T0192342

Refining Used Lubricating Oils

CHEMICAL INDUSTRIES

A Series of Reference Books and Textbooks

MOST RECENTLY PUBLISHED

Refining Used Lubricating Oils, James Speight and Douglas I. Exall

The Chemistry and Technology of Petroleum, Fifth Edition, James G. Speight

Educating Scientists and Engineers for Academic and Non-Academic Career Success, James Speight

Transport Phenomena Fundamentals, Third Edition, Joel Plawsky

Synthetics, Mineral Oils, and Bio-Based Lubricants: Chemistry and Technology, Second Edition, Leslie R. Rudnick

Modeling of Processes and Reactors for Upgrading of Heavy Petroleum, Jorge Ancheyta

Synthetics, Mineral Oils, and Bio-Based Lubricants: Chemistry and Technology, Second Edition, Leslie R. Rudnick

Fundamentals of Automatic Process Control, Uttam Ray Chaudhuri and Utpal Ray Chaudhuri

The Chemistry and Technology of Coal, Third Edition, James G. Speight

Practical Handbook on Biodiesel Production and Properties, Mushtaq Ahmad, Mir Ajab Khan, Muhammad Zafar, and Shazia Sultana

Introduction to Process Control, Second Edition, Jose A. Romagnoli and Ahmet Palazoglu

Fundamentals of Petroleum and Petrochemical Engineering, Uttam Ray Chaudhuri

Advances in Fluid Catalytic Cracking: Testing, Characterization, and Environmental Regulations, edited by Mario L. Occelli

Advances in Fischer-Tropsch Synthesis, Catalysts, and Catalysis, edited by Burton H. Davis and Mario L. Occelli

Transport Phenomena Fundamentals, Second Edition, Joel Plawsky

Asphaltenes: Chemical Transformation during Hydroprocessing of Heavy Oils, Jorge Ancheyta, Fernando Trejo, and Mohan Singh Rana

Refining Used Lubricating Oils

James G. Speight • Douglas I. Exall

CRC Press is an imprint of the
Taylor & Francis Group, an **informa** business

CRC Press
Taylor & Francis Group
6000 Broken Sound Parkway NW, Suite 300
Boca Raton, FL 33487-2742

First issued in paperback 2021

Version Date: 20140219

ISBN 13: 978-1-03-223596-7 (pbk)
ISBN 13: 978-1-4665-5149-7 (hbk)

Publisher's Note
The publisher has gone to great lengths to ensure the quality of this reprint but points out that some imperfections in the original copies may be apparent.

Library of Congress Cataloging-in-Publication Data

Speight, James G.
Refining used lubricating oils / authors, James G. Speight, Douglas I. Exall.
pages cm. -- (Chemical industries)
Summary: "Used lubricating oil is a valuable resource. This book examines recycling processes for a wide range of products with different properties and different criteria. It also compares the various recycling methods and resulting products to conventional products obtained from original refining processes. The reviews, data, and comparisons provided by the authors allow readers to identify which processes are likely to produce a product with specific properties, and enables them to combine this with an analysis of the economic data to identify attractive oil recycling propositions"-- Provided by publisher.
Includes bibliographical references and index.
ISBN 978-1-4665-5149-7 (hardback)
1. Petroleum waste--Recycling. 2. Mineral oils--Refining. 3. Lubricating oils--Recycling. 4. Petroleum waste--Management. I. Exall, Douglas I. II. Title.

TP687.S74 2014
665.5'389--dc23 2013049513

Visit the Taylor & Francis Web site at
http://www.taylorandfrancis.com

and the CRC Press Web site at
http://www.crcpress.com

Contents

Preface

Used lubricating oil is a valuable resource. However, during service, lubricating oil becomes contaminated by a number of additional components (from engine wear) such as iron and steel particles, copper, lead, zinc, barium and cadmium (two highly toxic metals), sulfur, water, and dirt. In addition, the organic constituents of lubricating oil will also undergo changes in use and produce undesirable contaminants. Because of these contaminants, disposal of used lubricating oil can be more environmentally damaging than crude oil pollution.

Thus, lubricating oil becomes unfit for further use for two main reasons: accumulation of contaminants in the oil and chemical changes in the oil. In addition, there is a shortage of prime lubricating oils that is difficult to make up and requires use of hard-to-replace petroleum for products other than fuels. Resource preservation and issues related to oil disposal are priorities that dictate the expansion of the re-refining option.

The economically and environmentally sound way to dispose of used lubricating oil is to recycle the used oil to its original use. A second recycling option is to clean up the used oil and use it as a fuel. In North America and Europe, as well as in other countries in the world, legislation is being introduced that requires waste materials to be recycled under a hierarchy of options that favor returning the materials to their original use where this is practicable and moving to a lower level of recycling, which includes energy recovery and could involve their disposal by destructive means such as use as a fuel, where this can be justified. Although the definition of what is practicable and what justifications are acceptable can be the subject of discussion, there is no doubt that the re-refining of used lubricating oils is becoming more popular and desirable.

Recent studies using life cycle assessment methods have highlighted the relative merits of the used oil recycling methods and include analyses of the energy use in recycling and in the original production of similar products from crude oil. Although crude oil prices fluctuate from year to year, valuable energy resources can be conserved by the use of recycled lubricating oil and fuel oil made from reclaimed motor oil. The energy saved by collecting and recycling used motor oil can also help reduce our dependence on foreign oil imports.

This book presents to the reader the properties of used lubricating oils and the means by which such materials can be re-refined and converted once more into useful lubricants as well as other products. This book will also strongly suggest the feasibility of such action and leave the reader with the knowledge of the currently feasible used lubricating oil recycling programs that are being introduced in many parts of the world.

Finally, at a time when many nations of the world have undergone a transferal to the metric system of measurement, there are still those disciplines that are based on such scales as the Fahrenheit temperature scale as well as the foot measure instead of the meter. Accordingly, and where appropriate, the text contains both the metric and nonmetric measures, but it should be noted that exact conversion is not often feasible, and thus conversion data are often taken to the nearest whole number. Indeed, conversions involving the two temperature scales—Fahrenheit and Celsius—are, at the high temperatures quoted in the text, often *rounded* to the nearest 5 degrees, especially when serious error would not arise from such a conversion.

Additional material is available from the CRC Web site: http://www.crcpress.com/product/isbn/9781466551497.

James G. Speight, PhD, DSc
Douglas I. Exall, PhD, P.Eng.

Authors

Dr. James G. Speight, C. CChem., FRSC, FCIC, FACS, earned his BSc and PhD degrees from the University of Manchester, U.K. He also holds a DSC degree in geological sciences (VINIGRI, St. Petersburg, Russia) and a PhD degree in petroleum engineering (Dubna International University, Moscow, Russia). Dr. Speight is the author of more than 50 books in petroleum science, petroleum engineering, and environmental sciences. Formerly the CEO of the Western Research Institute (now an independent consultant), he has served as an adjunct professor in the Department of Chemical and Fuels Engineering at the University of Utah and in the Departments of Chemistry and Chemical and Petroleum Engineering at the University of Wyoming. In addition, he has also been a visiting professor in chemical engineering at the following universities: the University of Missouri–Columbia, the Technical University of Denmark, and the University of Trinidad and Tobago.

Dr. Speight was elected to the Russian Academy of Sciences in 1996 and awarded the Gold Medal of Honor that same year for outstanding contributions to the field of petroleum sciences. He has also received the Scientists without Borders Medal of Honor of the Russian Academy of Sciences. In 2001, the Academy also awarded Dr. Speight the Einstein Medal for outstanding contributions and service in the field of geological sciences.

Dr. Douglas I. Exall, P.Eng., is an engineering consultant with more than 40 years of experience in oil and gas production and processing technologies with patents, conference presentations, journal publications, industrial R&D reports, and teaching experience in most areas of chemical engineering, including refinery operations, process modeling and simulation, and economic analysis. He received his engineering education at the University of Natal in South Africa and the Swiss Federal Institute of Technology in Zurich, holding BSc, MSc, and PhD degrees in chemical engineering from the University of Natal. He has worked as a research manager in the oil and gas industry and Alberta Research Council in Canada, and as a professor or adjunct professor in the University of Calgary, the University of Regina, the University of Natal, the University of Trinidad and Tobago, and the Universidad de America in Colombia, as well as coordinating the petroleum engineering programs at the Southern Alberta Institute of Technology in Calgary. He has taught courses in petroleum refining, process modeling and simulation, and economic evaluation, besides many of the chemical engineering courses covering heat, mass, and momentum transfer with chemical reaction kinetics and transformations.

Since 2007, consulting work has included reviewing options for the re-refining of lubricating oils, the available processes and technologies and their economic viability, as well as modeling oil and gas production and field production economics.

1 Manufacture of Lubricating Oil

1.1 INTRODUCTION

The constant demand for products such as liquid fuels is the main driving force behind the petroleum industry. Indeed, fuel products—for example, gasoline, kerosene, and diesel fuel—are the prime products of the current era. In addition, other products, such as lubricating oils, waxes, and asphalt, have also added to the popularity of petroleum as an important resource. Petroleum products are the basic materials used for the manufacture of synthetic fibers for clothing and in plastics, paints, fertilizers, insecticides, soaps, and synthetic rubber. In fact, the use of petroleum as a source of raw material in manufacturing is central to the functioning of modern industry (Speight and Ozum 2002; Hsu and Robinson 2006; Gary et al. 2007; Speight 2014).

After kerosene, the early petroleum refiners produced paraffin wax for the manufacture of candles, and lubricating oil was, at first, a by-product of wax manufacture—the preferred lubricants in the 1860s (at the dawn of the refining industry) were lard oil (oil from pig fat), sperm oil (whale oil), and tallow (a rendered form of beef or mutton fat, processed from suet, which is solid at room temperature). The demand that existed for kerosene did not also exist for petroleum-derived lubricating oils, and the demand for these high-boiling oils took time and market demand to develop, although petroleum-derived lubricating oil was used to supplement the animal-derived and vegetable-derived lubricating oils. However, industry evolved and the trend to heavier machinery increased the demand for mineral lubricating oils and, after the 1890s, petroleum displaced animal and vegetable oils as the predominant source of lubricants.

Oils derived from petroleum sources (mineral oils) are often used as lubricating oils but also have medicinal and food uses. A major type of hydraulic fluid is the mineral oil class of hydraulic fluids, which are produced from higher-boiling petroleum distillates. Carbon numbers ranging from C_{15} to C_{50} occur in the various types of mineral oils, with the higher-boiling distillates having higher percentages of the higher–carbon number compounds.

Crankcase oil (motor oil) may be either mineral-based or synthetic. The mineral-based oils are more widely used than the synthetic oils and may be used in automotive engines, railroad and truck diesel engines, marine equipment, jet and other aircraft engines, and most small two-stroke and four-stroke engines. The mineral-based oils contain hundreds to thousands of hydrocarbon compounds but unfortunately also include a substantial fraction of nitrogen-containing and sulfur-containing compounds. The hydrocarbons are mainly mixtures of straight and branched chain hydrocarbons (alkanes), cycloalkanes, and aromatic hydrocarbons. Polynuclear aromatic hydrocarbons (and the corresponding alkylated derivatives) and metal-containing constituents are components of motor oils and crankcase oils, with the used oils typically having higher concentrations of these substances than the new unused oils.

As a general note, lubricating oils produced from petroleum represent the majority of lubricating oils used each year in the United States. The use of synthetic lubricating oil is increasing in popularity and the market for synthetic lubricating oils is expanding because of their interesting properties, such as (1) high viscosity index due to the absence of aromatic compounds, (2) low volatility, (3) high thermal stability, and (4) low-temperature flow behavior.

In the current context, the major forces driving change in the automotive lubricant industry are higher fuel economy, smaller engines operating at higher rotational speeds, further low-temperature improvements, longer drain intervals, and emission control system durability. Increased use of lighter-viscosity multigrade oils to achieve better fuel economy and low-temperature performance has made base oil properties key factors in lubricant design. Currently, the more severe demand for volatility and low temperature properties requires the full integration of base oils and formulation technologies. Conventional mineral-based oils alone are no longer adequate to meet the top requirements among North American and European specifications, especially regarding volatility.

1.2 BASE OIL MANUFACTURE FROM PETROLEUM SOURCES

Base oil is the base stock or blend of base stocks used in American Petroleum Institute (API)-licensed oil and a base stock slate is a product line of base stocks that have different viscosities but are in the same base stock grouping and are from the same manufacturer.

Base oil (the raw material for lubricants) is the name given to lubrication grade oils initially produced from refining crude oil (mineral base oil) or through chemical synthesis (synthetic base oil). Base oil is typically defined as oil with a boiling point range between 300°C (550°F) and 565°C (1050°F), consisting of hydrocarbons with 18 to 40 carbon atoms. This oil can be either paraffinic or naphthenic in nature depending on the chemical structure of the constituent molecules.

Like many petroleum products, there are no longer many (if any) processes that are responsible for the direct manufacture of base oil. The base oil is typically a blend of products from several streams to which additives are added to adjust the properties to meet specifications and desired service life (Pillon 2007, 2010; Rudnick 2013).

1.2.1 BASE OIL MANUFACTURE

Lubricating oil manufacture was well established by 1880, and the method depended on whether the crude petroleum was processed primarily for kerosene or for lubricating oils. Usually, the crude oil was processed for kerosene, and primary distillation separated the crude into three fractions, naphtha, kerosene, and residuum. To increase the production of kerosene, the cracking distillation technique was used, and this converted a large part of the gas oils and lubricating oils into kerosene. The cracking reactions also produced coke products and asphalt-like materials, which gave the residuum a black color and, hence, it was often referred to as *tar* (Speight and Ozum 2002; Hsu and Robinson 2006; Gary et al. 2007; Speight 2014).

The rapid evolution of lubricating oil manufacture and use occurred during the early decades of the twentieth century. Petroleum-based oils first became available and as the demand for automobiles grew, so did the demand for better lubricants. By 1923, the U.S. Society of Automotive Engineers (SAE) classified engine oils by viscosity as light, medium, and heavy lubricating oils. However, engine oil contained no additives and had to be replaced every 800 to 1000 miles. In the 1920s, more lubrication manufacturers started "processing" base oils to improve their performance. Three popular processing routes were: (1) clay treatment, (2) acid treatment, and (3) sulfur dioxide treatment (Kramer et al. 2001).

Clay treatment was used to soak up and remove some of the worst undesirable components in the petroleum base oil. These compounds were usually aromatic and highly polar compounds containing sulfur and nitrogen (Speight and Ozum 2002; Hsu and Robinson 2006; Gary et al. 2007; Speight 2014).

Acid treatment with concentrated sulfuric acid was used to react with the worst components in the base oil and convert them into a sludge that could be removed. Although this process effectively cleaned up the oil, it was expensive and the technology is no longer used in many refineries due to environmental concerns about the acid and the acid sludge (a thick viscous material that separates when petroleum or petroleum products are treated with sulfuric acid) formed in the process (Speight and Ozum 2002; Hsu and Robinson 2006; Gary et al. 2007; Speight 2014).

Continuous acid treatment involves the same steps as batch refining with the exception that (1) the acid and feedstock oil and neutralizing agent are mixed with pumps or static mixers, (2) excess acid and sludge and excess neutralizing agent and soaps are removed using centrifuges or centrifugal extractors, (3) water washing is conducted using centrifugal extractors, and (4) drying of the oil is conducted in continuous strippers. The advantages for the continuous process over the batch process are (1) higher yields of oil, (2) lower chemical consumption, and (3) a reduction in air and water pollution.

Sulfur dioxide treatment was a primitive extraction process to remove the worst components in the lubricating oil using a recyclable solvent which, unfortunately, was highly toxic. Although it also has been virtually phased out, it was a useful stepping stone to conventional solvent extraction (Speight and Ozum 2002; Hsu and Robinson 2006; Gary et al. 2007; Speight 2014).

Currently, catalytic dewaxing and solvent dewaxing (the most prevalent) are processes commonly in use—older technologies include cold settling, pressure filtration, and centrifuge dewaxing (Sequeira 1994; Kramer et al. 2001; Speight and Ozum 2002; Hsu and Robinson 2006; Gary et al. 2007; Speight 2014).

1.2.1.1 Distillation

Within a naphthenic or paraffinic type, base stocks are distinguished by their viscosities and are produced to certain viscosity specifications. Because viscosity is approximately related to molecular weight, the first step in manufacturing is to separate out the lube precursor components that have the correct molecular weight range. Lower-boiling fuel products of such low viscosities and volatilities, which have no application in lubricants—naphtha, kerosene, jet, and diesel fuels—are distilled off in an atmospheric pressure distillation tower. The higher molecular weight components, which do not vaporize at atmospheric pressure, are then fractionated by distillation at reduced pressures of from 10 to 50 mm Hg (i.e., vacuum fractionation).

The constituents of the base oil are extracted from crude oil, which undergoes a preliminary purification process (sedimentation) before it is pumped into fractionating towers (Speight and Ozum 2002; Hsu and Robinson 2006; Gary et al. 2007; Speight 2014). A typical high-efficiency fractionating tower, 25 to 35 ft. in diameter and up to 400 ft. tall, is constructed of high-grade steels to resist the corrosive compounds present in crude oils. The tower internals consist of an ascending series of condensate collecting trays (Figure 1.1).

Within the first tower (atmospheric distillation tower), the constituents of petroleum are separated from each other (fractional distillation; Speight and Ozum 2002; Hsu and Robinson 2006; Gary et al. 2007; Speight 2014). As the vapors rise up through the tower, the various fractions cool, condense, and return to liquid form at different rates determined by their respective boiling points (the lower the boiling point of the fraction, the higher it rises before condensing).

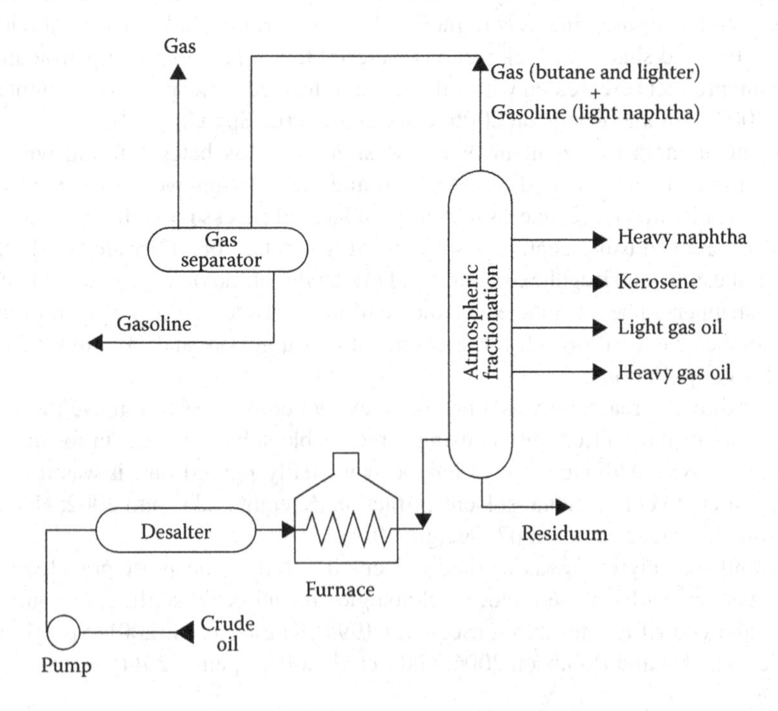

FIGURE 1.1 An atmospheric distillation unit. (From OSHA Technical Manual, Section IV, Chapter 2: Petroleum Refining Processes. http://www.osha.gov/dts/osta/otm/otm_iv/otm_iv_2.html.)

Thus, the atmospheric residuum (bottoms) from the atmospheric tower are fed to the vacuum tower, in which intermediate product streams such as light vacuum gas oil and heavy vacuum gas oil are produced. These may be either narrow cuts of specific viscosities destined for a solvent-refining step or broader cuts destined for hydrocracking to lubes and fuels.

The nonvolatile material from the atmospheric distillation tower is reheated and pumped into this second tower (vacuum distillation tower) wherein vacuum pressure lowers the residual oil's boiling point so that it can be made to vaporize at a lower temperature. The higher–molecular weight constituents with higher boiling points, such as residuum which contains inorganic contaminants originally in the crude oil, remain for further processing (Figure 1.2).

However, the nonvolatile residuum from the vacuum tower (vacuum tower bottoms) may contain valuable high-viscosity lube precursors (boiling point >510°C, >950°F) and these are separated from asphaltic components (nonvolatile highly aromatic components that are difficult to refine) in a deasphalting unit.

Deasphalting units separate asphalt from deasphalted oil using solubility differences, and this is usually soluble in propane, and further refining of the deasphalted oil by dewaxing and solvent refining or by hydrotreatment produces bright stock, which is a heavy (very viscous) base stock that is a residue (i.e., it is not a distillate overhead). The deasphalted oil can also be part of the feedstock to a lube hydrocracker to produce heavier base stocks.

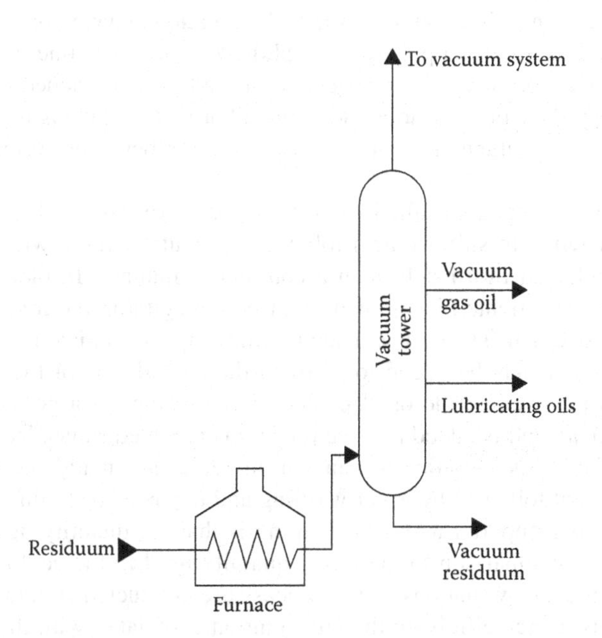

FIGURE 1.2 A vacuum distillation unit. (From OSHA Technical Manual, Section IV, Chapter 2: Petroleum Refining Processes. http://www.osha.gov/dts/osta/otm/otm_iv/otm_iv_2.html.)

The waxy distillates and deasphalted oil require three further processing steps to obtain acceptable base stock: (1) the viscosity–temperature relationship of the base stock (improvement of the viscosity index) has to be enhanced by the removal of aromatics to meet industry requirements for paraffinic stocks; (2) oxidation resistance and performance must be improved by the removal of aromatics, particularly polynuclear aromatic compounds, nitrogen-containing constituents, and some of the sulfur-containing compounds; and (3) the temperature at which the base stock "freezes" due to the crystallization of wax must be lowered by wax removal so that lubricated equipment can operate at winter temperatures.

After further processing to remove unwanted compounds, the lubricating oil fraction that has been collected in the two fractionating towers is passed through several ultrafine filters, which remove remaining impurities. Aromatics, one such contaminant, contain six carbon rings that would affect the lube oil's viscosity undesirably if they were not removed by solvent extraction. Solvent extraction is possible because aromatics are more soluble in the solvent compared with the lube oil fraction. When the base oil is treated with the solvent, the aromatic constituents dissolve and the insoluble fraction (aromatic constituents) is recovered by separation from the solvent.

1.2.1.2 Chemical Refining Processes

In chemical refining processes, constituents with chemical structures unsuitable for lubes are wholly or partially converted to acceptable base stock components. These processes all involve catalysts acting in the presence of hydrogen (catalytic hydroprocessing). Examples are the hydrogenation and ring opening of polynuclear aromatic systems to polycyclic naphthenes with the same or fewer rings and the isomerization of wax components to more highly branched isomers with lower freezing points. Furthermore, the chemical properties of existing useful constituents may be simultaneously altered such that even better performance can be achieved.

Acid-alkali refining, also called *wet refining*, is a process in which lubricating oils are contacted with sulfuric acid followed by neutralization with alkali. The process is conducted in a batch or in a continuous manner. In older batch processes, the oil to be treated is pumped to a treating agitator and mixed with acid of the desired strength. The oil and acid are mixed by mechanical means or by air blowing and water may be added to assist in the coagulation of the acid sludge. The sludge is removed or the oil decanted after settling for a period of several hours. Additional acid is added and the process is repeated as needed. The acidic oil (sour oil) from this operation is then neutralized using an aqueous or alcoholic neutralizing agent followed by water washing and drying. The main difference in acid finishing as compared with acid refining is that the quantity of acid used is usually lower in comparison to the larger quantity used in the acid refining processes. More modern variations of the process are conducted in totally enclosed treating vessels, which eliminate the air pollution associated with the older open treating units.

Acid-clay refining, also called *dry refining*, is similar to acid-alkali refining with the exception that clay and a neutralizing agent are used for neutralization. This process is used for oils that form emulsions during neutralization. Neutralization with aqueous and alcoholic caustic, soda ash lime, and other neutralizing agents, is used to remove organic acids from some feedstocks. This process is conducted to reduce organic acid corrosion in downstream units or to improve the refining response and color stability of lube feedstocks.

1.2.1.3 Hydroprocessing

Hydroprocessing (hydrotreating) was developed in the 1950s and first used in base oil manufacturing in the 1960s. It was used as an additional purification step after a conventional solvent refining process. In the process, hydrogen is added to the base oil at elevated temperatures in the presence of a catalyst to stabilize the most reactive components in the base oil, improve color, and increase the useful life of the base oil. This process removes some of the nitrogen-containing and sulfur-containing species but the process was not sufficiently severe to remove a significant amount of aromatic molecules.

Hydrocracking is a more severe form of hydroprocessing and requires higher temperatures and pressures than hydrotreating (Figure 1.3). Feedstock molecules are cracked (thermally decomposed) into lower molecular weight products. Most of the sulfur, nitrogen, and aromatics are removed and the molecular alteration of the remaining saturated species occurs as naphthenic rings are opened and paraffin isomers are redistributed, and driven by thermodynamics with reaction rates facilitated by catalysts.

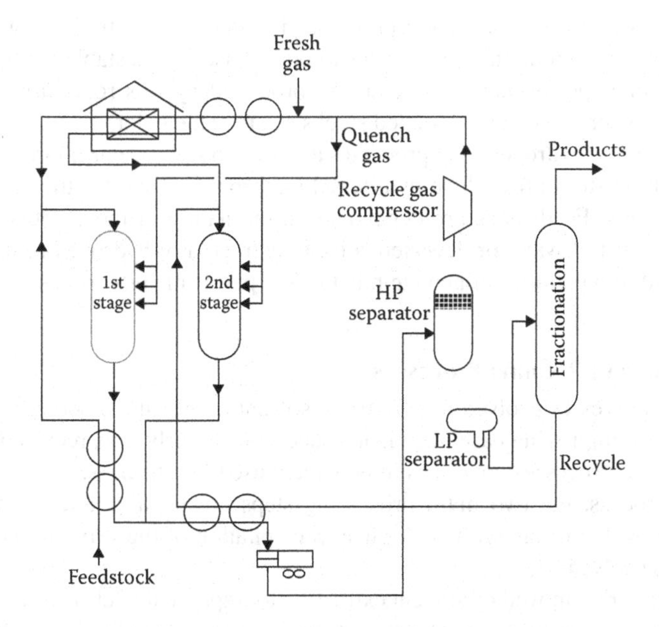

FIGURE 1.3 Two-stage hydrocracking. (From OSHA Technical Manual, Section IV, Chapter 2: Petroleum Refining Processes. http://www.osha.gov/dts/osta/otm/otm_iv/otm_iv_2.html.)

Hydroprocessing, which has been generally replaced with solvent refining, consists of lube hydrocracking as an alternative to solvent extraction, and hydrorefining to prepare specialty products or to stabilize hydrocracked base stocks. Hydrocracking catalysts consist of mixtures of cobalt, nickel, molybdenum, and tungsten on an alumina or silica–alumina-based carrier. Hydrotreating catalysts are proprietary, but usually consist of nickel–molybdenum on alumina. The hydrocracking catalysts are used to remove nitrogen, oxygen, and sulfur, and convert polynuclear aromatics and polynuclear naphthenes to mononuclear naphthenes, aromatics, and isoparaffins, which are typically desired in lube base stocks. Feedstocks consist of unrefined distillates and deasphalted oils, solvent-extracted distillates and deasphalted oils, cycle oils, hydrogen refined oils, and mixtures of these hydrocarbon fractions.

In the first stage of the catalytic hydrogenation process, part of the feedstock aromatic constituents are saturated by hydrogenation to form cycloparaffins. The process also promotes significant molecular reorganization by carbon–carbon bond breaking to improve the rheological (flow) properties of the base stock (by improving the viscosity index).

Typically, in this stage, feedstock sulfur and nitrogen are both essentially eliminated. Some of the carbon–carbon bond breaking produces overhead (distillable, volatile) products in the form of low-sulfur gasoline and middle distillates. The fractionated waxy lube streams, usually those boiling at approximately higher than 370°C (700°F), are then dewaxed, either by solvent dewaxing or more frequently by catalytic hydroprocessing (in which wax is either cracked to gasoline or isomerized to low melting point isoparaffins in high yields and which has a positive effect on the viscosity index).

The final step in the conversion processes is usually catalytic hydrogenation to saturate most of the remaining aromatics to make base stocks stable for storage and to improve their performance. Base stocks produced by this route are frequently water white, whereas solvent extracted stocks retain some color.

Lubricating oil hydrogenating processes are used to stabilize or improve the quality of lube base stocks from lube hydrocracking processes and for the manufacture of specialty oils. Feedstocks are dependent on the nature of the crude source, but generally consist of waxy or dewaxed solvent–extracted or hydrogen-refined paraffinic oils and refined or unrefined naphthenic and paraffinic oils from some selected crude oils.

1.2.1.4 Solvent Refining Processes

As refinery processes evolved in the 1930s, solvent processing emerged as a viable technology for improving base oil performance using a fairly safe, recyclable solvent and the solvent processes continues to be widely used in refineries.

In the process, the two main processing steps are (1) aromatics removal and (2) wax removal chilling, with subsequent precipitation of the wax, in the presence of a different solvent.

Aromatics are removed by solvent extraction to improve the lubricating quality of the oil. Aromatics make good solvents but they make poor quality base oils because they are among the most reactive components in the lubricating oil boiling range. Oxidation of aromatics can start a chain reaction that can dramatically shorten the useful life of base oil. The viscosity of aromatic components in base oil also responds relatively undesirably to changes in temperature.

Lubricants are often designed to provide a viscosity that is low enough for good cold weather starting and high enough to provide adequate film thickness and lubricity in hot, high-severity service. Therefore, when hot and cold performance is required, a small response to changes in temperature is desired, which is measured by the viscosity index—a higher viscosity index indicates a smaller, more favorable response to temperature. Correspondingly, many turbine manufacturers have a minimum viscosity index specification for their turbine oils. Base oil selection is the driver for meeting this specification because turbine oil additives do not normally contribute positively to the viscosity index in turbine oil formulations.

Aromatics are removed by feeding the raw lube distillate (vacuum gas oil) into a solvent extractor in which it is contacted countercurrently with a solvent, such as furfural, *N*-methyl-2-pyrrolidone (NMP), and DUO-SOL. Phenol was another popular solvent but it is rarely used today due to environmental concerns. Solvent extraction typically removes 50% to 80% of the impurities (aromatics, polar constituents, sulfur-containing and nitrogen-containing species). The second step is solvent dewaxing in which wax is removed from the oil to keep the lubricating oil from gelling (through wax deposition) at low temperatures (Figure 1.4). In the process, wax is removed by first diluting the raffinate with a solvent to lower its viscosity improving low-temperature filterability. Popular dewaxing solvents are methyl-ethyl ketone (MEK)/toluene, MEK/methyl-isobutyl ketone, or (rarely) propane. The diluted oil is then chilled to −10°C to −20°C (14°F to −4°F). Wax crystals form, precipitate, and are removed by filtration (Kramer et al. 2001).

After the undesirable chemical compounds (e.g., polynuclear aromatic constituents) are removed using the solvent-based separation process (solvent refining), the by-products (extracts) represent a yield loss in producing the base stock. The base

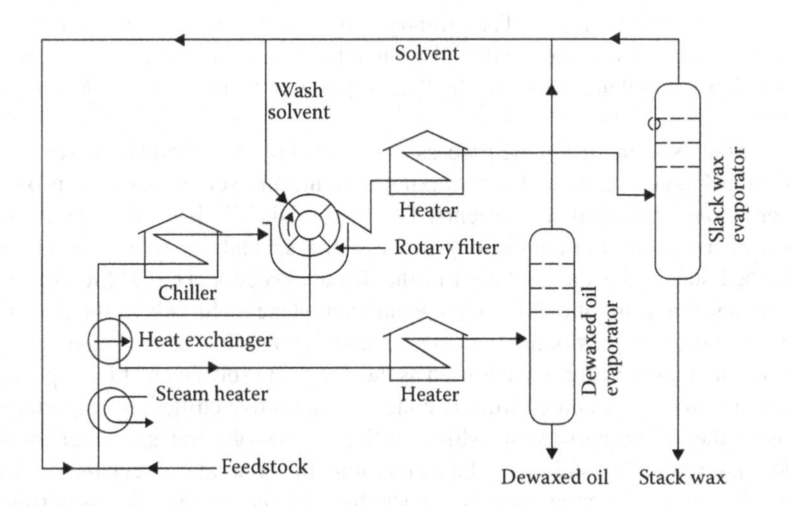

FIGURE 1.4 Solvent dewaxing. (From OSHA Technical Manual, Section IV, Chapter 2: Petroleum Refining Processes. http://www.osha.gov/dts/osta/otm/otm_iv/otm_iv_2.html.)

stock properties are determined by the chemical composition of the crude oil because the molecular species in the final base oil are unchanged from those in the feedstock.

Feedstocks for solvent-refining processes consist of paraffinic distillate, naphthenic distillate, hydrogen-refined distillates, deasphalted oil, cycle oil, and dewaxed oil. The products are refined oils destined for further processing or finished lube base stocks. The by-products are aromatic extracts, which are used in the manufacture of rubber, carbon black, petrochemicals, catalytic cracking feedstock, fuel oil, or asphalt. The major solvents in use are NMP and furfural, with phenol and liquid sulfur dioxide used to a lesser extent.

The solvents are typically recovered in a series of flash towers. Steam or inert gas strippers are used to remove traces of solvent, and a solvent purification system is used to remove water and other impurities from the recovered solvent.

1.2.1.5 Solvent Dewaxing

Solvent dewaxing is a variation of the solvent refining processes describe above (Speight and Ozum 2002; Hsu and Robinson 2006; Gary et al. 2007; Speight 2014) and consists of the following steps: crystallization, filtration, and solvent recovery. In the crystallization step, the feedstock is diluted with the solvent and chilled, solidifying the wax components. The filtration step removes the wax from the solution of dewaxed oil and solvent. Solvent recovery removes the solvent from the wax cake and filtrate for recycling by flash distillation and stripping. The major process in use today is the ketone dewaxing process.

Other processes that are used to a lesser degree include the dimethyl ketone process Di/Me process and the propane dewaxing process. The most widely used ketone processes are the Texaco solvent dewaxing process and the Exxon Dilchill Process. Both processes consist of diluting the waxy feedstock with solvent while chilling at a controlled rate to produce a slurry. This slurry is then filtered using rotary vacuum filters and the wax cake is washed with a cold solvent. The filtrate is used to chill the feedstock and solvent mixture. The primary wax cake is diluted with additional solvent and filtered again to reduce the oil content in the wax. The solvent is recovered from the dewaxed oil and wax cake by flash vaporization and recycled back into the process.

The Texaco solvent dewaxing process (also called the MEK process) uses a mixture of MEK and toluene as the dewaxing solvent, and sometimes uses mixtures of other ketones and aromatic solvents. The Exxon Dilchill dewaxing process uses a direct cold solvent dilution-chilling process in a special crystallizer in place of the scraped surface exchangers used in the Texaco process. The Di/Me dewaxing process uses a mixture of dichloroethane and methylene dichloride as the dewaxing solvent. The propane dewaxing process is essentially the same as the ketone process except for the following: propane is used as the dewaxing solvent and higher pressure equipment is required, and chilling is done in evaporative chillers by vaporizing a portion of the dewaxing solvent. Although this process generates a better product and does not require crystallizers, the temperature differential between the dewaxed oil and the filtration temperature is higher than for the ketone processes (higher energy costs), and dewaxing aids are required to get good filtration rates.

1.2.1.6 Catalytic Dewaxing and Hydroisomerization

The first catalytic dewaxing and wax hydroisomerization technologies were commercialized in the 1970s (Kramer et al. 2001). Catalytic dewaxing was a desirable alternative to solvent dewaxing especially for conventional neutral oils because it removed *n*-paraffins and waxy side chains from other molecules by catalytically cracking them into smaller molecules. This process lowered the pour point of the base oil so that it flowed at low temperatures, like solvent dewaxed oils. Hydroisomerization also saturated most of the remaining aromatics and removed most of the remaining sulfur and nitrogen species.

In 1993, the first modern wax hydroisomerization process was commercialized and was an improvement over earlier catalytic dewaxing because the pour point of the base oil was lowered by isomerizing (reshaping) the *n*-paraffins and other molecules with waxy side chains into very desirable branched compounds with superior lubricating qualities rather than cracking them away. Hydroisomerization was also an improvement over earlier wax hydroisomerization technologies because it eliminated the subsequent solvent dewaxing step, which was a requirement for earlier generation wax isomerization technologies to achieve adequate yield at standard pour points.

Current wax hydroisomerization processes yield products with exceptional purity and stability due to an extremely high degree of saturation. They are very distinctive because, unlike other base oils, they typically have no color.

Because solvent dewaxing is relatively expensive for the production of low pour point oils, various catalytic dewaxing (selective hydrocracking) processes have been developed for the manufacture of lube oil base stocks. The basic process consists of a reactor containing a proprietary dewaxing catalyst followed by a second reactor containing a hydrogen finishing catalyst to saturate olefins created by the dewaxing reaction and to improve stability, color, and to mitigate emulsion formation by the finished lube oil during service.

1.2.2 Older Processes

Older processes are those which were the prime means of lubricating oil and wax production—some of which are occasionally reported as still being operational in present-day refineries.

In older refineries, there was often cracking in the primary (atmospheric) distillation at the high temperatures used in the process, and the gas oil used as the precursor to the lubricating oil contained dark-colored, sludge-forming asphaltic materials. These undesirable materials were removed by treatment with sulfuric acid followed by lye washing (Speight and Ozum 2002; Speight 2014). This was followed by separation of the wax from the acid-treated paraffin distillate, which was achieved by chilling and filtration. The chilled, semisolid paraffin distillate was then squeezed in canvas bags in a knuckle or rack press (similar to a cider press) so that the oil would filter through the canvas, leaving the wax crystals in the bag. Later developments saw chilled paraffin distillate filtered in hydraulically operated plate and frame presses, and the use of these continued almost to the present time (Speight and Ozum 2002; Speight 2014).

The oil from the press (*pressed distillate*) was subdivided into three fractions by re-distillation. Two overhead fractions of increasing viscosity, the heavier with an SAE viscosity of approximately SAE 10 (*paraffin oils*) and the residue in the still (viscosity equivalent to a light SAE 30) was known as *red oil*. All three fractions were again acid-treated and lye (alkali)–treated and then washed with water. The treated oils were pumped into shallow pans in the bleacher house, where air was blown through the oil and exposure to the sun through the glass roof (of the bleacher house or pan) removed cloudiness or made the oils *bright*—with no obvious signs of sediment or other impurities.

Further treatment of the paraffin oil produced pale oil; thus, if the paraffin oil was filtered through bone charcoal, fuller's earth, clay, or similar absorptive material, the color changed from a deep yellow to a pale yellow. The filtered paraffin oil was called pale oil to differentiate it from the nonfiltered paraffin oil, which was considered of lower quality.

The wax separated from paraffin distillate by cold-pressing contained approximately 50% oil (*slack wax*), which was melted and cast into cakes that were pressed in a hot or hard press. This squeezed more oil from the wax (*scale wax*) which, by a process known as *sweating*, was subdivided into several paraffin wax products with different melting points.

In contrast, crude petroleum processed primarily as a source of lubricating oil was handled differently from crude oils processed primarily for kerosene. The primary distillation removed naphtha and kerosene fractions, but without using temperatures high enough to cause cracking. The yield of kerosene was thus much lower, but the absence of cracking reactions increased the yield of lubricating oil fractions. Furthermore, the residuum was distilled using steam, which eliminated the need for high distillation temperatures, and cracking reactions were thus prevented. Thus, various overhead fractions suitable for lubricating oils and known as neutral oils were obtained; many of these were so light that they did not contain wax and did not need dewaxing; the more viscous oils could be dewaxed by cold-pressing.

If the wax in the residual oil could not be removed by cold-pressing, it was removed by cold settling. This involved admixture of the residual oil with a large volume of naphtha, which was then allowed to stand for as long as necessary in a tank exposed to low temperatures, usually climatic cold (winter). This caused the waxy components to congeal and settle to the bottom of the tank. In the spring, the supernatant naphtha–oil mixture was pumped to a steam still, in which the naphtha was removed as an overhead stream; the bottom product was known as steam-refined stock. If the steam-refined stock is filtered through charcoal or a similar filter material, the improvement in color caused the oil to be known as *bright stock* (a high-viscosity base oil, which is refined from paraffinic crude oil and widely used in marine oil, mono-grade motor oil, gear oil, and grease; Haycock and Hiller 2004). Mixtures of steam-refined stock with the much lighter paraffin, pale, red, and neutral oils produced oils of any desired viscosity.

The wax material that settled to the bottom of the cold settling tank was crude *petrolatum*. This was removed from the tank, heated, and filtered through a vessel containing clay, which changed its red color to brown or yellow. Further treatment with sulfuric acid produced white grades of petrolatum.

If the crude oil used for the manufacture of lubricating oils contained asphalt, it was necessary to acid treat the steam-refined oil before cold settling. Acid-treated, settled steam-refined stock was widely used as steam cylinder oils.

The crude oils available in North America until about 1900 were either paraffin base or mixed base; hence, paraffin wax was always a component of the raw lubricating oil fraction. The mixed-base crude oils also contained asphalt, and this made acid treatment necessary in the manufacture of lubricating oils. However, the asphalt-base crude oils (also referred to as naphthene-base crude oils) that contained little or no wax yielded a different kind of lubricating oil. Because wax was not present, the oils would flow at much lower temperatures than the oils from paraffin- and mixed-base crude oils even when the latter had been dewaxed. Hence, lubricating oils from asphalt-base crude oils became known as low cold-test oils; furthermore, these lubricating oils boiled at a lower temperature than oils of similar viscosity from paraffin-base crude oils. Thus, higher-viscosity oils could be distilled from asphalt-base crude oils at relatively low temperatures, and the low cold-test oils were preferred because they left less carbon residue in gasoline engines.

The development of vacuum distillation led to a major improvement in both paraffinic and naphthenic (low cold-test) oils. By vacuum distillation, the more viscous paraffinic oils (even oils suitable for bright stocks) could be distilled overhead and could be separated completely from residual asphaltic components. Vacuum distillation provided the means of separating more suitable lubricating oil fractions with predetermined viscosity ranges and removed the limit on the maximum viscosity that might be obtained in distillate oil.

However, although vacuum distillation effectively prevented residual asphaltic material from contaminating lubricating oils, it did not remove other undesirable components. The naphthenic oils, for example, contained components (naphthenic acids) that caused the oil to form emulsions with water. In particular, naphthenic oils contained components that caused oil to thicken excessively when cold and become very thin when hot. The degree to which the viscosity of an oil is affected by temperature is measured on a scale that originally ranged from 0 to 100 and is called the *viscosity index* (Speight 2014). Oil that changes the least in viscosity when the temperature is changed has a high viscosity index. Naphthenic oils have viscosity indices of 35 or lower, compared with 70 or higher for paraffinic oils.

1.2.3 Finishing Processes

By combining three catalytic hydroprocessing steps (hydrocracking, hydroisomerization, and hydrotreating), molecular constituents with poor lubricating qualities are transformed and reshaped into higher quality base oil molecules. Pour point, viscosity index, and oxidation stability are controlled independently. All three steps convert undesirable constituents into desirable product, rather than having one, two, or all three steps rely on subtraction. Among the many benefits of this combination of processes is greater crude oil flexibility; that is, less reliance on a narrow range of crude oils from which to make high-quality base oils. In addition, the base oil performance is exceptionally favorable and substantially independent of crude source, unlike solvent-refined base oil.

The process consists of fixed bed catalytic reactors that typically use a nickel–molybdenum catalyst to neutralize, desulfurize, and denitrogenate lube base stocks. These processes do not saturate aromatics or break carbon–carbon bonds as in other hydrogen finishing processes. Sulfuric acid treatment is still used by some refiners for the manufacture of specialty oils and the reclamation of used oils. This process is typically conducted in batch or continuous processes similar to the chemical refining processes with the exception that the amount of acid used is much lower than that used in acid refining.

Clay contacting involves mixing the oil with fine bleaching clay at elevated temperature followed by separation of the oil and clay. In the process, the clay is mixed intimately with the oil at elevated temperatures for a short period followed by separation of the oil and clay. This process may be used alone or in combination with an acid treating process for the finishing and neutralization of lube base stocks.

The process is an adsorption process used to remove polar compounds from lubricating oils thus improving color and the chemical, thermal, and color stability of the lube base oil. The process variables include the type of clay, clay dosage, and high treatment temperatures (150°C–370°C, 300°F–700°F). Clay contacting has been replaced with hydrogen finishing in the manufacture of base oils with the exception that some manufacturers use the process for manufacturing specialty oils. This process improves color and chemical, thermal, and color stability of the lube base stock, and is often combined with acid finishing.

Clay percolation (a cyclic process consisting of adsorption and regeneration cycles) is a static bed adsorption process used to purify, decolorize, and finish lube stocks and waxes. It is still used in the manufacture of refrigeration oils, transformer oils, turbine oils, white oils, and waxes. Clay percolation has, in large part, been replaced by hydrogen finishing but is still in limited use for the manufacture of refrigeration oils, transformer oils, turbine oils, white oils, and waxes. Although Attapulgus clay can be used, the most frequently used clay is activated bauxite. The process variables include temperature, flow rate, throughput and type of clay.

Attapulgus clay is a hydrous magnesium-aluminum silicate and belongs to the naturally active class of adsorbents, that is, adsorptive properties are developed by thermal treatment alone. The degree of activity is determined by the amount of water of hydration retained after thermal treatment. Attapulgus clay may be used to decolorize and neutralize any petroleum oil. Due to the relatively large pores in Attapulgus clay, it is well adapted to the removal of high–molecular weight sulfonates, resin constituents, and asphaltene constituents. It is moderately effective in removing odorous compounds and trace metals, but does not strongly adsorb aromatics.

Activated bauxite is composed primarily of hydrated aluminum oxide (bauxite) with minor amounts of silica, titania (titanium dioxide, titanium oxide), kaolinite, and hematite. Like Attapulgus clay, activated bauxite belongs to the class of adsorbents activated by heat alone. In addition to decolorization, activated bauxite reduces organic acidity, affords oils of improved demulsibility, often improves oxidation stability, and deodorizes. It is a good refining agent for turbine and transformer oils. A major advantage in using activated bauxite is that although the efficiency declines with successive adsorption–regeneration cycles, it eventually reaches a constant value (typically 60%–70% of the new clay efficiency). Once this equilibrium efficiency is obtained, no additional reduction in efficiency occurs with further regeneration. Conversely, Attapulgus clay does not attain equilibrium efficiency. The efficiency of Attapulgus clay continues to decline with each successive adsorption–regeneration cycle.

Hydrofinishing processes (hydrorefining processes) have largely replaced the acid and clay finishing processes. Hydrofinishing processes are mild hydrogenation processes used to improve the color, odor, thermal, and oxidative stability, as well as the demulsibility of lube base stocks. Chemically, these processes do not saturate aromatics or break carbon–carbon bonds as in other hydrogen processes. Sulfuric acid treating is still used by some refiners for the manufacture of specialty oils and the reclamation of used oils.

The operating conditions are dependent on feedstock composition (related to crude source as well as type and severity of prior processing), catalyst, and product specifications. The effects of hydrogen finishing temperature and pressure are highly dependent on the quality of the feedstock, product specifications, and the type of catalyst used. An increase in temperature or pressure will generally improve desulfurization, denitrification, product color, and product stability (Speight and Ozum 2002; Hsu and Robinson 2006; Gary et al. 2007; Speight 2014). However, increasing the temperature above some maximum that is related to the catalyst and feedstock quality will degrade the color, color stability, oxidation stability, and other properties of the base oil.

Finally, although an increase in temperature will usually lead to an improvement in color, excessively high temperatures will darken the oil much like excessive acid contact time or high contact temperature leads to color degradation. Use of excessively high temperature will also lead to cracking (hydrocracking) of the feedstock. On the other hand, an increase in pressure will improve the color of the base oil.

1.3 BLENDING

In a blending operation, the base oil is mixed with other base oil fractions and with additives to give it the desired physical properties (such as the ability to withstand low temperatures). At this point, the product (lubricating oil) is subjected to a variety of quality control tests that assess its viscosity, specific gravity, color, flash and fire points. Oil that meets quality standards is then packaged for sale and distribution.

Most applications of lube oils require that they be nonresinous, pale-colored, odorless, and oxidation-resistant. Common physical tests and chemical tests used to classify and determine the grade of lubricating oils include measurements for viscosity, specific gravity, and color, whereas typical chemical tests include those for flash and fire points.

However, of all the properties, viscosity is often considered to be the most important property. The application and operating temperature range are key factors in determining the proper viscosity for lubricating oil. For example, if the oil is too viscous, it offers too much resistance to the metal parts moving against each other. On the other hand, if it is not viscous enough, it will be squeezed out from between the mating surfaces and will not be able to lubricate them sufficiently. The Saybolt Standard Universal Viscometer is the standard instrument for determining the viscosity of petroleum lubricants between 70°F and 210°F (21°C and 99°C). Viscosity is measured in the Saybolt universal second (SUS), which is the time in seconds required for 50 mL of oil to empty out of a Saybolt viscometer cup through a calibrated tube orifice at a given temperature (ASTM D2161, ASTM 2012).

The specific gravity of oil depends on the refining method and the types of additives present, such as lead, which give the lubricating oil the ability to resist extreme mating surface pressure and cold temperatures. The color of the lubricating oil indicates the uniformity of a particular grade or brand. The oil's flash and fire points vary with the crude oil's origin. The *flash point* is the temperature to which an oil has to be heated until sufficient flammable vapor is driven off so that it will flash when brought into contact with a flame. The *fire point* is the higher temperature at which the oil vapor will continue to burn when ignited.

Common engine oils are classified by viscosity and performance according to specifications established by the U.S. SAE. Performance factors include wear prevention, oil sludge deposit formation, and oil thickening.

The consistent performance of hydrocarbon lubricant base oils is a critical factor in a wide variety of applications such as engine oils, industrial lubricants, and metalworking fluids (ASTM D6074). In addition, in many of these applications, humans are exposed to the base oils as a component of a formulated product such that health or safety considerations may need to be addressed. This guide suggests a compilation of properties and potential contaminants that are understood by those knowledgeable in the manufacture and use of hydrocarbon lubricants to be of significance in some or all applications.

Potential sources of base oil variation include the raw material, manufacturing process, operating conditions, storage, transportation, and blending. The test methods, base oil properties, and potential contaminants suggested are those that would likely be useful in many common situations, although it is recognized that there are specific applications and situations that could have different requirements.

Some lubricants are incompatible because of differences in additive chemistry that lead to undesirable chemical reactions. If these oils are mixed, insoluble material may form and then deposit onto sensitive machine surfaces. For a hydraulic fluid, this could lead to lubricant starvation, valve failure, or increased wear. A second form of lubricant incompatibility is more difficult to handle because no visible changes occur when the products are mixed. The problem appears only after the mixture is used in a piece of equipment that consequently fails or loses performance. Optimum performance requires carefully balanced frictional and antiwear properties in the finished product that are upset when the lubricants are mixed.

Some incompatible lubricant mixtures may also affect synthetic rubber seals. Lubricants are formulated to be neutral to seals or cause them to swell slightly. Too much seal swell, seal shrinkage, or chemical deterioration may occur with some combinations of lubricants. Engine oils formulated with certain types of dispersants attack fluorocarbon seals.

Lubricant incompatibility is a chemistry problem—two oils made by the same manufacturer may be incompatible. The most common cause of lubricant incompatibility that results in the formation of harmful solids is the reaction of an acidic component in one oil with a basic component in another oil.

Base oil testing helps ensure that quality is maintained throughout the refining, blending, transportation, and storage process. Expertise in chemical analysis techniques is also available to help with tough problem-solving, troubleshooting, formulation, and trace contamination issues. Performance testing related to the specific application should serve as the basis for identifying any contaminants and proving the acceptability of the oil (ASTM 2012). And this is particularly true of any blended products.

1.4 COMPOSITION AND PROPERTIES

Base stocks are classified into two broad types: (1) naphthenic and (2) paraffinic depending on the crude types they are derived from. Naphthenic crudes are characterized by the absence of wax or have very low levels of wax so they are largely cycloparaffinic in nature and aromatic in composition. Therefore, naphthenic lube fractions are generally liquid at low temperatures without any dewaxing. On the other hand, paraffin-base petroleum contains wax, consisting largely of *n*-paraffins and isoparaffins, which have high melting points. Waxy paraffinic distillates have melting or pour points too high for winter use; therefore, the paraffins have to be removed by dewaxing. After dewaxing, the paraffinic base stocks may still solidify, but at higher temperatures compared with naphthenic ones because their molecular structures have a more paraffinic character.

Lubricating oils are distinguished from other fractions of crude oil by their usually high (>400°C, >750°F) boiling point, as well as their high viscosity. Materials suitable for the production of lubricating oils are comprised principally of hydrocarbons containing from 25 to 35 or even 40 carbon atoms per molecule, whereas residual stocks may contain hydrocarbons with 50 or more (up to 80 or so) carbon atoms per molecule. The composition of lubricating oil may be substantially different from the lubricant fraction from which it was derived because wax (normal paraffins) is removed by distillation or refining by solvent extraction, and adsorption preferentially removes nonhydrocarbon constituents as well as polynuclear aromatic compounds and the multiring cycloparaffins (Speight and Ozum 2002; Hsu and Robinson 2006; Gary et al. 2007; Speight 2014).

Normal paraffins up to C_{36} have been isolated from petroleum, but it is difficult to isolate any individual hydrocarbon from the lubricant fraction of petroleum. Various methods have been used in the analysis of products in the lubricating oil range, but the most successful procedure involves a technique based on the correlation of simple physical properties, such as refractive index, molecular weight or viscosity, and density. Results are obtained in the form of carbon distribution and the methods may also be applied to oils that have not been subjected to extensive fractionation. Although they are relatively rapid methods of analysis, the lack of information concerning the arrangement of the structural groups within the component molecules is a major disadvantage.

Nevertheless, there are general indications that the lubricant fraction contains a greater proportion of normal and branched paraffins than the lower boiling portions of petroleum. For the poly-cycloparaffin derivatives, a good proportion of the rings appear to be in condensed structures, and both cyclopentyl and cyclohexyl nuclei are present. The methylene groups appear principally in unsubstituted chains at least four carbon atoms in length, but the cycloparaffin rings are highly substituted with relatively short side chains.

Mononuclear, dinuclear, and trinuclear aromatic compounds seem to be the main constituents of the aromatic portion, but material with more aromatic nuclei per molecule may also be present. For the dinuclear aromatic constituents, most of the material consists of naphthalene types. For the trinuclear aromatic constituents, the phenanthrene type of structure predominates over the anthracene type. There are also indications that the greater part of the aromatic compounds occur as mixed aromatic–cycloparaffin compounds.

Base oils come from distillation or other refinery processes carried out on certain types of crude oil, and selected according to the characteristics required of the lubricants. During the process, paraffin wax is removed from the base oil because of the tendency of the wax to foul the base oil and render the lubricating oil unsuitable for the designated service use. The wax is not a waste product but is used to make a wide range of products such as candles. Paraffin wax is an inert, impermeable, shiny, and biodegradable product that burns without harmful or corrosive fumes being given off.

Once the purification process has been performed, base oils can be conveniently subdivided into five categories (API 1509 2012, Appendix E):

1. Group I base stocks contain less than 90% v/v saturated constituents or greater than 0.03% w/w sulfur and have a viscosity index greater than or equal to 80 and less than 120—the temperature range for the use of these oils is from 0°C to 65°C (32°F–150°F) and the base oils of this group are prepared by solvent refining and tend to be the cheapest base oils on the market.
2. Group II base stocks contain greater than or equal to 90% v/v saturated constituents and less than or equal to 0.03% w/w sulfur and have a viscosity index greater than or equal to 80 and less than 120—these base oils are often manufactured by hydrocracking, which is a more complex process than what is used for Group I base oils and because all of the hydrocarbon constituents of these oils are saturated, Group II base oils have better anti-oxidation properties; they also have a clearer color and cost more in comparison to Group I base oils. Group II base oils are becoming very common on the modern market and are priced very close to Group I oils.
3. Group III base stocks contain greater than or equal to 90% saturates and less than or equal to 0.03% w/w sulfur and have a viscosity index greater than or equal to 120—these oils are refined to a greater extent than Group II base oils and generally are severely hydrocracked (higher pressure and heat)—this more severe process is designed to achieve a purer base oil. Although made from petroleum, Group III base oils are sometimes described as synthesized hydrocarbons. Like Group II base oils, these oils are also becoming more prevalent on the modern market.

4. Group IV base stocks are composed of poly α-olefins (PAO)—these synthetic base oils have a much broader temperature application range and are very suitable for use in extreme cold conditions and high heat applications.
5. Group V base stocks, initially included all other base stocks not included in Groups I, II, III, and IV, but the group has been expanded to include base oils such as those produced from silicone polymers, phosphate ester, polyalkylene glycol, polyol-ester, and oils made from biomass sources (Speight 2008, 2011). These base oils are at times mixed with other base stocks to enhance the properties of the base oil. An example would be PAO-based compressor oil that is mixed with a polyol-ester. Esters are common Group V base oils used in different lubricant formulations to improve the properties of the existing base oil. Ester oils can take more abuse at higher temperatures and will provide superior detergency compared with PAO synthetic base oil, which in turn increases the hours of use.

To summarize again, almost every lubricant used in plants today started off as base oil. The API has categorized base oils into five categories. The first three groups are refined from petroleum crude oil. Group IV base oils are fully synthetic (PAO) oils. Group V is for all other base oils not included in Groups I through IV. Before all the additives are added to the mixture, lubricating oils begin as one or more of these five API groups.

On the other hand, the SAE (SAE International) viscosity grades constitute a classification for engine lubricating oils in rheological terms only and are intended for use by engine manufacturers in determining the engine oil viscosity grades to be recommended for use in their engines and by oil marketers in formulating and labeling their products.

Finally, it is noteworthy that the use of base oils in modern systems has changed dramatically in comparison to those in use two decades ago. For example, modern Group II base oils are the most commonly used base oils and make up 47% v/v of the plant capacity compared with 21% v/v for both Group II and III base oils two decades ago. Currently, Group III base oils account for less than 1% v/v of the plant capacity and Group I base oils, which previously made up 56% v/v of the plant capacity, now account for 28% v/v of modern plant capacity.

1.5 USES

Lubricating oils may be divided into many categories according to the types of service they are intended to perform. However, there are two main groups: (1) oils used in intermittent service, such as motor and aviation oils, and (2) oils designed for continuous service, such as turbine oils. Lubricating oil is distinguished from other fractions of crude oil by a high boiling point (>400°C, >750°F), as well as a high viscosity and, in fact, lubricating oil is identified by viscosity.

This classification is based on the SAE J 300 specification. The single-grade oils (e.g., SAE 20) correspond to a single class and have to be selected according to engine manufacturer specifications, operating conditions, and climatic conditions. At −20°C (68°F), multigrade lubricating oil such as SAE 10W-30 possesses the viscosity of a 10W oil and at 100°C (212°F) the multigrade oil possesses the viscosity of an SAE 30 oil.

Oils used in intermittent service must show the least possible change in viscosity with temperature; that is, their viscosity indices (Speight 2014) must be high. These oils must be changed at frequent intervals to remove the foreign matter collected during service. The stability of such oils is therefore of less importance than the stability of oils used in continuous service for prolonged periods without renewal. Oils used in continuous service must be extremely stable, but their viscosity indices may be low because the engines operate at fairly constant temperature without frequent shutdown.

The pour point is an indicator of the ability of lubricating oil to flow at cold operating temperatures. The pour point of a fluid can be lowered with additives, called pour point depressants, also known as cold flow improvers. As the lube oil sample is cooled, small wax crystals form, the temperature at which this occurs is the cloud point. As the sample is cooled further, the crystals agglomerate and grow in size until the entire sample solidifies. Most pour point depressants do not alter the initial formation temperature of the crystals and, thus, they do not generally affect the cloud point. Rather, they inhibit the crystals from combining and growing to a size large enough to plug filters.

Many researchers have tried to find a universal mechanism to explain the change of the wax crystals in habit and particle size, from which the product design of pour point depressants according to different kinds of diesel oils would benefit. Adsorption, co-crystallization, nucleation, and improved wax solubility have been accepted as the most widely used theories in explaining the mechanism (Zhang et al. 2004; Chen et al. 2010). Many kinds of polymers that can be used as pour point depressants have been developed to influence the behavior of paraffin crystallite formation (Zhang et al. 2003; Soldi et al. 2007; Al-Sabagh et al. 2009; Chen et al. 2009). Common pour point depressants include wax-alkylated naphthalene derivatives, phenol derivatives, polymethacrylate derivatives, and styrene–ester copolymers.

The most important single property of lubricating oil is its viscosity. Selected chemicals (Nehal 2008), could be blended with the lubricating oils to impart to them certain specific properties. Resistance of a lubricant to viscosity change with temperature is determined by its viscosity index, which is an arbitrary number calculated from the observed viscosities at two widely separated temperatures. The normal range of viscosity index is from 0 up to 100. Oils of high viscosity index could resist excessive thinning at high temperatures, whereas those of low viscosity index experience an extreme thinning at high temperatures.

The viscosity index is very important because it indicates the suitability of an oil to lubricate properly at the elevated temperature of the engine. At the same time, lubricating oil should be moderately viscous for operating during starting. Polyisobutylene and polymethacrylate are used as viscosity index improvers for lubricating oil.

Viscosity index improvers can be regarded as the key to high-performance multigrade oil. They are generally oil-soluble polymers, and oils containing viscosity index improvers can achieve a viscosity index up to 150. Three major families of viscosity index improvers are known: (1) olefin copolymers, (2) hydrogenated diene copolymers, and (3) acrylic acid–based copolymers.

1.6 SYNTHETIC LUBRICATING OIL

Synthetic lubricating oil is a lubricant consisting of chemical compounds that are artificially synthesized, rather than occurring naturally as petroleum constituents (Rudnick and Shubkin 1999). Synthetic oils may cover a wide range of chemicals but are generally found within the following categories: (1) synthetic hydrocarbons, (2) hydrocarbon esters, (3) phosphate esters, (4) glycols, (5) chlorinated hydrocarbons, and (6) silicone oils. Synthetic hydrocarbons are similar in composition to those found in petroleum base oils, but they are synthesized using chemical processes in which basic carbon and hydrogen compounds are combined. Generally, they are incorporated into oils to impart to the finished product enhanced levels of the properties possessed by the petroleum distillate feedstocks.

Synthetic oils were originally designed for the purpose of having very pure base oils with excellent properties. Thus, synthetic lubricating oil can be manufactured using chemically modified petroleum constituents rather than whole crude oil, but can also be synthesized from other raw materials. Synthetic lubricating oil is used as a substitute for lubricating oil produced from petroleum when operating in extremes of temperature because, in general, it provides superior mechanical and chemical properties than those found in traditional mineral oil. Aircraft jet engines, for example, require the use of synthetic oils, whereas aircraft piston engines generally do not.

Synthetic lubricating oils provide a more advanced fluid option, giving greater performance and reliability through more enhanced mechanical and chemical properties than traditional mineral oils. Several types of synthetic oils exist, all containing organic compounds or synthetic hydrocarbons. Contrary to what most people believe, engines that run with synthetic oils require as much maintenance as engines running with nonsynthetic oils.

Synthetic lubricating oils consist of the following classes of lubricants: (1) PAO—API Group IV base oils; (2) synthetic esters—API Group V base oils (non-PAO) synthetics such as diesters, polyol-esters, alkylated naphthalene derivatives, and alkylated benzenes; and (3) hydrocracked/hydroisomerized—API Group III base oils prepared by processes involving the catalytic conversion of feedstocks under pressure in the presence of hydrogen into high-quality mineral lubricating oil.

Semisynthetic lubricating oils (synthetic blends) are blends of mineral oil with less than 30% v/v synthetic oil and are designed to have many of the benefits of synthetic oil without matching the cost of pure synthetic oil. Lubricating oils that have synthetic base stocks even lower than 30% v/v, and high-performance additive packs consisting of esters can also be considered synthetic lubricants. In general, the ratio of the synthetic base stock to mineral oil base stock is used to define commodity codes among the customs declarations for tax purposes.

Group II and Group III–type base stocks help formulate more economic type semisynthetic lubricants. Groups I, II, II+, and III–type mineral-based oil stocks are widely used in combination with additive packages, performance packages, and esters or Group IV PAOs to formulate semisynthetic-based lubricants. Group III base oils are sometimes considered synthetic, but they are still classified as highest-top-level mineral-based stocks.

High-performance synthetic oils—technologically advanced motor fluids for enhanced performance in the most extreme conditions—contain customized additives that add to the engine performance. Friction modifiers, for example, provide more horsepower at better fuel economy. The resistance of the fluid to breakdown (from oxidation) and the formation of deposits make these oils effective lubricants that also protect the engine. In preparing the perfect high-performance synthetic oil, oil manufacturers have tried different compositions of additives and synthetic fluids to get the mixture with the most desirable attributes, modifying high-performance oils for viscosity grade and other engine dynamics by adding or subtracting these additives from the base fluid.

1.6.1 Poly Alpha-Olefins

PAOs are the most popular synthetic lubricants—the chemical structure and properties are similar to those of mineral oils. PAOs (synthetic hydrocarbons) are manufactured by polymerization of hydrocarbon molecules (α-olefins) in the presence of a catalyst.

A polyolefin (poly-alkene) is a polymer that is produced from a simple olefin (alkene, C_nH_{2n}) as a monomer. A more specific type of olefin is a PAO, a polymer made by polymerizing an α-olefin, which is an alkene where the carbon–carbon double bond starts at the α-carbon atom, that is, the double bond is between the #1 and #2 carbons in the olefin molecule. Furthermore, with increasing options of catalytic processes available, it is now possible to produce tailor-made PAOs and thus control the properties of the end products, such as viscosity and the viscosity index.

Many PAOs have flexible alkyl branching groups on every other carbon of their polymer backbone chain. These alkyl groups, which exist in numerous three-dimensional conformations (molecular shapes), make it very difficult for the polymer molecules to align themselves side-by-side in an orderly way. Therefore, many PAOs do not crystallize or solidify easily and are able to remain oily, viscous liquids even at lower temperatures. Because of this property, low–molecular weight PAOs are useful as synthetic lubricants such as synthetic motor oil for vehicles used in a wide range of temperatures (Ray et al. 2012).

The hydrogenated PAOs result from the oligomerization of decene-1 ($C_{10}H_{20}$, $CH_3CH_2CH_2CH_2CH_2CH_2CH_2CH_2CH{=}CH_2$) as well as from dodecene-1 ($C_{12}H_{24}$, $CH_3CH_2CH_2CH_2CH_2CH_2CH_2CH_2CH_2CH_2CH{=}CH_2$), which are obtained by the polymerization of ethylene ($CH_2{=}CH_2$). The resulting oligomers are hydrogenated and distilled in different fractions. In terms of manufacture, these base oils are completely paraffinic in nature and do not contain aromatics or various heteroatoms. They are characterized by a high viscosity index, ease of adjusting volatility by distillation, and good cold-flow behavior. In fact, with regard to the manufacture of conventional base oils, synthetic oils are tailor-made.

1.6.2 Poly Internal-Olefins

Poly internal-olefins are long-chain hydrocarbons, typically a linear backbone with some branching randomly attached and are obtained by oligomerization of internal *n*-olefins (e.g., $R^1CH=CHR^2$, where R^1 and R^2 are various alkyl groups; Corsico et al. 1999; Rudnick 2013).

The poly internal-olefins have good thermal and oxidative stability and are used extensively in synthetic lubricating oil formulations. Poly-internal olefins offer similar performance to PAO oils and are a high-quality base stock for modern lubricant formulations.

1.6.3 Poly-Butenes

Poly-butene is typically made from the C_4 stream (butene-1, butene-2, and isobutylene) from a catalytic cracker after the stream is treated to remove sulfur (Speight and Ozum 2002; Hsu and Robinson 2006; Gary et al. 2007; Speight 2014). The ethylene steam cracker C_4 stream is also used as supplemental feedstock for the production of poly-butene. On the other hand, poly-isobutylene is produced from pure isobutylene made in the petrochemical complex of a major refinery. The presence of isomers other than isobutylene can have several effects including: (1) lower reactivity due to steric hindrance at the terminal carbon in, for example, the manufacture of poly-isobutenyl succinic anhydride dispersant manufacture; (2) the molecular weight–viscosity relationships of the two materials may also be somewhat different (Decroocq and Casserino 2005).

The reaction is an acid-catalyzed carbocation polymerization typically using aluminum chloride ($AlCl_3$) or a hydrogen halide acid such as hydrofluoric acid. The repeat unit in case of 1-butene is $[-CH_2CH(CH_2CH_3)-]_n$ and the repeat unit in case of 2-butene is $[-CH(CH_3)CH(CH_3)-]_n$. One of the end units in the polymer chain contains a double bond, allowing reactivity with other compounds to provide functional chemistry mainly for lubricant additives for engine oils, fuels, and greases (Rudnick 2013).

1.6.4 Organic Esters

An ester oil is a synthetic base oil that has been chemically synthesized. Esters are derived from carboxylic acids (which contain the –COOH functional group) in which the hydrogen atom in the carboxylic acid group is replaced by a hydrocarbon group. This could be an alkyl group like methyl or ethyl, or one containing a benzene ring such as the phenyl group (C_6H_5–). Esters have been used successfully in lubrication for more than 60 years and are the preferred stock in many severe applications in which their benefits solve problems or bring value.

Ester oils are produced by the reaction of acids and alcohols with water. Ester oils are characterized by very good high-temperature and low-temperature tolerance.

There are many types of esters including polyolesters and complex esters. Esters are often used in jet engine lubrication due to their low temperature flow properties with clean, high-temperature capacity. Ester oils are often used in combination with

PAO lubricating oils in fully synthetic motor oils—ester content varies from 5% to 25% w/w depending on the ester and the desired properties. The high-performance properties and custom design versatility of esters is ideally suited to problem solving in the industrial arena. These oils are available in a wide range of viscosities and are more generally used under high temperatures (maximum from 180°C to 230°C, 355°F to 445°F; Rudnick 2013).

Ester lubricants are a broad and diverse family of synthetic lubricant base oils, which can be custom-designed to meet specific physical and performance properties. The inherent polarity of esters improves their performance in lubrication by reducing volatility, increasing lubricity, providing cleaner operation, and making the products biodegradable. A wide range of available raw materials allow an ester lubricant designer the ability to optimize a product over a wide range of variables to maximize the performance and value to the client. They may be used alone in very high-temperature applications for optimum performance or blended with PAOs or other synthetic base stocks in which their complementary properties improve the balance of the finished lubricant. The application of ester lubricants continues to grow as the drive for efficiency makes operating environments more severe. Because of the complexity involved in the design, selection, and blending of an ester base stock, the choice of the optimum ester should be left to a qualified ester engineer who can better balance the desired properties.

The primary structural difference between ester-based lubricating oil and petroleum-based lubricating oil (including PAO lubricating oils) is the presence of oxygen in the hydrocarbon molecules in the form of multiple ester linkages (COOR), which impart polarity to the constituent molecules of the lubricating oil. As a result, this affects the manner in which esters behave as lubricants, such as in the following properties: (1) volatility, (2) lubricating behavior, often referred to as lubricity, (3) detergent properties and ability to act as a dispersant, and (4) biodegradability.

In terms of volatility, the polarity of the ester molecules causes an intermolecular attraction (through hydrogen bonding) causing them to be attracted to one another, and this intermolecular attraction requires more energy (heat) for the esters to transfer from a liquid to a gaseous state. Therefore, at a given molecular weight or viscosity, the ester-based lubricating oils will exhibit a lower vapor pressure, which translates into a higher flash point and a lower rate of evaporation for the lubricant. Generally, the more ester linkages in a specific ester, the higher its flash point and the lower its volatility.

The lubricating behavior of ester-based lubricating oils is also subject to the polar effect of the ester group. Polarity also causes the ester moieties to be attracted to positively charged metal surfaces. As a result, the molecules tend to line up on the metal's surface, creating a film that requires additional energy (load) to wipe off. The result is a stronger film, which is reflected in better lubricating behavior and lower energy consumption in lubricant applications.

The polar nature of esters also introduces detergent and dispersant properties, which makes the ester-based oils good solvents and dispersants. This allows the esters to solubilize or disperse oil degradation by-products, which might otherwise be deposited as sludge, and is manifested as cleaner operation and improved additive solubility in service.

Although ester-based oils have enhanced stability (compared with petroleum-based oils) against oxidative and thermal breakdown, the ester linkage provides a vulnerable site for microbes and the rate of biodegradability of the ester-based oils is enhanced.

However, as with any product, there are also disadvantages to the use of ester-based lubricating oils. The most important issue is the compatibility of the ester base stocks with the elastomer material used in the machinery seals. Esters (and there are few if any exceptions) will swell and soften most elastomer seals but, fortunately, the degree to which the seals swell can be controlled through proper selection. When used as the exclusive base stock, the ester should be designed for compatibility with seals or the seals should be changed to those types that are more compatible with esters.

Another potential disadvantage with esters is their ability to react with water or hydrolyze under certain conditions. Generally this hydrolysis reaction requires the presence of water and heat with a relatively strong acid or base to act as a catalyst. Because esters are usually used in very high temperature applications, high amounts of water are usually not present and hydrolysis is rarely a problem in actual use. Where the application environment may lead to hydrolysis, the ester structure can be altered to greatly improve its hydrolytic stability and additives can be selected to minimize any effects.

1.6.5 Polyglycols

Polyglycols are produced by the oxidation of ethylene and propylene. The oxides are then polymerized, resulting in the formation of polyglycols, which are water soluble. Polyglycols are characterized by very low coefficients of friction. They are also able to withstand high pressures without extreme pressure additives (Rudnick 2013).

1.6.6 Silicones

Silicones are a group of inorganic polymers, molecules of which represent a backbone structure built from repeated chemical units (monomers) containing Si=O moieties. Two organic groups are attached to each Si=O moiety: for example, methyl + methyl [$(CH_3)_2$], methyl + phenyl (CH_3+ C_6H_5), phenyl + phenyl [$(C_6H_5)_2$]. The most popular silicone is polydimethylsiloxane—the monomer is $(CH_3)_2SiO$. Polydimethylsiloxane is produced from silicon and methyl chloride. Other examples of silicones are polymethylphenylsiloxane and polydiphenylsiloxane.

The viscosity of silicones depends on the length of the polymer molecules and on the degree of cross-linking. Short non–cross-linked molecules make fluid silicones, whereas long cross-linked molecules result in elastomeric silicones. Silicone lubricants (oils and greases) are characterized by a broad temperature range: –73°C to 300°C (–100°F to 570°F; Rudnick 2013).

1.6.7 Blended Oils

Blended oils (compounded oils) are lubricating oils in which the properties are improved by the addition of additives. Additives enhance certain desirable properties of the base lubricating oil or give it new desired properties. Additives that improve oiliness include fatty acids, castor oil, and coconut oil. Antioxidant additives include phenolic compounds and amino compounds. Antifoaming agents that are added to lubricating oils include glycerol and glycols, emulsifying additives include sulfuric acid and sodium salts, and viscosity index–improving additives include compounds with high molecular weights, such as hexanol ($C_6H_{13}OH$; *n*-hexanol, boiling point 158°C, 315°F).

However, modern high-performance lubricants are specifically formulated with a carefully selected balance of performance additives and base stocks to match the lubrication requirements of the equipment in which they are used. Thus, blending oils should be performed with extreme care—when lubricants are mixed, this balance is often upset. Mechanical problems leading to shorter equipment life can occur, sometimes catastrophically.

A general rule is that lubricants made with synthetic base stocks should not be mixed with products made with mineral oil, even if they are designed for the same application. The limited exceptions include some PAO and ester-based products. Even then, compatibility is often concentration-dependent. Deposits may form because of additive incompatibility or seal compatibility may be compromised.

1.6.8 Biolubricants

Vegetable lubricants are based on soybean, corn, castor, canola, cottonseed, and rapeseed oils. Vegetable oils are an environmentally friendly alternative to mineral oils because they are biodegradable. Lubrication properties of vegetable base oils are identical to those of mineral oils. The main disadvantages of vegetable lubricants are their low oxidation and temperature stabilities.

Animal lubricants are produced from animal fat. There are two main animal fats: hard fats (stearin) and soft fats (lard). Animal fats are mainly used for manufacturing greases. Animal oils and vegetable oils are extracted from animal and vegetable sources. They are typically triglycerides—the chains of fatty acids that are bound with a single molecule of glycerol. Animal and vegetable oils are high in viscosity and provide good lubrication. Animal oils include seal oil, whale oil, and lard oil. Vegetable oils include cottonseed oil, olive oil, mustard oil, and palm oil.

Animal and vegetable oils have some disadvantages: they oxidize easily to form acidic products, they are expensive, and they get hydrolyzed easily in the presence of moist air or another aqueous medium. Chemical additives are added to animal and vegetable oils to improve the properties.

Vegetable oils as new renewable raw materials for industrial applications such as lubricants have been of great importance because of the emphasis on environmentally friendly lubricants, and are attractive due to the rapid depletion of world fossil fuel reserves and increasing concern for environmental pollution from excessive mineral oil use (Chauhan and Chhibber 2013).

Vegetable oils with high oleic contents are considered to be the best alternative for substituting conventional mineral oil–based lubricating oils and synthetic esters. Vegetable oils are preferred over synthetic fluids because they are more ecofriendly, renewable resources and cheaper too. The properties of vegetable oils, fatty esters, chemically modified esters, and synthetic esters are relevant for performance as lubricants in various industrial applications such as hydraulic oils, refrigeration oils, chainsaw lubricants, metal-working fluids, engine oils, and two-stroke engine oils (Chauhan and Chhibber 2013).

Advantages such as high lubricity, appropriate viscosity–temperature relationship, low lubricant consumption, and energy efficiency combined with suitable public health, safety, and environmental contamination records, more than offset the disadvantages of initial costs in most of these applications. It has been suggested that modified and stabilized oils from wastes of forest origin and other nonedible oils and their chemically modified derivatives can be produced at relatively cheaper cost than similar oils marketed in the developed world and can be introduced in India with immense environmental and performance benefits.

1.7 BASE OIL PROPERTIES

Base oils are manufactured to specifications that place limitations on their physical and chemical properties, and these in turn establish parameters for refinery operations (Bartz 1993; Sequeira 1994). Because of the differences in petroleum composition and subtle differences in refining process parameters (Speight and Ozum 2002; Hsu and Robinson 2006; Gary et al. 2007; Speight 2014), base stocks from different refineries will generally not be identical, although they may have some properties (e.g., viscosity at a particular temperature) in common. At this point, it is worth briefly reviewing what measurements are involved in these specifications, what they mean, and where in the process they are controlled.

The various properties are listed below in alphabetical order for the convenience of the reader.

1.7.1 Aniline Point

The aniline point is a measure of the ability of the base stock to act as a solvent and is determined from the temperature at which equal volumes of aniline and the base stock are soluble (ASTM D611). High aniline points imply a paraffinic base stock, whereas low aniline points (<100°C) imply a naphthenic or aromatic stock.

1.7.2 Cloud Point

The cloud point is the temperature at which wax crystals first form as a cloud of microcrystals. It is therefore higher than the pour point, at which crystals are so numerous that flow is prevented. The long-standing ASTM method is D2500, with three new automated methods being ASTM D5771, ASTM D5772, and ASTM D5773. Many base stock inspection sheets no longer provide cloud points. Cloud points can be 3°C to 15°C above the corresponding pour points.

1.7.3 Color

Solvent-extracted/solvent-dewaxed stocks will retain some color as measured by ASTM D1500. Hydrocracked stocks, when hydrofinished at high pressures, are usually water white and their color is best measured on the Saybolt color scale (ASTM D156, ASTM 2012).

1.7.4 Density and API Gravity

Knowledge of density is essential when handling quantities of the stock and the values can also be seen to fit with the base stock types. An alternative measure is the API gravity scale where:

$$\text{API gravity} = (141.5/\text{specific gravity}) - 131.5$$

Density increases with viscosity, boiling range, and aromatic and naphthene content, and decreases as isoparaffin levels increase and as the viscosity index increases.

1.7.5 Distillation

At one time, this would have been carried out using an actual physical distillation using either ASTM D86, a method performed at atmospheric pressure and applicable to very light lubes, or by vacuum distillation according to ASTM D1160, for heavier ones. Neither of these methods is employed much for base stocks nowadays because of their time and manpower requirements. Distillation today is usually performed using gas chromatography and the method is commonly called either simulated distillation (SimDist) or gas chromatographic distillation using ASTM D2887. This method is capable of excellent accuracy, repeatability, and fast turnaround times and is normally automated. It is applicable to samples with final boiling points of less than 538°C (1000°F).

1.7.6 Flash Point

The flash point is the temperature at which there is sufficient vapor above a liquid sample to ignite when exposed to an open flame and is a significant feature in product applications in which it is used as a common safety specification. Flash points are a reflection of the boiling point of the material at the front end of the base stock's distillation curve. Flash points generally increase with viscosity grade and a high flash point for a given viscosity is desirable. Careful fractionation so that undesirable constituents are rejected and increased viscosity index of the base stock favor higher flash points. The Cleveland Open Cup test method (ASTM D92) is the most often cited for North American base stocks, although the Pensky–Martens test (ASTM D93) is sometimes used.

1.7.7 Pour Point

The pour point measures the highest temperature at which a base stock no longer flows, and for paraffinic base stocks, pour points are usually between −12°C (10°F) and −15°C (5°F), and are determined by operation of the dewaxing unit. For specialty purposes, pour points can be much lower. The pour points of naphthenic base stocks, which can have a very low wax content, may be much lower (−30°C to −50°C, −22°F to −58°F). For very viscous base stocks such as bright stocks, pour points may actually reflect a viscosity limit.

The pour point is measured traditionally by ASTM D97, but three new automated equivalent test methods are the "tilt" method (ASTM D5950), the pulse method (ASTM D5949), and the rotational method (ASTM D5985).

1.7.8 Refractive Index and Refractivity Intercept

The refractive index is used to characterize base stocks, with aromatic ones having higher values than paraffinic ones. The value increases with molecular weight.

The significance of refractive index data is the relationship between the refractive index (n) and the density (d) of hydrocarbons. Thus,

$$\text{Refractivity intercept} = n - d/2$$

This is characteristic of each hydrocarbon series and has been shown to represent a more constant relationship between the density and refractive index of the members of a homologous series of hydrocarbons than the usual equations for refractivity. This is a means of characterizing the composition of the sample—values range from 1.03 to 1.047.

1.7.9 Viscosity

Base stocks are primarily manufactured and sold according to their viscosities at either 40°C (104°F) or 100°C (212°F), using kinematic viscosity. Viscosity grades are now defined by kinematic viscosity in centistokes (cS) at 40°C (104°F); formerly, they were established on the SUS scale at 38°C (100°F). Higher viscosity base stocks are produced from heavier feedstocks (e.g., a 100 cS at 40°C oil is produced from heavy vacuum gas oil and cannot be made from a light vacuum gas oil because the molecular precursors are not present). As viscosity increases, so does the distillation midpoint.

1.7.10 Viscosity–Gravity Constant

The viscosity–gravity constant is an indicator of base stock composition and solvency that is calculated from the density and viscosity according to ASTM D2501; it usually has a value between 0.8 and 1.0. High values indicate higher solvency and therefore greater naphthenic or aromatic content. This is usually of interest for naphthenic stocks.

1.7.11 Viscosity Index

Different oils have different rates of change of viscosity with temperature. For example, distillate oil from naphthenic base crude would show a greater rate of change of viscosity with temperature than would a distillate oil from a paraffin crude. The viscosity index is a method of applying a numerical value to this rate of change, based on comparisons with the relative rates of change of two arbitrarily selected types of oil that differ widely in this characteristic. A high viscosity index indicates a relatively low rate of change of viscosity with temperature, whereas a low viscosity index indicates a relatively high rate of change of viscosity with temperature.

The viscosity index is a measure of the extent of viscosity change with temperature; the higher the viscosity index, the less the change, and generally, higher viscosity indices are preferred. The viscosity index is usually calculated from measurements at 40°C (104°F) and 100°C (212°F). The minimum viscosity index for a paraffinic base stock is 80 but, in practice, the norm is 95, established by automotive market needs. Naphthenic base stocks may have a viscosity index of approximately zero. The conventional solvent extraction/solvent dewaxing route produces base stocks with a viscosity index of approximately 95. Lower raffinate yields (higher extract yields) in solvent-refining mean a higher viscosity index, but it is difficult economically to go much higher than 105. In contrast, conversion processes enable a wide viscosity index range of 95 to 140 to be attained, with the final product's viscosity index depending on the feedstock viscosity index, first-stage reactor severity, and the dewaxing process. Dewaxing by hydroisomerization gives the same or higher viscosity index relative to the solvent dewaxing product. To obtain a viscosity index greater than 140, the feedstock generally must be either petroleum wax or Fischer–Tropsch wax (Chadeesingh 2011; Speight 2014).

1.7.12 Volatility

Volatility has emerged as a significant factor in automotive lubricant products from environmental and operational standpoints and again pertains predominantly to the distillation front end.

Low volatility (minimal losses at high temperatures) reduces emissions, is beneficial for emission control catalysts, reduces oil consumption, and helps prevent engine oil viscosity changes. Volatility is obviously affected by viscosity grade, but for a constant viscosity is established in part by sharper fractionation and in part by VI. It is measured either by the Noack method (ASTM D5800), using a thermogravimetric analyzer (TGA) method (namely ASTM D6375), or by gas chromatography (ASTM D6417) for engine oils.

1.8 ADDITIVES

Lubricating oil is not produced by any one refining process—the product of a refinery is, initially, base oil that might arise from two or more processes. The base oils are blended, depending on their respective properties, and the lubricating oil made by introducing proper additives to base oil(s) cuts manufactured from paraffin or mixed-base crude oils. The main functions of lubricating oil include reducing friction, carrying away heat, protecting against rust, protecting against wear, and removing contaminants from the engine. Additives are used to enhance the natural properties of the lubricating oil and to prevent undesirable properties (Rudnick 2009).

Thus, additives are chemical compounds added to lubricating oils to impart specific properties to the finished oils (Ludema 1996; Leslie 2003; Rizvi 2009; Ahmed and Nassar 2011). Some additives impart new and useful properties to the lubricant, some enhance properties already present, whereas some additives act to reduce the rate at which undesirable changes take place in the product during the service life of the lubricating oil.

The quality and performance of finished lubricating oil is improved by the addition of additives, which began to be used widely in 1947 when the API started to categorize engine oils by severity of service: regular, premium, and heavy duty. Additives were used to extend the life of premium and heavy-duty oils only. In 1950, multigrade oils were first introduced and its performance (viscosity index) was enhanced with the use of polymer additives, which improved the hot and cold performance of the oil. For several more decades, the lubricants industry continued to rely heavily on additive technology to improve the performance of finished oils. Lubricant quality improved significantly only when the additive chemistry improved, and was the predominant form of improving lubricating oil performance until a significant improvement in base oil technology became available.

Each additive is selected for its ability to perform one or more specific functions in combination with other additives. Selected additives are formulated into packages for use with a specific lubricant base stock and for a specified end-use application. The largest end use is in automotive engine crankcase lubricants. Other automotive applications include hydraulic fluids and gear oils. In addition, many industrial lubricants and metalworking oils also contain lubricating oil additives. The major functional additive types are dispersants, detergents, oxidation inhibitors, antiwear agents, extreme-pressure additives, and viscosity index improvers.

Typically, various additives are present in lubricating oil in concentrations on the order of 12% to 15% w/w and they play a major role in assuring that the finished oil meets the required specifications. Because there are differences between the various base oils, it is important to understand that small quality differences that occur between the base oils, either new or regenerated, become insignificant compared with the role played by additives. Without additives, base oil would undergo rapid changes (especially changes induced by oxidation) during its use, leading to a change in base oil viscosity, and the formation of corrosive (oxidation) products, leading to deposits. More precisely, if oxidation is the culprit, the rate of oxidation of a hydrocarbon doubles with each 10°C (18°F) increase in temperature.

Government regulations have had a major effect on the use of lubricating oil additives in the past and are likely to remain important in the future because upgrading lubricants is part of the effort to improve fuel economy and to meet more stringent emission-control requirements.

Additives impart special performance features to the finished oil. The choice of additives and the balance among them differentiate an antiwear hydraulic oil from a turbine oil, for example. Some additives affect the physical properties of the finished lubricant. Others change the lubricant's chemical properties or are added for cosmetic purposes.

Some lubricants are incompatible because of differences in additive chemistry that lead to undesirable chemical reactions. If these oils are mixed, insoluble material may form and then deposit onto sensitive machine surfaces. For a hydraulic fluid, this could lead to lubricant starvation, valve failure, or increased wear.

A second form of lubricant incompatibility is more difficult to handle because no visible changes occur when the products are mixed (Mushrush and Speight 1995). The problem appears only after the mixture is used in a piece of equipment that consequently fails or loses performance. Optimum performance requires carefully balanced frictional and antiwear properties in the finished product that can be changed when lubricants are mixed.

1.8.1 Antifoaming Agents

Agitation and aeration of lubricating oil occurring under certain conditions may result in the formation of air bubbles in the oil (foaming). Foaming not only enhances oil oxidation but also decreases the lubrication effect—causing oil starvation.

To decrease the tendency of oil foaming, mostly due to the presence of detergents and dispersing agents, a very small quantity (mg/kg) of antifoam additive is added. Products like silicone or alkyl polymethacrylate with low molecular weight are used: they are insoluble in oil and concentrate at the liquid–air interface. Their weak surface tension inhibits the formation of stable foam by rapid coalescence of air bubbles.

1.8.2 Antioxidants

Mineral oils react with oxygen from the air, forming organic acids. The oxidation reaction products cause an increase of the oil viscosity, formation of sludge and varnish, corrosion of metallic parts and foaming. Antioxidants inhibit the oxidation process of oils.

During the oxidation process, in a median period, oil viscosity is increased although the acidity remains constant because there are alcohols of unsaturated hydrocarbons produced from the decomposition of hydroperoxides process so it neutralizes the effect of acidity formed from another oxygen-bearing compound such as aldehydes, ketones, and acids. For most lubricant applications, trapping catalytic impurities and the destruction of hydrocarbon radicals, alkyl peroxy radicals, and hydroperoxides can be achieved through the use of radical scavengers, peroxide decomposers, and metal deactivators.

An antioxidant can act as a radical inhibitor of one of the steps of oxidation by neutralization of the free radicals. These are compounds such as phenol, alkaline earth phenolates and salicylates, and aromatic amines. In addition, an antioxidant destructive to oxidation products, such as hydroperoxide compounds that could initiate new oxidation chain reactions, is a necessary additive for most lubricating oils.

1.8.3 Antiwear Additives

Friction during service will cause wear on unprotected metal surfaces (Ludema 1996; Leslie 2003; Masabumi et al. 2008; Rizvi 2009). The surfaces of machinery components appear well-finished to the naked eye. When magnified, however, surface imperfections become readily apparent. These microscopic defects (asperities) rub, lock together, and break apart when dry surfaces move relative to one another. The resistance generated when these adjacent surfaces come in contact is called friction. The welding together and breaking apart of asperities is a form of adhesive wear. Another form of wear may occur when a hard contaminant particle becomes trapped between two opposing surfaces. When this occurs, the contaminant acts as a miniature lathe, cutting into the softer machinery surface (abrasive wear). Another consequence of friction is that the energy created by resistance is converted into heat. Therefore, the primary functions of a lubricant are the formation of a protective film between adjacent surfaces to reduce wear and the dissipation of heat generated at these wear surfaces.

To overcome this, additives are incorporated into the lubricating oil, which are then adsorbed onto surfaces in contact, thereby forming a solid protection film. Antiwear additives prevent direct metal-to-metal contact between the machine parts when the oil film is broken down and prolongs machine life due to higher wear and score resistance of the components. The mechanism of antiwear additives is simple: the additive reacts with the metal on the part surface and forms a film, which may slide over the friction surface.

1.8.4 Corrosion and Rust Inhibitors

Another role provided by a lubricant is the prevention of system corrosion. In environments where contamination of the system with water (especially seawater or petroleum brine) is likely, protection of machinery components from corrosion is of the utmost importance—salt water is considerably more corrosive than fresh water and water molecules may also diffuse through the lubricant and enter surface microcracks, causing hydrogen embrittlement and subsequent surface failure. It is thus imperative that water contamination of machinery systems be minimized. To achieve corrosion protection, lubricants must form a protective barrier on machinery surfaces. Modern-day lubricants often contain corrosion inhibitors, which chemically bond to the metallic surfaces of equipment components.

Corrosion inhibitors and rust inhibitors form a barrier film on the substrate surface, thereby reducing the rate of corrosion and the rate of rust formation. The inhibitors also adsorb onto the metal surface forming a film protecting the part from the attack of oxygen, water and other chemically active substances.

Corrosion by organic acids (produced by various oxidation processes) can occur, for example, in the bearing inserts used in internal combustion engines. Some of the metals used in these inserts, such as the lead in copper–lead or lead–bronze, are readily attacked by organic acids in oil. The corrosion inhibitors form a protective film on the bearing surfaces that prevents the corrosive materials from reaching or attacking the metal. The film may be either adsorbed on the metal or chemically bonded to it. It has been found that the inclusion of highly alkaline materials in the oil will help neutralize these strong acids as they are formed, greatly reducing corrosive wear.

These types of additives can be either (1) oxygenated inhibitors or (2) nitrogenous inhibitors. The oxygenated inhibitors are essentially carboxylic acids with long organic chains whereas the nitrogenous inhibitors are essentially fatty amines and their derived products. Detergent additives and dispersing agents also have antirust properties—rust is formed by the combined action of water and oxygen in the air on the iron, resulting in the formation of ferrous hydroxides and then ferric hydroxides. Corrosion is due more specifically to the action of the acidity of sulfur compounds and of acids resulting from the oxidation of oil or fuel.

An oxidation inhibitor, not to be confused with a rust inhibitor, is used to prevent or eliminate the oxidation reactions of lubricating oil. The conditions for oxidation reactions are present in the crankcase of the engine. These reaction conditions include high temperature, the presence of air, and the presence of metal in the engine—the metals act as a catalyst for the oxidation reactions. Furthermore, the products of oil oxidations act as the catalyst to further accelerate the rate of oxidation. Oil oxidation usually occurs in a series of steps in which substances are formed such as petroleum hydrocarbons, aldehydes, organic acids, resinous substances, varnish-like materials, lacquers, carbonaceous material, and coke (Speight 2014). In actual practice, the amount of lubricating oil that is lost in the oxidation reactions is small. However, the products of oil oxidation interfere with lubrication by forming sludge and corrosive by-products to the extent that the oil must be either replaced or purified. A variety of organic chemical compounds have been used as oxidation inhibitors such as sulfur compounds, phosphorus compounds, sulfur/phosphorus compounds, amines and phenolic derivatives.

1.8.5 Detergents and Dispersants

Detergent and dispersant additives are used to keep particulate matter in suspension or dispersion. Particulate matter consists of fuel soot, resins, and oil oxidation products. By remaining suspended, these products will neither adhere to metal surfaces (such as varnish or lacquers) nor settle out in the engine sump (such as sludge). Detergent additives also have antioxidant and anticorrosive properties, whereas the dispersant, being polymeric in nature, may also improve the viscosity index.

Detergents neutralize strong acids present in the lubricant (for example, sulfuric and nitric acid produced in internal combustion engines as a result of the combustion process) and remove the neutralization products from the metal's surface. Detergents also form a film on the part surface preventing high-temperature deposition of sludge.

The role of detergents is to prevent deposits on the surfaces of the engine at high temperature and to keep the lubricant distribution network clean. These additives can be made to have an alkaline reserve by incorporating colloidal calcium or magnesium carbonates. This colloidal dispersion is absolutely limpid and its solution in oil is completely stable despite the addition of a quantity of carbonate up to 35% w/w of the oil.

The reserve of alkalinity neutralizes the acids formed during the oxidation of oil or resulting from the combustion of the fuel. These additives are calcium or magnesium salts of organic acids. The most current ones are natural or synthetic sulfonate derivatives characterized by molecular weights high enough to confer on them a sufficient oleophilic character in the oil medium. The natural additives are generally obtained as by-products during the manufacture of white oil, whereas synthetic additives are often synthesized from aromatic heavy alkylate formed from the alkylation of aromatic hydrocarbons by heavy olefins. A second frequently used type contains phenolates in which the metal is either magnesium or calcium. Using phenol itself as a simple example:

$$C_6H_5OH + Mg^{2+} \rightarrow C_6H_5O\text{–}Mg^{2+}\text{–}OH_5C_6 + H_2O$$

$$C_6H_5OH + Ca^{2+} \rightarrow C_6H_5O\text{–}\ Ca^{2+}\text{–}OH_5C_6 + H_2O$$

Dispersants keep the foreign particles present in a lubricant in a dispersed form (finely divided and uniformly dispersed throughout the oil). Typically, foreign particles are materials such as sludge, dirt, oxidation products, and water.

Dispersing additives came into use owing to the necessity to maintain in fine suspension the materials susceptible to settling in the lubrication circuits. This property was improved owing to the development of additives without ash; the first ones of this type proposed on the market were alkenyl succinimides, which are surfactants with a polybutene moiety as the oleophilic part of the molecule and with molecular mass in the range of 800 to 1500.

The polar (often nitrogenous) part is adsorbed on particles (dust, water, soot, metals from wear, and solid residues of oxidation) and stabilizes the particles in the oil medium. Other molecules of dispersing additives without ashes, including the same oleophilic radical, are also marketed as a Mannich base (a beta-amino ketone, which is formed in the reaction of an amine [RNH_2], formaldehyde [$CH_2{=}O$], or an aldehyde [$RCH{=}O$] and a carbon acid) or succinic esters.

1.8.6 Pour Point Depressants

The pour point is the lowest temperature at which the lubricating oil may still flow. Wax crystals formed in mineral oils at low temperatures reduce their fluidity and pour point depressants inhibit the formation and agglomeration of wax particles, keeping the lubricant fluid at low temperatures. Thus, pour point depressants are used to lower the pour point temperature of the lubricating oil below the starting temperature of the engine.

Pour point depressants hinder the process of growth of the crystals of paraffin wax crystals, which form in the oil at low temperatures. Polymethacrylates with low molecular masses are used—the same effect is also obtained with some additives used to improve the viscosity index.

Petroleum-based lubricating oils could contain some dissolved wax and, because the oil is chilled, this wax begins to separate as crystals that interlock to form a rigid structure that traps the oil in small pockets in the structure. When this wax crystal structure becomes sufficiently complete, the oil will no longer flow under the conditions of the test. However, because mechanical agitation can break up the wax structure, it is possible to have an oil flow at temperatures considerably below its pour point. Cooling rates also affect wax crystallization; it is possible to cool an oil rapidly to a temperature below its pour point and still have it flow.

Although the pour point of most oils is related to the crystallization of wax, certain oils, which are essentially wax-free, have viscosity-limited pour points. In these oils, the viscosity becomes progressively higher as the temperature is lowered until at some temperature no flow can be observed. The pour points of such oils cannot be lowered with pour point depressants because these agents act by interfering with the growth and interlocking of the wax crystal structure.

1.8.7 Viscosity Index Improvement

The viscosity of lubricating oil sharply decreases at high temperatures. Low viscosity causes a decrease in the oil's lubrication ability. Viscosity index improvers keep the viscosity at acceptable levels, which provide a stable oil film even at increased temperatures. Viscosity improvers are widely used in multigrade oils, the viscosity of which is specified at both high and low temperatures.

Viscosity index improvers are long-chain, high–molecular weight polymers that function by causing the relative viscosity of an oil to increase more at high temperatures than at low temperatures. Generally, this result is due to a change in the polymer's physical configuration with increasing temperature of the mixture.

In practice, the lubricant oil fraction is obtained and separated into fractions in the vacuum distillation column, which is fed by the residue of the atmospheric column, also called *topping*. The remaining oil present in the residual fraction is obtained by propane deasphalting to get the bright stock, which corresponds to the most viscous fraction of the recoverable oil. The vacuum distillates and bright stock so obtained cannot be used as they are, and are subject to the following treatments: (1) a partial extraction of the aromatic compounds by a solvent (phenol, furfural, or NMP) is necessary to decrease the variation in viscosity of the oil by temperature change, which makes it possible to increase the viscosity index of the lubricating oil in the range of 95 to 105, instead of values lower than 50 for straight-run vacuum distillates, and (2) the elimination of long-chain high–molecular weight constituents, which are responsible for high pour points, by either dewaxing or by a catalytic treatment to crack and hydroisomerize these structures.

Different oils have different rates of change of viscosity with temperature. For example, distillate oil from naphthenic base crude would show a greater rate of change of viscosity with temperature than would a distillate oil from a paraffin crude. The viscosity index is a method of applying a numerical value to this rate of change, based on comparisons with the relative rates of change of two arbitrarily selected types of oil that differ widely in this characteristic. A high viscosity index indicates a relatively low rate of change of viscosity with temperature, whereas a low viscosity index indicates a relatively high rate of change of viscosity with temperature. For example, a high-viscosity index oil and a low-viscosity index oil having the same viscosity at the same temperature would, as the temperature increased, lead to a thicker high-viscosity index oil and, therefore, would have a higher viscosity than the low-viscosity index oil at higher temperatures.

Once the viscosity characteristics and a lower pour point are acquired, a finishing treatment is generally applied through a mild catalytic hydrogenation.

1.9 PERFORMANCE

Base oils made by hydrocracking and early wax isomerization technologies showed favorably differentiated performance, which prompted the API to categorize base oils by composition (API 1509 2012; Table 1.1). Group I base oils are differentiated from Group II base oils because the Group II base oils contain significantly lower levels of aromatic constituents (<10% w/w) and lower amounts of sulfur-containing constituents (<300 ppm w/w sulfur). In addition, Group II oils made using the Chevron hydroisomerization technology (Speight 2014) have almost no color, indicating a high level of purity, which is an indication of longer service life. More specifically, the oil is more inert and has a lower tendency to oxidize to form oxidation by-products that increase base oil viscosity and react with some of the additives, thereby nullifying the effect of the additives. More recently, the selective dewaxing process (ExxonMobil)—an all-hydroprocessing route—has been added to Group II production base oil production.

In summary, production of Group II base oils, along with specially designed additives, has had a significant effect on finished oil performance. In some applications, including turbine oils, lubricating oils formulated with Group II base oils can outperform synthetic oil bases made from PAOs (Kramer et al. 2001).

In terms of the remaining base oils, Group III base oils produced by solvent dewaxing, which are now produced by the Chevron all-hydroprocessing route, have greatly improved oxidation stability and low-temperature performance relative to the solvent-refined Group III base oils. From a processing standpoint, modern Group III base oils are manufactured by essentially the same processing route as modern Group II base oils—the higher viscosity index is achieved by increasing the temperature or time in the hydrocracker. Alternatively, the product viscosity index can be increased by increasing the viscosity index of the process feedstock by selecting the appropriate crude oil as the source of the feedstock (Kramer et al. 2001).

TABLE 1.1
Base Oil Categories as Published by the American Petroleum Institute

	Base Oil Category	Sulfur (%)		Saturates (%)	Viscosity Index
Mineral	Group I (solvent refined)	>0.03	and/or	<90	80–120
	Group II (hydrotreated)	<0.03	and	>90	80–120
	Group III (hydrocracked)	<0.03	and	>90	>120
Synthetic	Group IV	PAO Synthetic Lubricants			
	Group V	All other base oils not included in Groups I, II, III or IV			

Source: API 1509 2012.

These Group III base oils have properties that allow them to perform at a level that is significantly higher than petroleum-derived Group I and Group II base oils, and typically match existing levels of performance established by traditional synthetic base oils.

Generally, base oils produced from PAOs have had superior lubricating performance characteristics such as viscosity index, pour point, volatility, and oxidation stability that could not be achieved with conventional petroleum-derived base oils. However, because viscosity index, pour point, volatility, and oxidation stability can be independently controlled, Group III base oils can be designed and manufactured so that their performance closely matches base oils derived from PAOs in most commercially significant finished lube applications.

Generally, pour point is the one property in which Group III base oils do not meet the performance of base oils derived from PAOs. Although it is certainly true that the pour point of the neat Group III base oil is higher than that of base oil from PAOs of comparable viscosity, it is important to understand that the pour point of the fully formulated lubricant (base oils plus additives) is the critical property. Base oils manufactured with modern isomerization catalysts respond very well to pour point depressant additives.

Oxidation stability and thermal stability are among the most important advantages of synthetic base oil—better base oil stability means better additive stability and longer service life. Group III base oils routinely exceed the performance of base oil produced from PAOs in this area. The stability of modern Group III base oils depends mostly on the viscosity index, which is an indication of the fraction of highly stable isoparaffin structures in the base oil (Kramer et al. 2001). However, because modern Group III stocks also undergo additional severe hydrofinishing after hydrocracking and hydroisomerization, they achieve an additional increase in stability because only trace amounts of aromatics and other impurities remain in the finished base oils. On the other hand, the performance of PAO base oils seems to depend largely on residual olefin content, which contributes to instability.

REFERENCES

Ahmed, N.S. and Nassar, A.M. 2011. Lubricating Oil Additives, Chapter 10. In: *Tribology—Lubricants and Lubrication*, C.-H. Kuo (Editor). InTech—Open Access Company, New York. Available at http://www.intechopen.com/books/tribology-lubricants-and-lubrication. Accessed on May 16, 2013.

Al-Sabagh, A.M., Noor El-Din, M.R., Morsi, R.E., and Elsabee, M.Z. 2009. Styrene-Maleic Anhydride Copolymer Esters as Flow Improvers of Waxy Crude Oil. *Journal of Petroleum Science and Engineering*, 65(3–4): 139–146.

API 1509. 2012. *Engine Oil Licensing and Certification System*, 17th Edition. American Petroleum Institute, Washington DC. September.

ASTM. 2012. *Annual Book of Standards*. American Society for Testing and Materials, West Conshohocken, Pennsylvania.

ASTM D86. 2012. Standard Test Method for Distillation of Petroleum Products at Atmospheric Pressure. *Annual Book of Standards*. American Society for Testing and Materials, West Conshohocken, Pennsylvania.

ASTM D92. 2012. Standard Test Method for Flash and Fire Points by Cleveland Open Cup Tester. *Annual Book of Standards*. American Society for Testing and Materials, West Conshohocken, Pennsylvania.

ASTM D93. 2012. Standard Test Methods for Flash Point by Pensky–Martens Closed Cup Tester. *Annual Book of Standards*. American Society for Testing and Materials, West Conshohocken, Pennsylvania.

ASTM D97. 2012. Standard Test Method for Pour Point of Petroleum Products. *Annual Book of Standards*. American Society for Testing and Materials, West Conshohocken, Pennsylvania.

ASTM D156. 2012. Standard Test Method for Saybolt Color of Petroleum Products (Saybolt Chromometer Method). *Annual Book of Standards*. American Society for Testing and Materials, West Conshohocken, Pennsylvania.

ASTM D611. 2012. Standard Test Methods for Aniline Point and Mixed Aniline Point of Petroleum Products and Hydrocarbon Solvents. *Annual Book of Standards*. American Society for Testing and Materials, West Conshohocken, Pennsylvania.

ASTM D1160. 2012. Standard Test Method for Distillation of Petroleum Products at Reduced Pressure. *Annual Book of Standards*. American Society for Testing and Materials, West Conshohocken, Pennsylvania.

ASTM D1500. 2012. Standard Test Method for ASTM Color of Petroleum Products (ASTM Color Scale). *Annual Book of Standards*. American Society for Testing and Materials, West Conshohocken, Pennsylvania.

ASTM D2161. 2012. Standard Practice for Conversion of Kinematic Viscosity to Saybolt Universal Viscosity or to Saybolt Furol Viscosity. *Annual Book of Standards*. American Society for Testing and Materials, West Conshohocken, Pennsylvania.

ASTM D2501. 2012. Standard Test Method for Calculation of Viscosity–Gravity Constant (VGC) of Petroleum Oils. *Annual Book of Standards*. American Society for Testing and Materials, West Conshohocken, Pennsylvania.

ASTM D2887. 2012. Standard Test Method for Boiling Range Distribution of Petroleum Fractions by Gas Chromatography. *Annual Book of Standards*. American Society for Testing and Materials, West Conshohocken, Pennsylvania.

ASTM D5771. 2012. Standard Test Method for Cloud Point of Petroleum Products (Optical Detection Stepped Cooling Method). *Annual Book of Standards*. American Society for Testing and Materials, West Conshohocken, Pennsylvania.

ASTM D5772. 2012. Standard Test Method for Cloud Point of Petroleum Products (Linear Cooling Rate Method). *Annual Book of Standards*. American Society for Testing and Materials, West Conshohocken, Pennsylvania.

ASTM D5773. 2012. Standard Test Method for Cloud Point of Petroleum Products (Constant Cooling Rate Method). *Annual Book of Standards*. American Society for Testing and Materials, West Conshohocken, Pennsylvania.

ASTM D5800. 2012. Standard Test Method for Evaporation Loss of Lubricating Oils by the Noack Method. *Annual Book of Standards*. American Society for Testing and Materials, West Conshohocken, Pennsylvania.

ASTM D5949. 2012. Standard Test Method for Pour Point of Petroleum Products (Automatic Pressure Pulsing Method). *Annual Book of Standards*. American Society for Testing and Materials, West Conshohocken, Pennsylvania.

ASTM D5950. 2012. Standard Test Method for Pour Point of Petroleum Products (Automatic Tilt Method). *Annual Book of Standards*. American Society for Testing and Materials, West Conshohocken, Pennsylvania.

ASTM D5985. 2012. Standard Test Method for Pour Point of Petroleum Products (Rotational Method). *Annual Book of Standards*. American Society for Testing and Materials, West Conshohocken, Pennsylvania.

ASTM D6074. 2012. Standard Guide for Characterizing Hydrocarbon Lubricant Base Oils. *Annual Book of Standards*. American Society for Testing and Materials, West Conshohocken, Pennsylvania.

ASTM D6375. 2012. Standard Test Method for Evaporation Loss of Lubricating Oils by Thermogravimetric Analyzer (TGA) Noack Method. *Annual Book of Standards*. American Society for Testing and Materials, West Conshohocken, Pennsylvania.

ASTM D6417. 2012. Standard Test Method for Estimation of Engine Oil Volatility by Capillary Gas Chromatography. *Annual Book of Standards*. American Society for Testing and Materials, West Conshohocken, Pennsylvania.

Bartz, W.J. (Editor). 1993. *Engine Oils and Automotive Lubrication*. CRC Press, Taylor & Francis Group, Boca Raton, Florida.

Chadeesingh, R. 2011. The Fischer-Tropsch Process, Part 3, Chapter 5. In: *The Biofuels Handbook*, J.G. Speight (Editor). The Royal Society of Chemistry, London, United Kingdom, pp. 476–517.

Chauhan, P.S., and Chhibber, V.K. 2013. Non-Edible Oil as a Source of Bio-Lubricant for Industrial Applications: A Review. *International Journal of Engineering Science and Innovative Technology*, 2(1): 299–305.

Chen, W.H., Zhang, X.D., Zhao, Z.C., and Yin, C.Y. 2009. UNIQUAC Model for Wax Solution with Pour Point Depressant. *Fluid Phase Equilibria*, 280(1–2): 9–15.

Chen, W., Zhao, Z., and Yin, C. 2010. The Interaction of Waxes with Pour Point Depressants. *Fuel*, 89(5): 1127–1132.

Corsico, G., Mattei, L., Roselli, A., and Gommellini, C. 1999. *Poly(Internal Olefins)—Synthetic Lubricants and High-Performance Functional Fluids*, Chapter 2. Marcel Dekker Inc., New York, pp. 53–62.

Decroocq, S., and Casserino, M. 2005. Polybutenes, Chapter 17. In: *Synthetics, Mineral Oils, and Bio-Based Lubricants: Chemistry and Technology*, L.R. Rudnick (Editor). CRC Press, Taylor & Francis Group, Boca Raton, Florida.

Gary, J.G., Handwerk, G.E., and Kaiser, M.J. 2007. *Petroleum Refining: Technology and Economics*, 5th Edition. CRC Press, Taylor & Francis Group, Boca Raton, Florida.

Haycock, T.F., and Hiller, J.E. 2004. *Automotive Lubricants Reference Book*, 2nd Edition. Society of Automotive Engineers (SAE International), Warrendale, Pennsylvania.

Hsu, C.S., and Robinson, P.R. (Editors). 2006. *Practical Advances in Petroleum Processing*, Volume 1 and Volume 2. Springer Science, New York.

Kramer, D.C., Lok, B.K., and Krug, R.R. 2001. The Evolution of Base Oil Technology. In: *Turbine Lubrication in the 21st Century*, ASTM Special Publication No. ASTM STP #1407, W.R. Herguth and T.M. Warne (Editors). American Society for Testing and Materials, West Conshohocken, Pennsylvania.

Leslie, R.R. 2003. *Lubricant Additives: Chemistry and Applications*. Marcel Dekker, Inc., New York, pp. 293–254.

Ludema, K.C. 1996. *Friction, Wear, Lubrication: A Textbook in Tribology*. CRC Press, Boca Raton, Florida, pp. 124–134.

Masabumi, M., Hiroyasu, S., Akihito, S., and Osamu, K. 2008. Prevention of Oxidative Degradation of ZnDTP by Microcapsulation and Verification of Its Antiwear Performance. *Tribology International*, 41: 1097–1102.

Mushrush, G.W., and Speight, J.G. 1995. *Petroleum Products: Instability and Incompatibility*. Taylor & Francis, Washington, DC.

Nehal, S.A. 2008. Lubricants Additives from Maleate Copolymers. *Petroleum Science and Technology*, 26(3): 298–306.

Pillon, L.Z. 2007. *Interfacial Properties of Petorleum Products*. CRC Press, Taylor and Francis Group, Boca Raton, Florida.

Pillon, L.Z. 2010. *Surface Activity of Petroleum Derived Lubricants*. CRC Press, Taylor and Francis Group, Boca Raton, Florida.

Ray, S., Rao, P.V.C., and Choudhary, N.V. 2012. Poly-α-Olefin-Based Synthetic Lubricants: A Short Review on Various Synthetic Routes. *Lubrication Science*, 24(1): 23–44.

Rizvi, S.Q.A. 2009. *A Comprehensive Review of Lubricant Chemistry, Technology, Selection, and Design*, Monograph No. MNL59-EB. ASTM International, West Conshohocken, Pennsylvania, pp. 100–112.

Rudnick, L.R. (Editor). 2009. *Lubricant Additives: Chemistry and Applications*, 2nd Edition. CRC Press, Taylor & Francis Group, Boca Raton, Florida.

Rudnick, L.R. (Editor). 2013. *Synthetics, Mineral Oils, and Bio-Based Lubricants*, 2nd Edition. CRC Press, Taylor and Francis Group, Boca Raton, Florida.

Rudnick, L., and Shubkin, R. 1999. *Synthetic Lubricants and High-Performance Functional Fluids*, 2nd Edition Revised and Expanded. CRC Press, Boca Raton, Florida.

Sequeira, A. 1994. *Lubricant Base Oil and Wax Processing*. Marcel Dekker Inc., New York.

Soldi, R.A., Oliveira, A.R.S., Barbosa, R.V., and César-Oliveira, M.A.F. 2007. Polymethacrylates: Pour Point Depressants in Diesel Oil. *European Polymer Journal*, 43(8): 3671–3678.

Speight, J.G., and Ozum, B. 2002. *Petroleum Refining Processes*. Marcel Dekker Inc., New York.

Speight, J.G. 2008. *Synthetic Fuels Handbook: Properties, Processes, and Performance*. McGraw-Hill, New York.

Speight, J.G. (Editor). 2011. *The Biofuels Handbook*. Royal Society of Chemistry, London, United Kingdom.

Speight, J.G. 2014. *The Chemistry and Technology of Petroleum*, 5th Edition. CRC Press, Taylor and Francis Group, Boca Raton, Florida.

Zhang, J., Wu, C., Li, W., Wang, Y., and Han, Z. 2003. Study on the Performance Mechanism of Pour Point Depressants with Differential Scanning Calorimeter and X-Ray Diffraction Methods. *Fuel*, 82(11): 1419–1426.

Zhang, J., Wu, C., Li, W., Wang, Y., and Cao, H. 2004. DFT and MM calculation: The Performance Mechanism of Pour Point Depressants. *Fuel*, 83(3): 315–326.

2 Types and Properties of Lubricating Oils

2.1 TYPES OF LUBRICATING OILS

There are two main types of lubricants: (1) those that are petroleum-based and (2) those that are manufactured as a synthetic product (Chapter 1). Each of these is suited for particular purposes and conditions. The different types are also subject to varying levels of oxidation and degradation, and are compatible with only certain types of machinery components, demands, and environments.

The motor oil in an automobile's engine is commonly a petroleum-based lubricant (Speight and Ozum 2002; Hsu and Robinson 2006; Gary et al. 2007; Speight 2014). The hydrocarbon-based or petroleum-based lubricating motor oil is designed to protect the various moving parts of the engine, whereas gasoline, which is also a petroleum product, is formulated to produce the explosive heat needed to power the engine.

Lubricants may be liquid (such as motor oil and hydraulic oil), semisolid or solid (such as grease or Teflon tape), or dry or powdered (such as graphite or molybdenum disulfide). All lubricating materials for mechanized equipment are designed to form some sort of protective coating between moving parts of machinery to protect these parts from undue wear, contamination, and oxidation.

Synthetic lubricants have precisely engineered chemical reactions on particular components. These reactions are created by specifically applying varying amounts of heat and pressure on the components. Synthetic motor oil is gaining popularity among automobile owners who use it instead of petroleum-based motor oil. This type is also used more extensively in industry because, although costlier to use originally, they are better suited to the demands of modern engine and machine technologies. Because synthetic motor and machinery oils don't have to be changed as frequently, consumers actually save in the long run.

There are petroleum-based and synthetic hydraulic lubricants, also known as hydraulic oils, that are formulated to be lighter and more free-flowing. They are used not only for lubrication but also for the actual operation of hydraulic machinery. Hydraulic oils must be able to flow freely through the pumps that compress the oil for the operation of the machinery, and at the same time, must have film-forming additives to lubricate the moving parts of the pumping equipment.

Although most modern lubricants are petroleum-based, bases such as vegetable oil and esters are gaining increased popularity for this purpose. The base of a particular lubricating fluid is the primary determinant as to whether it is petroleum-based or synthetic oil.

The used lubricating oil that is collected for recycling may contain a mixture of oils of different types, particularly when a service station contains a tank for liquids disposal and brake fluid, antifreeze, gear oil, parts washer solvent, engine oil, and possibly grease are all dumped into this tank. Although efforts have been made to encourage the separate collection of these types of fluids, this type of discipline will probably only be respected when an efficient recycling program has been introduced and popularized. In most jurisdictions, transformer oils have to be recycled separately due to the presence of polychlorobiphenyls in the oil, and there is generally a legislated limit to the content of the polychlorobiphenyls in an oil that may be accepted for re-refining, due to the potential formation of carcinogenic substances in unsuitable processing facilities and the release of potentially harmful substances to the environment. The bio-based lubricants are also not suitable for mixing with the mineral oil–derived lubricants, but these constitute a small fraction of the lubricants industry.

The lubricant oils that a re-refining plant has to deal with on a general basis will consist of mineral oil derivatives (petroleum-derived oils) or synthetic oils, such as the poly α-olefins, or mixtures of these two components, formulated from a base oil with the performance-enhancing additives and application-related contaminants (Chapter 1).

2.1.1 Lubricating Oils Derived from Mineral Oils

The base oils derived from mineral oils are either paraffinic or naphthenic in nature, depending on the source of the oil from which they are made (Chapter 1). Paraffin-based oils have a high content of waxy hydrocarbons such as *n*-paraffins and iso-paraffins, whereas the naphthenic oils are high in cycloparaffins and aromatics and are low in wax content. Because the waxes tend to solidify at low temperatures, naphthenic base oils do not show the same problems as a waxy paraffinic base stock. For low-temperature use, therefore, paraffinic base oils have to be dewaxed, but their other properties make them a better choice for most lubricating oil applications than naphthenic oils, paraffinic base oils making up approximately 85% of the world's base stock market.

Lubricating oil base stocks produced from petroleum sources are obtained from the gas oil fractions coming from the vacuum distillation columns of the petroleum refinery, or from deasphalted residues from the atmospheric distillation columns (Thiault 1995; Speight and Ozum 2002; Hsu and Robinson 2006; Gary et al. 2007; Lynch 2008; Speight 2014).

2.1.2 Lubricating Oils Derived from Synthetic Sources

The lubricating oils containing synthetic components, such as the poly α-olefins, generally have better low-temperature performance in not showing as much increase in viscosity as a mineral oil–based lubricant, a desirable property for cold engine starts, as well as better high-temperature performance in being more stable at comparable higher temperatures. These desirable properties are balanced by the significantly higher price obtained for the synthetic oils due to the additional processing required in their production.

2.2 SOLID LUBRICANTS AND GREASES

Solid lubricants and greases are lubricants that exist in solid form or in semisolid form and are key ingredients in high-performance anti-seize pastes and anti-friction coatings, used as additives in some greases and oils. In service, because of temperature effects, the solid lubricant or semi-solid grease may exist in liquid form. These special lubricants and additives fill in and smooth surface asperity peaks and valleys as they adhere to the substrate and cohere to each other. The solids provide effective boundary lubrication, optimizing friction reduction, and reducing wear under extreme operating conditions. Boundary films formed by solid lubricants can maintain a steady thickness that is unaffected by load, temperature, or speed, unlike oil or grease fluid films for hydrodynamic lubrication.

2.2.1 SOLID LUBRICANTS

Solid lubricants vary widely in terms of composition and properties. The most common types of solid lubricants are used in anti-seize pastes and anti-friction coatings. Typically, an anti-seize paste will contain 40% to 60% solids in a base-oil carrier, whereas an anti-friction coating will contain approximately 30% solids blended with a solvent carrier and resin binder. Solids (up to 10%) may be used in greases and oils for lubrication during start-up, shutdown, and shock-load conditions.

Graphite has a layered lattice structure with weak bonding between layers, providing excellent lubricity as long as moisture is available. Graphite solids provide: (1) high-temperature stability, (2) good lubrication in high humidity, (3) low coefficient of friction under high loads, and (4) protection against fretting corrosion.

Molybdenum disulfide (MoS_2) also has a lamellar structure, easily sheared in the direction of motion. Particle size and film thickness can be matched to surface roughness and these types of lubricants offer: (1) high load-carrying capacity, (2) wide service temperature range, (3) excellent adhesion, (4) protection against fretting corrosion, (5) decreased friction with increasing loads, and (6) stick-slip prevention.

Polytetrafluoroethylene (PTFE) consists of carbon and fluorine atoms. Low surface tension makes this one of the most slippery man-made materials known. This lubricant provides: (1) colorless film lubricity, (2) low load-carrying capacity, (3) low coefficient of friction at low loads, (4) good chemical resistance, and (5) good sliding-friction reduction.

2.2.2 GREASES

Grease is lubricating oil to which a thickening agent has been added for the purpose of holding the oil to surfaces that must be lubricated (Speight and Ozum 2002; Hsu and Robinson 2006; Gary et al. 2007; Speight 2014). The development of the chemistry of grease formulations is closely linked to an understanding of the physics at the interfaces between the machinery and the grease. With this insight, it is possible to formulate greases that are capable of operating in increasingly demanding and wide-ranging conditions.

A wide range of lubricant base fluids is used in grease technology. However, the largest segment consists of a variety of products derived from the refining of crude oil and downstream petroleum raw materials. These mineral oils can contain a very wide spectrum of chemical components, depending on the origin and composition of the crude oil as well as the refining processes to which they have been submitted.

There are three basic groups of mineral oils: (1) aromatic base, (2) naphthene base, and (3) paraffin base (Chapter 1). Historically, the first two have represented the principal volumes used in grease formulation, largely due to availability but also due to their solubility characteristics. However, concerns about the carcinogenic aspects of molecules containing aromatic and polynuclear aromatic ring structures have led to their replacement by paraffinic oils as the mineral fluids of choice.

There are three basic components that contribute to the multiphase structure of lubricating grease: (1) a base fluid, (2) a thickener, and (3) very frequently, in modern grease, a group of additives. The function of the thickener is to provide a physical matrix to hold the base fluid in a solid structure until operating conditions, such as load, shear, and temperature, initiate viscoelastic flow in the grease. To achieve this matrix, a careful balance of solubility between the base fluid and the thickener is required.

The most widely used thickening agents are soaps of various kinds, and grease manufacture is essentially the mixing of soaps with lubricating oils. Until relatively recently, grease-making was considered an art. To stir hot soap into hot oil is a simple business, but to do so in such a manner as to form grease is much more difficult, and the early grease-makers needed much experience to learn the essentials of the trade. Therefore, it is not surprising that grease-making is still a complex operation. The signs that *told* the grease-maker that the soap was *cooked* and that the batch of grease was ready to run have been replaced by scientific tests that follow the process of manufacture precisely.

Modern base oils in lubricating greases are therefore often a blend of severely refined paraffinic and naphthenic oils, designed to provide the final product with the appropriate characteristics of mechanical stability, lubricity, and dropping point.

Finally, the key to providing a grease matrix that is stable, both over time and under the operating shear within machine components, can be found in the thickener system. The thickeners themselves also contribute significantly to the extreme pressure and antiwear characteristics of grease and, additionally, thickeners provide a grease gel capable of carrying additives which in turn extends performance in these areas.

Water resistance, surface adhesion and tackiness, dropping point, and compatibility with other greases are all properties in which the selection of the right thickener is important. Increasingly, for centralized lubrication systems, pumpability is an additional prerequisite.

The early grease-makers made grease in batches in barrels or pans, and the batch method is still the chief method of making grease. Oil and soap are mixed in kettles that have double walls between which steam and water may be circulated to maintain the desired temperature. When temperatures higher than 150°C (300°F) are required, a kettle heated by a ring of gas burners is used. Mixing is usually accomplished in each kettle by horizontal paddles radiating from a central shaft.

The soaps used in grease-making are usually made in the grease plant, usually in a grease-making kettle. Soap is made by chemically combining a metal hydroxide with a fat or fatty acid:

$$\underset{\text{Fatty acid}}{R\,CO_2H} + NaOH \rightarrow \underset{\text{soap}}{R\,CO_2^-Na^+} + H_2O$$

The most common metal hydroxides used for this purpose are calcium hydroxide, lye, lithium hydroxide, and barium hydroxide. Fats are chemical combinations of fatty acids and glycerin. If a metal hydroxide is reacted with a fat, a soap containing glycerin is formed. Frequently, a fat is separated into its fatty acid and glycerin components, and only the fatty acid portion is used to make soap. Commonly used fats for grease-making soaps are cottonseed oil, tallow, and lard. Among the fatty acids used are stearic acid (from tallow), oleic acid (from cottonseed oil), and animal fatty acids (from lard).

To make grease, the soap is dispersed in the oil as fibers of such a size that it may be possible to detect them only by microscopy. The fibers form a matrix for the oil, and the type, amount, size, shape, and distribution of the soap fibers dictate the consistency, texture, and bleeding characteristics, as well as the other properties of grease. Greases may contain from 50% to 30% soap, and although the fatty acid influences the properties of grease, the metal in the soap has the most important effect. For example, calcium soaps form smooth buttery greases that are resistant to water, but are limited in use to temperatures under approximately 95°C (200°F).

Soda (sodium) salts form fibrous greases that disperse in water but can be used at temperatures well over 95°C (200°F). Barium and lithium soaps form greases similar to those from calcium soaps, but they can be used at both high temperatures and very low temperatures; hence, barium and lithium soap greases are known as multipurpose greases.

The soaps may be combined with any lubricating oil from a light distillate to a heavy residual oil. The lubricating value of the grease is chiefly dependent on the quality and viscosity of the oil. In addition to soap and oil, greases may also contain various additives that are used to improve the ability of the grease to stand up under extreme bearing pressures, to act as a rust preventative, and to reduce the tendency of oil to seep or bleed from grease. Graphite, mica, talc, or fibrous material may be added to greases that are used to lubricate machinery subject to shock loads to absorb the shock of impact. Other chemicals can make grease more resistant to oxidation or modify the structure of the grease.

The older, more common method of grease-making is the batch method, but grease is also made using a continuous method. The process involves soap manufacture in a series (usually three) of retorts. Soap-making ingredients are charged into one retort while soap is made in the second retort. The third retort contains finished soap, which is pumped through a mixing device in which the soap and the oil are brought together and blended. The mixer continuously discharges finished grease into suitable containers.

As with the conventional type of lubricating oil (Chapter 1), viscosity in service is the most important property of solid lubricants and greases because it determines the amount of friction that will be encountered between sliding surfaces and whether a thick enough film can be built up to avoid wear from solid-to-solid contact.

There is no single formulation that can satisfy all the requirements of a solid lubricant or grease on a cost-effective basis. Properties that should be considered are coefficient of friction, load-carrying capacity, corrosion resistance (susceptibility to galvanic corrosion), and electrical conductivity. Furthermore, one must consider the environment in which the solid-film lubricant or grease must perform. Such materials must be anticipated to be contained (inadvertently or advertently) in the same disposal pot as used lubricating oil and therefore contained in the used lubricating oil shipped to a re-refining unit.

2.3 USE AND APPLICATIONS

Although lubricating oils are typically composed of petroleum-derived base oils or synthetic-derived base oils (or both), their uses are many. In fact, it is the varied uses that cause issues during refining if an understanding of the constituents of these oils is not available. Knowing the character of the oils and the use to which they are put is a forward step in understanding the nature of the impurities and, therefore, the steps that will be needed to refine the used oil.

It is the purpose of this section to advance this understanding and allow the used oil refiner to design the necessary steps for re-refining the used oil. However, there is no substitute for an assiduous method of testing the used oil to confirm the nature of the foreign constituents that have been introduced or created during service.

2.3.1 Automotive Engines

Automotive oil (motor oil, engine oil) is oil used for lubrication of the internal combustion engine. Its main function is to lubricate the moving parts of the engine and prevent breakdown. The lubricating oil also cleans the engine parts, inhibits corrosion, improves sealing, and also improves engine cooling by dispersing the heat away from moving parts (Klamann and Rost 1984).

Automotive oil is derived from petroleum-based and non-petroleum-based chemical compounds (Chapter 1). Modern automotive oil is typically a blended product from base oils composed of hydrocarbons, poly α-olefins, and poly internal olefins. The base oils of some high-performance motor oils, however, contain up to 20% by weight of esters. Additives are also present to improve certain properties (Chapter 1).

The bulk of typical automotive oil consists of hydrocarbons with between 18 and 34 carbon atoms per molecule (Speight and Ozum 2002; Hsu and Robinson 2006; Gary et al. 2007; Speight 2014). One of the most important properties of motor oil in maintaining a lubricating film between moving parts is its viscosity. The viscosity of a liquid can be thought of as its thickness or a measure of its resistance to flow. The viscosity must be high enough to maintain a lubricating film, but low enough that the oil can flow around the engine parts under all conditions. The viscosity index is a measure of how much the oil's viscosity changes as temperature changes. A higher viscosity index indicates that the viscosity changes less with temperature than a lower viscosity index.

Motor oils must be able to flow adequately at the lowest temperature it is expected to experience to minimize metal-to-metal contact between moving parts upon starting up the engine. The *pour point* first defined this property of motor oil (ASTM D97), an index of the lowest temperature of its utility for a given application, but the cold-cranking simulator (ASTM D5293) and mini-rotary viscometer (ASTM D3829; ASTM D4684) are the properties required in motor oil specifications and at present define the Society of Automotive Engineers' (SAE) classifications.

Oil is largely composed of hydrocarbons that can burn if ignited. Still another important property of motor oil is its flash point, the lowest temperature at which the oil gives off vapors that can ignite. It is dangerous for the oil in a motor to ignite and burn, so a high flash point is desirable. At a petroleum refinery, fractional distillation separates a motor oil fraction from other crude oil fractions, removing the more volatile components, and therefore increasing the oil's flash point (reducing its tendency to burn).

Another manipulated property of motor oil is its total base number (TBN), which is a measurement of the reserve alkalinity of oil, meaning its ability to neutralize acids. The resulting quantity is determined as the amount of potassium hydroxide (in milligrams per gram of lubricant). Analogously, total acid number (TAN) is the measure of a lubricant's acidity (ASTM D664; ASTM D974). Other tests include zinc, phosphorus, or sulfur content, and testing for excessive foaming. The NOACK volatility test (ASTM D5800) determines the physical evaporation loss of lubricants in high temperature service. A maximum of 15% evaporation loss is allowable to meet American Petroleum Institute (API) SL and International Lubricant Standardization and Approval Committee (ILSAC) GF-3 specifications. Some original equipment manufacturers' (OEM) automotive oil specifications require less than 10%.

2.3.2 Diesel Engines

Generally, gasoline and diesel engine oils have the same anatomy or makeup. They are formulated from the blending of base oils and additives to achieve a set of desired performance characteristics. From this simple definition, the character of the lubricating oil changes because of the required performance for each engine type.

Proper lubrication of the inside of a heavy-duty engine such as a diesel engine requires more than just creating a protective film. Such engine oils must also disperse soot and control sludge to extend the life of the engine.

Diesel engine oils have a higher antiwear load in the form of zinc dialkyldithiophosphate. The catalytic converters in diesel systems are designed to be able to deal with this problem, whereas the gasoline systems are not. This is one of the main reasons you don't want to use diesel engine oil in your gasoline engine.

Diesel engine oils are now exposed to a higher level of contamination that can degrade the oil and damage engine parts. There is concern that exhaust gas recirculation can have a detrimental effect on engine durability and its effects on the oil. Oils exposed to the exhaust gas recirculation environment show an increase in soot content, acid number, and viscosity, whereas the engine and oil are both exposed to corrosive/acidic gases and particle buildup.

Viscosity is the single most important property of a lubricant, and oil having the right viscosity is of the utmost importance. The selected viscosity needs to be pumpable at the lowest start-up temperature while still protecting the components at in-service temperatures.

Typically, diesel engine oil will have a higher viscosity than automotive lubricating oil but the low-temperature pumpability of this higher viscosity is an issue. During cold starts, the oil may be very thick and difficult for the oil pump to deliver to the vital engine components in the lifter valley. This most certainly will lead to premature wear, as the components will be interacting without the benefit of lubrication.

Diesel engine lubricating oil has more additives per volume than gasoline engine lubricating oil. The most prevalent are over-base detergent additives, which function by neutralizing acids keeping the diesel engine clean. Diesel engines create a great deal more soot and combustion byproducts. Through blow-by, these find their way into the crankcase, forcing the oil to deal with them. When this lubricating oil with the extra additives is used in a gasoline engine, the effects can be devastating to engine performance. The detergent will work as it is designed and try to clean the cylinder walls, which can have an adverse effect on the seal between the rings and cylinder, resulting in lost compression and efficiency.

2.3.3 Tractors and Other Engines

The lubricating oils required for use in tractors and agricultural machinery fall into two categories (Harperscheid and Omeis 2007): the universal tractor transmission oils (UTTO) and super tractor oils universal (STOU). The first type is used in the hydraulic systems, the gearbox, and wet brake systems, whereas the second can be used in these applications as well as in the engine. This means that the STOU oils have to have similar additives to the automotive engine oils, including antifoaming agents, detergents, and dispersants, as well as oxidation and corrosion inhibitors.

The viscosity grades correspond to the automotive oils of 5W-40, 10W-40, and 15W-40. The oils used in gearboxes usually have to meet the API GL-4 specifications, but certain manufacturers specify their own requirements, which have to be met for their equipment. In general, a normally aspirated engine requires at least an API CE lubricating oil, whereas a turbo-charged engine needs an API CF or CF-4 oil (Harperscheid and Omeis 2007). In fact, most tractor and agricultural manufacturers now issue their own specifications for all their products after testing the oils in their equipment.

Some of the latest tractor oils are also claimed to be more environmentally friendly than older oils, but this often means that they can contain bio-based oils such as rapeseed or sunflower seed oils, or synthetic esters that are rapidly biodegradable. If these oils are mixed with the other used lubricating oils that are recycled to a refinery for regeneration, this can complicate the refining process.

2.3.4 Aviation Oils

Aviation-derivative gas turbines present unique turbine oil challenges that call for oils with much higher oxidation stability. Of primary concern is the fact that the lube oil in aero-derivative turbines is in direct contact with metal surfaces ranging from 204°C to 316°C (400°F–600°F). Sump lube oil temperatures can range from 71°C to 121°C (160°F–250°F). These compact gas turbines utilize the oil to lubricate and to transfer heat back to the lube oil sump. In addition, their cyclical operation imparts significant thermal and oxidative stress on the lubricating oil. These most challenging conditions dictate the use of high-purity synthetic lubricating oils. Average lube oil makeup rates of 0.15 gallons per hour (0.57 L/h) will help rejuvenate the turbine oil under these difficult conditions.

2.3.5 Turbine Oils

Steam turbines, gas turbines, and hydro turbines operate on lubricating oil known as rust and oxidation inhibited oil (R&O oil). Turbine equipment geometry, operating cycles, maintenance practices, operating temperatures and the potential for system contamination present unique lubricating oil demands versus other lubricating oils such as gasoline and diesel engine applications.

Utility steam turbine and gas turbine sump capacities can range in size from 1000 to 20,000 gallons (3785–75,700 L), which drives the economic incentive for long-life lubricating oil. Low turbine oil makeup rates (~5% per year) also contribute to the need for high-quality, long-life lubricants. Without significant oil contamination issues, turbine oil life is primarily dictated by oxidation stability. Oxidation stability is adversely affected by heat, water, aeration, and particulate contamination. Antioxidants, rust inhibitors, and demulsification additives are blended with premium quality base stock oil to extend oil life. Lube oil coolers, water removal systems, and filters are installed in turbine lubrication systems for the same purpose.

Unlike most gasoline and diesel engine oil applications, turbine oil is formulated to shed water and allow solid particles to settle where they can be removed through sump drains or kidney loop filtration systems during operation. To aid in contaminant separation, most turbine oils do not have high levels of added detergents or dispersants that clean and carry away contaminants. Turbine oils are not exposed to fuel or soot and therefore do not need to be drained and replaced on a frequent basis.

Well-maintained steam turbine oil with moderate makeup rates should last 20 to 30 years. When steam turbine oil fails early through oxidation, it is often due to water contamination. Water reduces oxidation stability and supports rust formation, which among other negative effects, acts as an oxidation catalyst.

Varying amounts of water will constantly be introduced to the steam turbine lubrication systems through gland seal leakage. Because the turbine shaft passes through the turbine casing, low-pressure steam seals are needed to minimize steam leakage or air ingress leakage to the vacuum condenser. Water or condensed steam is generally channeled away from the lubrication system but, inevitably, some water will penetrate the casing and enter the lube oil system. Gland seal condition, gland sealing steam pressure, and the condition of the gland seal exhauster will affect the amount of water introduced into the lubrication system. Typically, vapor extraction systems and high-velocity downward flowing oil create a vacuum that can draw steam past shaft seals into the bearing and oil system. Water can also be introduced through lube oil cooler failures, improper powerhouse cleaning practices, water contamination of makeup oil, and condensed ambient moisture.

In many cases, the effect of poor oil–water separation can be offset with the right combination and quality of additives including antioxidants, rust inhibitors, and improvers to prevent emulsification. Excess water may also be removed on a continuous basis through the use of water traps, centrifuges, coalescers, tank headspace dehydrators, or vacuum dehydrators. If the improvers preventing emulsification fail, exposure to water-related lube oil oxidation is then tied to the performance of water separation systems.

Heat will also cause reduced turbine oil life through increased oxidation. In utility steam turbine applications, it is common to experience bearing temperatures of 49°C to 71°C (120°F–160°F) and lube oil sump temperatures of 49°C (120°F). The effect of heat is generally understood to double the oxidation rate for every 10°C above 60°C (every 18°F above 140°F).

A conventional mineral oil will start to rapidly oxidize at temperatures above 82°C (180°F). Most tin-babbited (a thin surface layer in a complex, multimetal structure) journal bearings will begin to fail at 121°C (250°F), which is well above the temperature limit of conventional turbine oils. High-quality antioxidants can delay thermal oxidation, but excess heat and water must be minimized to gain long turbine oil life.

For most large gas turbine frame units, high operating temperature is the leading cause of premature turbine oil failure. The drive for higher turbine efficiencies and firing temperatures in gas turbines has been the main incentive for the trend toward more thermally robust turbine oils. Today's large-frame units operate with bearing temperatures in the range of 71°C to 121°C (160°F–250°F). Next-generation frame units are reported to operate at even higher temperatures.

As new-generation gas turbines are introduced into the utility market, changes in operating cycles are also introducing new lubrication hurdles. Lubrication issues specific to gas turbines that operate in cyclic service started to appear in the mid-1990s. Higher bearing temperatures and cyclic operation led to fouling of system hydraulics that delayed equipment start-up. Properly formulated hydrocracked turbine oils were developed to remedy this problem and to extend gas turbine oil drain intervals.

Hydro turbines typically use ISO46 or ISO68 rust- and oxidation-inhibited oils. Prevention of emulsion formation and hydrolytic stability are the key performance parameters that affect turbine oil life due to the constant presence of water. Ambient temperature swings in hydroelectric service also make viscosity stability, as measured by viscosity index, an important performance criterion.

Current technology turbine oils for land-based power generation turbines are described as 5 cSt turbo oils. Aero-derivative turbines operate with much smaller lube oil sumps, typically 50 gallons (189 L) or less. The turbine rotor is run at higher speeds, 8000 to 20,000 rpm, and is supported by rolling element bearings.

Synthetic turbo oils are formulated to meet the demands of military aircraft gas turbo engines identified in military specification (MIL) format. These specifications are written to ensure that similar quality and fully compatible oils are available throughout the world and as referenced in original equipmret manufacturers (OEM) lubrication specifications.

Type II turbo oils were commercialized in the early 1960s to meet demands from the U.S. Navy for improved performance, which created MIL-L (PRF)-23699. The majority of aero-derivatives in power generation today deploy these type II, MIL-L (PRF)-23699, polyol ester base stock, synthetic turbo oils. These type II oils offer significant performance advantages over the earlier type I diester-based synthetic turbo oils.

Enhanced type II turbo oils were commercialized in the early 1980s to meet the demands from the U.S. Navy for better high-temperature stability. This led to the creation of the new specification MIL-L (PRF)-23699 HTS. In 1993, Mobil jet oil 291 was commercialized as the first fourth-generation turbo oil to satisfy present and advanced high-temperature and high-load conditions of jet oils. Improvements continue to be made in turbo oil lubricant technology.

2.3.6 Compressor Oils

Compressed air is a critical part of many manufacturing facilities; without it, production would cease. Lubrication is key to keeping air compressors running and the compressor lubricants are produced by many lubricant manufacturers, ranging in quality from poor to excellent. Poor air compressor oil could cause the compressor to have a very short life, but excellent quality air compressor oil reduces maintenance and can extend compressor life.

The API governs the minimum quality standards for engine oils. Air compressor oils are not governed by any organization, so no official performance standards exist. This leaves the responsibility for producing a satisfactory product to the individual lubricant manufacturers. Air compressor OEM help eliminate some confusion by publishing minimum oil specifications required for their individual air compressors. These minimum oil specifications ensure minimum lubricant performance.

Any oil that meets or exceeds the minimum specifications can be used without voiding the standard or regular compressor warranty (usually 1 or 2 years). Although OEMs do not manufacture their own compressor oil, they frequently market their own brand of compressor oil and have often been able to tie separately purchased extended warranty requirements to the use of their own branded oil.

Compressor oils have many attributes and functions, including: (1) oxidation stability, (2) hydrolytic stability, (3) rust protection, (4) foam resistance, (5) copper corrosion resistance, and (6) antiwear performance. Different chemistries are required to achieve a particular performance parameter, and sacrifices are often made in other areas. For example, chemistry that is good for rust protection may cause foaming, and chemistry that is good for antiwear may not be good for oxidation resistance. Although there is a wide variance in performance, it is generally recognized that no single oil will perform perfectly in all categories. However, it is important to have a balanced air compressor oil that will perform all functions well that will deliver consistent, dependable air compressor operation.

2.3.7 Industrial Oils

Industrial machinery runs on oil, and the successful outcome of manufacturing depends on that oil being maintained properly. Hence, oil maintenance programs, when in place at all, have historically depended on a time-based change program (often at an annual shutdown). Although this is better than nothing, roughly 70% of the oils subjected to a time-based change are removed from service unnecessarily. But the time-based change also does not guarantee that lubricating oils that are well beyond the end of their useful lives are removed before they damage the machine.

Industrial oils run cold compared with other oils (such as automotive oils) and they tend to accumulate moisture. The moisture comes from humidity in the air, or in some cases, it's directly introduced to the oil from coolants and related systems. Moisture affects the lubricity of the oil, decreasing its effectiveness. Moisture in the oil can cause a variety of problems, such as poorly running hydraulic rams, machines seizing up, and chatter.

Another negative effect of moisture in oil is acidity. Oil, by its molecular nature, cannot become an acid, but there is always a little moisture present in oils operating at relatively cool temperatures, and that moisture can turn acidic. Acids in a machine's oil sump will corrosively attack internal parts, not only the metallic parts but also the seals as well. Corroded valves become ineffective. Many issues arising during machinery operation can be directly attributed to oil condition. Although oils do not respond to the pH test, there is a neutralization test (TAN test) that can easily spot oil that is becoming problematic.

Industrial oil becomes abrasive from wear in metals, abrasive dirt, and particle contamination. The most serious result of abrasive oil is the detrimental effect it has on seals. Machine seals are lubricated by the system's oil, and they will last a long time if the oils are maintained effectively. If they are not maintained properly, the seals will degrade and cause leakage. Leaking machines require pans under them, which need to be vacuumed regularly, and the used oils pose a disposal problem. Fresh oil is purchased needlessly, running up maintenance costs. Machines that leak oil also run the risk of being run low on oil and having improper oils used as replacement. All these expensive problems can be eliminated by keeping machine oils in a serviceable condition.

Many industrial operations hire filtration companies to filter insoluble materials and abrasive contaminants from their oil. Some plants operate their own filtration equipment. Filtering oil that's currently in use is a good idea, and it helps companies avoid needlessly purchasing virgin oil products, but it has limits. Oil that is filtered too many times can contain damaged additives. If the additives are damaged, the oil can't function effectively: the oil loses lubricity and becomes oxidized. There is a point at which the additives either need to be restored or the oil needs to be replaced, and oil analysis is useful in determining this point. It can also help to rate the effectiveness of a filtration program.

Not all wear metals and abrasive contaminants can be filtered out of the oil; they tend to accumulate and eventually reach levels that leave the oil unserviceable.

2.3.8 Marine Oils

Marine oil (also known as marine diesel oil) is a type of fuel oil and is a blend of gas oil and heavy fuel oil, with less gas oil than intermediate fuel oil used in the maritime field.

2.4 PROPERTIES OF LUBRICATING OILS

The main requirements for lubricants are that they are able to: (1) keep surfaces separate under all loads, temperatures, and speeds, thus minimizing friction and wear; (2) act as a cooling fluid by removing the heat produced by friction or from external sources; (3) remain adequately stable to guarantee constant behavior over the forecasted useful life; (4) protect surfaces from the attack of aggressive products formed during operation; and (5) fulfill detersive and dispersive functions to remove residue and debris that may form during operation.

This is achieved by determining a series of selective properties (properties that are adequate to the task depending on the type of lubricating oil) and interpreting the analytical data accordingly. The main properties of lubricants, which are usually indicated in the technical characteristics of the product are (1) viscosity, (2) viscosity index, (3) pour point, and (4) flash point. However, a large number of standard test methods are available (ASTM 2012) and should be chosen to match the desired data and use of the oil.

Below is a list of the various properties (including the four properties listed above) that are often used to portray the quality and performance capabilities of a new lubricating oil. Physical and chemical properties such as viscosity, acid number, flash point, elemental analysis, and contamination performance tests work by testing a lubricant in a way that is similar to the actual working environment of the equipment. The reported results characterize or measure how the lubricant responded to the challenge.

2.4.1 Viscosity and Its Temperature Dependence

The most important property of lubricating oil is its viscosity and how this varies with changes in temperature under the operational conditions to which it is subjected in the lubricated equipment. It is the characteristic of a liquid that relates an applied shearing stress to the velocity gradient it produces in the liquid.

Viscosity is strongly dependent on the temperature and is also a function of pressure and density. With increasing temperature, the viscosity has to be stated for a certain temperature.

Viscosity testing can indicate the presence of contamination in used engine oil. The oxidized and polymerized products dissolved and suspended in the oil may cause an increase in the oil's viscosity, whereas decreases in the viscosity of engine oils indicate fuel contamination. Oxidation of base oils during use in an engine environment produces corrosive oxidized products, deposits, and varnishes that lead to an increase in viscosity.

Lubricating oils are identified by their SAE number. The SAE viscosity numbers are used by most automotive equipment manufacturers to describe the viscosity of the oil they recommend for use in their products. The greater or higher the SAE viscosity number, the more viscous or heavier the lubricating oil (Mang 2007). Viscosity numbers are often presented in terms of Saybolt second universal (SSU), or in centistokes. Viscosity is strongly dependent on the temperature—with increasing temperature, the viscosity of the oil can decrease rapidly. The addition of certain additives is for the improvement of viscosity–temperature characteristics.

Figure 2.1 shows two flat plates, each of area A m^2, one stationary and the other with an applied force of F newtons moving with a velocity v in a direction parallel to the faces of the plates. The space between the plates is filled with the lubricating oil. The oil is said to have Newtonian fluid behavior if the velocity increases in a linear manner as shown in Figure 2.1 so that the velocity gradient $\frac{dv}{dy}$ is a straight line, where y is the distance from the bottom plate surface and v is the velocity at position y. Almost all lubricating base oils show Newtonian behavior, the exception being where waxes start to precipitate out at low temperature and the velocity gradient can then be curved.

The shear stress τ is defined as the applied force per unit area of the plates, or

$$\tau = \frac{F}{A}$$

τ has units of newtons/m^2, equivalent to Pascals or kg/m · s^2

The relationship between this applied shear stress and the velocity gradient produced between the plates is given by

$$\tau = \mu \cdot \frac{dv}{dy}$$

μ is the dynamic viscosity of the fluid, having units of kg/m · s or Pa · s in the SI unit system. Another common unit for the dynamic viscosity is the centipoise, defined as

$$1 \text{ cP} = 10^{-3} \text{ Pa} \cdot \text{s}$$

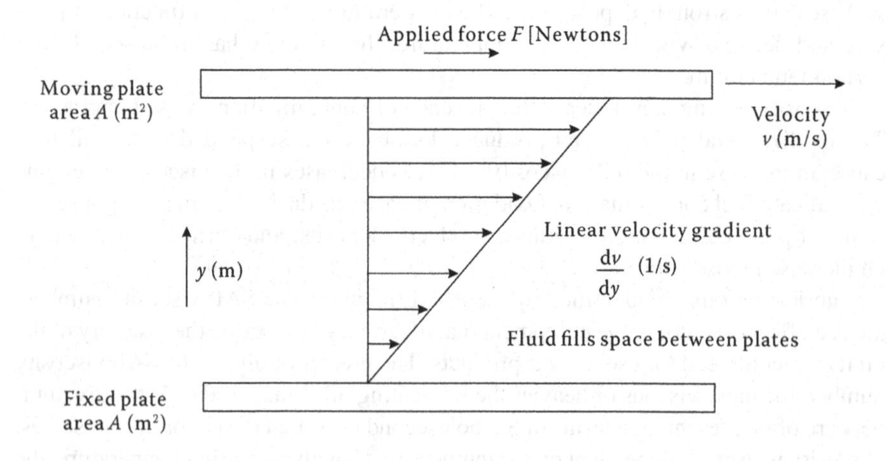

FIGURE 2.1 Newtonian fluid behavior in shear flow.

The kinematic viscosity KV of the fluid is defined as

$$KV = \frac{\mu}{\rho}$$

ρ is the density of the fluid in kg/m^3 and KV has units of m/s. The common unit of kinematic viscosity is the centistoke, defined as

$$1 \text{ cS} = 10^{-6} \text{ m/s or } 1 \text{ mm/s}$$

The ASTM D445 standard describes the procedure for measuring the kinematic viscosity of a fluid such as lubricating oil, and this involves timing the flow of a fixed volume of the test oil through a calibrated capillary viscometer under a head, which is accurately reproducible, in a liquid bath that has an accurately controlled temperature. Samples of standard viscosity oils are available and the time taken for a standard oil to run through the same viscometer provides the calibration data required to calculate the kinetic viscosity of the test oil.

As the temperature of lubricating oil is increased, its viscosity decreases. This is what is expected in an engine in which the lubricating oil may have to function at very low winter temperatures when the engine is cold, and at very high temperatures in different parts of the engine when it is operating or in very hot ambient conditions. Because the engine is designed to have oil between the parts to be lubricated, which meets certain viscosity performance parameters, it is important to check that the change in viscosity of the oil with changes in temperature is within acceptable limits and it does not become too "thin" at higher temperatures, or too "thick" at low temperatures. In general, the less the viscosity changes with change in temperature, the more preferable that oil will be for lubricating purposes. It has been found that the paraffinic lubricating oils usually have a smaller change in viscosity with temperature change than the naphthenic oils, and the viscosities of the synthetic lubricating oils are generally changed less by temperature changes than either paraffinic or naphthenic oils.

The change in viscosity of oil with change in temperature can be described by the Walther equation (Lynch 2008), also known as the Ubbelohde–Walther equation (Mang 2007):

$$\log \log (KV + C) = A - B \cdot \log T$$

where KV is the kinetic viscosity in centistokes at temperature T in degrees Kelvin, A and B are constants, and C for mineral oils, is a constant between 0.6 and 0.9, often taken as 0.7 because it does not affect the results significantly in most cases.

Plotting this equation on the appropriate logarithmic scale axes produces a straight line with a slope of B. However, the use of this equation has not proved to be as popular as the viscosity index concept for describing the temperature sensitivity of the viscosity of a lubricating oil.

2.4.2 Viscosity Index

The viscosity index is strictly an empirical number and indicates the effect of change in temperature on viscosity. A high viscosity index indicates a small change in viscosity with temperature, which also means better protection of an engine that operates under vast temperature variations. Viscosity index improvers are among the common additives that improve the efficiency of the oil. However, engine oil with a high addition level of viscosity index improvers tends to degrade more rapidly. A high viscosity index is due to the absence of aromatic and volatile compounds and is also an indicator of good thermal stability and low temperature fluidity.

The early work of Dean and Davis in developing the concept of a viscosity index is described in detail by Lynch (2008). The viscosity index represents the temperature dependence of the viscosity of an oil, the original method assigning a value of 100 to a Pennsylvania crude oil whose viscosity changed relatively little with change in temperature, and a value of 0 assigned to a U.S. Gulf Coast crude oil whose viscosity changed by a greater amount with change in temperature. The viscosities of these two reference samples were originally measured at temperatures of 100°F and 210°F (37.8°C and 98.9°C), but these temperatures were changed to 40°C and 100°C as described in the standard ASTM D2270.

Figure 2.2 shows a representation of the viscosities of the two reference oils, labeled H for the VI = 100 oil, and L for the VI = 0 oil, both having the same viscosity as an unknown oil at 100°C, together with the measured viscosities of the unknown oil at 40°C and 100°C. The viscosities are in units of centistokes and the viscosity index of the unknown oil is calculated from the expression:

$$\text{VI} = \frac{(L-U)}{(L-H)} \times 100$$

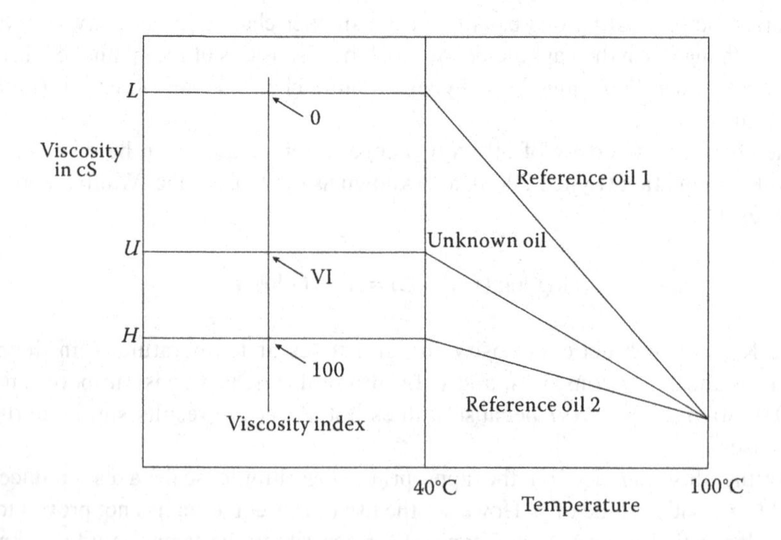

FIGURE 2.2 Viscosity index evaluation.

At the time of development, it was expected that the viscosity index for a lubricating oil would lie between 0 and 100, and this equation is used to calculate viscosity indices in this range. However, modern oils show viscosity index values ranging up to as high as 200 and negative viscosity indices have also been encountered. To overcome difficulties with the low temperature and high temperature ranges, and to cope with the extended VI range, for a viscosity index of greater than 100, and for oils with a viscosity at 100°C between 2 and 70 cSt, the following formula is used to calculate the viscosity index of an unknown oil:

$$\mathrm{VI} = \frac{\left[(\text{antilog } N) - 1\right]}{0.00715} + 100$$

where

$$\mathrm{N} = (\log H - \log U)/\log Y$$

and Y is the viscosity in centistokes at 40°C of the unknown oil.

For samples of oil with a viscosity at 100°C (212°F) greater than 70 cSt, L and H are obtained from (Lynch 2008):

$$L = 0.8353\ Y^2 + 14.67\ Y - 216$$

$$H = 0.1684\ Y^2 + 11.85\ Y - 97$$

There are a number of approximate formulas available for calculating the viscosity index of an oil for which the viscosity at 40° and 100°C are known, but the standard ASTM D2270 contains tables for the L and H values of the reference oils having the same 100°C viscosity as the unknown oil. Interpolation between the values in the table may be necessary, and the Lagrangian interpolation routine may be downloaded as a Microsoft Excel spreadsheet from the publisher of this book to provide a reproducible and accurate estimate of the required parameters. It is necessary to enable macro operation within the Excel spreadsheet to run this routine.

For use in vehicle engines, a viscosity index value of greater than 95 is generally considered to be desirable. Synthetic base oils often have a viscosity index around 120 or higher.

2.4.3 Pressure Dependence of Viscosity

The viscosity of an oil increases rapidly with increasing pressure (Mang 2007) and the relationship can be represented by the following equation:

$$\mu_p = \mu_1 \cdot e^{\alpha(p - p_1)}$$

μ_p is the dynamic viscosity of the oil at pressure p
μ_1 is the dynamic viscosity of the oil at a pressure of 1 bar (100 kPa)

p is the pressure and p_1 is the baseline pressure of 1 bar
α is the viscosity–pressure coefficient

2.4.4 Density

The density of a substance is equal to the mass of a substance divided by the volume of the substance (ASTM D941):

$$\text{Density } (D) = \text{mass } (m)/\text{volume } (v)$$

Specific gravity is influenced by the chemical composition of the oil. An increase in the amount of aromatic compounds in the oil results in an increase in the specific gravity, whereas an increase in the saturated compounds results in a decrease in the specific gravity. An approximate correlation exists between the specific gravity, sulfur content, carbon residues, viscosity, and nitrogen content. Used engine oils' specific gravity increases with the presence of increasing amounts of solids in the used engine oil. One percent of weight of solids in the sample can raise the specific gravity by 0.007—used engine oil is contaminated with oxidized and condensed products rich in carbon or soot (or both).

2.4.5 Pour Point and Cloud Point

The cloud point (ASTM D2500) is the temperature at which paraffinic wax first starts to precipitate as the oil is cooled at a standard rate (Chapter 1). The pour point (ASTM D97) is the temperature at which the oil ceases to flow under defined conditions, again as the temperature is decreased at a standard rate, and is an important property of lubricating oil in low-temperature environments.

2.4.6 Total Acid Number and Total Base Number

The TAN is a measure of the acidic and basic components present in the oil (ASTM D664; ASTM 2012)—it is the weight (in milligrams) of potassium hydroxide required to neutralize 1 g of the materials in the oil that will react with potassium hydroxide under specific test conditions. In new lubricating oils, the TAN is essentially determined by the composition and quantity of the various additives in the oil. For used lubricating oil, the TAN is of interest as an indicator of the extent of oxidation in the oil.

The usual major components of such materials are organic acids or soaps of heavy metals. Because engine oils are subjected to elevated temperatures, oxidation occurs, which leads to the formation of organic acids in the oil. TAN has been considered to be an important indicator for engine oil quality, specifically in terms of defining oxidation states. The presence of oxygen (in most engine oil environments) and hydrocarbons (which make up the base oil) leads to some reactions. This reaction may lead to the formation of carbonyl-containing products (primary oxidation products); subsequently, these undergo further oxidation to produce carboxylic acids (secondary products) that result in an increase in the TAN value. In addition, with time and elevated temperatures, the oxidation products formed then polymerize, leading to the precipitation of sludge, which decreases the efficiency of engine oil and causes excessive wear.

The TBN is an indicator of the reserve basic content of the oil, which will decrease with use as the acidic components formed by combustion neutralize the alkaline reserves (Noll and Müller 2007).

Internal combustion engine oils are formulated with highly alkaline additives to neutralize the acidic products' composition. The TBN is a measure of the highly alkaline additives and it may be used as an indication for the engine oil's replacement time. This is because TBN is depleted with time in service. Higher TBNs are more effective at neutralizing acids for longer periods of time. The rate of consumption of the additives is an indication of the projected service life of the oil.

2.4.7 Ash Formation

The ash-forming propensity (derived from the mineral matter content) of oil shows the level of impurities in the oil (ASTM D482), and these can be dust or oxidation products, or arise from the additives that contain metals (Noll and Müller 2007). In used oils, the metal particles resulting from wear on the lubricated surfaces accumulate in the oil and increase the ash content. Onstream monitoring of the metal content of the lubricating oil during its use can also indicate when an oil change is recommended.

For the refinery dealing with used lubricating oils, the metals content can be important if some of these metals find their way into the catalytic treatment operation as they can have a deleterious effect on the catalysts. Fortunately, most will be separated in the asphaltic, high boiling point residue stream from the refinery.

However, metals do occur in used oil and are regarded as heteroatoms found in engine oil mixtures. The amounts of metals are in the range of a few hundred to thousands of parts per million, and their amounts increase with an increase in the boiling points or decrease in the API gravity of the engine oil. The metallic constituents of lubricating oil are associated with high molecular weight constituents and they mainly appear in the residues. Base lubricating oils have very little metal content, which indicates their purity. Some metals present in virgin oils in high concentrations are in the form of various additives that improve the performance of the engine oil. Many others are introduced into the oils after use due to depletion of various additives, engine bearings, or bushings, and dilution of the engine oil with fuel containing metal additives. Metals are found in used engine oil in two forms: (1) metal particulate contamination and (2) elemental metals.

Metallic particulates enter the engine oil as a consequence of the breakdown of oil-wetted surfaces due to ineffective lubrication, mechanical working, abrasion erosion, or corrosion. Metallic particles from deteriorating component surfaces are generally hard and increase the wear rate as their concentration in the oil increases.

Elemental metals are found in many oil constituents containing metallic elements that have been added to enhance the oil's efficiency. In general, metals in engine oils are regarded as contaminants that should be removed completely to produce suitable base oil for producing new virgin oil. Copper (Cu) is introduced to engine oils after use from bearings, wearing of metal parts during service, and valve guides. Engine oil coolers can also be contributing to copper content along with some oil additives.

2.4.8 Flash Point and Fire Point

The flash point is the minimum temperature at which an oil gives off sufficient vapor to form an explosive mixture with air. The flash point test gives an indication of the presence of volatile compounds in oil and is the temperature to which the oil must be heated under specific conditions to give off sufficient vapor to form a flammable mixture with air.

The flash point of lubricating oil is an indication of the contamination of the oil. A substantially low flash point of lubricating oil is a reliable indicator that the oil has become contaminated with volatile products such as gasoline. In the presence of 3.5% v/v fuel, or greater in used engine oils, the flash point will be potentially reduced to less than 55°C (131°F). The flash point is also an aid in establishing the identity of a particular petroleum product. The flash point increases with increasing molecular mass of the oil and oxidation would result in the formation of volatile components, which leads to a decrease in the flash point.

The flash point (Cleveland open cup method; ASTM D92) is the temperature at which a flash appears on the surface of the sample when a small flame of specified size is passed across the cup at regular temperature intervals while the oil in the cup is being heated at a specified rate.

The fire point (Cleveland open cup method; ASTM D92) is the temperature at which the oil ignites and continues to burn for at least 5 s. Fire point is obtained as a continuation of the flash point test.

2.4.9 Volatility or Noack Evaporation Loss

The evaporation loss (ASTM D5800) is of particular importance in engine lubrication. Where high temperatures occur, portions of oil can evaporate, which may contribute to oil consumption in an engine and can lead to a change in the properties of an oil. Many engine manufacturers specify a maximum allowable evaporation loss for elevated temperatures. Some engine manufacturers, when specifying a maximum allowable evaporation loss, quote this test method along with the specifications.

Procedure C (ASTM D5800), using the Selby–Noack apparatus, also permits the collection of volatile oil vapors for the determination of their physical and chemical properties. Elemental analysis of the collected volatiles may be helpful in identifying components such as phosphorus, which has been linked to premature degradation of the automobile emission control system catalyst.

2.4.10 PONA Analysis

The determination of the paraffins, olefins, naphthenes, and aromatics (PONA) in lubricating oil is of interest in providing a "fingerprint" of the oil composition. It can also provide an indication of the potential stability or reactivity of the oil as the double bonds in the olefins make these generally more reactive than the saturated paraffins, particularly when exposed to oxygen in the air.

2.4.11 API Lubricant Base Stock Oil Groups

The lubricating oil base stocks are divided into groups by the API, as shown in Table 2.1. Groups II+ and III+ are not formal API groups, but the lubricants industry often makes use of these additional categories to distinguish an oil with better performance.

2.4.12 Additives to Base Stock Oils

The base stock components in the lubricating oils are subject to high rates of shear and high temperatures in a modern internal combustion engine, coming into contact with oxygen from the air, and these conditions are conducive to oxidation of the hydrocarbon components, causing the oil to age and change its properties. In addition, exposure to high shear forces and elevated temperatures can produce degradation of the base stock oils and the formation of product chemicals that have a negative effect on the properties of the oil. These and other undesirable changes can, to some extent, be alleviated by using various additives to the base oil, these additives often amounting in total to around 20% w/w of the final lubricating oil, each additive being used in a range from parts per million to much larger quantities (Braun 2007). The additives themselves have two main mechanisms in their operation: (1) they can change the physical and chemical properties of the base oil, such as the viscosity–temperature behavior, emulsifying properties, and stability to oxidation processes, and (2) they can change the surface properties of the metals in the engine to inhibit corrosion or reduce friction.

TABLE 2.1
API Lubricant Base Stock Oil Groups

Group	Sulfur (Mass %)	Saturates (Mass %)	Viscosity Index	Remarks
I	>0.03	and/or <90%	80–119	
II	<0.03	and >90%	80–119	
II+	<0.03	and >90%	110–119	
III	<0.03	and >90%	≥120	
III+	<0.03	and >90%	≥130	
IV	Poly α-olefins			
V	All others (naphthenics, polyalkylene glycols, esters, etc.)			Normally used as additives

Note: Groups II+ and III+ are not API categories but are commonly used in the industry. In some cases, Group II+ is defined as having a VI between 115 and 119 (Henderson 2006) and Group III+ as having a VI of ≥130 and <150 (Lynch 2008).

The range of materials used in these additives is large, with zinc, sulfur, and other elements and compounds being used for various purposes to improve the properties of the oil as a lubricant. The presence of these additives in used lubricating oil adds to the complexity of the re-refining process as they are often seen as undesirable components in the refining process.

2.4.12.1 Antioxidant Additives

The oxidation of the hydrocarbons in the base oil proceeds through a chain reaction through free radical intermediates to intermediate hydroperoxides and dialkylperoxides, and then to the oxygen-containing products of acids, peroxides, alcohols, aldehydes, ketones, and esters (Lynch 2008; Braun 2007). These products tend to increase the viscosity of the lubricating oil, and the insoluble polymer products tend to form sludge and varnish deposits on the internal metal surfaces in the engine. The inhibition of the oxidation of the hydrocarbon components in the oil is thus of great interest in extending the life of the lubricating oil.

The antioxidant additives fall into two categories: (1) primary antioxidants, which act as radical scavengers and (2) secondary antioxidants, which act as peroxide decomposers to form innocuous end products, thus interfering with the degrading oxidation process.

The radical scavenger types include the phenol derivatives, which are the most effective antioxidants, and the oil-soluble secondary aromatic amines such as the alkylated diphenylamines, which are effective scavengers but tend to form greater amounts of sludge compared with the phenol derivative types. The additives containing sulfur and phosphorus are also effective, particularly the zinc dithiophosphate derivatives, which also have antiwear and corrosion inhibitor properties.

The peroxide decomposing additives include organosulfur compounds and organophosphorus compounds such as triaryl and trialkylphosphites. Other compounds include organocopper chemicals in combination with other peroxide decomposers, and antioxidants for higher temperature use include basic phenates and salicylates of magnesium and calcium.

The antioxidants are often used in synergistic mixtures by combining amine and phenolic antioxidants of several types, combining radical scavengers with peroxide decomposers, a combination of phenolic antioxidant with phosphites has been found to be very effective for hydrotreater base stocks (Braun 2007). As metals often act as catalysts in the oxidation process, combining rust inhibitors and metal passivators in the additive mixture can provide a synergistic effect.

Lynch (2008) also provides a discussion of the mechanisms and importance of the aromatic and sulfur components in the oxidation stability of base stock oils.

2.4.12.2 Viscosity Modifier Additives

The additives used to improve the viscosity–temperature behavior of the lubricating oils are known as viscosity modifiers or viscosity index improvers, and generally consist of polymers whose solubility in the base oil decreases as the temperature decreases. As the temperature increases, the polymer chains unravel and increase the viscosity of the oil at higher temperatures. Because these polymer chains are subject to damage under high shear forces and elevated temperatures, the stability under shear is an important property of these additives. If the polymer chain is broken, there tends to be a permanent degradation of the additive, but a temporary loss of viscosity can be produced if the polymer chain aligns itself with the shear force field, returning to the random alignment when the shear force field is removed.

For engine oils, the additive polymer types include olefin copolymers, hydrotreated styrene-isoprene copolymers, and hydrotreated styrene-butadiene polymers. For gear oils, polyalkyl methacrylate derivatives and polyisobutylene derivatives are used.

2.4.12.3 Antiwear and Extreme Pressure Additives

When an engine is started cold, it takes a finite time for the hydrodynamic lubrication film to form between the metal parts of the engine. During this period, it is very important to have some additional protection on these surfaces to avoid metal-to-metal contact and subsequent wear. It is also necessary to have additional lubricating potential under conditions of high load when the pressures involved are higher than normal. These materials are generally polar compounds that are adsorbed or chemisorbed on the metal surfaces to form protective layers of iron phosphides, sulfides, sulfates, oxides, or carbides.

Friction modifiers such as fatty acids and fatty oils are only physically adsorbed on the metal's surfaces and have limited effectiveness under low to moderate pressure conditions. The chemically adsorbed and active additives perform better, particularly under higher pressure conditions, and include organic phosphorus compounds for use under medium pressure conditions. They are mostly neutral or acidic phosphoric acid ester derivatives, their metal or amine salts, or amides.

Sulfur-containing and phosphorus-containing additives such as zinc dialkyldithiophosphate derivatives are important because their properties are tailored by adjusting the alkyl groups, and also act as antioxidants and metal passivators. Other dialkyldithiophosphate derivatives using ammonium, antimony, and molybdenum groups have also been used, as well as ashless esters of dialkyldithiophosphoric acid. For high-temperature use, triphenylphosphorothionate has been found to function well.

Sulfur-containing and nitrogen-containing additives such as the dithiocarbamate derivatives of zinc or methylene have been used in greases and gear oils as antiwear agents. Sulfur alone was used for a long time as an additive to improve metal-cutting oil performance, oil-soluble organic sulfur carriers can be used to provide sulfur to the metal surfaces, and have been used in gear oils and cutting fluids together with calcium sulfonate or sodium sulfonate.

Chlorine compounds have also been used, but are falling out of favor due to the environmental problems involved with recycling chlorine-containing fluids: incineration at temperatures that are too low results in the potential formation of carcinogenic dioxins. The potential formation of hydrogen chloride is also problematic and requires the use of neutralizers to avoid excessive corrosion in the equipment.

Solid lubricants have also been used and these include finely powdered graphite and suspensions of molybdenum disulfide; however, these have practically disappeared from modern lubricating oils although they are still being used in some greases and specialty applications.

2.4.12.4 Friction Modifiers

These additives are also active in the period after start-up, before the hydrodynamic lubricating films have been established, and consist of monomolecular layers of adsorbed polar oil–soluble compounds that have lower friction properties than the antiwear or extreme pressure additives. They may be grouped into the following functional groups:

- Those with mechanical effects, the solids such as molybdenum disulfide, graphite, and polytetrafluoroethylene (Teflon)
- Those with adsorption layers such as the long-chain carboxylic acids, fatty acid esters, ethers, alcohols, amines, amides, and imides
- Those with tribochemical reaction layers—such reactions occur when tribological sliding contact leads to a chemical reaction—including saturated fatty acids, phosphoric, and thiophosphoric acid esters
- Friction-reducing polymer-forming modifiers such as glycol dicarboxylic acid partial esters, unsaturated fatty acids, and sulfurized olefins
- Organometallic compounds such as molybdenum compounds like molybdenum dithiophosphates and copper-containing organics

These additives are used in many modern lubricants marketed as delivering better gas mileage, in automatic transmission fluids, and limited-slip differential oils.

2.4.12.5 Detergents and Dispersants

These additives keep the oil-insoluble combustion products in suspension and stop the particles from agglomerating into clumps, and the basic metal-containing compounds can neutralize acidic combustion and oxidation products with their reserve of alkalinity. This then prevents the deposition of varnishes and sludge on the internal metal parts of the engine, combats corrosion, and slows down the increase in viscosity of the oil. Particulate products of wear and soot particles are effectively dispersed and held in suspension.

These additives generally have a large oleophilic hydrocarbon tail and a polar hydrophilic head group. When the additive encounters a target particle, for instance, the head groups attach to the particle and the tail groups, which are soluble in the oil, radiate out from the particle to form a layer that is soluble in the base oil; this process is known as peptization. The particles are thus held apart and also prevented from depositing on the surrounding metal surfaces.

The metal-containing detergents include the phenates, with calcium phenate being the most widely used. Other additives of this type include the salicylates, particularly for diesel engine applications, the thiophosphates derivatives, which could be the calcium salts of thiopyrophosphonate derivatives, thiophosphates derivatives, phosphonate derivatives, and the sulfonate derivatives that are metal salts of long-chain alkylarylsulfonic acids. In particular, calcium sulfonate is relatively cheap and shows good performance characteristics, whereas magnesium compounds have very good anticorrosion properties but tend to form undesirable ash products when they thermally decompose.

The dispersants that do not form ash on combustion are metal-free detergents made from hydrocarbon polymers, with the most popular and cheapest being the poly-butenes, which are combined with maleic anhydride to form poly-isobutene succinic acid anhydride, which is then added to amino-alkylene oligomers to form imides such as poly-isobutene succinimide. Some of the constituent groups in this product tend to react with the fluorocarbon elastomers used in seals in the engine and various inhibitors have been used to control this undesirable property.

2.4.12.6 Corrosion Inhibitors

Corrosion inhibition is achieved with molecules with long alkyl chains and polar groups that adsorb onto the metal surface and form a hydrophobic protective film. In doing so, they compete with other polar additives such as the antiwear and extreme pressure chemicals for the metal surfaces. They can be classified into the following two groups:

1. *Antirust additives.* Petroleum sulfonates were very popular but are being displaced by new synthetic alkylbenzene sulfonates. In particular, sulfonates with high alkalinity reserves are important in engine oils and show detergent properties while also neutralizing acidic products of oxidation. Carboxylic acid derivatives are also used where the carboxylic group adsorbs onto the metal surface, whereas alkylated succinic acids, with their ester and amide products, are used in machine and hydraulic fluids. Other widely used materials include amides and imides produced by reacting saturated fatty acids with alkylamines and alkanolamines. For industrial oils, amine neutralized alkylphosphoric acid partial esters are popular. Vapor phase corrosion inhibitors are also used in some applications in which the metal surface does not come into intimate contact with the liquid lubricating fluid, and these are mostly of the amine type.
2. *Metal passivators.* These fall into the categories of film-forming passivators, complex-forming chelating agents, and sulfur scavengers. Combining these with antioxidant additives prevents the formation of copper ions and inhibits oxidation processes, making this a very popular additive type in modern lubricating oils. Benzotriazole and tolyltriazole and their alkylated derivatives are the most widely used. Examples of film-forming passivators are 2,5-dimercapto-1,3,4-thiadiazole derivatives, which also act as sulfur scavengers. Chelating agent examples include *N,N'*-disalicylidenealkylendiamines, in which the "alkylen" is ethylene or propylene, zinc dialkyldithiophosphates, and dialkyldithiocarbamates (Braun 2007).

2.4.12.7 Antifoaming Agents

The formation of foam in a lubricating oil is very undesirable because this can accelerate the oxidation of the oil and also interfere with the lubricating performance of the oil. The dispersion of the air in the lubricating oil can take three different forms:

1. Surface foam formation—this can be counteracted with antifoaming agents
2. Internal foam in which the bubbles of gas are dispersed throughout the oil—this cannot be treated with the antifoaming agents, which tend to stabilize the dispersion
3. Dissolved air in the oil—which can cause problems with cavitation caused by pressure variations as the oil flows through the engine or mechanical equipment

The most efficient defoamers are those containing silicon as liquid silicones, such as the cyclic poly-dimethylsiloxane derivatives, and these are dissolved in aromatic solvents and have a strong attraction to polar metal surfaces. Increasing use is being made of defoamer additives that are free of silicones and consist of polymers of organic compounds.

2.4.12.8 Pour Point Depressants

The pour point of lubricating oil is the lowest temperature at which the oil will remain in a flowing state. Most engine base oils contain waxes and paraffins that solidify at cold temperatures. Engine oils with high wax and paraffin content will have a higher pour point. Pour point is highly affected by the viscosity of the oil, and lubricating oils with high viscosity are characterized by having high pour points. The pour point is an important variable, especially when starting the engine in cold weather. The oil must have the ability to flow into the oil pump and then be pumped to the various part of the engine, even at low temperatures.

As the temperature of lubricating oil decreases, the paraffinic components tend to crystallize out and increase the viscosity of the oil. This crystallization cannot be prevented, but an additive can be used to co-crystallize with the paraffins and change the crystal lattice structure and properties. For mineral-based oils, polyalkyl methylacrylates can be used to prevent the formation of the usual needle-shaped crystals, which cause a rapid increase in the viscosity of the oil, and instead promote the formation of round crystals that have a much smaller effect on the viscosity of the oil at low temperatures.

2.4.12.9 Demulsifiers and Emulsifiers

The demulsifiers are particularly important in industrial oils such as gear, hydraulic, compressor, turbine, and metal-working oils in which any water in the oil must be easily separated. This is done with surface-active agents, which are usually alkaline earth metal salts of organic sulfonic acids, for example, the barium or calcium dinonylnaphthene sulfonate derivatives. Other applications use polyethylene glycol derivatives and other ethoxylated compounds.

Emulsifiers are used mainly in water-based metal-working fluids and are generally anionic or nonionic soaps, which can contain sodium or amine groups or organic derivatives.

2.4.12.10 Dyes

Dyes may be added to some lubricating oils in which leak detection is important and these may find their way into the recycled oil disposal containers. The dyes are generally oil-soluble azo dyes or fluorescent materials.

2.4.13 SAE/API Classification

When the base oil has been blended with the required additives to form the marketable product, it is suitable for application in various types of machinery. The American SAE, together with the API, developed a classification system based on a series of performance tests that certified the lubricating oil as being suitable for the defined application. For gasoline engines, for instance, the oil type is indicated by a code starting with S, and a second letter being assigned sequentially as the requirements for the performance of the oil were raised or modified: the SN designation indicates lubricating oil that meets newer requirements compared with lubricating oil designated previously as SM or SL.

The test requirements include measurements of the oil's foaming tendencies, emulsion retention capability, sulfur content, phosphorus content, elastomer or seal compatibility, varnish deposition tendencies, oxidation susceptibility, and other properties before the classification is approved.

Besides the general performance characteristics of the lubricating oil, the viscosity characteristics of the oil and its low temperature behavior have also been summarized in a grade designation described in the publication SAE J300. For instance, it is required that the oil have a low viscosity at low temperatures to facilitate starting of a cold engine, but an oil starting with a low viscosity at that point might have a viscosity at the engine operating temperature that is too low to protect the engine's lubricated moving parts. Oils that are suitable for winter or low temperature use are designated in the single-grade type as 0W, 5W, and so on up to 25W, and for each of these, a maximum viscosity at low temperatures is specified, together with a minimum viscosity at 100°C. The higher temperature designation of 20, 30, up to 60, contains a specification of the minimum viscosity at 100°C and a maximum viscosity at the same temperature. The multigrade oil specifications, for instance 5W30, indicate an oil that meets the 5W requirements for low-temperature use as well as the requirements for the 30 grade oil at 100°C.

2.4.14 ILSAC Classification

A similar general classification of lubricating oils is provided by the International Lubricant Standardization and Approval Committee (ILSAC), and this seal of approval is also found on most lubricating oil containers sold in North America. In this case, the GF-5 classification replaced the GF-4 designation after September 30, 2011, and indicates an oil having better fuel economy characteristics than earlier classifications of lubricating oils.

2.4.15 Classification by the Association des Constructeurs Européens d'Automobiles

The Association des Constructeurs Européens d'Automobiles (ACEA) classification for marketed lubricating oils consists of four categories:

1. "A" indicates a gasoline engine lubricating oil with categories from A1 through A5 with A4 being reserved for a future classification, and the higher numbers indicating better performance characteristics and better fuel economy

2. "B" indicates a passenger car diesel engine oil, again with categories up to B5
3. "C" indicates a catalyst compatible oil for use in vehicles with a diesel particulate filter or three-way catalysts in a gasoline engine application
4. "E" indicates an oil for use on heavy duty diesel engines

2.4.16 Original Equipment Manufacturers (OEM) Classifications

A number of vehicle manufacturers have developed their own specifications for the lubricating oils they require to be used in their products to maintain the manufacturer's warranty on the engine or other lubricated units in the vehicle in which the lubricating oils have been tested to these specifications and have been approved. This may also be indicated on the container in which the lubricating oil is sold.

2.5 USE OF TEST DATA

For many years, lubricant inspection and testing has been used to help diagnose the internal condition of oil-wetted components and provide valuable information about lubricant serviceability. The first test methods used for this purpose included such simple procedures as smelling used oil for the sour odor of excess acid, checking visually for obvious signs of contamination, or placing a drop of sample on absorbent paper to detect contaminants and monitor additive effectiveness. As basic research and technology expanded, progress in lubricant testing kept pace. An increasingly large number of tests were developed to assess lubricant physical properties and detect contaminants (Section 2.4 and Chapter 3).

Routine testing and analysis can show you how the condition of a particular lubricant can affect equipment performance and, ultimately, equipment life by helping identify minor problems before they become major failures, thereby maximizing equipment reliability.

The analytical data provided by the application of various test methods is an indication of the manner by which the lubricating oil will perform during service or is an indication of the manner in which the lubricating oil has behaved in service. The analytical data provides a check on (1) the properties of the oil, (2) the presence of suspended contaminants, and (3) the presence of wear debris due to machinery wear (often referred to as *tribology*). In addition, analysis of lubricating oil is necessary for the oil to be certified in compliance with various standards and specifications.

Analysis of lubricating oil should be performed during routine preventive maintenance, which can be used to avoid or mitigate the consequences of failure of equipment and to provide meaningful and accurate information on lubricant and machine condition. By tracking oil analysis sample results over the life of a particular machine, trends can be established that can help eliminate costly repairs.

Thus, lubricating oil analysis can be divided into three categories: (1) analysis of oil properties including those of the base oil and its additives, (2) analysis of contaminants, and (3) analysis of wear debris from machinery. In addition to monitoring oil contamination and wear metals, modern usage of lubricating oil analysis includes the analysis of the additives in oils to determine if an extended drain interval may be used. Maintenance costs can be reduced using oil analysis to determine the remaining useful life of additives in the oil. By comparing the oil analysis results of new and used lubricating oil (Chapter 3), it is possible to determine when oil has passed its useful service life and must be replaced. Careful analysis might even allow the oil to be restored (*sweetened* must not be confused with sulfur removal as used in the refining industry) to the original additive levels by either adding fresh oil or replenishing additives that were depleted.

REFERENCES

ASTM. 2012. *Annual Book of Standards*. American Society for Testing and Materials, West Conshohocken, Pennsylvania.

ASTM D92. 2012. Standard Test Method for Flash and Fire Points by Cleveland Open Cup Tester. *Annual Book of Standards*. American Society for Testing and Materials, West Conshohocken, Pennsylvania.

ASTM D97. 2012. Standard Test Method for Pour Point of Petroleum Products. *Annual Book of Standards*. American Society for Testing and Materials, West Conshohocken, Pennsylvania.

ASTM D445. 2012. Standard Test Method for Kinematic Viscosity of Transparent and Opaque Liquids (and Calculation of Dynamic Viscosity). *Annual Book of Standards*. American Society for Testing and Materials, West Conshohocken, Pennsylvania.

ASTM D482. 2012. Standard Test Method for Ash from Petroleum Products. *Annual Book of Standards*. American Society for Testing and Materials, West Conshohocken, Pennsylvania.

ASTM D664. 2012. Standard Test Method for Acid Number of Petroleum Products by Potentiometric Titration. *Annual Book of Standards*. American Society for Testing and Materials, West Conshohocken, Pennsylvania.

ASTM D941. 2012. Standard Test Method for Density and Relative Density (Specific Gravity) of Liquids by Lipkin Bicapillary Pycnometer. *Annual Book of Standards*. American Society for Testing and Materials, West Conshohocken, Pennsylvania.

ASTM D974. 2012. Standard Test Method for Acid and Base Number by Color-Indicator Titration. *Annual Book of Standards*. American Society for Testing and Materials, West Conshohocken, Pennsylvania.

ASTM D2270. 2012. Standard Practice for Calculating Viscosity Index from Kinematic Viscosity at 40 and 100°C. *Annual Book of Standards*. American Society for Testing and Materials, West Conshohocken, Pennsylvania.

ASTM D2500. 2012. Standard Test Method for Cloud Point of Petroleum Products. *Annual Book of Standards*. American Society for Testing and Materials, West Conshohocken, Pennsylvania.

ASTM D3829. 2012. Standard Test Method for Predicting the Borderline Pumping Temperature of Engine Oil. *Annual Book of Standards*. American Society for Testing and Materials, West Conshohocken, Pennsylvania.

ASTM D4684. 2012. Standard Test Method for Determination of Yield Stress and Apparent Viscosity of Engine Oils at Low Temperature. *Annual Book of Standards*. American Society for Testing and Materials, West Conshohocken, Pennsylvania.

ASTM D5293. 2012. Standard Test Method for Apparent Viscosity of Engine Oils and Base Stocks between –5 and –35°C Using Cold-Cranking Simulator. *Annual Book of Standards*. American Society for Testing and Materials, West Conshohocken, Pennsylvania.

ASTM D5800. 2012. Standard Test Method for Evaporation Loss of Lubricating Oils by the Noack Method. *Annual Book of Standards*. American Society for Testing and Materials, West Conshohocken, Pennsylvania.

Braun, J. 2007. Additives. In: *Lubricants and Lubrication*, 2nd Edition, T. Mang and W. Dresel (Editors). Wiley-VCH Verlag GmbH & Co. KGaA, Weinheim, Germany, pp. 88–118.

Gary, J.G., Handwerk, G.E., and Kaiser, M.J. 2007. *Petroleum Refining: Technology and Economics*, 5th Edition. CRC Press, Taylor & Francis Group, Boca Raton, Florida.

Harperscheid, M., and Omeis, J. 2007. Lubricants for Internal Combustion Engines. In: *Lubricants and Lubrication*, 2nd Edition, T. Mang and W. Dresel (Editors). Wiley-VCH Verlag GmbH & Co. KGaA, Weinheim, Germany, pp. 191–229.

Henderson, H.E. 2006. Chemically Modified Mineral Oils. In: *Synthetics, Mineral Oils, and Bio-based Lubricants*, L.R. Rudnick (Editor). CRC Press, Taylor and Francis Group, Boca Raton, Florida, pp. 287–315.

Hsu, C.S., and Robinson, P.R. (Editors). 2006. *Practical Advances in Petroleum Processing*, Volume 1 and Volume 2. Springer Science, New York.

Klamann, D., and Rost, R.R. 1984. *Lubricants and Related Products: Synthesis, Properties, Applications, International Standards*. Verlag Chemie, Leipzig, Germany.

Lynch, T.R. 2008. *Process Chemistry of Lubricant Base Stocks*. CRC Press, Taylor & Francis Group, Boca Raton, Florida.

Mang, T. 2007. Rheology of Lubricants. In: *Lubricants and Lubrication*, 2nd Edition, T. Mang and W. Dresel (Editors). Wiley-VCH Verlag GmbH & Co. KGaA, Weinheim, Germany, pp. 23–33.

Noll, S., and Müller, R. 2007. Laboratory Methods for Testing Lubricants. In: *Lubricants and Lubrication*, 2nd Edition, T. Mang and W. Dresel (Editors). Wiley-VCH Verlag GmbH & Co. KGaA, Weinheim, Germany, pp. 715–735.

Speight, J.G., and Ozum, B. 2002. *Petroleum Refining Processes*. Marcel Dekker Inc., New York.

Speight, J.G. 2014. *The Chemistry and Technology of Petroleum*, 5th Edition. CRC Press, Taylor & Francis Group, Boca Raton, Florida.

Thiault, B. 1995. Characteristics of Non-Fuel Petroleum Products. In: *Crude Oil, Petroleum Products, Process Flowsheets*, J.-P. Wauquier (Editor). Part 1 of the Series "Petroleum Refining", Éditions Technip, Institut Français du Pétrole, Paris, France, pp. 271–292.

3 Used Lubricating Oils

3.1 INTRODUCTION

Used lubricating oil—often referred to as *waste oil* without further qualification—is any lubricating oil, whether refined from crude or synthetic components, which has been contaminated by physical or chemical impurities as a result of use.

Lubricating oil loses its effectiveness during operation due to the presence of certain types of contaminants. These contaminants can be divided into: (1) extraneous contaminants and (2) products of oil deterioration.

Extraneous contaminants are introduced from the surrounding air and by metallic particles from the engine. Contaminants from the air are dust, dirt, and moisture—in fact, air itself may be considered as a contaminant because it can cause foaming of the oil. The contaminants from the engine are (1) metallic particles resulting from wear of the engine, (2) carbonaceous particles due to incomplete fuel combustion, (3) metallic oxides present as corrosion products of metals, (4) water from leakage of the cooling system, (5) water as a product of fuel combustion, and (6) fuel or fuel additives or their by-products, which might enter the crankcase of engines.

In terms of the products of oil deterioration, many products are formed during oil deterioration. Some of these important products are (1) sludge: a mixture of oil, water, dust, dirt, and carbon particles that results from the incomplete combustion of the fuels. Sludge may be deposited on various parts of the engine or remain in colloidal dispersion in the oil; (2) lacquer: a hard or gummy substance that gets deposited on engine parts as a result of subjecting sludge in the oil to high temperature operation; and (3) oil-soluble products: the result of oil oxidation products that remain in the oil and cannot be filtered out and are thus deposited on the engine parts.

The quantity and distribution of engine deposits vary widely depending on the conditions at which the engine is operated. At low temperatures, carbonaceous deposits originate mainly from incomplete combustion products of the fuel and not from the lubricating oil. At high temperature, the increase in lacquer and sludge deposits may be caused by the lubricating oil.

Used mineral-based crankcase oil is the brown-to-black, oily liquid removed from the engine of a motor vehicle when the oil is changed (U.S. Department of Health and Human Services 1997). It is similar to a heavy fraction of virgin mineral oil, except that it contains additional chemicals from its use as an engine lubricant. The chemicals in oil include hydrocarbons, which are distilled from crude oil to form a base oil stock, and various additives that improve the oil's performance. Used oil also contains chemicals formed when the oil is exposed to high temperatures and pressures inside an engine. It also contains some metals from engine parts and small amounts of gasoline, antifreeze, and chemicals that come from gasoline when it burns inside the engine. The chemicals found in used mineral-based crankcase oil vary, depending on the brand and type of oil, whether gasoline or diesel fuel was used, the mechanical condition of the engine that the oil came from, and the amount of use between oil changes. Used oil is not naturally found in the environment.

Lubrication technology has different types of oils (thus, different compositions) that are currently available. For example, solvent-refining technology emerged and displaced naturally occurring petroleum distillates due to its improved refined properties. In the 1960s, hydroprocessing technologies were introduced, which further improved base oil purity and performance. In the 1970s and 1980s, Group II base oils were manufactured and recognized as a separate American Petroleum Institute (API) category due to their positive differentiation over conventional stocks. Modern hydroisomerization technologies, such as isodewaxing, became widely accepted and grew rapidly since it was first commercialized in 1993. Widespread licensing of this technology has created an abundant supply of Group II oils that have exceptional stability and low temperature performance relative to their Group I and Group II predecessors (Kramer et al. 1999).

Group III base oils, especially those made using modern hydroisomerization processes, offer most of the performance advantages of traditional poly α-olefin–based synthetic lubricating oils. Furthermore, the widespread availability of modern Group II and III mineral oils is accelerating the rate of change and new improved base oils are helping meet increasing demands for better, cleaner lubricants from the engine and equipment manufacturers. This is particularly true for turbine oils, which typically contain more than 99% v/v base oil. Turbine oils made from hydroisomerized Group II base oils and the appropriate additives have demonstrated significantly longer service life than turbine oils made with Group I base oils. In addition, lubrication performance that currently can be achieved only in small-volume niche applications will be more widely available using the new generation of Group II and Group III oils. Furthermore, the use of additives has increased the useful life and range of applications of lubricant products.

These changes complicate the issue of understanding the chemistry and technology of used lubricating oil even further. As base oil technology continues to evolve and improve, used lubricating oil will contain a variety of products that present to the re-refiner a more complex solution to the re-refining process (Chapter 7). For example, although the increased use of additives in lubricating oil offers enhanced economic and environmental benefits during use, additional difficulties are presented for processors wishing to re-refine the lubricants after use. Although re-refined base oil will be improved by greater use of synthetic oils from mineral oil sources, a particularly difficult issue for some re-refining processes is the increasing use of esters in synthetic lubricating oil.

In addition, used oil originates from diverse sources, including petroleum refining operations (such as sludge containing appreciable amounts of oil originating from the various parts of petroleum plants such as sumps, gravity separators, and the cleaning of storage tanks), the forming and machining of metals, small generators (do-it-yourself [DIY] car and other equipment maintenance), and industrial sources, and the rural farming population. Collecting used oil from nonindustrial sources and local/small generators is very difficult and requires a well-established and efficient infrastructure to accomplish the task. In this regard, it is important to develop adequate reuse or recycling options, to properly handle the collected volume of oil, to address the specific properties of the concerned waste, and to assess the degree to which used oils could be treated. A major source of oily wastes arising in some parts of the world is the sludge recovered from tanks used for the storage of leaded gasoline. This sludge, which is usually produced by high-pressure water jet cleaning of storage tanks, consists of iron oxide corrosion products and sediments, onto which organic and inorganic lead compounds have been absorbed or adsorbed mixed with fuel. The free fuel is usually readily removed by gravity or mechanical separation and used as an energy source. The highly toxic organic lead compounds associated with the sludge have to be chemically or thermally oxidized (calcined) into inorganic lead compounds to facilitate its disposal.

For the purposes of this book, used oil must have been refined from crude oil or made from synthetic materials (i.e., derived from coal, shale, or polymers). Examples of crude oil–derived oils and synthetic oils are motor oil, mineral oil, laminating surface agents, and metal working oils. Thus, animal and vegetable oils as well as bottom clean-out from virgin fuel oil storage tanks or virgin oil recovered from a spill, as well as products solely used as cleaning agents or for their solvent properties, and certain petroleum-derived products such as antifreeze and kerosene are not included.

The oil must have been used as a lubricant, coolant, heat (noncontact) transfer fluid, hydraulic fluid, heat transfer fluid, or for a similar use. Lubricants include, but are not limited to, used motor oil, metal-working lubricants, and emulsions. An example of a hydraulic fluid is transmission fluid. Heat transfer fluids can be materials such as coolants, heating media, refrigeration oils, and electrical insulation oils. The used oil must be contaminated by physical (e.g., high water content, metal shavings, or dirt) or chemical (e.g., lead, halogens, solvents, or other hazardous constituents) impurities as a result of use.

Among the facilities generating used lubricating oil are vehicle repair shops, fleet maintenance facilities, industries, nonprofit organizations, government agencies, and educational institutions. Used lubricating oil is also generated by private citizens who change the motor oil on their own vehicles and are known as household DIY oil changers.

Many countries follow specified standards regulating all used lubricating oil facility types while at the same time state law provides numerous locations where a DIY operator/vehicle owner (who is not always otherwise exempt from regulation) can return used lubricating oil for recycling (Harrison 1994). Generally, in the United States, under Article 23, Title 23 of the Environmental Conservation Law, any service establishment that sells at least 500 US gallons (1893 L) per year of new oil and performs vehicle servicing must accept from the public at no charge up to 5 U.S. gallons (18.93 L) of used lubricating oil per person per day. Retailers who do not service vehicles but sell at least 1000 US gallons (3785 L) per year of new oil must either accept used lubricating oil from the public, as service establishments do, or contract to have another service or retail establishment accept it on their behalf. Some municipalities in the state also collect DIY used lubricating oil as part of a hazardous waste program.

Used lubricating oil that acquired hazardous characteristics during its use as an oil (as a lubricant, for example) is exempt from hazardous used materials regulation on the presumption that the oil will eventually be recycled. Mixing the oil with hazardous wastes or disposing of the oil instead of recycling it invalidates the presumption (40 CFR 279.10). Provided that no mixing or disposal occurs, the oil is subject to its own set of regulatory standards, whether or not it displays any hazardous characteristics.

Therefore, unlike other wastes, there is generally little to no need for a used lubricating oil generator (unless suspicions of used oil quality have been aroused) to perform a hazardous waste determination on the used lubricating oil being generated.

In summary, re-refining used oil (Chapters 4, 6, and 7) is complicated by the fact that the oil is difficult to characterize. The re-refining process is very much dependent on (1) the source of the used oil, (2) the types of used oil, (3) the method of collection, and (4) the intended end-use of the oil. The highly variable quality of used oils drives the complexity of the facilities necessary to treat it for reuse. As already noted, used oil is made up of base oils and various additives that impart the desired quality characteristics to the finished product. High-tech motor oil, for example, contains up to 30% v/v of additive components in the finished product. This includes additives such as viscosity improvers, demulsifiers, detergents, and antiwear additives.

On the other hand, used oils comprise the original base oils and additives plus water, noncombustible ash, heavy metal compounds, sulfur, and solids such as dirt and grit from blowby carbon in diesel engines. Water can exist in free form or in emulsified form with the chemical additives in motor oils that are designed to prevent free water from accumulating in engine sumps. Heavy metals come from certain additives and from the engine itself whereas sulfur is an important part of the antiwear and detergent additives. Finally, used oil can also contain chlorinated solvents, polychlorobiphenyl derivatives, paints, solvents, and other extraneous materials that may require special handling.

3.2 CAUSES OF OIL DEGRADATION

Lubricating oil degradation can be responsible for many types of equipment failure. A lubricant in service is subjected to a wide range of conditions that can degrade its base oil and additive system. Such factors include heat, entrained air, incompatible gases, moisture, internal or external contamination, process constituents, radiation, and inadvertent mixing of a different fluid. The processes presented below give indications of the means by which they can (or will) cause degradation of the oil.

Oxidation is the reaction of materials with oxygen. It can be responsible for increased viscosity, varnish formation, sludge and sediment formation, additive depletion, base oil breakdown, filter plugging, loss in antifoam properties, increased acid number, rust, and corrosion. The oxidation and polymerization products that were dissolved and suspended in the oil will cause the increase of oil viscosity—a decrease in the viscosity of lubricating oil indicates fuel contamination. On the other hand, oxidation will produce acidic products and an increase in acid number is caused by the oxidation of lubricating oil. Thus, controlling oxidation is a significant challenge in trying to extend the service life of a lubricant.

Thermal breakdown occurs in a mechanical working environment in which temperature can increase beyond the point of the thermal stability of the oil. In addition to separating the moving parts of the machinery, the lubricant must also dissipate heat and, thus, the lubricant will sometimes be heated above its recommended stable temperature. Overheating can cause the light ends of the lubricant to vaporize or the lubricant itself to decompose. This can cause certain additives to be removed from the system without performing their job, or the viscosity of the lubricant may increase.

At temperatures that greatly exceed the thermal stability point of the lubricant, cracking (thermal decomposition) will occur—producing lower molecular weight volatile products. Thermal cracking can initiate side reactions, induce polymerization, produce gaseous by-products, destroy additives, and generate insoluble by-products. In some cases, thermal degradation will cause a decrease in viscosity.

Micro-dieseling (pressure-induced thermal degradation) is a process in which an air bubble transitions from a low-pressure to a high-pressure zone, resulting in adiabatic compression. This may produce localized temperatures of more than 1000°C (1830°F), resulting in the formation of carbonaceous by-products and accelerated oil degradation.

Additive depletion occurs because most additive systems are designed to be sacrificial. Monitoring additive levels is important not only to assess the condition of the lubricant for continued service life but also because it may provide clues related to specific degradation mechanisms. Monitoring additive depletion can be complex depending on the chemistry of the additive component.

Electrostatic spark discharge occurs when clean, dry oil rapidly flows through tight clearances and internal (molecular level) friction within the oil generates static electricity. This may accumulate to the point in which it produces a sudden discharge or spark. The temperature of these sparks have been estimated (but not measured) to be on the order of between 1000°C and 2000°C (1800°F and 3600°F) and typically occur in mechanical filters.

Contamination is the occurrence of foreign substances in lubricating oil, which greatly influences the type and rate of lubricant degradation. Metals such as copper and iron are catalysts in the degradation process. Water and air can provide a large source of oxygen to react with the oil. In addition, water, even in small amounts, causes rusting of iron or steel. The water also results in the formation of water sludge (emulsions), which may clog oil passages, pump valves, and other oil-handling equipment. Water also contributes to foaming problems.

Therefore, a contaminant-free lubricant is ideal and monitoring a fluid's contamination levels provides significant insight to the service life of the lubricant and the operability of the machine.

3.3 AMOUNT OF USED OILS

The United States Environmental Protection Agency (U.S. EPA) estimates that approximately 1.37 billion gallons (1.37×10^9 gal. or 5.19×10^9 L) of used oil are generated each year in the United States, of which 780 million gallons (2953 ML) are used as fuel, 165 million gallons (625 ML) are re-refined, and 426 million gallons (1613 ML) are disposed of in landfills or other improper locations (U.S. EPA 2011).

A key issue has been the DIY oil changers and their reluctance to recycle oils in a proper manner. It has been estimated that more than 80% of the used oil generated by DIY activities ends up being disposed of in an improper manner (U.S. Department of Energy [U.S. DOE] 2006). However, one of the positive observations is that since 1996, the volume percentage of automobile oil changes that take place in the increasingly popular *do it for me* (DIFM) quick lube shops, garages, and auto dealerships has increased and that the volume percentage of DIY changes has decreased. This increase in DIFM oil change volumes would lead one to conclude that the annual volume of improperly disposed of oil has decreased from 426 million gallons (1613 ML) in 1996 to an estimated 348 million gallons (1317 ML) in 2004. From the information that is available on the used oil recycling progress in individual states, it is believed that there have been improvements in the DIY oil changers' behavior, but little authoritative data is available nationwide. All in all, this is a positive message and reflects the effective programs that have been implemented at the state and local government levels plus the efforts of the quick lube shops, service stations, auto dealerships, and used oil collectors/recyclers.

Furthermore, because of the amounts of used oil that become available each year, there must also be storage facilities that can accommodate the oil at any given site until transportation is available. However, as anticipated, there are regulations that must be adhered to.

Used oil may be stored in tanks (stationary) or containers (portable), including drums. All aboveground tanks and containers must be kept in good condition and must not leak. Aboveground tanks, containers, and the fill pipes of underground tanks must be labeled *Used Oil*, and aboveground tanks must be labeled with the tank's design and working capacity. The fill pipes of underground tanks must also be labeled with the tank's capacity. Underground tanks are subject to the requirements of the federal underground storage tank (UST) regulations (40 CFR 280).

All aboveground and underground used oil tanks, regardless of capacity, are subject to the standards of any existing (state or federal) petroleum bulk storage regulations, including the registration of all used oil tanks, regular (typically) monthly visual inspection of aboveground tanks, and standards for new tank installations. Secondary containment is required for all containers and aboveground tanks at transfer, processing, re-refining, and burning facilities. At generator facilities, secondary containment is required for aboveground tanks whose capacity is 10,000 gal. (37,854 L) or greater and for smaller tanks if it can be reasonably expected that any leakage from them could contaminate local water course or local ground water. Even when not required, secondary containment at generator sites is often necessary—and very strongly recommended.

3.4 COLLECTION

Collection centers and transportation centers (often the same as the collection centers) are the so-called midpoint agents between the generators and the end consumers of used oils. They perform a service for the industry and their business is very competitive. Both re-refiners and processors also act as collectors, and thus there is competition between the end users of the recycled oil, those that seek to combust the oil for its heating value and the processors or re-refiners who seek to rejuvenate or regenerate the used oil into base oils similar in quality to virgin base oils.

Certain oily wastes can be managed as used oil even when they do not fit the used oil definition. For example, a material that is not dangerous used oil, but is contaminated with or contains used oil, may be managed under the used oil rule. A good example of this is oily rags. A rag (a product) is used for its intended purpose (to wipe up an oil spill) and, after use, may be managed under the used oil regulations. Other wastes that may (but in many cases should not) fit into this category include oily bilge water, oily used water from oil/water separators, and oily waste water.

Used oil that is removed from material containing or contaminated with used oil can also be managed under the used oil regulations. Examples may include used oil drained from an oil filter, engine oil pan, or used oil removed from oil/water separators (Khawaja and Aban 1996).

Any type of oily used material that does not meet the definition or exceptions mentioned above cannot be managed under the used oil regulations. Some examples include brake cleaner, solvents used for degreasing (including petroleum distillates and mineral spirits), paint and oil paint, oil-based inks, antifreeze, degreasers, and carburetor cleaners. These products would most likely be designated as *hazardous used* or *dangerous used materials* when disposed of. No solid used or hazardous used materials should be mixed into or with used oil. If this occurs, the mixture is no longer considered used oil, and could be classed as a hazardous used material.

Used oil is a recyclable commodity and must be handled in accordance with the regulations. The term *used oil* is often applied to *waste oil* but, in many cases, waste oil is contaminated by constituents other than those that occur in lubricating oil during service and such waste oil must be handled by a permitted treatment, storage, and disposal facility.

The principles and regulations laid down in many countries for the collection of used oil include the following aspects that can be used as a model for setting up a used oil management scheme: (1) regulation on the character/properties/composition of the oil that will be accepted, (2) reprocessing procedures, (3) stipulation of limit values for polychlorobiphenyl derivatives and total halogens, (4) introduction of a labeling and take-back obligation, (5) priority given to materials recycling over disposal, (6) positioning of used oil collection stations in the vicinity of sales outlets and places of use, and (7) prohibition of the admixture of foreign substances (such as used oils containing polychlorobiphenyl derivatives, other hazardous wastes, and solvents).

In addition, some industrial lubricant products, particularly cutting fluids, hydraulic oils, gear oils, and quenching oils are cleaned on-site or at merchant facilities so that contaminants such as water and various solids are removed before additive packages are replaced and the lubricant is returned to use. Only when the lubricant is no longer fit for the simple cleaning or reconditioning process is it sent for disposal. In the past, companies cleaned a range of industrial lubricants on behalf of major lubricant vendors. However, many of the larger vendors no longer offer this service as part of their total fluid care services to industry. The commercial drivers for this change in strategy are (1) used oil can be disposed of free of charge, (2) maintaining lubricant quality at remote industrial sites is overreliant on a few highly skilled individuals, (3) the requirement for specialized equipment on-site, and (4) virgin lubricant is often more readily available. Thus, used oil generators are seeking alternate methods of oil disposal.

For used oil generators, typically, transportation of more than 55 gal. (208 L) of used oil must be delivered to an approved used oil collection center. The center must collect and retain the following information about the used oil generator: (1) name, (2) address, (3) telephone number, (4) date of delivery, and (5) amount of used oil delivered to the center.

Throughout many countries, various levels of government encourage businesses that change motor oil to volunteer to serve as used oil collection sites—a place where private citizens and DIY practitioners can drop off used motor oil in an appropriately designed drum or tank. Typically, the collection center is well marked to ensure that it is used for only uncontaminated lubricating oil and it should be serviced regularly by a used oil transporter to make sure there is always room for more oil.

Participation in such a program helps fight pollution, conserves a valuable natural resource, and serves a very critical need by (1) keeping many tons of used oil out of landfills, and (2) keeping used oil from contaminating waters that may serve as drinking supplies and ensure clean drinking water. Recycling used oil can also significantly decrease the amount of damage done to natural habitats such as wetlands and waterways, which ensures the preservation of wildlife and sport fishing areas.

One final aspect of collection that is worthy of note is that many industrial customers make no effort to segregate types of lubricant and it is not unusual for the mixed waste to contain less than 10% v/v of oil and have up to 20% w/w of solids. Specialist oil recovery companies collect wastes with a higher proportion of oil and typically avoid collecting the more heavily contaminated oily wastes and oil–water mixtures.

Such oil is usually cleaned to an extent that is environmentally and purchaser acceptable and marketed as recovered fuel oil, which can be used as fuel for electricity-generating stations and other industrial combustors. However, it has been reported by some purchasers that recovered fuel oil causes greater levels of corrosion in spray heads and pipe work than would be the case with virgin product.

In some countries (mostly developing countries), used oil is classified as hazardous waste and must be disposed of accordingly. However, the necessary data on the quantity and type of used oils arising, or on their sources, is just as unlikely to be available as a formal collection and reprocessing system. For used oil, complete data should be available. Furthermore, regulated and monitored collection systems are not always available and used oil tends to be collected on a random basis and put to various uses. Otherwise, excess used oil is disposed of in an uncontrolled manner (tipping).

Currently, hazardous waste is generally subject to statutory regulation in most developing countries, and the types of hazardous waste are frequently defined and measures for handling and disposal are prescribed; however, in many cases, these are not specific enough to provide a useable basis to allow meaningful implementation. The legislation is not always complete nor is it easy to enforce in practice.

A significant shortcoming in many countries is the absence of collection systems and the lack of adequate treatment plants and disposal sites for hazardous waste, in particular, for waste in liquid form. Another negative factor in many places is the lack of equipment for the analysis and measurement of the amounts of used oil. The dissemination of information and the raising of awareness of the subject of used oil in industry (especially in small industries and micro-industries) are indispensable prerequisites for future improvements.

3.5 TYPES OF USED OILS AND ANALYSES

Examples of used oil include: (1) engine oil (typically includes gasoline and diesel engine crankcase oil, piston-engine oils from automobiles, trucks, boats, airplanes, trains, and heavy equipment), (2) transmission fluid, (3) refrigeration oil, (4) turbine and compressor oil, (5) metalworking fluids and oils, (6) laminating oils,

(7) industrial hydraulic fluid, (8) electrical insulating oil, (9) industrial process oil, (10) oils used as buoyant, and (11) synthetic oil (usually derived from a polymer-based starting material) (Chapter 1).

However, used oil, as defined for this text, is not (1) used oil from bottom clean-out of virgin fuel storage tanks, virgin fuel oil spill cleanups, or other oil wastes that have not actually been used in service; (2) products such as antifreeze and kerosene; (3) vegetable and animal oils, even when used as a lubricant; or (4) petroleum distillates used as solvents.

Oil analysis is the laboratory analysis of the properties of a lubricating oil or a used lubricating oil to determine (in the latter case) properties, suspended contaminants, and wear debris. The analysis should be performed during routine preventative maintenance to provide meaningful and accurate information on lubricant and engine or machine condition. By tracking oil analysis using a well-defined chain of custody protocol (API 2012), with sample results over the life of a particular machine, trends can be established, which can help eliminate costly repairs and potential unhealthy environmental issues.

Before embarking on a testing regime to determine the quality and the nature of the contaminants in used lubricating oil, it is necessary to follow a specific *sampling* protocol. This will allow any purchaser of the used oil to determine the potential suitability of the used oil for re-use.

The importance of adhering to a rigorous sampling protocol to ensure that samples are representative of the bulk product, or appropriate to the specific requirement, cannot be overemphasized. Moreover, it is important that the sampling procedure does not introduce any contaminant into the sample or otherwise alter the sample so that subsequent test results are affected. Because of the importance of proper sampling procedures in establishing used oil properties (and quality), it is imperative that appropriate procedures be used (ASTM D4057). In many liquid manual sampling applications, the material to be sampled (such as used lubricating oil) may contain heavy components (such as metals), which tend to separate from the main component and unless certain conditions can be met to allow for this, the sample may not be truly representative of the used oil. In addition, adequate records of the circumstances and conditions during sampling have to be made.

Thus, a proper oil sampling protocol is critical to an effective oil analysis program. Without a representative sample, further oil analysis endeavors are likely to produce inaccurate data. There are two primary goals in obtaining a representative oil sample.

The first goal is to maximize data density—the sample should be taken in a way that ensures there is as much information per milliliter of oil as possible. This information relates to such criteria as cleanliness and dryness of the oil, depletion of additives, and the presence of wear particles being generated by the machine.

The second goal is to minimize data disturbance—the sample should be extracted so that the concentration of information is uniform, consistent, and representative. It is important to make sure that the sample does not become contaminated during the sampling process. This can distort the data by making it difficult to distinguish the contaminants that were originally in the used oil from contaminants that possibly were introduced into the used oil during the sampling process.

To ensure good data density and minimum data disturbance in oil sampling, the sampling procedure, sampling device, and sampling location should be considered. The procedure by which a sample is drawn is critical to the success of oil analysis. Sampling procedures should be documented and followed uniformly by all members of the oil analysis team. This ensures consistency in oil analysis data and helps to institutionalize oil analysis within the organization. It also provides a recipe for success to new members of the team.

As with all petroleum and petroleum product sampling methods, sampling records for any procedure must be complete and should include, but is not restricted to these items, information such as

1. The precise (geographic or other) location (or site or refinery or process) from which the sample was obtained
2. The identification of the location (or site or refinery or process) by name
3. The character of the bulk material (solid, liquid, or gas present) at the time of sampling
4. The means by which the sample was obtained
5. The means and protocols that were used to obtain the sample
6. The date and the amount of sample that was originally placed into storage
7. Any chemical analyses (elemental analyses, fractionation by adsorbents or by liquids, functional type analyses) that have been determined to date
8. Any physical analyses (API gravity, viscosity, distillation profile) that have been determined to date
9. The date of any such analyses included in items 7 and 8
10. The methods used for analyses that were employed in items 7 and 8
11. The analysts who carried out the work in items 7 and 8
12. A log sheet showing the names of the persons (with the date and the reason for the removal of an aliquot) who removed the samples from storage and the amount of each sample (aliquot) that was removed for testing

In summary, there must be a means to identify the sample history as carefully as possible so that each sample is tracked and defined in terms of source and activity. Thus, the accuracy of the data from any subsequent procedures and tests for which the sample is used will be placed beyond a reasonable doubt (Speight 2001, 2002).

For example, automotive lubricating oil undergoes a range of chemical and physical transformations during routine engine operation, and oil transformations continue after being released to the environment. Some components of motor oil, particularly used oils, are considered a threat to public health and the environment. Gas chromatography–mass spectroscopy (GC-MS), Fourier-transformed infrared spectroscopy (FTIR), nuclear magnetic resonance spectroscopy (NMR), and flame atomic absorption spectrophotometry (FAAS) techniques were employed to characterize the chemical composition of fresh, used, and weathered used oil samples. Used oil was weathered by adding it to soil at a rate of 1.5% w/w, seeding with mixed grass and legume species, and incubating. Soxhlet-extractable oil was analyzed over 150 days. Compared with fresh motor oil, used oil contained new aliphatic and aromatic hydrocarbon compounds such as 1,3,5-trimethylbenzene, p-xylene, and methyl ester undecanoic acid. FTIR bands at 1704 to 1603 cm^{-1} were related to the presence of carbonyl groups, and bands at 869, 813, and 1603 cm^{-1} were associated with new aromatic hydrocarbons including polycyclic aromatic hydrocarbons. NMR analysis revealed new peaks in the used oil in the ranges of 2.1 to 2.7 and 6.8 to 7.2 parts per million (ppm), which were associated with new aliphatic and aromatic hydrocarbon products, respectively. FTIR and GC-MS analysis of weathered used oil indicated the presence of various alcohols, aldehydes, and ketones, indicating substantial biological and chemical decomposition (Dominguez-Rosado and Pichtel 2003).

The study of wear in machinery (tribology) requires frequent performance evaluation and interpretation of oil analysis data from (1) analysis of oil properties including those of the base oil and its additives (Table 3.1), (2) analysis of contaminants, and (3) analysis of wear debris from machinery.

TABLE 3.1
Common Additives Found in Used Oil

Purpose of Additive	Additives
Anticorrosion	Zinc dithiophosphates, metal phenolates, fatty acids, and amines
Antifoamant	Silicone polymers and organic copolymers
Antiodorant	Perfumes and essential oils
Antioxidant	Zinc dithiophosphates, hindered phenols, aromatic amines, and sulfurized phenols
Antiwear additive	Chlorinated waxes, alkyl phosphites and phosphates, lead naphthenate, metal triborates, and metal and ashless dithiophosphates
Color stabilizer	Aromatic amine compounds
Corrosion inhibitor	Metal dithiophosphates, metal dithiocarbamates, metal sulfonates, thiodiazoles, and sulfurized terpenes
Detergent	Alkyl sulfonates, phosphonates, alkyl phenates, alkyl phenolates, alkyl carboxylates, and alkyl-substituted salicylates
Dispersant	Alkylsuccinimides and alkylsuccinic esters
Emulsifier	Fatty acids, fatty amides, and fatty alcohols
Extreme pressure additives	Alkyl sulfides, polysulfides, sulfurized fatty oils, alkyl phosphites and phosphates, metal and ashless dithiophosphates and carboxylates, metal dithiocarbamates, and metal triborates
Friction modifier	Organic fatty acids, lard oil, and phosphorus-based compounds
Metal deactivator	Metal deactivator organic complexes containing nitrogen and sulfur amines, sulfides, and phosphates
Pour point depressant	Alkylated naphthalene and phenolic polymers, and polymethacrylates
Rust inhibitor	Metal alkylsulfonates, alkylamines, alkyl amine phosphates, alkenylsuccinic acids, fatty acids, alkylphenol ethoxylates, and acid phosphate esters
Seal swell agent organic	Organic phosphate aromatic hydrocarbons
Tackiness agent	Polyacrylates and polybutenes
Viscosity	Polymers of olefins, methacrylates, dienes, or alkylated styrenes

3.6 RECYCLING AND RE-REFINING CAPACITIES

Used lubricating oil is insoluble, persistent, and can (typically, more than likely does) contain toxic chemicals and heavy metals. The used oil is slow to biodegrade and is a major source of oil contamination of waterways, which also leads to the pollution of drinking water sources.

In the United States, approximately four million people reuse motor oil as a lubricant for other equipment or take it to a recycling facility. Recycled used motor oil can be (1) re-refined into new (re-refined) oil, (2) processed into fuel oils, and (3) used as raw materials for the petroleum industry. As a measure of the re-refining effort, 1 gallon (3.785 L) of used lubricating oil will provide the same 2.5 US quarts (2.37 L) of lubricating oil as one barrel (42 US gal. or 159 L) of crude oil.

Recycling or re-refining used lubricating oil is typically based on petroleum refining-type processes such as distillation, adsorption, and hydrotreating processes (Kim et al. 1997). During the process, all efforts are made to preserve or rejuvenate oil quality and prevent corrosion and fouling of equipment. The product is a high-quality premium product that can be used as base oil for constituting saleable lubricating product (Chapter 2).

3.6.1 Re-Refining Capacities

Although it has been debatable that significant new investment in re-refining capacity (Chapter 7) was uneconomic in the past, the economics are not as unfavorable as has been believed (Chapter 8) and the choice of a suitable process can make the investment attractive (Chapter 9). In addition, benefits such as the regeneration (rather than waste) of a prime product such as lubricating oil and environmental benefits favor re-refining because the toxic heavy metals (zinc, cadmium, chromium, and lead, among others) are extracted from the used oil and are not to be ignored.

Compared with combustion (Chapter 5), the re-refining process separates heavy metal compounds, which are then solidified and stabilized into solids, therefore posing minimal environmental risk. In the combustion process, the metals in the combustion flue gases must be treated with air pollution abatement equipment before being released to the atmosphere. Combustion processes in cement kilns, steel mills, and other large-scale industrial combustion processes featuring state-of-the-art flue gas treatment are effective in addressing these environmental issues.

On the other hand, combustion of used motor oils in space heaters does not provide for similar levels of pollutant reduction. To achieve maximum energy conservation and environmental benefit, it is generally preferable to re-refine used oils into regenerated base oils that can be blended with additives to make finished lube oil products rather than combustion for heating value recovery. However, the environmental benefit of combusting used oil depends on the fuel that is displaced. Displacement of high–environmental impact fuels such as coal or petroleum coke would make the combustion of used oil rank higher from an environmental perspective (U.S. DOE 2006).

The more elaborate re-refining process (Chapter 7) includes flash evaporation to remove light ends and water, a defueling step to separate gas oils, lube distillate separation from heavy residue, and a hydrofinishing step (Pyziak and Brinkman 1993). These steps require energy and material inputs, such as natural gas for heating, electricity for pressurization and pumping, and hydrogen for hydrofinishing. The hydrofinishing step also requires the use of a catalyst. Sodium hydroxide is used in several process steps (including used-water treatment).

Re-refining results in the recovery of a high-purity lubricating base oil that displaces virgin lube base oil. The heavy metals and other contaminants in used oil are concentrated in the residual (asphalt type) by-product of the re-refining process. This material can be used as a roofing tar, asphalt concrete additive, or for other traditional asphalt bitumen uses. Use of the asphalt product could result in some risk due to leaching of heavy metals; however, leaching tests (Speight and Lee 2000; Speight 2005) will be necessary to show (1) whether or not heavy metals are present, and (2) if the heavy metals are bound within the asphalt matrix and insignificant leaching occurs.

3.6.2 Used Oil Management

Generally, *used oil* is any oil that has been refined from crude oil or any synthetic oil that has been used and, as a result of such use, is contaminated by physical or chemical impurities (40 CFR 279.1). Sources of used oil include: (1) automotive lubricating oil, (2) industrial oil, (3) hydraulic oil, and (4) oily sludge.

Used oil is regulated by the U.S. EPA as well as state environmental agencies. The federal regulations are set forth in 40 CFR Part 279 (Standards for the Management of Used Oil), although some states, such as California, have adopted regulations that are more stringent than the federal regulations.

The Resource Conservation and Recovery Act (RCRA), is the United States federal law that was originally enacted by Congress in 1976. The primary goals of the law are to protect human health and the environment from the potential hazards of used oil disposal, to conserve energy and natural resources, to reduce the amount of used oil generated, and to ensure that wastes are managed in an environmentally sound manner. The law was amended in 1984 and one of the amendments specifically directed the U.S. EPA to regulate used oil so as to encourage recycling and protect the environment.

Under the RCRA, used oil may be considered hazardous if it is ignitable, corrosive, reactive (e.g., explosive), or contains toxic chemicals. In addition to these characteristic wastes, the U.S. EPA has also developed a list of more than 500 specific hazardous wastes. However, the used oil management regulations establish that recycled used oils and used oil destined for recycling are not hazardous wastes (Boughton and Horvath 2004). In fact, the U.S. EPA regulations presume that if used oil will be recycled, it no longer has the benefit of being regulated under the used oil management standards, and must be tested to determine if it is a hazardous used material.

Halogens are a group of nonmetal elements that exhibit similar properties. They include fluorine, chlorine, bromine, iodine, and astatine. Sources of halogens include: (1) chlorinated solvents, (2) pesticides, and (3) polychlorinated biphenyls (PCBs). It is worth noting that used oil containing PCBs in concentrations of 50 ppm or greater are subject to the U.S. Toxic Substances Control Act (40 CFR 279.10).

In fact, identifying the sources of chlorine in a facility will allow the segregation of spent solvents from used oil storage tanks.

Furthermore, according to the U.S. EPA's Rebuttable Presumption Rule, any used oil containing 1000 ppm of total halogens is presumed to have been mixed with hazardous used material and, consequently, the entire used oil/used mixture constitutes a hazardous used material [40 CFR 261.3(a)(2)(v)]. The rule also allows generators to rebut the presumption in several different ways.

For example, when using a field test designed for the identification and quantification of contaminants, it is clear if the result shows that the used oil contains a concentration of more than 1000 ppm total halogens, a sample of the oil can be sent for laboratory analysis to determine the source of the halogens. If laboratory analysis shows that

- RCRA-listed solvents in the used oil were not in significant concentrations (<100 ppm): then the used oil can be classified as nonhazardous and regulated as used oil. (See 509 U.S. Federal Regulation 49176, November 29, 1985.) If the total concentration of halogen is less than 4000 ppm, then the used oil can be classified as *on-specification used oil* provided four special metals (lead, arsenic, cadmium, and chromium) are all below the required concentrations. Also, the flashpoint must be 100°F (37.8°C) or greater: see 40 CFR 279.11. A decrease in the flash point indicates contamination by dilution of lubricating oil with unburned fuel—an increase in the flash point indicates the evaporation of some of the lower boiling constituents during service.
- RCRA-listed solvents were in significant amounts: then the used oil is classified as *hazardous used oil* unless the source of the halogens can be conclusively shown not to be hazardous used oil.
- RCRA-listed solvents were not in significant amounts, but the total halogens were more than 4000 ppm: then the used oil can be classified as *off-specification used oil.* If the used oil contains more than 1000 ppm total halogens, the used oil is presumed to be mixed with listed hazardous used material and is classified as hazardous used oil—unless the presumption can be rebutted.

Thus, for the purposes of regulating used oil burned for energy recovery, the U.S. EPA makes a distinction between on-specification used oil and off-specification used oil based on a finding that certain contaminants in used oil pose a significant threat to human health or the environment. As a result, the U.S. EPA has established maximum concentration limits for these constituents of concern. If the used oil is shown not to exceed any allowable level, it is on-specification and can be burned in any combustion device. Otherwise, it is considered off-specification used oil.

Some used oils can be managed as *used oil* even when they do not fit the used oil definition. For example, a material that is not a dangerous waste, but is contaminated with or contains used oil, may be managed under the used oil rule. Wastes, although not included in this text, which may fit into this category include oil bilge water, oily waste from oil/water separators, and oily wastewaters. Used oil that is removed from material containing or contaminated with used oil can also be managed under the used oil regulations. Examples may include used oil drained from an oil filter, engine oil pan, or oil waste removed from oil/water separators.

Any other type of oily waste that does not meet the definition or exceptions mentioned above cannot be managed under the used oil regulations. Some examples include brake cleaner, solvents used for degreasing (including petroleum distillates and mineral spirits), paint and oil paint waste, oil-based inks, antifreeze, degreasers, and carburetor cleaners. These products would typically be designated as dangerous waste when disposed of.

In addition, even though the following wastes fit the definition of used oil, they cannot be managed as used oil for the purpose of burning: (1) used oil which is designated as extremely hazardous waste, usually for high levels of halogenated organic compounds, (2) certain metal-working oils with any amount of chlorinated compounds in the formulation, and (3) transformer oils such as polychlorobiphenyl oils.

In summary, it is essential that the used oil, before it leaves the facility for disposal as used oil or for any other use, be subjected to a thorough series of tests. If the used oil tests positive for high levels of halogens, and the presumption cannot be rebutted, the contaminated used oil must be classified as hazardous used oil and a licensed hazardous used oil company must be used for transportation and disposal of the used material. In fact, it is often a rule of thumb as an initial precautionary measure that any used oil (except edible oil) that is unclassified in terms of analysis and inspection should be considered to be hazardous until it is proven otherwise.

3.7 ENVIRONMENTAL ASPECTS OF USED OIL MANAGEMENT

Used oil is any oil that has been refined from crude oil or any synthetic oil that has been used and as a result of such use is contaminated by physical or chemical impurities. In brief, used oil is exactly what its name implies—any petroleum-based or synthetic oil that has been used (Chapters 1 and 2). The additive content of lubricants and motor oils in particular, gives rise to the environmental concerns involved in the combustion of used oils. It is common that up to 30% v/v of typical motor oil blends are made up of additives that are used to improve the quality, stability, and longevity of motor oils in combustion engine applications. Although industrial oils are also components of used oils, they do not contain the same level of additives and thus they tend to dilute the effect of automotive lubricating oil additives.

Because of the complexity and composition of used oil, danger to the environment and health from inappropriate management of used oil, and in view of the possible content of problematic foreign substances (such as heavy metals, PCBs, or other halogen compounds), used oils require special handling.

Generally, used oil is semiliquid or liquid and consists entirely or partially of mineral oil or synthetic oils (Chapter 1). Accordingly, used oil is oil that has taken up foreign substances or impurities and can no longer be used for lubrication purposes.

It may contain small quantities of toxic substances that are liable to degrade the quality of air, soil, and ground water if it is not handled and disposed of in an appropriate manner.

In developing countries, most of the used oil that is generated originates from lubricating oil in the transport sector; the quantities of hydraulic oil and transformer oil are relatively small. Agriculture (irrigation pumps, etc.) and the energy sector (small-scale power stations with diesel engines) are the second most significant source. Most used oil arises in urban centers and along major roads at filling stations and motor vehicle repair shops, and in some cases also at DIY oil-change stations on sites adjacent to major roads.

Mismanagement of used lube oil is a serious environmental problem. Almost all types of used oil have the potential to be recycled safely, saving a precious non-renewable resource and at the same time minimizing environmental pollution. Unfortunately, in some countries, most of the used oil is handled improperly. Some is emptied into sewers going directly into water waste, adversely affecting water treatment plants; some is dumped directly onto the ground to kill weeds or is poured onto dusty roads or is dumped in deserts, where it can contaminate surface water and ground water.

The disposal of used lubricating oil into the ecosystem creates environmental hazards. Laws have been enacted throughout the world for the disposal of waste petroleum products, and every effort should be made for reuse of these products. In most cases, used oil can be reused after reconditioning with or without the addition of any additives, resulting in the conservation of a prime product. Thus, regeneration, reclamation, or recycling of spent lubricating oils has become an important process industry, adopting various techniques for oil purification.

If used oil is not disposed of properly so that the risk to the environment is at an absolute minimum, there is a risk that it and any other substances that it may contain will enter natural cycles and the food chain via water, soil, and the air. In this way, hydrocarbons, heavy metals, polychlorobiphenyl derivatives, and other halogen compounds that may be contained in used oil pose a risk to human health and impede the growth of plants and their ability to take up water. Less than 1 gallon (3.785 L) of used oil has the potential to contaminate up to 1 M gal. (3.785 ML) of ground water or surface water—the concentration of oil in the water is then 1 ppm, which is the level that is seen as the upper limit of oil in water that can still be considered tolerable for general use.

The quality of used oil is determined mainly (apart from the content of middle distillates and highly volatile components) by the treatment of lubricating oils with additives and the conditions under which the oils are used. Additives serve the purpose of improving viscosity or flow properties, for example, or of reducing wear (Chapter 1). Furthermore, the growing use of synthetic oils is a significant factor. The infiltration of oil into soil as a separate phase or in dissolved form constitutes above all a long-term threat to groundwater and hence potentially to drinking water quality. Both the biodegradation of oils in soil and the discharge of the substances into ground water proceed very slowly (Speight and Arjoon 2012).

Thus, it is of fundamental importance to obtain as much information as possible about how used oil becomes contaminated and the nature of the contaminants. This knowledge will provide information about the type of oil or the other dangerous substances in the used oil. In some cases, it will be straightforward because the nature of the contamination will be obvious. Other cases, such as with contaminated land remediation, will be more difficult, and the history of the site will require additional investigation.

To meet the U.S. EPA definition of used oil, it must fulfill each of three criteria: (1) origin—must have been refined from crude oil or made of synthetic materials; (2) used—must have been used as a lubricant, hydraulic fluid, heat transfer fluid, buoyant, or other similar purpose; and (3) contaminated—by either physical or chemical impurities generated from handling, storing, and processing of used oil and could include foreign material such as metal shavings, sawdust, dirt, solvents, and halogens.

Finally, as a result of re-refining (Chapter 7) the environmentally noxious compounds are solidified and stabilized and, due to their disposition as a solid, pose a much lower environmental risk. On the other hand, combustion of used oil (Chapter 5) results in air emissions, the magnitude dependent on the quality of the air pollution control equipment utilized.

3.8 CONVERTING USED OIL INTO USEABLE OIL

Although this topic is dealt with in much more detail later (Chapters 4 through 7), it is appropriate at this point to summarize the processing options for the production of clean used oil. The first step is analytical testing to ensure quality because used lubricating oil from a wide variety of sources may contain chemical or physical contaminants that are hazardous or prevent that oil from being used in the re-refining process. Constant testing ensures that used lubricating feedstocks are capable of producing the highest quality end products with the result of premium products (base oils) and environmentally correct recycling.

For example, water is a common constituent of used oil and it is typically removed using an evaporation (or distillation) process, after which the water is collected, treated to be chemically and biologically safe, and then discharged. Any low molecular weight constituents that evaporate (distill) with the water are extracted for reuse. If the glycol that may be present in the used oil is not removed during the evaporation, the dehydrated used oil goes to tall recovery (distillation) towers where glycol and low-boiling fuel constituents are separated and collected. The glycol can be reprocessed and recycled as a finished automotive grade product. Higher molecular weight constituents are removed in a vacuum distillation unit—these constituents can be blended with distillate fuel oil. The remaining oil is treated with hydrogen to remove sulfur, nitrogen, chlorine, heavy metals, and other impurities. This step also corrects any issues with odor, color, and corrosion performance. The purified oil is tested before being returned to the marketplace as base oil.

Depending on the analytical data after each stage, the used oil product can be selected for burner fuel alone or as a blend with heavy fuel oil or the ultimate product—base oil for reuse in lubricating oil production.

REFERENCES

American Petroleum Institute (API). 2012. *Bulk Engine Oil Chain of Custody and Quality Documentation*. Report No. API 1525A. American Petroleum Institute, Washington, DC, January.

ASTM D4057. 2012. Practice for Manual Sampling of Petroleum and Petroleum Products. *Annual Book of Standards*. American Society for Testing and Materials, West Conshohocken, Pennsylvania.

Boughton, R., and Horvath, A. 2004. Environmental Assessment of Used Oil Management Methods. *Environmental Science & Technology*, 38(2): 353–358.

Dominguez-Rosado, E., and Pichtel, J. 2003. Chemical Characterization of Fresh, Used and Weathered Motor Oil via GC/MS, NMR and FTIR Techniques. *Proceedings of the Indiana Academy of Science*, 112(2): 109–116.

Harrison, C. 1994. The Engineering Aspects of a Used Oil Recycling Project. *Waste Management*, 14(3–4): 231–235.

Khawaja, M.A., and Aban, M.M. 1996. Characteristics of Used Lubricating Oils: Their Environmental Impact and Survey of Disposal Methods. *Environmental Management and Health*, 7(1): 23–32.

Kim, M.-S., Hwang, J.-S., and Kim, H.-R. 1997. Re-Refining of Waste Lube Oils by Vacuum Distillation with Petroleum Atmospheric Residuum. *Journal of Environmental Science & Health, Part A: Environmental Science & Engineering*, A32(4): 1013–1024.

Kramer, D.C., Ziemer, J.N., Cheng, M.T., Fry, C.E., Reynolds, R.N., Lok, B.K., Krug, R.R., and Sztenderowicz, M.L. 1999. Influence of Group II and III Base Oil Composition on VI and Oxidation Stability. Proceedings, AIChE Spring Meeting, Houston, Texas, March 14–18.

Pyziak, T., and Brinkman, D.W.J. 1993. Recycling and Re-Refining Used Lubricating Oils. *Lubrication Engineering*, 49(5): 339–346.

Speight, J.G. 2001. *Handbook of Petroleum Analysis*. John Wiley & Sons Inc., Hoboken, New Jersey.

Speight, J.G. 2002. *Handbook of Petroleum Product Analysis*. John Wiley & Sons Inc., Hoboken, New Jersey.

Speight, J.G. 2005. *Environmental Analysis and Technology for the Refining Industry*. John Wiley & Sons Inc., Hoboken, New Jersey.

Speight, J.G., and Arjoon, K.K. 2012. *Bioremediation of Petroleum and Petroleum Products*. Scrivener Publishing, Salem, Massachusetts.

Speight, J.G., and Lee, S. 2000. *Environmental Technology Handbook*, 2nd Edition. Taylor & Francis, New York.

U.S. Department of Health and Human Services. 1997. *Toxicological Profile for Used Mineral-Based Crankcase Oil*. Agency for Toxic Substances and Disease Registry, Public Health Service, United States Department of Health and Human Services, Washington DC. September.

U.S. Department of Energy (U.S. DOE). 2006. *Used Oil Re-Refining Study to Address Energy Policy Act of 2005 Section 1838*. Office of Oil and Natural Gas, Office of Fossil Energy, United States Department of Energy, Washington, DC.

U.S. Environmental Protection Agency (U.S. EPA). 2011. Materials Characterization Paper in Support of the Final Rulemaking: Identification of Nonhazardous Secondary Materials that are Solid Waste Used Oil, February 3. Available at http://www.epa.gov/epawaste/nonhaz/define/index.htm.

4 Composition and Treatment

4.1 INTRODUCTION

Large quantities of used oil are generated each year, which, if properly collected and processed, can be a valuable energy source or refined to produce usable products such as new lubricating oil. However, used oil is contaminated with water and other liquids, halogens, and other elements including heavy metals. It is not surprising that, in most countries, it is regarded as potentially hazardous waste and must be handled, processed, and stored appropriately. The transport, storage, and ultimate use of used oil is governed by a variety of direct and indirect national and international legislation and industry standards.

Because of the danger to the environment and health from inappropriate management of used oil and in view of the possible content of problematic foreign substances (such as heavy metals, polychlorinated biphenyls [PCBs], or other halogen compounds), used oils require special handling. In fact, in most countries, used oil is classified as hazardous waste and the collective term *used oil* includes used and contaminated mineral oils, oily residues from containers, emulsions, and water–oil mixtures (Chapter 1). In fact, the quality of used oils is determined mainly, apart from their content of middle distillates and highly volatile components (ASTM D86; ASTM D1160; Chapter 1), by the treatment of lubricating oils with additives and the conditions under which the oils are used.

In industrialized countries, the term used oil may also be defined according to the means of disposal rather than according to the composition of the material. Whatever the meaning of the term, used oil is more than likely to contain small quantities of toxic substances that are liable to degrade the quality of air, soil, and ground water if it is not handled and disposed of in an appropriate manner.

There are a number of legally approved routes for the disposal of used oil, each subject to legislation and, as a result of evolving regulations, some of these uses may become more restrictive as legislation changes. Currently, the main legal disposal routes are (1) direct combustion/use as fuel, (2) processing to produce secondary fuels, and (3) re-refining to produce new base oil and other petroleum products—the reuse route chosen in different countries varies greatly and depends on local energy policies.

The management of used oil is particularly important because of the large quantities generated globally, their potential for direct reuse, reprocessing, reclamation and regeneration, and because they may cause detrimental effects on the environment if not properly handled, treated, or disposed of. Used lubricating and other oils represent a significant portion of the volume of organic waste liquids generated worldwide. The three most important aspects of used oils in this context are (1) contaminant content, (2) energy value, and (3) hydrocarbon properties.

The processing of used oils has been practiced for many years, with organized recycling of engine lubricating oil from vehicle fleets being well established by the 1930s. Certain used oil streams arising from oil refinery sites have been fed into so-called *crude ponds*. A portion of the materials that have accumulated in these ponds have been recycled.

Briefly, it must not be forgotten that the prime objective in the production of lubricating oil is the separation of wax distillate and cylinder stock without any decomposition or cracking of the lubrication oil fractions; thus, a vacuum distillation unit is used to separate the wax distillate and the bottom stock at a lower temperature (Chapter 1). The properties that make the high-boiling paraffin hydrocarbons suitable for lubricating oil manufacture include (1) stability at high temperatures, (2) fluidity at low temperature, (3) only a moderate change in viscosity over a broad temperature range, and (4) sufficient adhesiveness to keep the oil in place under high shear forces. The desired fractions for the manufacture of the lubricating oil have high boiling points, and its separation into various boiling point range cuts must be accomplished under reduced pressure (Chapter 1). The vacuum tower produces some fuel oil overhead, which is sold as a separate product or sent to another area of the refinery for further processing and blending. The two main products from the vacuum tower are wax distillate and cylinder stock, which is the bottom product—each stream contains a number of desirable lubricating oil constituents as well as by-products. The wax distillate is charged directly to the dewaxing unit. The vacuum tower bottoms, or cylinder stock, are charged to the deasphalting unit. These two fractions form the basic stock for lubricating oil manufacture (Speight and Ozum 2002; Hsu and Robinson 2006; Gary et al. 2007; Speight 2014).

In the context of this book, *used oil* is any semisolid or liquid used product consisting totally or partially of mineral oil or synthesized hydrocarbons (synthetic oils; Chapters 1 and 3). Thus, used oil is oil arising from industrial and nonindustrial sources in which the oil has been used for lubricating, hydraulic, heat transfer, electrical insulating (dielectric), or other purposes and whose original characteristics have changed during use, thereby rendering the oil unsuitable for further use for the purpose for which the oil was originally intended.

In terms of the terminology used for used oil regeneration, there are four commonly used terms: (1) recycling, (2) reprocessing, (3) reclamation, and (4) regeneration.

Recycling is the commonly used generic term for the reprocessing, reclaiming, and regeneration (re-refining) of used oils using an appropriate selection of physical and chemical methods of treatment.

Reprocessing usually involves treatment to remove insoluble contaminants and oxidation products from used oils such as by heating, settling, filtering, dehydrating, and centrifuging. Depending on the quality of the resultant material, this can be followed by blending with base oils and additives to bring the oil back to its original or an equivalent specification. Reprocessed oil is generally returned to its original use.

Reclamation usually involves treatment to separate solids and water from a variety of used oils. The methods used may include heating, filtering, dehydrating, and centrifuging. Reclaimed oil is generally used as a fuel or fuel extender.

Regeneration involves the production of base oils from used oils as a result of processes which remove contaminants, oxidation products, and additives, i.e., re-refining involving the production of base oils for the manufacture of lubricating products. These processes include predistillation, treatment with acids, solvent extraction, contact with activated clay, and hydrotreating (Audibert 2006; Udonne 2011; Emam and Shoaib 2012). These methods should not be confused with the simpler methods of treating oils, such as those described under reclamation.

Used oils originate from diverse sources. These include petroleum refining, the forming and machining of metals, small generators (do-it-yourself car and other equipment maintenance) and industrial sources, oil and gas well drilling operations, and the rural farming population. Collecting used oil from nonindustrial sources and local/small generators is very difficult and requires a well-established and efficient infrastructure to accomplish the task. In this regard, it is important to develop adequate reuse or recycling options, to properly handle the collected volume of oil, to address the specific properties of the concerned waste, and to assess the degree to which used oils could be treated.

Used lubricating oil is routinely collected from gasoline/petrol-fueled engines during maintenance operations. Used motor oils consist of the base petroleum-derived oil contaminated with low levels of combustion products, which appear over time during the course of engine use. Used lubricating oil may become mixed with a wide variety of materials. How the resulting mixture is regulated can vary greatly depending on the type of material that is mixed with the used oil.

In general, used oils are not volatile and do not present a significant inhalation health hazard. Some components are hazardous, including the type of base oil used to formulate the unused product, as well as accumulated polynuclear aromatic hydrocarbons. Prolonged and repeated skin contact must be avoided. Used oils may also contain residual amounts of additives that are eye and skin irritants, and possibly sensitizers.

Used lubricating oils retain a high energy potential. However, the hazards and costs associated with collecting, storing, transporting, and general handling of the used oil has limited the efforts to collect used motor oil for disposal or recycling. Although previous methods provide limited processing of used motor oil for other petroleum products, there remains a need to improve the procedure of converting used motor oil to a high-quality energy source.

Lubricating oil undergoes a range of chemical and physical transformations during routine in-service operation, and oil transformations continue after being released into the environment. Some components of lubricating oil, particularly used oils, are considered a threat to public health and the environment.

In fact, used lubricating oil is a complex mixture of paraffinic, naphthenic, and aromatic petroleum hydrocarbons that may contain one or more of the following: (1) carbon deposits, (2) sludge, (3) aromatic and nonaromatic solvents, (4) water, as a water-in-oil emulsion, (5) glycols, (6) wear metals and metallic salts, (7) silicon-based antifoaming compounds, (8) fuels, (9) polynuclear aromatic hydrocarbons, and (10) miscellaneous lubricating oil additives. In the unlikely event that used transformer oils are mixed with other used oils, then PCBs and polychlorinated terphenyl derivatives may also be present.

In addition, polynuclear aromatic hydrocarbons have the potential to cause skin cancer. Used lubricating oil may contain components that (1) are harmful to aquatic organisms or (2) can cause long-term adverse effects in the aquatic environment. Furthermore, because of thermal changes to the original oil during service and the leakage of unburned fuel into the oil sump, used lubricating oils may contain fuel that could reduce the flash point and make the material flammable.

Thus, if used lubricating oil is not managed properly, it can result in the pollution of soil, ground water, and surface water. Just a quart (0.95 L) of oil can make more than a million gallons (3.79 ML) of drinking water unfit to drink. One pint (0.47 L) of used oil can create an oil sheen more than an acre (0.405 ha.) in size, which can block necessary oxygen and sunlight from moving through the surface of the water. Hydrocarbons in oil can harm young fish, upset fish reproduction, and interfere with the growth and reproduction of bottom-dwelling aquatic organisms. Used oil spilled on the ground could also kill plants and could be toxic to pets and wildlife.

The task of the re-refiner is to remove all the aforementioned contaminants and restore the oil to its original condition. The important point to note is that, in many cases (Chapter 7), the technology used by the re-refiner is virtually identical to that used to refine crude petroleum; the difference being that the level of contamination in used oil is much lower than that in crude oil.

The recycling of used engine oil and other lubricating oils and used oil management was conceived as early as the 1930s. Initially, the used lubricating oils were combusted to produce energy and later these oils were reblended with engine oils after treatment. Due to the increasing necessity of environmental protection and more and stricter environmental legislation, the disposal and recycling of used oils have become very important and can be accomplished with different methods (Gorman 2005).

Another option for recycling used lubricating oil and other oils is reclamation, which is carried out without earlier pretreatment; therefore, harmful components (such as heavy metals, noncombusted halogenated, hydrocarbons, and sulfur oxides) are concentrated in the exhaust gas of the power plant. However, slurry, metal particulates, and other impurities have to be separated from used oil by physical separation.

The composition of used oils is the main element determining its recycle/disposal processing options because this could determine the possibility of being recycled as a fuel in power plants or in re-refining operations.

Thus, after initial clean-up of the used oil, re-refining uses the same basic techniques that are used in crude oil refining, namely, distillation, hydroprocessing, and various finishing processes (Speight and Ozum 2002; Hsu and Robinson 2006; Gary et al. 2007; Speight 2014). Generally, used oil has not lost the basic lubricating properties through its use, but suffers from fuel dilution, additive depletion, contamination by soot and other combustion by-products, and metals produced by wear on the machinery. The objective of re-refining is to recover the base oils that can then be blended with additives to produce new lubricating oils (Chapter 7).

Re-refining results in the recovery of high-purity lubricating base oil that displaces virgin lube base oil (Chapter 7). The heavy metals and other contaminants in used oil are concentrated in the asphalt by-product of the re-refining process. This material can be used as a roofing tar, asphalt concrete additive, or for other traditional asphalt uses. However, the asphalt product could be a risk because it could result in the leaching of heavy metals—although leaching tests have shown that the heavy metals are usually bound within the asphalt matrix and any leaching is insignificant. The phenomenon of insignificant leaching is more than likely product-dependent, and before use the asphalt product should be subjected to the toxicity characteristic leaching procedures (Speight 2005; Speight and Arjoon 2012).

Compared with other disposal routes, re-refining involves considerable capital cost and, unless favorably treated by local tax regimes, can be more expensive than lubricating oil derived from crude. However, re-refining yields energy benefits compared with the refining of new crude oil and generates less pollution than burning the waste as fuel. These factors, combined with volatile oil prices and a desire by many countries to reduce dependence on imported oil, suggests that re-refining may (or should) become a more popular disposal route in the future. Furthermore, in some countries, new legislation stipulates that re-refining is the preferred disposal route to prevent the export of waste oils for incineration elsewhere.

Finally, re-refined base oil is used motor oil that has undergone an extensive refining process to remove contaminants, distill the used oil, and reform the molecular constituents to produce new base oils that have the same performance characteristics as virgin base oils made from crude oil (Chapters 1 and 7).

Although a major source of used oil is used automotive lubricants, the oil recycling industry has developed so that a wide variety of wastes are treated to produce usable product. There are a number of legally approved routes for the disposal of used oil, each subject to legislation; regulations are also being reviewed continually in many countries and some of these (for used oil) uses may become restricted as legislation changes. The main *legal* disposal routes are (1) direct combustion/use as fuel (Chapter 5), (2) processing to produce secondary fuels, and (3) re-refining to produce new base oil and other petroleum products. The routes chosen in different countries vary greatly (depending on the operative legislation) and of the used oil produced—approximately 50% or more is, on average, used as fuel. Re-refining involves a significant investment in the power plant and, in some countries, a premium price for the product. Although some countries are claiming re-refining rates of more than 70% of waste oil collected, others are close to zero. A significant proportion of used oil is disposed of illegally—as much as 25% of all used oil produced is disposed of in this way.

4.2 PRIMARY TREATMENT

The primary step in re-refining used lubricating oil is dehydration, in which the oil is stored to allow water and solids to separate out from the oil, then the oil is heated to at least 120°C (248°F) in a closed vessel to boil off any emulsified water and some of the light hydrocarbons and fuel diluents.

The dehydrated oil can then be fed continuously into a vacuum distillation plant for fractionation in exactly the same fashion as crude petroleum. The fractions obtained are as follows: (1) light fuel and diesel, which can be used as fuel to the burners and boilers; (2) lubricating oil, which is a preferential product insofar as base oils are premium products to recycle as lubricating oil stock; and (3) residuum, which is the nonvolatile portion of the feedstock and contains all the carbon, wear metals, degraded additives, and most of the lead and oxidation products. The residuum might be suitable as an asphalt additive for road paving.

During this step, other contaminates such as ethylene glycol (antifreeze; ASTM D2982) that has leaked into the lubricating oil or has been disposed of incorrectly and included in the used lubricating oil are also removed. Ethylene glycol (boiling point: 197°C, 355°F) boils sufficiently lower than lubricating oil (boiling point: >285°C, 545°F, or more typically >300°C, 570°F) to be separated during the dehydration step. The conditions employed for the dehydration depend on the amount of water, gasoline contaminants, and other contaminants (such as ethylene glycol) in the used oil. Subsequently, the glycol–water mixtures can be separated using membrane technology (Srinivasa Rao et al. 2007).

In one processing option, the lubricating oil fractions are then passed through an extraction tower in the presence of *N*-methyl-pyrrolidone (NMP)—a selective solvent that, in addition to removing some color and odor, is able to extract all unwanted aromatic contaminants present in the paraffinic lubricating oil fraction, subsequent to fractional distillation.

Currently, because of the high value of base oils, recycling technology is preferably based on the re-refining process, although cracking the used oil to produce diesel and other distillates is still considered an option in some countries. The cracking process of choice is often a pyrolysis process in which the used oil is thermally decomposed at staged temperatures up to 500°C (930°F) to remove water and other volatile constituents before thermal decompositon commences (Kim and Kim 2000; Ramasamy and Raissi 2007; Lam et al. 2011, 2012; Manasomboonphan and Junyapoon 2012).

Various processes allow high-quality base oil blending components to be produced. In most cases, solvent or heterogeneous catalytic hydrogenation is applied for the re-refining of used lubricating oil. The hydrogenation processes might spread in a wide extent due to the increasing demand for modern, high-performance Group III base oils.

4.3 SEPARATION

The re-refining of used lubricating oils is a physical and chemical conversion that consists of the removal of mechanical and oxidation contaminants. The properties of re-refined used engine oils are similar to the new oils. Currently, four different re-refining technologies are applied, which have different yield and product properties, construction, and operational costs. These are (1) clay treatment and acidic or solvent re-refining, (2) clay treatment and vacuum distillation, (3) vacuum distillation and solvent re-refining, and (4) vacuum distillation and hydrogenation.

4.3.1 Chemical Separation

An example of chemical separation is the Vaxon process (Gorman 2005), which uses chemical treatment, vacuum distillation, and solvent refining. The advantage of this process is the special vacuum distillation unit (detailed below) in which the cracking of oil is minimized.

In the first aspect of this technology, chemical treatment is carried out with alkali-hydroxides (sodium hydroxide and potassium hydroxide) for removing chlorides, metals, additives, and acidic compounds. Alkoxides are formed on the catalyst surface from the insoluble alkali-hydroxides, which are soluble in oil. The impurities can be bonded with asphaltene molecules by these reactants; therefore, these impurities can be easily separated from the oil.

After chemical treatment, the feedstock is separated into light products, catalyst, base oil, and residue. The feedstock is distilled into two parts by a cyclonic column. Because of the formation of a tangentially flowing thin film, the light hydrocarbons are easily and quickly distilled. The polycyclic aromatic hydrocarbons are separated by solvent refining with polar solvents (such as dimethyl-formamide, NMP), which is carried out in a multistage extractor followed by solvent recovery from both phases. The raffinate contains the wide boiling range base oil, which can be separated by vacuum distillation into different viscosity grade base oils. The polycyclic aromatic hydrocarbons, which are concentrated in the extract, are used for heat energy production or as an asphalt blending component. Used oil containing high amounts of metals can also be processed using these methods (Gorman 2005). More recently, acetic acid has been proposed as a solvent for washing impurities out of used oil (Hamawand et al. 2013).

4.3.2 Physical–Chemical Separation

In the Universal Oil Products direct contact hydrogenation (DCH) process, the feedstock is mixed with hydrogen in the first separation stage, where metals and other contaminants are removed. After this step, the feedstock is flowed to a fixed bed hydrogenation reactor, which is followed by cooling and the second separation step. The hydrogen flow from the separation step contains light hydrocarbons, hydrogen sulfide, ammonia, and steam, which have to be separated before recycling. The hydrogenated oil is fractionated to gasoline, gas oil, and base oil fractions.

4.3.3 Physical Separation

The Dominion Oil process is a solvent-refining process in which the feedstock is dewatered in the first stage, and then the light hydrocarbons (naphtha, diesel, gas oil) and residue are separated by vacuum distillation.

The main product of the vacuum distillation unit is the base oil, which is refined with *N*-methyl-2-pyrrolidone in the second stage. In the solvent-refining stage, the removal of polycyclic aromatic hydrocarbons is carried out. The contaminants which are larger than 1 μm are filtered from the re-refined oil. The vacuum residue is used as an asphalt blending component.

The Fetherstonhaug process for used oil re-refining consists of four steps. In the first step, asphaltene constituents and other heavy impurities are dissolved in propane. After this step, the re-refined oil is recovered by distillation. The solvent-refined oil is fractionated to low-boiling hydrocarbons, vacuum gas oil, and base oil by atmospheric distillation (200°C–250°C, 390°F–480°F) and vacuum distillation (310°C–340°C, 590°F–645°F) under low pressure in a packed tower. The final step is hydrogenation to form base oil.

4.4 FINISHING

Finishing processes are those processes used in the treatment of used lubricating oils to obtain a purified oil product. Finishing processes typically follow the removal of contaminants from used lubricating oils by a series of treatments. The result is the production of finished oil from demetallized used oil.

Materials contained in a typical used crankcase oil that are undesirable if the oil is to be reused include submicron-sized carbon particles, atmospheric dust, metal and metal particles, detergents, pour point depressants, oxidation inhibitors, viscosity index improvers, and resins. Besides lead, which can be present at concentrations of 1% to 2.5% w/w, particularly in the decreasing number of areas where leaded gasoline is still in use, appreciable amounts of zinc, barium, calcium, phosphorus, and iron are also present in the used crankcase oil. Once these contaminants have been removed (Chapters 4 and 6), the oil is treated in a finishing process, which may involve clay treatment or hydrotreatment in the presence of a catalyst.

4.4.1 Clay Treatment

Contacting the hydrocarbons with some natural clay minerals activated by roasting or treatment with steam or acids is an established refining process for removing traces of impurities from petroleum and petroleum-derived products (Speight and Ozum 2002; Hsu and Robinson 2006; Gary et al. 2007; Speight 2014). The phenomenon is similar to that described under the adsorption process—the clay retains the longer chain molecules within its highly porous structure.

In the clay treatment process, which many scientists and engineers consider to be the primary treatment process (after water removal), the used lubricating oil is mixed with clay in a reactor. The mixture is preferably heated to between 105°C and 200°C (220°F–390°F), which is sufficiently low to avoid cracking (thermal decomposition) of the oil (Emam and Shoaib 2012; Martin de Julian et al. 2012). After a measured period, depending on the quality of the used feedstock oil, the mixture is pumped through filters. Cakes of clay and contaminants remain in the filters, while the oil emerges without the contaminants. In a second stage of the process—for removing ash or soot, very fine carbon particles, and other organic compounds from used motor oil—the oil–clay mixture is passed through the filters after a centrifuge is used to remove most of the clay contaminated with soot so that the filters will not be clogged.

Used oils, which have not deteriorated to any great extent, are often clay-contacted or treated with adsorbents without any acid treatment (Emam and Shoaib 2012). They are generally given a preliminary settling, filtering, centrifuging, or vacuum dehydrating treatment. Insulating oils and transformer oils are often treated in this way.

In the conventional acid-treating process, dewatered oil is treated with sulfuric acid (96%) and the acid-refined oil is vacuum-distilled to separate lube base oil from the low-boiling spindle oil and gas oil (Speight 2014). With sulfuric acid treatment, it is necessary to dehydrate the feedstock completely before subjecting it to acid treatment to prevent dilution of the concentrated sulfuric acid.

In the sulfuric acid–treating process, lubricating oil is contacted with sulfuric acid (85%–104% w/w) to (1) improve color and color stability, (2) improve oxidation stability, and (3) remove sulfur, nitrogen, and the more active aromatic compounds. The use of fuming acid has been desirable when the complete removal of aromatics is required. Treating at higher temperatures (50°C–65°C, 120°F–150°F) may be advisable because sulfonation of aromatics proceeds much faster and the acid is more efficiently used. However, care must be taken to prevent burning and darkening, which will occur if the temperature is too high.

The action of sulfuric acid on lubricating distillates is complex and is generally both chemical and physical in nature. Strong acid attacks almost all of the constituents present in the oil including saturated and aromatic hydrocarbons and sulfur, nitrogen, and oxygen compounds. Selective action of the acid on these compounds may be obtained by varying treating conditions, such as reaction temperature, acid concentration, and contact time between the oil and acid.

In general, there is no need to remove crankcase dilution or fuel components ahead of the acid-treating step because these could be conveniently stripped from the hot oil in the subsequent clay contacting step. The presence of these components during acid treatment reduces the viscosity of the oil and thereby increases the ease of separating the acid sludge. However, sulfuric acid treatment and the addition of clay produce waste streams such as acid tar and spent clay, resulting in the problem of waste disposal.

However, in terms of environmental hazards from by-product and waste streams, the acid–clay process may be the least environmentally sound of the re-refining processes. The principal reason for this is the large quantity of by-product acidic tar produced, which presents difficulties in disposal. Therefore, such a process should not be used if there is no (or inadequate) capacity or facility to treat and dispose of the acid sludge from the process (Teixeira and de A. Santos 2008).

Thus, the classic chemical (sulfuric acid and clay refining) processes that were originally used for the refining of lube oil base stocks have been or are being replaced by solvent extraction (solvent refining) and hydrotreating (hydrogen refining) processes because they are (1) more effective for the upgrading of feedstocks, (2) more cost-effective, and (3) environmentally more acceptable. Although some refiners still use chemical refining processes, chemical refining is most often used for the reclamation of used lubricating oils or in combination with solvent or hydrogen refining processes for the manufacture of specialty oils such as refrigeration, transformer, and white oils (Speight 2014).

Vacuum distillation involves the distillation of oils under sub-atmospheric pressure, which lowers the necessary operating temperature and reduces the occurrence of problems such as thermal breakdown. The use of wiped film equipment, which allows material with a significant amount of solid contents to be more readily processed with reduced thermal breakdown, is increasing. Clay with high adsorptive capacity is used to remove impurities such as heavy metals and breakdown products arising from the use of the oil. The clay may also be used before distillation to provide a cleaner feedstock and also to give the recovered product oil a final cleaning (polish).

However, recycled base oils using the acid–clay and the clay filtration processes are not the same as re-refined base oils because they do not have the typical base oil viscosity fractions/grades. In addition, these recycled base oils may not have the same performance and oxidation stability, and the quality of such base oils may be very variable depending on the feedstock.

Finally, the yield of base oil from the acid–clay process is low because of losses from the acid sludge (some degree of sulfonation takes place). The two distillation processes do not recover bright stocks and this is reflected in their moderate lube oil recovery. However, bright stocks are recovered only in the acid–clay process and the process would be favored in any situation in which used oils contain extremely high proportions of bright stocks.

4.4.2 Catalytic Processes

The problem of dealing with spent clay, now designated as a hazardous waste in many places, has led many refiners to replace clay treatment facilities with a mild hydrotreating process or a hydrocracking process.

Catalytic treatment (hydroprocessing) of used oils provides a commercially viable alternative to high-temperature incineration or chemical treatment. Selective hydrogenation could be utilized to remove contaminants such as PCB derivatives or heavy metals from used oils.

Catalytic hydrogenation of contaminated organic waste streams is carried out at moderate temperatures and pressures. The treated organic phase is generally suitable for reuse as a fuel oil. The use of this technology is primarily constrained by economics, although disposal costs of organochlorine-contaminated oil could be substantially reduced. Thus, the use of this type of technology could have positive economic benefits.

Hydrotreating was developed in the 1950s and was first used in base oil manufacturing in the 1960s as an additional cleanup step that was added at the end of a conventional solvent refining process (Zakarian et al. 1987; Howell 2000). Hydrotreating is a process for adding hydrogen to the base oil at elevated temperatures in the presence of catalyst (1) to stabilize the most reactive components in the base oil, (2) to improve color, and (3) to increase the useful life of the base oil. This process removed some of the nitrogen-containing and sulfur-containing molecules but was not severe enough to remove (hydrogenate) all of the aromatic molecule constituents.

Hydrocracking is a more severe form of hydroprocessing and is accomplished by adding hydrogen to the feedstock at even higher temperatures and pressures compared with simple hydrotreating (Speight 2014). Feedstock constituents are cracked (thermally decomposed) into lower molecular weight products and most of the sulfur, nitrogen, and aromatics are removed. In addition, naphthene rings are opened and paraffin isomers are redistributed, driven by thermodynamics with reaction rates facilitated by catalysts.

The challenge in a hydroprocessing operation (hydrotreating or hydrocracking) of used lubricating oil is that it is very difficult to sustain operations without frequent catalyst change-out. Some re-refineries use guard beds to function as sacrificial reactors to remove deleterious constituents before the feedstock passes into the main catalyst reactor (Speight 2014). These guard beds typically tend to last 1 to 3 months at a maximum (depending on the contaminants in the used oil feedstock) before switching over to another guard bed reactor. The first guard bed is then cleaned and fresh guard-catalyst is added before switching back onstream and repeating the process with the second guard bed reactor.

Hydrotreating (hydroprocessing) is a common process for removing sulfur and nitrogen impurities from oils as well as hydrogenating unsaturated moieties in the oil (Speight and Ozum 2002; Hsu and Robinson 2006; Gary et al. 2007; Speight 2014). The oil is combined with high-purity hydrogen and either vaporized or retained in the liquid phase, and then contacted with a catalyst such as tungsten, nickel, or a mixture of cobalt and molybdenum oxides supported on an alumina base. Operating temperatures are usually between 260°C and 425°C (500°F and 800°F) at pressures of 200 to 1000 psi (1.38–6.90 MPa). Operating conditions are set to facilitate the desired level of sulfur removal without promoting any change to the other properties of the oil.

The sulfur in the oil is converted to hydrogen sulfide (H_2S) and the nitrogen to ammonia (NH_3). Aromatic and unsaturated constituents are hydrogenated into saturated compounds and remain in the oil. The hydrogen sulfide is removed from the circulating hydrogen stream by absorption in a solution such as diethanolamine. The solution can then be heated to remove the sulfide and can then be reused. The hydrogen sulfide recovered is useful for manufacturing elemental sulfur of high purity. The ammonia is recovered and can be either converted to elemental nitrogen and hydrogen, burned in the refinery fuel-gas system, or processed into agricultural fertilizers.

If a more severe process is desired, used lubricating oil from internal combustion engines can be cracked (hydrocracking) to yield a distillate that approximates no. 2 diesel fuel. The starting hydrocarbon material, used motor oil, is characterized by a mixture of paraffinic constituents, naphthenic constituents, aromatic constituents, and olefin constituents. The rate of the cracking reaction is high, with more than 30% distilled, and cracking becomes vigorous between 215°C and 345°C (600°F and 650°F). The condensation temperature of the distilled material never exceeds the maximum boiling point temperature for no. 2 oil (diesel fuel). Metal particles are typically in the micron and submicron range, and represent particles sufficiently small such that they pass through standard oil filters and may require the application of an ultrafiltration procedure for near-to-complete removal (Rodriguez et al. 2002).

4.5 TEST METHODS

Elemental analysis is an essential part of the environmental protection and quality control procedures associated with the recycling of used oil. In fact, lubricating oils made from re-refined base stocks must undergo the same testing and meet the same standards as those from new base stock to receive the certification mark of approval of the American Petroleum Institute (API). Vehicle and engine manufacturers have issued warranty statements that allow the use of re-refined oil as long as it meets API standards.

The ultimate assessment of the performance of lubricating oil includes a variety of vehicle fleet tests that simulate the full range of customer driving conditions. Specific tests are selected to mimic challenging field conditions and must be predictive of and applicable to a variety of vehicle tests under similar field conditions. The relationships between engine sequence tests and vehicle fleet tests are judged to be valid based only on the range of base oils and additive technologies investigated—generally, those that have been proven to have satisfactory performance in service and that are in widespread use.

Furthermore, no matter what process is chosen to accommodate used oil for reuse, re-refined lubricating oil must meet the same specifications as unused lubricating oil for the re-refined product to be of value (ASTM D4485). There are a number of standard test methods and specifications that are used to ensure the quality of lubricating oils (Table 4.1; Speight 2002; Ogbeide 2010). Deviations from these standard specifications are expected when the test methods are applied to used lubricating oil after a suitable chain of custody has been applied (API 1525A).

Properties such as density, relative density, and API gravity (ASTM D4052), viscosity (ASTM D445), flash and fire points (ASTM D92; ASTM D93), corrosivity (ASTM D130), mineral matter content (measured as mineral ash; ASTM D482), water and sediment (ASTM D1796), and foaming tendency (ASTM D892) are the minimal properties to be considered as important aspects when comparing unused base oils and used oils. The boiling behavior (i.e., the distillation curve) of the oil is typically determined by a gas chromatographic method that calculates the distillation curve boiling temperatures based on standardized retention times (ASTM D2887). The distillation curve is one of the most important properties that can be measured for any complex fluid because it is the only practical avenue to assess the volatility or the vapor liquid equilibrium.

As a corollary to the distillation test protocol, advanced distillation curve metrology has been used to examine four unused automotive crankcase oils and four used lubricant oils, including used automotive crankcase oil (Ott et al. 2010). The distillation curves of the three unused petroleum crankcase oils followed the expected trends based on the stated weights of the oils. Three out of the four used lubricant oils had distillation curves that indicated initial boiling at much lower temperatures than the remainder of the distillation curves, indicating that these three used oils contained lower boiling components that would require removal during re-refining of the oil if the oil were to be reused. The composition-explicit data channel was used to probe the composition of a low-boiling distillate fraction of the used automotive oil—several components of automotive gasoline were present.

Fourier transform infrared spectrometry (FTIR) provides information on compounds, rather than elements, found in oil (ASTM E2412). The method can be used to measure several useful degradation parameters, and thus is particularly useful in engine oil samples. Infrared analysis detects the presence of water and can also be used to identify oil base stocks. Although inductively coupled plasma (ICP) spectroscopy measures emissions of radiation of specific wavelength in the visible and ultraviolet regions of the electromagnetic spectrum, infrared analysis measures the specific wavelengths of radiation in the infrared region. The various degradation by-products and contaminants found in the oil cause characteristic absorptions in

TABLE 4.1
Standard Test Methods used to Determine the Quality of Lubricating Oil

ASTM D92, Standard Test Method for Flash and Fire Points by Cleveland Open Cup
ASTM D93, Standard Test Methods for Flash Point by Pensky-Martens Closed Cup Tester
ASTM D445, Standard Test Method for Kinematic Viscosity of Transparent and Opaque Liquids (and the Calculation of Dynamic Viscosity)
ASTM D892, Standard Test Method for Foaming Characteristics of Lubricating Oils
ASTM D1552, Standard Test Method for Sulfur in Petroleum Products (High-Temperature Method)
ASTM D2007, Standard Test Method for Characteristic Groups in Rubber Extender and Processing Oils and Other Petroleum Derived Oils by the Clay-Gel Absorption Chromatographic Method
ASTM D2270, Standard Practice for Calculating Viscosity Index from Kinematic Viscosity at 40 and 100°C
ASTM D2622, Standard Test Method for Sulfur in Petroleum Products by Wavelength Dispersive X-Ray Fluorescence Spectrometry
ASTM D2887, Standard Test Method for Boiling Range Distribution of Petroleum Fractions by Gas Chromatography
ASTM D3120, Standard Test Method for Trace Quantities of Sulfur in Light Liquid Petroleum Hydrocarbons by Oxidative Microcoulometry
ASTM D3244, Standard Practice for Utilization of Test Data to Determine Conformance with Specifications
ASTM D4294, Standard Test Method for Sulfur in Petroleum and Petroleum Products by Energy-Dispersive X-Ray Fluorescence Spectroscopy
ASTM D4485, Standard Specification for Performance of Engine Oils
ASTM D4683, Standard Test Method for Measuring Viscosity at High Shear Rate and High Temperature by Tapered Bearing Simulator
ASTM D4684, Standard Test Method for Determination of Yield Stress and Apparent Viscosity of Engine Oils at Low Temperature
ASTM D4741, Standard Test Method for Measuring Viscosity at High Temperature and High Shear Rate by Tapered-Plug Viscometer
ASTM D4927, Standard Test Method for Elemental Analysis of Lubricant and Additive Components, Barium, Calcium, Phosphorus, Sulfur, and Zinc, by Wavelength Dispersive X-Ray Fluorescence Spectroscopy
ASTM D4951, Standard Test Method for Determination of Additive Elements in Lubricating Oils by Inductively Coupled Plasma Atomic Emission Spectrometry
ASTM D5119, Standard Test Method for Evaluation of Automotive Engine Oils in CRC L-38 Spark Ignition Engine
ASTM D5133, Standard Test Method for Low Temperature, Low Shear Rate, Viscosity/Temperature Dependence of Lubricating Oils Using a Temperature-Scanning Technique
ASTM D5185, Standard Test Method for Determination of Additive Elements, Wear Metals, and Contaminants in Used Lubricating Oils and Determination of Selected Elements in Base Oils by Inductively Coupled Plasma Atomic Emission Spectrometry (ICP-AES)
ASTM D5293, Standard Test Method for Apparent Viscosity of Engine Oils between –5 and –30°C Using the Cold-Cranking Simulator
ASTM D5302, Standard Test Method for Evaluation of Automotive Engine Oils for Inhibition of Deposit Formation and Wear in a Spark-Ignition Internal Combustion Engine Fueled with Gasoline and Operated Under Low-Temperature Light-Duty Conditions

(continued)

TABLE 4.1 (Continued)

Standard Test Methods used to Determine the Quality of Lubricating Oil

ASTM D5480, Standard Test Method for Motor Oil Volatility by Gas Chromatography

ASTM D5481, Standard Test Method for Measuring Apparent Viscosity at High-Temperature and High-Shear Rate by Multicell Capillary Viscometer

ASTM D5533, Standard Test Method for Evaluation of Automotive Engine Oils in the Sequence IIIE Spark Ignition Engine

ASTM D5800, Standard Test Method for Evaporation Loss of Lubricating Oils by the NOACK Method

ASTM D5844, Standard Test Method for Evaluation of Automotive Engine Oils for Inhibition of Rusting (Sequence IID)

ASTM D6082, Standard Test Method for High Temperature Foaming Characteristics of Lubricating Oils

ASTM D6202, Standard Test Method for Automotive Engine Oils on the Fuel Economy of Passenger Cars and Light-Duty Trucks in the Sequence VIA Spark Ignition Engine

ASTM D6335, Standard Test Method for Determination of High Temperature Deposits by Thermo-Oxidation Engine Oil Simulation Test

ASTM D6417, Standard Test Method for Estimation of Engine Oil Volatility by Capillary Gas Chromatography

ASTM D6557, Standard Test Method For Evaluation of Rust Preventative Characteristics of Automotive Engine Oils

ASTM D6593, Standard Test Method for Evaluation of Automotive Engine Oils for Inhibition of Deposit Formation in a Spark-Ignition Internal Combustion Engine Fueled with Gasoline and Operated Under Low-Temperature Light-Duty Conditions

ASTM D6616, Standard Test Method for Measuring Viscosity at High Shear Rate by Tapered Bearing Simulator Viscometer at 100°C

ASTM D6837, Standard Test Method for Measurement of Effects of Automotive Engine Oils on Fuel Economy of Passenger Cars and Light-Duty Trucks in Sequence VIB Spark Ignition Engine

ASTM D6794, Standard Test Method for Measuring the Effect on Filterability of Engine Oils After Treatment with Various Amounts of Water and a Long (6-h) Heating Time

ASTM D6795, Standard Test Method for Measuring the Effect on Filterability of Engine Oils After Treatment with Water and Dry Ice and a Short (30-min) Heating Time

ASTM D6891, Standard Test Method for Evaluation of Automotive Engine Oils in the Sequence IVA Spark-Ignition Engine

ASTM D6922, Standard Test Method for Determination of Homogeneity and Miscibility in Automotive Engine Oils

ASTM D7097, Standard Test Method for Determination of Moderately High Temperature Piston Deposits by Thermo-Oxidation Engine Oil Simulation Test-TEOST MHT

ASTM D7320, Standard Test Method for Evaluation of Automotive Engine Oils in the Sequence IIIG, Spark-Ignition Engine

Source: ASTM 2012.

specific regions of the infrared spectrum. The higher the level of contamination in the sample, the higher the degree of absorption in the characteristic region.

A plot of absorbance, transmittance, or concentration versus wave number is generated during the analysis of an oil sample, and is called the infrared spectrum. This spectrum is subsequently analyzed by specialized oil analysis software that usually yields measurements for soot, oxidation, sulfates, nitrates, and water. Other compounds, such as additives, fuel, and glycol (ASTM D2982) can also be measured, but for these, an accurate sample of the new oil is needed as a reference. If no such reference sample has been supplied, then readings of the latter parameters should be regarded with suspicion.

The soot index (also determined by FTIR) is a linear measurement that measures the extent to which the lubricating oil has become contaminated by fuel soot, an unwanted by-product of combustion. Dispersant additives are formulated in engine oils to hold soot in suspension, but there is a limit to the amount of soot a lubricant can hold in suspension. When the maximum amount is exceeded, deposits (sludge) form and the effects of sludge formation are manifested by an increase in the viscosity of the oil. This usually occurs rapidly to the point where the oil can no longer be pumped and engine failure ensues.

When interpreting the severity of the soot index measurement, soot readings on previous samples from the engine should be taken into account, as well as the magnitude of the change in the viscosity of the oil. Furthermore, high soot loading can negatively affect the accuracy of other infrared measurements.

As lubricating oil oxidizes, the ability of the oil to provide lubrication is reduced and, in cases of severe oxidation, noticeable changes occur: (1) the oil becomes darker and emits odor, (2) sludge is formed—the oxidation products are frequently referred to as varnishes or lacquers, and resins are formed, and (3) in the advanced stages of oxidation, viscosity increases at a rate dependent on the temperature of the oil—the chemical reaction between oxygen and lubricant at room temperature is slow and oxidative degradation is not an issue under these conditions. The situation changes when reaction conditions are altered to favor a more rapid reaction rate—because the rate of a chemical reaction (in this case, oxidation) doubles for every 10°C (18°F) increase in temperature, an excessively high operating temperature (overheating) is generally accompanied by increased wear (producing lead, copper, tin, and iron from bearing materials and lubricated surfaces) and increased baseline viscosity.

On occasion, overheating can also lead to the evaporation of volatile fractions in the oil, making regular top-ups necessary. In this case, the sump oil will exhibit increased additive levels (concentration of nonvolatile components) and an increased viscosity as a direct result of light end loss. As this lost oil is replaced with fresh oil, the antioxidants are replaced and oxidation is often not immediately evident. The extent of the oxidation can be determined by the acid number (ASTM D974; ASTM D5770) and by viscosity measurement for confirmation.

Determination of the acid number typically involves a titration method in which the total acid content of the oil dissolved in a mixed solvent is completely neutralized by the gradual addition of an alcoholic solution of potassium hydroxide (KOH). A colorimetric method of determining the end point is used—a chemical indicator changes color as soon as the acid is completely neutralized. Alternatively, a potentiometric method may also be used.

The acid number test is performed on oil samples to quantify the acid buildup in the oil—an increased acid number is a result of oxidation of the oil, perhaps caused by overheating, overextended oil service, or water or air contamination. The acid number limits for lubricating oils vary considerably—in some cases, an acid number exceeding 0.05 is unacceptable, whereas in other cases, an acid number higher than 4.00 may be acceptable. As with all other readings, trend analysis is the best indication of the health of both the oil and the machinery.

The measurement of the base number (ASTM D974; ASTM D2896; ASTM D5984) is determined as the total alkaline reserve of 1 g of oil dissolved in a mixed solvent after reaction by the gradual addition of a known excess of an acid solution. Typical starting values for diesel engine oils are between 8 and 12. However, marine engines burning heavy fuel oil need a much higher base number (possibly as high as 80) to handle the harsh combustion conditions from fuels containing a high concentration of sulfur. A general rule is to discard the lubricating oil when the base number drops below half its beginning value.

Another test method is often referred to as debris analysis, which involves the use of particle counting and is actually a test for particle contaminant levels and not specifically for wear debris. It does not distinguish between wear and dirt particles, but if it can be determined that nonferrous contamination has remained stable, then an increase in the particle count must be attributable to wear. A magnet can be used to modify the particle count to count ferrous debris only. There are various ways of doing this—essentially, a magnet holds back the ferrous debris while the nonferrous debris is flushed from the sample, after which a ferrous debris particle count is performed. The particle count is an easy test to interpret, assuming the test has been correctly performed—caution is advised because there are many factors that can negatively affect a particle count. An increasing count is simply an indication of an increased number of particles in the oil.

Although there are several test methods that should be applied to the used oil (Chapters 1 and 2), the two additional analytical techniques that should be used as part of the analytical protocols are energy dispersive x-ray fluorescence (EDXRF) spectrometry and ICP-optical emission spectrometry (ICP-OES) (Schulz 2011).

The EDXRF method involves irradiating the sample with a beam of x-rays to induce fluorescence in the atoms in the sample, which is then emitted as x-rays of a lower energy. Each element emits fluorescent x-rays of different and unique energies or wavelengths, whose intensity is proportional to the concentration of that element in the sample.

The ICP-OES method relies on the fact that every element has a unique atomic structure, with a massive positively charged nucleus and a number of electrons in orbits surrounding it. Each of these electrons exists at a precise energy level and under normal conditions, with no external influences (the ground state). However, if the atom is subjected to a source of energy, for example, in the form of heat or collisions with other atoms, then that energy can be absorbed and the electron(s) raised to a higher or excited state. This is an unstable condition, and the electrons quickly return to the ground state, at which point the previously absorbed energy is re-emitted. The intensity of this radiation is proportional to the number of atoms present (i.e., the concentration), which is the basis of the quantitative analysis by this method.

4.6 ENVIRONMENTAL ASPECTS

Modern lubricating oil mainly consists of two materials, which are the base oil and chemical additives (Chapter 1). With the addition of specific chemical additives, the properties of lubricating oil are enhanced and the rate of undesirable changes taking place during operation is reduced. Various types of additives are blended with base oil according to its grade and specific duty, and typical lubricating oil consists of minor amounts (several parts per million, up to 30% w/w) of additives (Table 4.2) and the composition differs considerably from used lubricating oil (Table 4.3).

These additives consist of antioxidants and metal deactivators, detergents, dispersants, corrosion inhibitors, and rust inhibitors, which also have a deleterious effect on the environment (Farrington and Slater 1997). The carbon number distribution and, by extension, the viscosity in hydraulic fluids varies depending on the anticipated application of the fluid. Generally, however, C_{15} to C_{50} is a reasonable range. Also, the higher the average carbon number, the higher the viscosity of the fluid in question.

In terms of the environmental aspects of used lubricating oil, used oil does not evaporate and is less subject to biodegradation (Speight and Arjoon 2012). Biodegradation, which provides an indication of the persistence of a particular substance in the environment, is the yardstick for assessing the ecofriendliness of substances. Due to the poor biodegradation rates observable for oils, various methods of bioremediation are currently being researched, with the isolation of various microbial species with the ability to consume lubricating oils as a carbon and energy source. Ultimately, vegetable oil–based hydraulic fluids will come to the fore as a

TABLE 4.2
Typical Composition of Lubricating Oil

Component	% w/w
Base oil	71.5–96.2
Metallic detergents	2.0–10.2
Dispersants	1.0–9.0
Zinc dithiophosphate	0.5–3.0
Antioxidant/antiwear	0.1–2.0
Friction modifier	0.1–3.0
Pour point depressant	0.1–1.5
Antifoam	2–15 ppm

Source: Gergel, W.C., *Lubricant Additive Chemistry*. Lubrizol Additive Company, Lubrizol Corporation, Wickliffe, Ohio, 1992.

TABLE 4.3
Comparison of Virgin and Waste Lubricating Oil Properties

Properties	Virgin Lube Oil	Used Lube Oil
Physical Properties		
Specific gravity	0.882	0.910
Dynamic viscosity SUS @ 100°F	–	324.0
Bottom sediment and water (BS&W), % v/v	0	12.3
Carbon residue, % w/w	0.82	3.00
Ash yield, % w/w	0.94	1.30
Flash point, °F	–	348.0
Pour point, °F	–35.0	–35.0
Chemical Properties		
Saponification number	3.94	12.7
Total acid number	2.20	4.40
Total base number	4.70	1.70
Nitrogen, % w/w	0.05	0.08
Sulfur, % w/w	0.32	0.42
Lead, ppm	0	7535
Calcium, ppm	1210	4468
Zinc, ppm	1664	1097
Phosphorus, ppm	1397	931
Magnesium, ppm	675	309
Barium, ppm	37	297
Iron, ppm	3	205
Sodium, ppm	4	118
Potassium, ppm	<1	31
Copper, ppm	0	29

Source: Gergel, W.C., *Lubricant Additive Chemistry*. Lubrizol Additive Company, Lubrizol Corporation, Wickliffe, Ohio, 1992.

suitable and more environmentally friendly solution to the demand for hydraulic fluids.

The materials that fall into the lubricating oil category are complex petroleum mixtures composed primarily of saturated hydrocarbons with carbon numbers ranging from C_{15} to C_{50} (some studies give ranges from C_{15} to C_{40}). At ambient temperatures, lubricating base oils are liquids of varying viscosities, with negligible vapor pressures. Base oils are produced by first distilling crude oil at atmospheric pressure to remove lighter components (e.g., gasoline and distillate fuel components), leaving a residue (residuum) that contains base oil precursors. This atmospheric residuum is then distilled under vacuum to yield a range of distillate fractions (unrefined distillate base oils) and a vacuum residuum (Chapter 1). Removal of the asphalt components of the vacuum residuum results in unrefined residual base oils. These distillate and residual base oil fractions may then undergo a series of extractive or transforming processes that improve the base oils' performance characteristics and reduce or eliminate undesirable components.

Distillate base oils contain components whose boiling points typically range from 300°C to 600°C (570°F–1110°F), which are often described as either naphthenic (saturated ring hydrocarbons) or paraffinic (straight or branched chain hydrocarbons) depending on the original petroleum from which they are derived or on the dominant hydrocarbons present (or both). The difference between naphthenic and paraffinic base oils is one of relative percentage because naphthenes and paraffins are present in both types of oils.

Thus, base oil might be called *paraffinic base oil* if it is 60% w/w paraffins and 30% w/w naphthenes or *naphthenic base oil* if it is 60% w/w naphthenes and 30% w/w paraffins. Base oils are also often described as either *light* (viscosity less than 19 cSt at 40°C, 104°F) or *heavy* (viscosity greater than 19 cSt at 40°C, 104°F). The naphthenic/paraffinic and light/heavy nomenclatures are primarily used to distinguish product application and lubricant quality parameters rather than health and safety characteristics because a significant amount of toxicology data exists that shows little differentiation between these four classifications.

Refined distillate base oils are produced from unrefined base oil fractions by undergoing additional processing designed to reduce or transform the undesirable components. In general, each additional step of processing (increasingly severe processing) results in: (1) lower levels of unwanted components; including aromatics, metals, waxes, and trace components causing unwanted colors or odors; (2) a narrower range of hydrocarbon molecules (increasing concentrations of paraffins and naphthenes); and (3) lower, if any, carcinogenic or mutagenic activity. Some distillate base oils are destined for use in food, food contact, cosmetic, pharmaceutical, and related applications. Known as white oils, these very severely refined distillate base oils undergo numerous processing steps that essentially eliminate or transform all undesired components, including unsaturated hydrocarbons and aromatics.

Synthetic base oils can be substituted for conventional petroleum-based oils. Most synthetic motor oils are fabricated by polymerizing short chain hydrocarbon molecules called α-olefins into longer chain hydrocarbon polymers called poly α-olefins. The degree of variation in molecular size, chain length, and branching in synthetically produced fluids is much less than occurs in base stocks extracted from crude oil. Although they seem chemically similar to mineral oils refined from crude oil, poly α-olefins do not contain the impurities or waxes inherent in conventional mineral oils. Poly α-olefins constitute the most widely used synthetic motor oils in the United States and Europe.

These various products have been applied ever since the discovery of crude oil. The recent worldwide emphasis on environmental preservation now raises questions that concern the long-term effects of the use, accidental spills, and intentional disposal of industrial products such as lubricating oils. Pressure is mounting to use more environmentally friendly products, especially in ecologically sensitive situations (Khawaja and Aban 1996). Hence, the biodegradability of lubricating oil is of interest because the environmental persistence of possibly toxic constituents of these lubricating oils is unacceptable (Speight and Arjoon 2012).

In terms of the environmental effects of lubricating oils, each type of crude oil or refined product has distinct physical properties that affect the way the oil spreads and breaks down, and ultimately determines the hazard it may pose to marine and human life, and the likelihood that it would pose a threat to natural and man-made resources. Lubricating oils are persistent and present a greater remediation challenge than lighter petroleum products (Khawaja and Aban 1996).

This prospective hazard is recognized by many regulatory bodies that stipulate, for example, that used oils in liquid form cannot be disposed of by any of the following methods: (1) disposed of into a drain or sewer or into surface water or ground water, (2) disposed of in a landfill, (3) burned in a municipal solid waste incinerator without any form of gas cleaning technology, and (4) used as dust control or weed control by spraying directly (or indirectly) on the ground.

The United States' Environmental Protection Agency (EPA) has laid down guidelines stipulating that accessible oil should be recovered in the event of a spill, and all contaminated soil be removed. Additionally, the EPA also requires that no visible oil sheen be evident downstream from facilities close to waterways. Another regulation requires that point discharges into waterways should not exceed 10 ppm of lubricating-based oils (U.S. Army Corps of Engineers 1999).

Residual base oils are derived from the residuum of the vacuum distillation tower and may contain components boiling as high as 800°C (1470°F). The residual oils have molecular weights that are much higher than the distillate base oils. Residual base oils are primarily used in situations requiring oils with a high viscosity, for example, gear oils. Detailed analytical testing indicated that the polynuclear aromatic constituents they contain are predominantly one to three rings that are highly alkylated (paraffinic and naphthenic). Because they are found in such a high-boiling material (>575°C, >1070°F), it is estimated that the alkyl side chains of the ring aromatics would be approximately 13 to 25 carbons in length. These highly alkylated aromatic ring materials are either devoid of the biological activity necessary to cause mutagenesis and carcinogenesis, or are largely non-bioavailable to the organisms (Roy et al. 1988).

For these reasons, the biodegradability and toxicity of petroleum-based lubricants are important factors with respect to environmental management. Although during the last decades, the increased public attention to the protection of the environment has stimulated the development of lubricants that show more or less compatibility with the environment, environmental compliance of lubricants and bioremediation of lubricant-polluted sites have become topics of interested research (Erhan and Asadauskas 2000; Boyde 2002; Willing 2001; Lee et al. 2007; Cecutti and Agius 2008; Matos Lopes and Bidoia 2009; Speight and Arjoon 2012).

Like fuels, lubricating oils are very important products in the industrialized twenty-first century with such numerous applications that they may be described as indispensable (Speight 2014). However, increasing environmental awareness has placed the spotlight on the persistence of the harmful components of lubricating oils in the environment, and their cumulative effects on ecosystems, especially in the light of their established poor biodegradability.

Spills are inevitable, and the development of bioremediation processes utilizing isolated hydrocarbon degrading bacteria has gone a long way in improving the efficiency of spill cleanups. The poor biodegradability of lubricating oils, coupled with the fact that remediation efforts always tend to be expensive, has made the option of the development of environmentally friendly oils based primarily on vegetable oils an attractive one. The future will bring with it the improvement of the efficiency of bio-based fluids in various applications, leading to the widespread use of these products; a situation which will no doubt be in the best interests of the global ecosystem.

REFERENCES

API 1525A. 2012. *Bulk Engine Oil Chain of Custody and Quality Documentation*. American Petroleum Institute, Washington DC, January.

ASTM. 2012. *Annual Book of Standards*. American Society for Testing and Materials, West Conshohocken, Pennsylvania.

ASTM D86. 2012. Standard Test Method for Distillation of Petroleum Products at Atmospheric Pressure. *Annual Book of Standards*. American Society for Testing and Materials, West Conshohocken, Pennsylvania.

ASTM D92. 2012. Standard Test Method for Flash and Fire Points by Cleveland Open Cup Tester. *Annual Book of Standards*. American Society for Testing and Materials, West Conshohocken, Pennsylvania.

ASTM D93. 2012. Standard Test Method for Flash Point by Pensky-Martens Closed Cup Tester. *Annual Book of Standards*. American Society for Testing and Materials, West Conshohocken, Pennsylvania.

ASTM D130. 2012. Standard Test Method for Corrosiveness to Copper from Petroleum Products by Copper Strip Test. *Annual Book of Standards*. American Society for Testing and Materials, West Conshohocken, Pennsylvania.

ASTM D445. 2012. Standard Test Method for Kinematic Viscosity of Transparent and Opaque Liquids (and Calculation of Dynamic Viscosity). *Annual Book of Standards*. American Society for Testing and Materials, West Conshohocken, Pennsylvania.

ASTM D482. 2012. Standard Test Method for Ash from Petroleum Products. *Annual Book of Standards*. American Society for Testing and Materials, West Conshohocken, Pennsylvania.

ASTM D892. 2012. Standard Test Method for Foaming Characteristics of Lubricating Oils. *Annual Book of Standards*. American Society for Testing and Materials, West Conshohocken, Pennsylvania.

ASTM D974. 2012. Standard Test Method for Acid and Base Number by Color-Indicator Titration. *Annual Book of Standards*. American Society for Testing and Materials, West Conshohocken, Pennsylvania.

ASTM D1160. 2012. Standard Test Method for Distillation of Petroleum Products at Reduced Pressure. *Annual Book of Standards*. American Society for Testing and Materials, West Conshohocken, Pennsylvania.

ASTM D1796. 2012. Standard Test Method Water and Sediment in Fuel oil by the Centrifuge Method (Laboratory Procedure). *Annual Book of Standards*. American Society for Testing and Materials, West Conshohocken, Pennsylvania.

ASTM D2887. 2012. Standard Test Method for Boiling Range Distribution of Petroleum Fractions by Gas Chromatography. *Annual Book of Standards*. American Society for Testing and Materials, West Conshohocken, Pennsylvania.

ASTM D2896. 2012. Standard Test Method for Base Number of Petroleum Products by Potentiometric Perchloric Acid Titration. *Annual Book of Standards*. American Society for Testing and Materials, West Conshohocken, Pennsylvania.

ASTM D2982. 2012. Standard Test Methods for Detecting Glycol-Base Antifreeze in Used Lubricating Oils. *Annual Book of Standards*. American Society for Testing and Materials, West Conshohocken, Pennsylvania.

ASTM D4052. 2012. Standard Test Method for Density, Relative Density, and API Gravity of Liquids by Digital Density Meter. *Annual Book of Standards*. American Society for Testing and Materials, West Conshohocken, Pennsylvania.

ASTM D4485. 2012. Standard Specification for Performance of Active API Service Category Engine Oils. *Annual Book of Standards*. American Society for Testing and Materials, West Conshohocken, Pennsylvania.

ASTM D5770. 2012. Standard Test Method for Semi-Quantitative Micro Determination of Acid Number of Lubricating Oils during Oxidation Testing. *Annual Book of Standards*. American Society for Testing and Materials, West Conshohocken, Pennsylvania.

ASTM D5984. 2012. Standard Test Method for Semi-Quantitative Field Test Method for Base Number in New and Used Lubricants by Color-Indicator Titration. *Annual Book of Standards*. American Society for Testing and Materials, West Conshohocken, Pennsylvania.

ASTM E2412. 2012. Standard Practice for Condition Monitoring of In-Service Lubricants by Trend Analysis Using Fourier Transform Infrared (FTIR) Spectrometry. *Annual Book of Standards*. American Society for Testing and Materials, West Conshohocken, Pennsylvania.

Audibert, F. 2006. *Waste Engine Oils: Rerefining and Energy Recovery*. Elsevier, Amsterdam, Netherlands.

Boyde, S. 2002. Green lubricants: Environmental benefits and impacts on lubrication. *Green Chemistry*, 12(4): 293–307.

Cecutti, C., and Agius, D. 2008. Ecotoxicity and biodegradability in soil and aqueous media of lubricants used in forestry applications. *Bioresource Technology*, 99: 8492–8496.

Emam, E.A., and Shoaib, A.M. 2012. Re-refining of used lube oil, II-by solvent/clay and acid/clay-percolation processes. *ARPN Journal of Science and Technology*, 2(11): 1034–1041.

Erhan, S.Z., and Asadauskas, S. 2000. Lubricant basestocks from vegetable oils. *Industrial Crops and Products*, 11(2): 277–282.

Farrington, A.M., and Slater, J.M. 1997. Monitoring of engine oil degradation by voltammetric methods utilizing disposable solid wire microelectrodes. *Analyst*, 122: 593–596.

Gary, J.G., Handwerk, G.E., and Kaiser, M.J. 2007. *Petroleum Refining: Technology and Economics*, 5th Edition. CRC Press, Taylor & Francis Group, Boca Raton, Florida.

Gergel, W.C. 1992. *Lubricant Additive Chemistry*. Lubrizol Additive Company, Lubrizol Corporation, Wickliffe, Ohio.

Gorman, W.A. 2005. Recovering base oils from lubricants. *Petroleum Technology Quarterly*, Q4: 85–88.

Hamawand, I., Yusaf, T., and Sardasht Rafat, S. 2013. Recycling of waste engine oils using a new washing agent. *Energies*, 6: 1023–1049.

Howell, R.L. 2000. Hydroprocessing routes to improved base oil quality and refining economics. Proceedings, 6th Annual Fuels and Lubes Conference, Singapore, January 25–28.

Hsu, C.S., and Robinson, P.R. (Editors). 2006. *Practical Advances in Petroleum Processing*, Volume 1 and Volume 2. Springer Science, New York.

Khawaja, M.A., and Aban, M.M. 1996. Characteristics of used lubricating oils: Their environmental impact and survey of disposal methods. *Environmental Management and Health*, 7(1): 23–32.

Kim, S.S., and Kim, S.H. 2000. Pyrolysis kinetics of waste automobile lubricating oil. *Fuel*, 79: 1943–1949.

Lam, S.S., Russell, A.D., Lee, C.L., and Chase, H.A. 2011. Production of hydrogen and light hydrocarbons as a potential gaseous fuel from microwave-heated pyrolysis of waste automotive engine oil. *International Journal of Hydrogen Energy*, 1–2: 1–11.

Lam, S.S., Russell, A.D., Lee, C.L., and Chase, H.A. 2012. Microwave-heated pyrolysis of waste automotive engine oils. Fuel: Properties of pyrolysis oil. *Fuel*, 91: 327–339.

Lee, S.H., Lee, S., and Kim, D.Y. 2007. Degradation characteristics of waste lubricants under different nutrient conditions. *Journal of Hazardous Materials*, 143: 65–72.

Manasomboonphan, W., and Junyapoon, S. 2012. Production of liquid fuels from waste lube oils used by pyrolysis process. Proceedings, 2nd International Conference on Biomedical Engineering and Technology, IPCBEE, IACSIT Press, Singapore, Volume 34, pp. 130–133.

Martin de Julian, P., Padrino Torres, L.R., and Torres Fonsec, P.A. 2012. Process for recovering used lubricating oils using clay and centrifugation. United States Patent Application US20120289441 A1, November 15.

Matos Lopes, P.R., and Bidoia, E.D. 2009. Evaluation of the biodegradation of different types of lubricant oils in liquid medium. *Brazilian Archives of Biology and Technology*, 52(5): 1285–1290.

Ogbeide, S.O. 2010. An investigation to the recycling of spent engine oil. *Journal of Engineering Science and Technology Review*, 3(1): 32–35.

Ott, L.S., Smith, B.L., and Bruno, T.J. 2010. Composition-explicit distillation curves of waste lubricant oils and resourced crude oil: A diagnostic for re-refining and evaluation. *American Journal of Environmental Sciences*, 6(6): 523–534.

Ramasamy, K.K., and Raissi, A. 2007. Hydrogen production from used lubricating oils. *Coal Today*, 129: 365–371.

Rodriguez, C., Sarrade, S., Schrive, L., Dresch-Bazile, M., Paolucci, D., and Rios, G.M. 2002. Membrane fouling in cross-flow ultrafiltration of mineral oil assisted by pressurized CO_2. *Desalination*, 144: 173–178.

Roy, T.A., Johnson, S.W., Blackburn, G.R., and Mackerer, C.R. 1988. Correlation of mutagenic and dermal carcinogenic activities of mineral oils with polycyclic aromatic compound content. *Fundamental and Applied Toxicology*, 10: 466–476.

Schulz, O. 2011. Elemental analysis in waste oil recovery and recycling, January. Available at http://www.digitalrefining.com/article/1000641. Accessed on March 30, 2013.

Speight, J.G. 2002. *Handbook of Petroleum Product Analysis*. John Wiley & Sons Inc., Hoboken, New Jersey.

Speight, J.G. 2005. *Environmental Analysis and Technology for the Refining Industry*. John Wiley & Sons Inc., Hoboken, New Jersey.

Speight, J.G. 2014. *The Chemistry and Technology of Petroleum*, 5th Edition. CRC Press, Taylor and Francis Group, Boca Raton, Florida.

Speight, J.G., and Arjoon, K.K. 2012. *Bioremediation of Petroleum and Petroleum Products*. Scrivener Publishing, Salem, Massachusetts.

Speight, J.G., and Ozum, B. 2002. *Petroleum Refining Processes*. Marcel Dekker Inc., New York.

Srinivasa Rao, P., Sridhar, S., Wey, M.Y., and Krishnaiah, A. 2007. Pervaporative separation of ethylene glycol/water mixtures by using cross-linked chitosan membranes. *Industrial & Engineering Chemistry Research*, 46: 2155–2163.

Teixeira, S.R., and de A. Santos, G.T. 2008. Incorporation of waste from used lube oil re-refining industry in ceramic body: Characterization and properties. *Revista Ciências Exatas—Universidade De Taubaté (Unitau)—Brasil*, 2(1): 1–6.

Udonne, J.D. 2011. A comparative study of recycling of used lubrication oils using distillation, acid and activated charcoal with clay methods. *Journal of Petroleum and Gas Engineering*, 2(2): 12–19.

U.S. Army Corps of Engineers. 1999. U.S. Army Manual EM1110-2-1424, Chapter 8. Available at http://www.publications.usace.army.mil/Portals/76/Publications/EngineerManuals/EM_1110-2-1424.pdf.

Willing, A. 2001. Lubricants based on renewable resources: An environmentally compatible alternative to mineral oil products. *Chemosphere*, 43: 89–98.

Zakarian, J.A., Robson, R.J., and Farrell, T.R. 1987. All-hydroprocessing route for high-viscosity index lubes. *Energy Progress*, 7(1): 59–64.

5 Combustion of Used Engine Oil

5.1 INTRODUCTION

Although most petroleum products can be used as fuels, the term *fuel oil*, if used without qualification, may be interpreted in various ways. For example, the term fuel oil may be associated with the black, viscous, residual material that remains as the result of refinery distillation of crude oil, either alone or in a blend with lower-boiling components, and which is used for steam generation for large slow-speed diesel engine operation and industrial heating and processing. Alternatively, the term fuel oil is applied to both residual and middle distillate type products such as domestic heating oil, kerosene, and burner fuel oils.

Because fuel oils are complex mixtures of compounds of carbon and hydrogen, they cannot be classified rigidly or defined exactly by chemical formulas or definite physical properties. For the purposes of this chapter, the term fuel oil will include all petroleum oils higher boiling than naphtha (gasoline) or kerosene (diesel fuel) that is used in burners. Because of the wide variety of petroleum fuel oils, the arbitrary divisions or classifications, which have become widely accepted in industry, are based more on their application than on their chemical or physical properties—two broad classifications are generally recognized: (1) distillate fuel oil and (2) residual fuel oil.

Middle distillate fuel oils are produced in the refinery by a distillation process in which petroleum is separated into fractions according to their boiling range. These middle distillate fuel oils may be produced directly from crude oil (straight-run fuel oil) as well as from subsequent refinery conversion processes, such as thermal or catalytic cracking. On the other hand, residual or heavy fuel oils are composed wholly or in part of nondistillable petroleum fractions from crude oil distillation (atmospheric or vacuum tower bottoms), visbreaking, or other refinery operations (Speight and Ozum 2002; Hsu and Robinson 2006; Gary et al. 2007; Speight 2014). The various grades of heavy fuel oil are generally produced to meet definite specifications to ensure suitability for their intended purpose. Residual oils are usually classified according to viscosity, in contrast with distillates, which are normally defined by boiling range.

An increasing trend in the fuel oil arena is the blending of used lubricating oils in burner oils to make industrial fuel oils. In fact, four grades of burner fuels are defined (ASTM D6448) that comprise (in whole or in part) with hydrocarbonbased used or reprocessed lubricating oil, or functional fluids such as preservative and hydraulic fluids. The grades of fuel are intended for use in various types of fuel oil-burning industrial equipment under various climatic and operating conditions. These fuels are not intended for use in residential heaters, small commercial boilers, or combustion engines. Grades RFO4, RFO5L, RFO5H, and RFO6 (ASTM D6448) are used lubricating oil blends (with or without distillate or residual fuel oil, or both), of increasing viscosity and are intended for use in industrial boilers equipped to handle these types of recycled fuels. The designation RFO identifies them as reprocessed fuel oils—L indicates light and H indicates heavy.

In fact, the major option, even before re-refining the used oil (Chapters 4 and 7), is the recovery of the energy in the oil, focusing principally on use of the oil as a fuel to recover the inherent heating value of used oil in a combustion process. A large volume of used oil is used solely for its energy content, as a secondary or substitute fuel (under controlled combustion conditions). The inherent high energy content of many used oil streams may encourage their direct use as fuels, without any pretreatment and processing, and without any quality control or product specification. Such direct uses do not always constitute good practice, unless it can be demonstrated that combustion of the used oil can be undertaken in an environmentally sound manner. The use of used oil as fuel is possible because contaminants do not present problems upon combustion, or it can be burned in an environmentally sound manner without modification of the equipment in which it is being burned.

However, used oils for use as fuel need to be subjected to treatments involving some form of settlement to remove sludge and suspended matter. Simple treatment of this type can substantially improve the quality of the material by removing sludge and suspended matter, carbon, and to varying degrees, heavy metals.

Although much of the used oil used as a secondary fuel receives only such basic pretreatment, every encouragement should be given to measures that improve the quality and control of this type of activity. Where fuels are to be marketed broadly, it is certainly desirable that used oils be subjected to both source and quality screening, and that products be supplied to a specification, even if only rudimentary. Where activities of this type are subject to a licence, permit, or authorization system, conditions should be specified to ensure that a minimum level of control is established, and that equipment for blending, separation, and any other processes are provided, used as necessary, and maintained properly.

The partial replacement of fuel with used oils is a technique that is widely applied around the world, in particular under controlled conditions in cement kilns. Other uses as fuel include in stone quarries for stone-drying purposes or in asphalt coating plants; in smelters handling iron, lead, tin, aluminum, and some precious metals; in chemical incinerators; in coking plants; and in brickworks, electricity generating stations, and steam raising boilers. In fact, an estimated 70% v/v of used oil is either used as a fuel (i.e., burned for energy recovery) or re-refined into lubricating oil. However, depending on the year, estimates indicate that as much as 90% v/v of all collected used oil is burned for energy recovery, leaving 10% v/v or less to be re-refined into lubricating oil to recover not only its inherent heating value but also its unique chemical composition so it can be reused in producing fresh lubricating oil products.

When used lubricating oil is to be blended with fuel oil for fuel use, there may (depending on the processes used to purify the used oil) be constituents within the used oil—contaminants such as water and rust or other particulate matter (PM)—that are introduced to the used oil during service as well as oil degradation products that form during extended storage as a manifestation of instability. In addition, instability (incompatibility) of the used oil and heavy fuel oil mixtures may also be an issue.

Lubricant incompatibility with heavy fuel oil is more difficult to determine if no visible changes occur when the products are mixed (Mushrush and Speight 1995). The problem appears only after the mixture is used in the combustion equipment that consequently fails or loses performance. Optimal lubrication performance requires carefully balanced frictional and antiwear properties in the finished product that can be changed when lubricants are mixed. It would be wise to institute a series of simple test protocols to determine the stability of used oils–heavy fuel oil mixtures before admixture (Mushrush and Speight 1995; Speight 2001, 2002, 2014).

The primary examples of oil combustion involve burning either fuel or used oil for the purposes of electricity generation or heating. These processes typically make use of petrochemical oils. Fuel oil heaters may be used to heat individual homes or much larger industrial settings, whereas used oil is more commonly used in commercial and industrial contexts. Residual fuel oil is sometimes used for electricity generation, and has also been used in vehicles like steamships and locomotives. Another type of oil combustion uses vegetable oil or animal fat to run diesel engines (Speight 2008).

Fuel oil is produced via the same fractional distillation process as gasoline, and is typically categorized into six different types (Speight 2014). These range from kerosene—a precursor to diesel fuel—to heavy residual fuel oil. After kerosene boils off in a fractional distillation, the next lightest type is heating oil.

Used oil typically refers to oil that has already been used for its intended purpose (Chapter 1), with the usage and precise definition of the terms varying from place to place. One example of used oil is the motor oil (lubricating oil) that is changed at regular intervals. After a certain amount of use, it may no longer be able to perform its lubrication functions, although it may still be useful as a combustion fuel. Used oil is typically combusted as a heat source, although it is sometimes used in boilers for various industrial applications.

The recycled oil standard limits the levels of certain heavy metals, total halogens, and halogenated compounds as contaminants. Used lubricating oil contains significantly higher concentrations of heavy metals, sulfur, phosphorus, and total halogens compared with low-sulfur crude-based heavy fuel oils (Kirk and Othmer 1996; Boughton and Horvath 2004). Because of a generally low quality as fuel, used oil is commonly blended with other fuel oils before use (Speight 2014). With blending, the specific level of contaminants in the finished fuel is lowered to an acceptable level for equipment specifications and temporal emission limits for any given user. Although this may satisfy the requirements for the levels of certain contaminants, it does not always satisfy the specifications for types of contaminants.

In addition, combustion of a blended fuel is assumed to not affect the net release of emissions with time—that is, from a life cycle perspective, the net emissions per unit of used oil consumed remain the same regardless of dilution. Mass of emissions alone does not convey a degree of significance to human health or the environment.

Because each elemental and chemical species has different specific effects, a weighing method is used to evaluate and compare the cases for different disposal options.

Currently, used oil marketed as fuel constitutes only a small amount (<10%) of the total fuel oil market in California. Because this amount is relatively small, it is assumed in a typical study in California that there is no significant displacement in crude refining capacity or in the fuel oil market by using used oil as a fuel.

Finally, facilities burn both on-specification and off-specification used oil. However, used oil that is off-specification may only be burned in specified combustion devices (U.S. Environmental Protection Agency 2008a,b): (1) industrial furnaces, as defined in 40 Code of Federal Regulation (CFR) §260.10 and §279.61(a), and industrial boilers, as defined in 40 CFR §260.20, are located at facilities that are engaged in a manufacturing process in which substances are transformed into new products; (2) utility boilers, as defined in 40 CFR §260.20, are used to produce electric power, steam, heated or cooled air, or other gases or fluids for sale; (3) used oil–fired space heaters, provided that the burner meets the provisions of 40 CFR §279.23; and (4) hazardous waste incinerators subject to regulation under 40 CFR parts 264 and 265, subpart O.

However, before used oil is reused (for example, in fuel applications), it must be subjected to a series of standard test methods (Chapter 3). Used oil is generated from many different sources, and then consolidated at key collection points (e.g., commercial and industrial operations, automotive repair shops, branch collection networks, and fleets) through well-established channels before being shipped to processing facilities and end users (Chapter 3).

Thus, the steps that are essential before used oil is employed as a fuel are as follows: (1) test the oil to determine whether it's on-specification or off-specification; (2) remove as much water and other liquids from the oil as possible by allowing the water to settle to the bottom, or also by using solvents to remove water and, on occasion, antifreeze; and (3) filter out metal scraps and other larger particles using screens—the filtering step can be performed when the oil is collected, or in the burner when the oil is added.

In addition to the processing steps outlined above, processors typically store used oil in tanks, allowing oil to separate from other denser liquids and rise to the top of the tank. These steps allow processors to provide users with relatively high-quality used oils. Before it is used as fuel, standard industry practice is to extract water from the used oil and to filter used oil.

5.2 DIRECT COMBUSTION: USE AS A FUEL

Used oil includes used crankcase oils from automobiles and trucks, used industrial lubricating oils (such as metal-working oils), and other used industrial oils (such as heat transfer fluids) (Chapter 2). When discarded, these oils become used oils due to a breakdown of physical properties and contamination by the materials they come in contact with. The different types of used oils may be burned as mixtures or as single fuels where supplies allow.

In terms of use as a fuel in power plants, hot asphalt mix plants, cement works, steelworks, and waste incinerators, there are directives that should be followed, and these directives set limits on the emissions to the atmosphere of substances generated during such combustion, and on emissions to water and any slag or ash generated. In fact, any slag or ash produced may itself be classified as hazardous waste and be subject to its own disposal controls. These emissions often depend on the design of the combustion system. Such plants may be licensed individually (rather than a blanket license for any group of similar plants) and are subject to routine monitoring and environmental audits to ensure that emission limits are met. It is, therefore, necessary to ensure that used oil sent as a fuel to these plants remains within the limits of impurities that the design can handle—and that the legislation will allow—so it is necessary to use known sources or to analyze the waste upon receipt before it is sent to the combustion chamber.

Co-incineration, when mixtures of used oil and other fuels are burned together, is also practiced, which can mean that highly contaminated material can be diluted with lesser contaminated fuel to keep emissions within limits—but this is merely skirting the real issue. It is not only the amount of used oil that is blended with another fuel but it is also the character of the contaminants in the used oil that must be taken into account. Contaminants such as halogens, heavy metals, and trace elements (to mention only three possible contaminants) are banned by many environmental laws in various countries, and dilution does not remove such contaminants.

To reduce the contaminants in used oil and before use as a fuel, the standard industry practice is to extract water from and to filter the used oil. Together with the gravity separation of the oil in settling tanks described previously, these processes do not always remove the contaminants to the satisfaction of the relevant environmental legislation. As stated previously, the operative watchwords are not only *quantity* but also *quality* in terms of defining the contaminants present.

Clearly, it will be necessary to ensure that the contaminants present in the used oil sent to various plants as fuel oil need to remain within the limits of impurities that the design can handle and within the limits of allowable types of contaminants; therefore, it is usually necessary to use known sources or to analyze the waste upon receipt. Co-incineration may be acceptable, providing there are no constituents in the blend (from any one of the blended materials) that would otherwise be classified as hazardous. Used oil can also be burned in space heaters and so-called small used oil burners and is used as a heat source in processing road aggregate and asphalt.

All such uses of used oil, no matter what the source of the oil, are subject to local or federal licensing practices and emission limits.

5.3 COMBUSTION CHEMISTRY AND EQUIPMENT

Most oil-fired domestic, commercial, industrial, or utility boilers can burn used oils. However, combustion of used oil not subject to strict product specifications or under uncontrolled conditions can lead to serious pollution of all environmental media. Also, combustion plants will suffer corrosion when oils with halogen contents are burned. Large industrial and utility boilers are generally considered to pose relatively low environmental risks because of their combustion efficiency, use of consistent quality fuel, and, in some cases, pollution control devices (e.g., electrostatic precipitators, bag filters, and high-energy venturi scrubbers), tall dispersive chimneys, and their location (often away from high population density areas).

5.3.1 Fuels

The terms *used oil* (as applied to use in a combustor), *recovered fuel oil,* or *processed fuel oil* are often used interchangeably. Processing typically involves basic cleanup such as dewatering and processes such as filtering, settling, or centrifuging to remove solid contaminants.

Used fuel oil has been in use for many years as a fuel for combustors, and various commercial specifications apply, with limits on elements such as (but not limited to) sulfur, halogens, nickel, vanadium, and sodium. More recently, environmental concerns have led to proposals for tighter controls on the use of used oil as a combustion fuel. For example, in the United States, used oil (and those oils categorized as recovered fuels by some vendors/purchasers) will be categorized as either *on-spec* (on-specification, i.e., the oil meets the desired fuel oil specifications) or *off-spec* (off-specification, i.e., the oil does not meet the desired fuel oil specifications).

Briefly, on-specification used oil (1) does not exceed any specification limits, has not been mixed with hazardous waste; or (2) has had ignitable characteristics typical of lubricating oil and meets the performance standard. On-specification oil generally may be managed or burned by anyone for any legitimate oil-burning purpose (e.g., space heaters, boilers, oil furnaces) regardless of whether it is generated on-site or not. On occasion, used oil may exhibit characteristics of hazardous waste because of normal use and may still qualify as on-specification used oil if hazardous waste has not been added to it and if the used oil generator can certify this claim.

The environmental laws in various countries (for example, the United States) may consider that only certain contaminants in used oil pose a significant threat to human health or the environment, and have established maximum concentration limits for these contaminants. Used oil with a low concentration of contaminants is on-spec and can be burned in any combustion device; however, various laws might specify not only the amounts of specific contaminants but also the mere presence of these contaminants at any level in the used oil disqualifies the oil from being used as a fuel. Off-specification used oil is regarded as waste and must be treated accordingly—it will be classified as hazardous waste and must be disposed of at a facility designated for that purpose.

5.3.2 Combustion Chemistry

Combustion is the rapid oxidation of a substance accompanied by a high temperature and usually a flame. It can produce a number of different products, depending on the materials available in the reaction (Gómez-Rico et al. 2003). The products of clean combustion between a hydrocarbon and oxygen are carbon dioxide (CO_2), water (H_2O), and energy. Incomplete or partial combustion can also form carbon monoxide (CO), free carbon or soot, nitrogen oxides, hydrogen cyanide (HCN), and ammonia (NH_3).

Fuel is the substance that burns during the combustion process. All fuels contain chemical potential energy; this is the amount of energy that will be released during a chemical reaction. A wide variety of substances can be used as fuels, but hydrocarbons are some of the most common. These include methane, propane, gasoline, and jet fuel, fuel oil, coal, natural gas, as well as biomass, such as wood.

The term *complete combustion* is generally used for the burning of hydrocarbons. If the hydrocarbon contains sulfur, sulfur dioxide will also be present. On the other hand, *incomplete combustion* results in some of the carbon atoms combining with only one oxygen atom to form carbon monoxide and other potentially harmful by-products. Complete combustion occurs when the fuel and oxygen are in the perfect combination, or ratio, to completely burn the fuel. This condition is also referred to as stoichiometric or zero excess air combustion. Incomplete combustion, as the term implies, may leave some of the fuel unused.

Usually, the combustion process is initiated by heating a hydrocarbon above its ignition temperature in an oxygen-rich environment. When the compound is heated, the chemical bonds of the hydrocarbon are split. The elements of the hydrocarbon then combine with the oxygen to form oxygen-containing compounds known as oxides. This rearrangement of hydrocarbon elements into oxides is accompanied by the release of energy and heat.

The combustion products of clean combustion include CO_2, H_2O, and energy. No other gases or solid particulates are formed as combustion products in this type of reaction. The following balanced reaction is the propane–oxygen reaction:

$$C_3H_8 + 5O_2 \rightarrow 3CO_2 + 4H_2O$$

Propane combustion serves as an example of a commonly burned hydrocarbon in household use. Usually, propane combustion will occur when the gas in the air mixture is between 2.2% and 9.6% v/v (flammability limits of propane). A properly functioning propane appliance producing an ideal burn will give off a blue flame and should present no danger of carbon monoxide poisoning, which is a deadly by-product of incomplete propane combustion.

Incomplete combustion of propane occurs when the mix ratio is higher or lower than the ideal ratio, but still occurs within the limits of flammability. If the ratio of propane to air is less than the ideal ratio, a lean burn will occur, as evidenced by flames that appear to lift away from the burner or go out. A rich burn occurs when the ratio of propane to air is greater than the ideal ratio, and can be recognized by larger flames that are yellow rather than blue. Incomplete combustion of propane or other hydrocarbons typically will result in carbon monoxide release, an extremely serious environmental and health hazard for humans as well as for most animals.

When combustion takes place in an oxygen-deprived environment (incomplete combustion), different combustion products can be made. Free carbon (soot) and carbon monoxide are produced along with carbon dioxide, water, and energy. The formation of soot as a combustion product is the reason why incomplete combustion (soot-forming combustion) is also known as *dirty combustion*. In the chemical industry, gasifiers burn flammable materials in oxygen-deprived environments to produce synthesis gas, which consists of hydrogen and carbon monoxide. Outside of chemical industries, incomplete combustion often occurs in internal combustion engines and poorly ventilated furnaces.

Oxygen from the air is the most common source of oxygen for most combustion reactions. Air is mostly composed of nitrogen, however, and during combustion, nitrogen is capable of producing a number of its own combustion products. Nitrogen oxides (NO_x) can be formed in a combustion reaction—the most common NO_x is the toxic nitrogen dioxide (NO_2), but ammonia (NH_3) and the extremely lethal hydrogen cyanide (HCN) may also be formed.

Halogen species, sulfur derivatives, and phosphorus derivatives in the fuel may also produce specific combustion products. Halogens such as chlorine can react with free radical hydrogen to form chemicals such as hydrogen chloride (HCl). Sulfur can produce the toxic and foul-smelling chemicals, sulfur dioxide (SO_2) and hydrogen sulfide (H_2S). When phosphorus is present in a combustion reaction, it produces phosphorus pentoxide (P_2O_5) as a white, solid particulate.

The tendency of a hydrocarbon fuel to favor clean or dirty combustion products can be estimated by examining the heat output potential of the reaction and the energy necessary to initiate the reaction. Increasing the heat output potential increases the tendency of the fuel to undergo incomplete combustion. Propane, which does not require much energy to initiate combustion, tends to burn cleanly. Aromatic compounds such as benzene, toluene, and their respective derivatives, on the other hand, tend to produce a lot of soot when burned.

5.3.3 Combustion Equipment

Combustion equipment consists of machinery designed to burn a fuel source in combination with the oxygen in air to produce heat and energy. The different types of combustion equipment include furnaces, boilers, turbines, and engines. The design can be further identified by the fuel source such as oil, natural gas, coal, or biomass (e.g., wood).

A furnace, or direct-fired heater, is a type of enclosed chamber in which heat is generated by the combustion of some type of fuel. A blower introduces air into the heating chamber where it is combined with the fuel and burned. A furnace can be used to produce heat energy for the heat itself, converted into mechanical energy to drive a process, or into electrical energy as in a power plant.

In coal-fired electricity-generating plants, recovered fuel oil is sprayed onto the coal at start-up and when adding significant new volumes of coal. Unless oil was added, heat output would decrease as the coal is added, making the power generation less smooth. The heat generated can be transferred away from the furnace through tubes filled with water, air, or oil, which is used as the heat transfer medium.

As a cautionary note, suppliers of recovered fuel oil may find some difficulty in consistently meeting the chlorine specification. Generally, used oils have been reported as being higher in chlorine content because of the contamination of used oils from other waste materials such as certain transformer oils or chlorinated grease.

5.3.3.1 Boilers

Boilers are combustion equipment units that produce steam or heated water to run industrial processes or to provide heat for buildings. The major types of industrial boilers are water-tube and fire-tube. Water-tube boilers move water through tubes running through a furnace chamber where fuel is combusted. The heat of combustion transfers through the tube walls to heat the water and create steam. Fire-tube boilers send heated gases generated by fuel combustion through tubes that pass through a water-filled chamber, where the heat transfer causes the water to vaporize into steam.

Boiler combustion is the means by which fuel is burned in boilers that heat water for steam. There are many applications for steam boilers, including chemical process heating, steam heat for buildings and hot water, and steam to drive electrical turbine generators. Combustion is the reaction of fuels with oxygen in air to create heat that is used for steam production.

The combustion area of a boiler normally has tubes containing water and steam passing through an open box that may contain burners and controls. Tube design can improve efficiency by using multipass systems. Water tubes entering the boiler may first pass through the flue gas zone, which takes some of the waste heat and preheats the water. Tubes can then pass through the combustion zone more than once to fully utilize combustion heat, which also improves efficiency.

Boiler combustion efficiency for air and fuel mixtures is critical to proper boiler operation. A molecule of fuel requires a theoretical amount of oxygen to burn completely, but in reality, excess oxygen is needed due to various losses in the combustion zone. Air is approximately 21% oxygen, so unburned nitrogen in the air must also be heated in the boiler and vented by the flue. This further affects boiler efficiency and produces nitrogen oxides (which do not originate from the fuel) that lead to the formation of acid rain and smog.

Too much oxygen reduces the boiler combustion temperature, creates some undesirable pollutants, and requires fuel to heat oxygen and nitrogen that are not used. Lack of oxygen can reduce boiler efficiency and create soot and other by-products that can damage the boiler over time. Monitoring oxygen and combustion gas concentrations in the flue gas, and maintaining a correct flue temperature (which is typically boiler dependent and fuel dependent), can optimize boiler performance.

Smaller boilers can be adjusted manually using flue gas sensors and flue gas thermometers, but many boilers can benefit from automatic controls. Boilers may not operate at a single operating point, but will have varying steam demands or operating conditions, which makes manual efficiency settings impractical. Older boilers can be retrofitted with electronic controls that provide feedback to air and fuel input pumps to give the best ratio for combustion.

A recovery boiler is a power plant subsystem used to capture energy that would otherwise be lost in waste products that were not completely combusted. This energy is recycled to the primary energy generating system to produce more power. For this reason, this equipment may alternatively be referred to as a heat recovery boiler or as a waste heat recovery boiler. The main requirement for its use is that the system's waste products have sufficient recoverable energy to drive the primary power-generating equipment. Recovery boilers may be used in the power systems for a variety of industrial equipment.

Waste recovered for further combustion in a recovery boiler is usually a gas, a liquid, or a combination of the two. Processes generating these types of wastes include metal refineries, petrochemical processing plants, and other industrial plants operating at high temperatures. Solid material may also be used, although this is less common. The most notable example of solids used in recovery boilers is in papermaking.

A variety of fuels can be used for boiler combustion, including natural gas, fuel oil, and biofuels produced from plants or animal wastes (Speight 2011). When fuel is sprayed or atomized into a boiler with air, an ignition coil or small pilot flame can ignite the mixture. Combustion releases a great deal of heat, some of which heats water to steam, and some is lost due to fuel losses (heated gases that are vented from the boiler through its flue or vent) and radiation (infrared heat loss)—the latter occurs from a hot boiler into a cooler room or environment.

Various fuels are used by different types of combustion equipment. Liquid petroleum products are the most common fuel sources in engines and turbines, and may include gasoline, kerosene, diesel, and heavier residual products such as bunker fuel. A large number of boilers, especially those producing building heat, are fired with natural gas. Power plants burning coal produce a significant percentage of the energy used worldwide. Biomass-fired (wood-fired) furnaces are less common in industrial applications and are more likely to be found in residential use.

Used oil can be burned in a variety of combustion systems including industrial boilers, commercial/institutional boilers, space heaters, asphalt plants, cement and lime kilns, other types of dryers and calciners, and steel production blast furnaces. Boilers and space heaters consume the bulk of the used oil burned. Space heaters are small combustion units (generally less than 250,000 British thermal units per hour [Btu/h] or 264 MJ/h input) that are common in automobile service stations and automotive repair shops where supplies of used crankcase oil are available.

Boilers designed to burn no. 6 (residual) fuel oils or one of the distillate fuel oils can be used to burn used oil, with or without modifications for optimizing combustion. As an alternative to boiler modification, the properties of used oil can be modified by blending it with fuel oil, to the extent required to achieve a clean-burning fuel mixture.

5.3.3.2 Turbines

Turbines are machines that create power through the rotation of blades or vanes produced by the movement of gas or liquid. A gas turbine uses a compressor to produce pressurized air. Combustion of the compressed air mixed with fuel generates a high-pressure gas that drives a rotor. This type of combustion equipment is found in smaller capacity power plants and in jet engines.

5.3.3.3 Internal Combustion Engines

An internal combustion engine produces energy by igniting fuel under high pressure in an enclosed space. The gas expansion created by the burning of fuel is turned into mechanical energy to move a crankshaft or rotor. These engines drive automobiles, electrical generators, pumps, forklifts, and other industrial machinery. Used oil is generally not suitable for use as a fuel in turbines or internal combustion engines.

5.4 FUEL OIL PROPERTIES

Heavy fuel oil can have high levels of contaminant metals, sulfur, and asphaltene constituents that can lower fuel quality (Chapter 3). Poor quality fuel causes ash fouling, slag, and corrosion. In fact, any property prescribed in a product's specification should be related to product performance or of value in the refining or handling of the product (Martin and Hicks 2010). Specific procedures commonly applied to heating and power generation fuels and their significance are varied but must be applied to assure product quality and suitability for utilization in the designated manner.

5.4.1 API AND SPECIFIC GRAVITY OR DENSITY

The gravity of a fuel oil is an index of the weight of a measured volume of the product. There are two scales in use in the petroleum industry: specific gravity or density (Speight 2001, 2002). On a weight basis, the heating value of petroleum fuels decreases with increasing specific gravity or density (decreasing API gravity), because the weight ratio of carbon (low-heating value) to hydrogen (high-heating value) increases as the specific gravity increases. On a volume basis, the increasing specific gravity more than compensates for the decreasing heating value per unit weight, with the net result that fuels having high specific gravity yield more heat energy per unit volume than those of low specific gravity. Because many fuels are purchased on a volume basis, the relationship of higher density fuels containing more energy is very significant to some purchasers.

Although gravity is not solely suitable for determining if a fuel oil or used oil is suitable for use in burners, it can certainly indicate the potential for the presence of lower boiling or higher boiling constituents. Use of a complementary test method will confirm or bring into question the data from this test.

5.4.2 FLASH POINT AND FIRE POINT

The *flash point* of a fuel is a measure of the temperature at which the fuel must be heated to produce an ignitable vapor–air mixture above the liquid fuel when exposed to an open flame. Flash point data are normally included in industry specifications for fuel oil. The minimum flash point for many fuel oils is 38°C or 40°C (100°F or 104°F), for safety reasons. The *firepoint* of a fuel is the temperature at which oil in an open container gives off vapor at a sufficient rate to continue to burn after a flame is applied.

All fuel oils should have a minimum flash point of 38°C (100°F) or higher—the purpose of this minimum flash point restriction is to reduce fire or explosion hazards during handling, transportation, and in the event of a leak or spill in the vicinity of furnaces or flames. Fuel oils with a flash point less than 38°C (100°F) introduce the risk of explosion, especially during initial ignition of the burner. Thus, fuel oil contaminated with a small amount of low-boiling material, resulting in a low flash point, may present a serious safety hazard in handling and use.

Similarly, used lubricating oil contaminated with low-boiling material, resulting in a low flash point, may be detectable with the use of this test. The test data may also confirm what had been obtained from the density/gravity test method. Therefore, determination of flash point can be useful in detecting such contamination and, thereby, avoiding serious safety hazards.

5.4.3 Viscosity

The viscosity of a fluid is a measure of its resistance to flow and is generally expressed as kinematic viscosity—alternate units include Saybolt universal seconds and Saybolt Furol seconds.

The determination of residual fuel oil viscosity is complicated by the fact that some fuel oils containing significant quantities of wax do not behave as simple Newtonian liquids in which the rate of shear is directly proportional to the shearing stress applied. At temperatures in the region of 38°C (100°F), residual fuels tend to deposit wax from solution. This wax deposition exerts an adverse effect on the accuracy of the viscosity result, unless the test temperature is increased sufficiently high enough for all wax to remain in solution. By using the kinematic system, the petroleum industry today has moved toward 100°C (212°F) as the standard test temperature for most residual fuels for viscosity determination.

Viscosity is one of the more important heating oil characteristics—it is indicative of the rate at which the oil will flow in fuel systems and the ease with which it can be atomized in a given type of burner. Because the viscosities of heavier residual fuel oils are high, this property tends to be particularly relevant to handling and use. The viscosity of a heavy fuel decreases rapidly with increasing temperature. For this reason, heavy fuels can be handled easily and atomized properly by preheating before use.

If no preheating facilities are available, lighter or less viscous oils must be used, and if the preheating equipment is inadequate, it may be necessary to burn lighter oil during cold weather. Overly viscous oil can produce problems throughout the system. Besides being difficult to pump, the burner may be hard to start, and flashback or erratic operation may be encountered.

Viscosity also affects the output or delivery of a spray nozzle and the angle of spray. With improper viscosity at the burner tip—or the presence and particulate matter that was not filtered out correctly as might occur in poorly prepared used lubricating oil—poor atomization can result in carbonization of the tip, carbon deposition on the walls of the fire box, or other conditions leading to poor combustion.

5.4.4 Pour Point

The pour point is the lowest temperature at which the oil will just flow under standard test conditions. Anticipated storage conditions and fuel application are usually the primary considerations in the establishment of pour point limits. Storage of higher-viscosity fuel oils in heated tanks will permit higher pour points than would otherwise be possible. The pour point is important for the deliverability of lighter fuel oils that are not normally heated, such as home heating oil. Although it does take time for a large volume of oil to cool in very cold weather, trying to deliver furnace fuel oil at temperatures below the pour point of the fuel could result in the fuel waxing or gelling in the tank truck or manifold, so that it cannot be pumped off a delivery truck into a fuel storage tank.

The failure to flow at the pour point normally is attributed to the separation of wax from the fuel and can be due to the effect of high viscosity in the case of very viscous oils—some used lubricating oils would fit this description. In addition, pour points, particularly in the case of residual fuel oil and used fuel oil, may be influenced by the previous thermal history of the oils. As an example, any loosely knit wax structure built upon cooling of the oil can normally be broken by the application of relatively little pressure. Thus, the usefulness of the pour point test in relation to residual fuel oil and used lubricating oil is open to question, and the tendency to regard the pour point as the limiting temperature at which a fuel will flow can be misleading, unless correlation is made with low temperature viscosity or with the data from other test methods.

5.4.5 Cloud Point

Middle distillate fuel oils, especially, begin to form wax crystals and become cloudy in appearance as they are cooled toward the pour point. Cloud points often occur at 4°C to 5°C (7°F–9°F) above the pour point, and temperature differentials of 8°C (15°F) or more are not uncommon. The temperature differential between cloud point and pour point depends on the nature of the fuel components, but the use of wax crystal modifiers or pour depressants tends to accentuate these differences.

As the temperature continues to decrease below the cloud point, the formation of wax crystals is accelerated. These crystals may clog fuel filters and lines and, thus, reduce the supply of fuel to the burner. Because the cloud point is at a higher temperature than the pour point, it can be considered even more important than the pour point in establishing distillate fuel specifications for cold weather usage, especially with newer, high-pressure burners equipped with fine filtration (e.g., 10 μm filters).

The application of the cloud point test method to used lubricating oil is often influenced by the color of the oil and the presence of unfiltered debris that may still exist in the used oil. The potential for anomalous data should be taken into account *before* the test method is applied to used lubricating oil.

5.4.6 Sediment and Water Content

The tests for water and insoluble solid content of fuel oils are important because contamination by water and sediment can lead to filter and burner problems and the production of emulsions, which are removable only with difficulty. The corrosion of storage tanks may also be associated with water bottoms that accumulate from atmospheric condensation and water contamination. In addition, the presence of water promotes microbial contamination.

Water in fuel oil or used lubricating oil can lead to delivery or flow problems when the temperature of fuel drops below 0°C (32°F). Free water will freeze, and may freeze a valve shut so it cannot be opened, or it may form ice crystals that will plug a screen in the off-loading line. Similarly, water that was dissolved in fuel oil or used lubricating oil at a warmer temperature (e.g., such as 5°C–15°C, 9°F–59°F) may separate out as ice crystals as the fuel is cooled to low temperatures, such as –20°C or –30°C (–4°F to –22°F).

5.4.7 Ash Content

Fuel oil and used lubricating oil do not contain ash but they do contain ash precursors that yield ash upon combustion—ash is the inorganic residue that remains after the combustion of oil in air at a specified, high temperature.

Ash-forming materials found in heavy fuel oil are normally derived from the metallic salts and organometallic compounds found in crude oils. Because the ash-forming constituents of crude oil ultimately concentrate in the distillation residue (Speight and Ozum 2002; Hsu and Robinson 2006; Gary et al. 2007; Speight 2014), distillate fuels tend to contain only negligible amounts of ash. However, both distillate and residual fuels may pick up ash contributors during transportation from the refinery.

On the other hand, used lubricating oil typically contains metals and other ash-forming constituents produced in the oil during service. Unless such materials are removed, ash formation will ensue during combustion and, depending on the use of the fuel oil and used lubricating oil, ash composition can have a considerable bearing on whether detrimental effects will occur. For example, ash precursors and ash itself can cause slagging or deposits, whereas high-temperature corrosion in boilers may also attack refractories in high-temperature furnaces and kilns. Therefore, methods have been developed for counteracting the effects of ash and include the use of additives (Chapter 1), modifications in equipment design, and the application of, in the current context, used oil processing methods (Chapter 7).

5.4.8 Carbon Residue

The carbon residue is the carbonaceous residue remaining after combustion and measures the coke-forming properties of fuel oil or used oil—the carbon residue also includes the mineral matter remaining after destructive distillation of the oil under specified conditions. The carbon residue of burner fuel—and the used lubricating oil that is to be used as fuel oil—serves as an approximation of the tendency of the fuel to form deposits in vaporizing pot-type and sieve-type burners, in which the fuel is vaporized in an air-deficient atmosphere.

During service, lubricating oil produces carbon residue precursors in the form of polynuclear aromatic systems that were not present in the original lubricating oil. These polynuclear aromatic systems cannot be detected and typically remain soluble in the used oil. Depending on the character of the fuel oil to be blended with the used lubricating oil, the polynuclear aromatic systems may separate out as a solid phase with the potential of interfering with the performance of the injector nozzles. As a component of a high-carbon residue, the polynuclear aromatic systems can be part of the cause of rapid carbon buildup and nozzle fouling.

5.4.9 Sulfur Content

Fuel oils contain varying amounts of sulfur (as organic compounds of sulfur), depending on the crude source, refining processes, and fuel grade (Speight and Ozum 2002; Hsu and Robinson 2006; Gary et al. 2007; Speight 2014). The high boiling range fractions and the residual fuels usually contain higher amounts of sulfur, which is generally regarded as an undesirable constituent because of its potential to create corrosion and pollution problems. Typically, lubricating oil does not contain sulfur or, at worst, contains very little sulfur, unless the sulfur is present in an additive.

In boiler systems, the conversion of even a small fraction of the sulfur to sulfur trioxide during the combustion of the fuel can cause low-temperature corrosion problems if this gas is allowed to condense and form corrosive sulfuric acid on the cool metal surfaces of the equipment (exhaust piping). In combination with sodium (also present in some lubricating oil additives), the sulfur from the blended fuel contributes to the formation of deposits on external surfaces of superheater tubes, which cause the corrosion of equipment and loss of thermal efficiency (heat transfer).

Combustion of sulfur-containing fuel oils produces sulfur oxides, which have been identified as atmospheric pollutants. To meet clean air standards (ASTM D396) in densely populated industrial areas, stack emission control devices and sulfur scrubbing procedures will be required.

5.4.10 Distillation

More generally, the distillation test is significant for the distillate fuel oil, but does not have the same relevance for residual fuel oil. The same might be said for used lubricating oil, unless the used fuel oil contains a considerable proportion of volatile material produced during service. In any case, a more appropriate test would be an evaporation test (ASTM D5800), which would give a better representation of the volatile constituents of used lubricating oil. However, the data from this test method must be compared with the data produced when the test method is applied to the original virgin lubricating oil.

5.4.11 Corrosion

Tests for corrosion are of a qualitative type and are made to ascertain whether fuel oils are free of a tendency to corrode copper fuel lines and brass or bronze parts used in the burner assemblies. Such test methods are also essential when blending used lubricating oil with fuel oil for used in burners.

The test method (ASTM D130/IP 154) is simplicity itself—the copper strip corrosion test is conducted by immersing a polished copper strip in a sample of fuel oil or used lubricating oil contained in a chemically clean test tube. The tube is then placed in a bath maintained at a temperature of 50°C (122°F) for 3 h and, after washing, the strip is then examined for evidence of corrosion and judged by comparison with the corrosion scale in the standard.

5.4.12 Heat Content

The heat content or thermal value of any fuel oil is the amount of heat given off as a result of its complete combustion. The energy content of petroleum-based fuels is measured by the heat of combustion and used lubricating oil has a heat content on

TABLE 5.1
General Ranges for the Heat Content of the Various Grades of Fuel Oil

Grade	Heating Value (Btu/US gallon)	Comments
Fuel Oil no. 1	132,900–137,000	Small space heaters
Fuel Oil no. 2	137,000–141,800	Residential heating
Fuel Oil no. 4	143,100–148,100	Industrial burners
Fuel Oil no. 5 (light)	146,800–150,000	Preheating is required
Fuel Oil no. 5 (heavy)	149,400–152,000	Preheating required
Fuel Oil no. 6	151,300–155,900	Bunker C

Note: Used Lubricating oil = 140,000–150,000 Btu/US gallon. 1 Btu/US gal = 278.7 J/L. No. 1 and No. 2 fuel are both used for residential heating purposes—No 2 fuel oil may be slightly more expensive, but the fuel gives more heat per gallon used. No. 1 fuel oil is used in vaporizing pot-type burners; No. 2 fuel oil is used in atomizing gun-type and rotary fuel oil burners. The heavier the grade of fuel used in an oil burner, the greater the care must be taken to ensure that oil is supplied the combustion process at the proper atomizing temperature; if the temperature is too low the fuel oil will not atomize and evaporate and deposits will be left on the injectors—the burner will not operate efficiently.

the order of 140,000 to 150,000 Btu/US gallon (39,022–41,809 kJ/L), which is comparable to the heat content of the heavier fuel oils (Table 5.1).

For most applications, such as fuels used in engines or furnaces, the net heat of combustion is the appropriate measure of energy. This value is sometimes called the lower heating value, net heating value, net calorific value, or specific energy and assumes that water produced from combustion goes "up the stack" as water vapor. The gross heat of combustion, also called the higher heating value, requires that water vapor be condensed back to liquid water to recover the "heat of vaporization" of water.

For most fuel oils and used lubricating oils, the net heat of combustion can be estimated directly from the density of the fuel, providing that there is a correction for water, ash, and sulfur content (Speight 2001, 2002, 2014).

5.4.13 Stability

In essence, the stability of a fuel oil or used fuel is the ability of the oil to resist change in composition. Instability is manifested by a change in color, the formation of gummy materials or insoluble solids, waxy sludge, or asphaltic deposition on the bottom of storage tanks or flow lines. The storage stability of fuel oils may be influenced by many factors. Among these are (1) origin, (2) composition, (3) water, and (4) other contaminants.

Fuel oils containing unsaturated hydrocarbons (olefins) and catalytically cracked components are inherently less stable chemically and have a greater tendency to form sediment upon aging compared with straight-run fuel oils. In addition, the presence of various contaminants in used lubricating oil can render the used oil unstable and, at times, incompatible with fuel oil. In addition to changes caused by in-service conditions, this may also be a consequence of factors such as oxidation, polymerization, and the method of production of the fuel oil, which can cause the formation of insoluble compounds that eventually settle to the tank bottom and form sediment (sludge). This can result in the clogging of external cold filters, blockage or restriction of lines, deposit formation on burner nozzles, and combustion difficulties.

5.4.14 Spot Test for Compatibility

A method for measuring the compatibility of residual fuel oil and used lubricating oil is the spot test (ASTM D4740). This test is often ignored, but in its simplicity is well worthy of consideration when used lubricating oil is utilized as a fuel or as a blend with heavy fuel oil. The two procedures in this test method are used alone or in combination to identify fuels or blends (especially blends with used lubricating oil that may contain contaminants that have not been removed during the processing) that could result in excessive centrifuge loading, strainer plugging, tank sludge formation, or similar operating problems.

A spot rating of no. 3 or higher on a finished fuel oil according to the cleanliness procedure indicates that the fuel contains an excessive amount of suspended solids and is likely to cause operating problems. However, although a fuel may test clean when subjected to the cleanliness procedure, suspended solids (which occur in most used lubricating oils) may precipitate when the fuel is mixed with a blend stock. Evidence of such incompatibility is indicated by a spot rating of no. 3 or higher in the compatibility procedure.

5.5 ENVIRONMENTAL RISKS

The environmental risks associated with burning used oil in any size boiler could be reduced through the application of comprehensive and integrated management strategies including: (1) pretreatment of used oil to meet established quality specifications, such as settling, centrifugation, vacuum distillation, solvent extraction; (2) dilution of used oil by blending with virgin fuels; and (3) installation of flue gas emission control devices.

Generally, used oil can be burned in cement kilns without many of the negative air quality effects normally associated with burning used oil in small-sized to medium-sized boilers. Unless boilers or other combustion processes are equipped with burners of high combustion temperatures and contaminant destruction efficiencies or with flue gas treatment devices, used oil burning should be strongly discouraged or prohibited if practicable. This implies that efforts should be made to bring adequate treatment and disposal technologies to remote communities and sparsely populated areas in need of such technologies, and implement a program providing the necessary infrastructure for establishing used oil collection/storage/transport/treatment systems for such small-volume used oil generators.

5.6 POSTCOMBUSTION CAPTURE

Many industrial operations, such as power plants, which use combustion equipment, produce substantial amounts of carbon dioxide from combusting hydrocarbon fuels and hydrocarbonaceous fuels, and that carbon dioxide can be captured in a process called *postcombustion capture* so that it does not negatively affect the environment. The most common method for postcombustion capture is to pass the carbon through a solvent that absorbs the carbon and the capture unit itself is relatively simple. The problem with using this process is that it has high running costs, and plants could use 10% to 30% more energy just to capture the carbon.

The flue gas is passed through a capturing unit, which combines the carbon gas with a solvent. Commonly, amine solvents (olamines) are used (Mokhatab et al. 2006; Speight 2007, 2014). The amine solvent absorbs and captures the carbon from the flue gas, so that the carbon can be transported and stored later.

Postcombustion capture has been used since the 1940s. One reason for its popularity is that scientists and industry workers have a lot of experience using this system. Another reason is that the capture unit can be easily retrofitted, or added to an existing plant. The postcombustion capture unit is usually integrated with the combustion chamber. Air and hydrocarbons are pumped into an area where, under high temperature, they combust and create energy. The exhaust gas is pumped into an amine tower, where it is instantly mixed with nitrogen and pushed below ground for storage.

5.7 USED OIL COMBUSTION

Used engine oil, essentially composed of hydrocarbons, is a potentially high-energy fuel that does not contain a heavy residual fraction characteristic of heavy fuels. Some of the physical properties of used engine oils improve its handling and combustion, and also includes the following advantages: (1) low viscosity, which allows the oil to be injected into a standard burner; (2) sufficient fluidity, which allows it to be stored and pumped at ambient temperature; and (3) relatively low sulfur content, which makes the oil comparable to low-sulfur fuel oil ($S < 1\%$ w/w).

However, not all used oils are the same and the emissions from burning used oils reflect its compositional variations. There are disadvantages to using used lubricating oil as fuel oil because it can contain polluting solvents that must be eliminated. Metallic and metalloid compounds must also be eliminated because they are transformed into oxides carried by the combustion gas and are partially transformed into sulfate species that are then deposited in the combustion chamber.

However, used oils contain significantly higher concentrations of heavy metals, sulfur, phosphorus, and total halogens compared with low-sulfur petroleum-based heavy fuel oils. Because of a generally low quality as fuel, used oil is commonly blended with other fuel oils before use. As a result of blending, the specific level of contaminants in the blended product is lowered to an acceptable level for equipment specifications and temporal emission limits for any given user. Combustion of a blended fuel is assumed to not affect the net release of emissions with time—the net emissions per unit of used oil consumed remain the same regardless of dilution. Unfortunately, this easy-to-justify rationale does not take into account certain contaminants that should not be in the fuel oil in any quantity whatsoever.

Additives and the corrosion of engine parts are responsible for this level of contamination (~5000 ppm w/w). Potential pollutants include carbon monoxide (CO), sulfur oxides (SO_x), nitrogen oxides (NO_x), particulate matter (PM), particles less than 10 μm (PM_{-10}), toxic metals, organic compounds, hydrogen chloride, and gases such as carbon dioxide (CO_2) and methane (CH_4).

5.7.1 Particulate Matter

Ash levels in used oils are normally much higher than ash levels in either distillate oils or residual oils. Used oils have substantially higher concentrations of most of the trace elements reported relative to those concentrations found in virgin fuel oils. Without air pollution emission controls, higher concentrations of ash and trace metals in the used oil translate to higher emission levels of PM and trace metals compared with virgin fuel oils.

5.7.2 Sulfur Oxides

Emissions of sulfur oxides (SO_x) are a function of the sulfur content of the fuel. The sulfur content varies but some data suggest that uncontrolled emissions of sulfur oxides will increase when used oil is substituted for a distillate oil, although it will decrease when residual oil is replaced.

5.7.3 Chlorinated Organics

Constituent chlorine in used oil typically exceeds the concentration of chlorine in virgin distillate and residual oils. High levels of halogenated solvents are often found in used oil as a result of the inadvertent or deliberate addition of contaminant solvents to the used oils in the collection tanks. Many efficient combustors can destroy more than 99.99% of the chlorinated solvents present in the fuel. However, given the wide array of combustor types that burn used oils, the presence of these compounds in the emission stream cannot be ruled out.

5.7.4 Flue Gas Treatment

Emissions can be controlled by the pretreatment of the used oil to remove pollutant precursors or with emission controls to remove the air pollutants. Reduction of emission levels is not the only purpose of pretreatment of the used oil. Improvement in combustion efficiency and reduction of erosion and corrosion of the combustor internal surfaces are also important considerations.

The most common pretreatment scheme uses sedimentation followed by filtration. Water and large particles (>10 μm in diameter) are removed without having much effect on sulfur, nitrogen, or chlorine contents. Other methods of pretreatment involve clay contacting; demetallization by acid, solvent, or chemical contacting; and thermal processing to remove residual water and light ends. These latter processes might be attractive as waste reduction schemes or to recycle the used oil, but the added costs probably hinder their use as part of a combustion process.

Blending of used oil with a virgin fuel oil is practised frequently and has the same effect as some of the other pretreatment processes. However, for the purpose of improving emission control factors, blending (by itself) is in the uncontrolled category.

Used oil serves as a substitute fuel for combustors designed to burn distillate or residual oils. Therefore, the emission controls used are usually those that were in place when the original fuels were first burned. For an asphalt plant, emissions of particulate matter, which include the dust from drying of the aggregate, are controlled with a fabric filter.

5.7.5 Control Technologies

There are three basic ways to control and stop the combustion process: (1) remove fuel, (2) remove the oxygen, or (3) remove the heat. Combustion may also be stopped by stopping the chemical chain reaction that creates flames. This is especially important when certain metals, such as magnesium, burn because adding water to the fire will only make it stronger. In such cases, dry chemicals or halomethanes are used to terminate the reaction. The best method to use is dependent on the type and the size of the fire.

5.8 COMBUSTION WITH HEAVY FUEL OIL

Used oil includes used crankcase oils from automobiles and trucks, used industrial lubricating oil (such as metal-working oil), and other used industrial oils (such as heat transfer fluids). When discarded, these oils become used oils due to a breakdown of the physical properties and contamination by the materials they come in contact with.

If the objective is to recover the heating value of the used oil by reprocessing it into a fuel oil, the following steps are necessary: (1) removal of water and sediment by settling, (2) filtration to remove particulates, and (3) blending to control the amount of ash (i.e., solid contaminants and metals present in the fuel in soluble compounds) produced in the combustor. Reprocessing typically does not remove heavy metals and halogens in the used oil.

Thus, unless combusted under controlled conditions (e.g., stacks with "scrubbers" designed to capture contaminant emissions), use of the reprocessed fuel is likely to release toxic constituents. In large-volume reprocessing operations, water, light-end fuels, and chlorinated solvents are removed by distillation. Centrifuges are used for more efficient particulate removal or a chemical treatment step may be included to break emulsions and reduce the content of ash and sulfur.

Heavy fuel oil (sometimes referred to as *residual fuel oil*) is the higher-boiling fraction resulting from the refinery distillation column—it may be blended with residuum and hence contains a portion of nonvolatile material. This type of oil can be one of the least expensive liquid fuels available, although they typically require special heating apparatuses to combust. There may also be a number of pollution concerns involved with the residual fuel oil combustion process. Large ships and power plants can both utilize the type of combustion equipment required by residual fuel oil.

Before dealing with the combustion of the mixtures, it is necessary to determine the range of miscibility of these two kinds of fuels, which varies owing to the structure of the hydrocarbons they contain. The used engine oil structure is paraffinic because of its multigrade character, which is achieved during its refinery manufacture (extraction of polynuclear aromatic compounds). On the other hand, heavy fuel oil, mainly composed of the petroleum crude residual fraction, contains three classes of well-differentiated products: (1) asphaltene constituents, more or less agglomerated, condensed compounds of high molecular weight; (2) resins constituents, with structures intermediate between asphaltene constituents and oil, the dispersing role of which has been clearly demonstrated (Speight 2014); and (3) oil constituents, which represent the continuous phase of the fuel oil.

In addition, when heavy fuel oil has been stored (rather than used immediately after production) a certain amount of settlement of solids and sludge will occur in tanks over time. This can cause problems when the fuel oil is introduced into pipes and the furnace injectors. Progressively finer strainers should be provided at various points in the oil supply system to filter away finer contaminants such as external dust and dirt, sludge, or free carbon. It is advisable to provide these filters in duplicate to enable one filter to be cleaned while oil supply is maintained through the other.

Furthermore, it is also necessary to investigate the miscibility/immiscibility of the used lubricating oil and the heavy fuel oil, which is dependent on (1) the used oil to heavy fuel ratio in mixtures, (2) the concentration of the dispersing additive in used oil, and (3) the resin content in the heavy fuel oil.

5.8.1 Particulate Matter Emissions

The amount of mineral matter (the ash-forming propensity) in used oil is higher than ash in the distillate oils and is often higher than the amount of mineral matter in residual fuel oil. In addition, used oil typically has substantially higher concentrations of most of the trace elements reported than those concentrations found in the virgin fuel oil. Thus, without controls, higher concentrations of mineral matter and trace metals in the used oil extrapolate to higher emission levels of total PM and trace metals in the flue gas emissions.

Low-efficiency pretreatment steps, such as large particle removal with screens or coarse filters, are commonly preferred procedures for oil to be used in oil-fired boilers. Reductions in total PM emissions are expected from these techniques, but little or no effect will be noticed in the level of small particles that are less than 10 μm in size (PM_{-10}).

5.8.2 Sulfur Dioxide Emissions

Lubricant-derived sulfur emissions are under increased scrutiny because of their potential to affect emissions subject to environmental air quality standards. Typical sulfur contents of used oil, distillate fuel oil, and residual fuel oil are on the order of 0.5% w/w (used oil), 0.25% w/w (distillate fuel oil), and 0.5% to 1.0% w/w (residual fuel oil). These data indicate that uncontrolled sulfur dioxide emissions will increase when used oil is substituted for distillate fuel oil, but will decrease when used oil is substituted for residual fuel oil.

Combustors that already burn distillate or residual oils are those that are most amenable for fuel substitution with used oil or with used oil added to the virgin fuel oil.

5.8.3 Chlorinated Organic Emissions

High levels of halogenated solvents are often found in used oil as a result of inadvertent or deliberate mixing of the contaminant solvents with the used oils. Although high-efficiency combustors can destroy almost all of these halogenated solvents, combusting used oil in lower-efficiency equipment can result in their partial release to the environment in the emissions stream. Also, the halogenated solvents potentially increase the level of hydrochloric acid (HCl) in the emissions stream.

5.8.4 Other Organic Emissions

The flue gases from used oil combustion need to be monitored for organic compounds other than chlorinated solvents. At parts per million by weight (ppm w/w) levels, some of the 170 organic compounds and organic classifications listed as hazardous under Title III of the U.S. Clean Air Act have been found in used oil. Benzene and toluene have been reported at concentrations of more than 5% v/v. Polychlorinated biphenyls (PCBs) and polychlorinated dibenzo dioxins (dioxins) have been detected in used oil samples. Additionally, these hazardous compounds may be formed in the combustion process as products of incomplete combustion.

5.8.5 Control Technologies

As discussed in Section 5.7, emissions from a used oil combustion process can be controlled by a combination of pretreatment processes to remove the potential pollutants before the combustion process and in subsequent emissions control equipment on the material streams leaving the combustor. These treatments will increase the cost of the used oil and thus cost considerations might impose limitations on the use of used oil as a fuel in many cases, particularly where stringent environmental protection legislation is in place.

5.8.6 Conclusions

Used oil can serve as a substitute fuel for combustors designed to burn distillate or residual oils. Although the recovery of the heat energy in the used oil by using it as a fuel is a simple application (in principle), the contamination of the oil by undesirable components during its use makes it necessary to either remove the contaminants before its use or to have extensive postcombustion treatments to avoid their emission into the environment. Many jurisdictions around the world are now implementing environmental protection legislation that affects the economics of this application of used oil. In addition, newer studies based on life cycle analysis methods show that there are often strong incentives to favor re-refining of used lubricating oils over its use as a combustion fuel. Because local circumstances, including the fuel that is displaced by the used oil, can have a strong effect on the outcomes of such studies, as will be seen in later chapters, it will be important to carefully analyze a particular case that examines the use of used oil as a combustion fuel.

There are, however, two applications of used oil as a fuel that are still very popular in various countries, and these include combustion in a cement kiln and in hot-mix asphalt plants, both of which operate at high temperatures and have fairly long residence times to enable the destruction of most of the undesirable chemical compounds in the oil, as well as efficient cleaning processes on the gaseous emission streams.

5.9 COMBUSTION IN CEMENT KILNS

High temperatures are required in the furnaces at cement manufacturing works to convert the raw materials into cement. Polluting substances such as polynuclear aromatic hydrocarbons, chlorinated hydrocarbons, and heavy metals are destroyed in cement production furnaces. Therefore, they offer an ideal situation for energy recovery from used oils in conditions that are respectful of the environment.

Used engine oil can be considered as a good fuel, unfortunately polluted by additive by-products, unburned particles from internal combustion engines, metallic particles from wear, etc. The oil, owing to its initial refining, is a product free from the petroleum residual fraction that characterizes heavy fuel oils and that is responsible for unburned carbon emissions. The oil burns with a bright flame and has a high heat value and thus cement plants may be an outlet for used oil combustion. The cement industry predominantly uses this fuel, which, in addition, is very easy to handle. Furthermore, the locations of the cement plants may allow a balanced distribution of used engine oil disposal in any country, depending on local circumstances.

With combustion of the used oil in a suitable cement kiln, there is apparently no appreciable increase of particle emissions into the atmosphere, particularly of organic, dioxin, furan, and other compounds. The alkalinity of the raw materials neutralizes compounds such as sulfur oxides, nitrogen oxides, and hydrogen chloride, and emissions of these gases into the atmosphere are therefore reduced. Other smaller particles are collected by electrostatic precipitation or filtering and then returned to the furnaces. The resultant ash of incombustible compounds (heavy metals in the used oils) is subjected to encapsulation processes.

In fact, with respect to current environmental issues, incineration in cement kilns seems to be more favorable than other forms of combustion. However, the primary reason for this is the substitution of the use of coal and pet coke, which have a relatively high carbon content and low heat value, as the primary fuel in cement kilns. The choice of fuel used is not, however, limited to technical reasons but is related to fuel prices on the world market.

5.10 COMBUSTION IN HOT-MIX ASPHALT PLANTS

Hot-mix asphalt paving materials are a mixture of size-graded, high-quality aggregate (which can include reclaimed asphalt pavement), and liquid asphalt cement, which is heated and mixed in measured quantities to produce hot-mix asphalt. Aggregate and reclaimed asphalt pavement (if used) constitute more than 92% w/w of the total mixture. Aside from the amount and grade of asphalt cement used, mix characteristics are determined by the relative amounts and types of aggregate and reclaimed asphalt pavement used.

Hot-mix asphalt paving materials can be manufactured by (1) batch mix plants, (2) continuous mix (mix outside dryer drum) plants, (3) parallel flow drum mix plants, and (4) counterflow drum mix plants. The plant can be constructed as a permanent plant, a skid-mounted (easily relocated) plant, or a portable plant. Most plants have the capability to use either gaseous fuels (natural gas) or fuel oil.

The fine PM_{-10} particulates and dust, which also includes fine solids from the drying of the aggregate in the asphalt plant, can be controlled using bag filters.

As in the case where used oil is combusted in a cement plant, much of the undesirable components in the oil end up in the asphalt product, and as long as they are shown to not leach out into the environment, this can be an acceptable disposal method that avoids the emission of these chemicals into the environment.

REFERENCES

ASTM D130. 2012. Standard Test Method for Corrosiveness to Copper from Petroleum Products by Copper Strip Test. *Annual Book of Standards*. American Society for Testing and Materials, West Conshohocken, Pennsylvania.

ASTM D396. 2012. Standard Specification for Fuel Oils. *Annual Book of Standards*. American Society for Testing and Materials, West Conshohocken, Pennsylvania.

ASTM D4740. 2012. Standard Test Method for Cleanliness and Compatibility of Residual Fuel by Spot Test. *Annual Book of Standards*. American Society for Testing and Materials, West Conshohocken, Pennsylvania.

ASTM D5800. 2012. Standard Test Method for Evaporation Loss of Lubricating Oils by the Noack Method. *Annual Book of Standards*. American Society for Testing and Materials, West Conshohocken, Pennsylvania.

ASTM D6448. 2012. Standard Specification for Industrial Burner Fuels from Used Lubricating Oils. *Annual Book of Standards*. American Society for Testing and Materials, West Conshohocken, Pennsylvania.

Boughton, R., and Horvath, A. 2004. Environmental Assessment of Used Oil Management Methods. *Environmental Science & Technology*, 38(2): 353–358.

Gómez-Rico, M.F., Martín-Gullón, I., Fullana, A., Conesa, J.A., and Font, R. 2003. Pyrolysis and Combustion Kinetics and Emissions of Waste Lube Oils. *Journal of Analytical and Applied Pyrolysis*, 68–69: 527–546.

Gary, J.G., Handwerk, G.E., and Kaiser, M.J. 2007. *Petroleum Refining: Technology and Economics*, 5th Edition. CRC Press, Taylor & Francis Group, Boca Raton, Florida.

Hsu, C.S., and Robinson, P.R. (Editors). 2006. *Practical Advances in Petroleum Processing*, Volume 1 and Volume 2. Springer Science, New York.

Kirk, R.E., and Othmer, D.F. 1996. *Kirk-Othmer Encyclopedia of Chemical Technology*, 4th Edition, Volume 21. John Wiley & Sons Inc., New York.

Martin, C.J., and Hicks, L. 2010. Burner, Heating, and Lighting Fuels, Chapter 6. In: *Significance of Tests for Petroleum Products*, 8th Edition, S.J. Rand (Editor). ASTM International, West Conshohocken, Pennsylvanian, pp. 65–79.

Mokhatab, S., Poe, W.A., and Speight, J.G. 2006. *Handbook of Natural Gas Transmission and Processing*. Elsevier, Amsterdam, Netherlands.

Mushrush, G.W., and Speight, J.G. 1995. *Petroleum Products: Instability and Incompatibility.* Taylor & Francis, Washington, DC.

Speight, J.G. 2001. *Handbook of Petroleum Analysis.* John Wiley & Sons Inc., Hoboken, New Jersey.

Speight, J.G. 2002. *Handbook of Petroleum Product Analysis.* John Wiley & Sons Inc., Hoboken, New Jersey.

Speight, J.G., and Ozum, B. 2002. *Petroleum Refining Processes.* Marcel Dekker Inc., New York.

Speight, J.G. 2007. *Natural Gas: A Basic Handbook.* GPC Books, Gulf Publishing Company, Houston, Texas.

Speight, J.G. 2008. *Synthetic Fuels Handbook: Properties, Processes, and Performance.* McGraw-Hill, New York.

Speight, J.G. (Editor). 2011. *The Biofuels Handbook.* Royal Society of Chemistry, London, United Kingdom.

Speight, J.G. 2014. *The Chemistry and Technology of Petroleum*, 5th Edition. CRC Press, Taylor and Francis Group, Boca Raton, Florida.

U.S. Environmental Protection Agency. 2008a. Code of Federal Regulations: Standards for the Management of Used Oil. Available at http://www.gpoaccess.gov/index.html.

U.S. Environmental Protection Agency. 2008b. Managing Used Oil: Advice for Small Businesses. Available at http://www.epa.gov/epaoswer/hazwaste/usedoil/usedoil.htm.

6 Alternative Processing Options

6.1 INTRODUCTION

Used lubricating oil (Chapter 3) is exactly what its name implies—any petroleum-based lubricating oil or synthetic lubricating oil that has been used. During normal use, impurities such as dirt, metal scrapings, water, or chemicals can get mixed in with the oil so that, in time, the oil no longer performs well. Eventually, this used oil must be replaced with virgin lubricating oil or re-refined lubricating oil to do the job correctly because used lubricating oil (1) is insoluble, persistent, and could contain toxic chemicals and heavy metals; (2) is slow to degrade; and (3) is a major source of oil contamination of waterways and can result in the pollution of drinking water sources.

Thus, the environmentally conscious design of processes and products is now increasingly viewed as an integral strategy in the sustainable development of new refining and chemical processes. Although re-refining of spent lubricating oils such as used motor oils has been practiced with varying technical and commercial success over the past several decades (Harrison 1994), but with questionable economics (Wolfe 1992), a sustainable processing technology is now a necessity (Chapter 7), and poor onstream efficiency, inconsistent product quality, and careless management of feedstock contaminants and by-products have often resulted in widespread environmental problems and poor economics. In addition, the economic aspects of the recycling (re-refining) process are now much more favorable (Chapter 8).

Thus, it is essential that recycling processes using nontoxic and cost-effective materials can be an optimal solution. For example, clay-based processes have been used as recycling methods for used lubricating oil during the past several decades; however, these processes do have disadvantages: (1) they produce a large quantity of pollutants, (2) they are not always suitable for modern multigrade oils, and (3) they may not always produce complete removal of asphaltic impurities.

The management of used lubricating oil is particularly important because of the large quantities generated globally; their potential for direct reuse, reprocessing, reclamation, and regeneration; and because they may cause detrimental effects on the environment if not properly handled, treated, or disposed of. Used lubricating oils and other oils represent a significant portion of the volume of organic waste liquids generated worldwide. The three most important aspects of used oils in this context are contaminant content, energy value, and hydrocarbon properties.

Conventional base oil production methods are energy-intensive, consume a diminishing fossil fuel resource, and place a large burden on the environment. The current trend of including higher percentages of hydrocracked base oils or ethylene-based synthetic products (or both) in the lubricant blend further increases the overall life cycle burden of the finished product. Spent lubricating oil containing a large quantity of high-viscosity index (and low pour point) base oil represents a valuable resource that should be properly managed. Although simple blending of used lubricating oil with low-quality fuel oil will recover the energetic value of this material (Chapter 5), the latent energy value of an engineered material containing very low amounts of aromatic constituents and waxy constituents is not recovered and recycling or preservation for reuse is becoming a major part in the process of lubricating oil disposal.

However, a used oil processing program must be based on used oil analysis (Chapter 3). In fact, analysis of lubricating oil in service should be a part of any maintenance program to give a fast and accurate assessment of the condition of the oil. Effective monitoring of lubricating oil in service allows maintenance to be scheduled efficiently, minimizing the risk of damage to expensive plants and avoiding unscheduled downtime. In this case, by the time the oil has served a useful life, the condition of the oil for reprocessing is not restricted to any form of guesswork. In addition, analysis of the wear debris will not only provide important information about the condition of the internal parts of a machine or engine but also provide information about the condition of the lubricant itself and the steps necessary for processing. From this information, the processing steps that are necessary to yield a product that will meet base oil specifications, such as viscosity, water content, particulate matter content, as well as the presence of undesirable chemical compounds, can be deduced.

Undesirable chemical compounds arise from the presence of the original additives that are added to base oil to provide new and desirable properties not originally present in the base oil (Chapter 1). Some of the most common additives include oxidation inhibitors, alkaline buffering agents to increase total base number, corrosion inhibitors, detergents, dispersants, antiwear agents, viscosity index improvers, pour point depressants, emulsifiers, and antifoaming agents.

Once the analytical data are available, used lubricating oil can be recycled or reused in a variety of ways. The first option in the waste management hierarchy is to conserve the original properties of the oil allowing for direct reuse. The second option is to recover its heating value (Table 6.1). It is important, first of all, to recycle the hydrocarbon content of used oils. Re-refining could be seen as one of the preferred methods for disposal of used oil (Ali and Hamdan 1995; Yee et al. 2002). It has the beneficial effect of reducing the consumption of virgin oils. However, it is very sensitive to the scale and economics of the operation (for instance, thousands of barrels per year of used oils would be required to sustain such a refinery operation).

Recycling is the commonly used generic term for the reprocessing, reclaiming, and regeneration (re-refining) of used oils by use of an appropriate selection of physical and chemical methods of treatment. A large range of used oils can be recycled and recovered, either directly in the case of high oil content wastes, or after some form of separation and concentration from high aqueous content materials. Certain types of used oils, lubricants in particular, can be processed allowing for direct reuse. The used oils, after treatment, can be used either as a high energy content, or a clean burning fuel, or a lubricating oil base stock (Chapter 1) comparable to a highly refined virgin oil.

TABLE 6.1
General Features of Some Used Oil Re-Refining Processes

Acid/Clay Process
Proven technology
Can be set-up for very small capacity
Low capital investment—cost-effective for small-scale and tiny-scale plants
Simple process
Simple to operate
Environmental pollution due to generation of acid sludge
Causes corrosion of equipment
Lower yield due to loss of oil in sludge as well as clay
Not compatible with new pollution control regulations
Activated Clay Process
No acid is required
Simple process
Suitable for small capacity plant
High clay consumption, low yield, inconsistent quality
Disposal of large quantity of spent clay is an environmental problem
Dependent on a particular type of clay that may not be readily available
Thin/Wiped Film Evaporator
Suitable for high-capacity plants
Thin film evaporator is capable of operating at high vacuum and normally used for high-value and heat-sensitive products
Does not cause pollution
Produces good quality base oils
Operates at high temperature and high vacuum
Plant has to be of a higher capacity to make it economically viable
Solvent Extraction Process
Propane is used as a solvent to remove additives, metals, and tar
Solvent is recyclable
Does not cause pollution
Produces good quality base oils
Typically operates at higher pressure (10 atm) and ambient temperature (27°C, 81°F)
Can involve operational solvent losses
May not be economical for low-capacity plants
Propane can cause fire and is an explosion hazard

Reprocessing typically involves treatment to remove insoluble contaminants and oxidation products from used oils such as by heating, settling, filtering, dehydrating, and centrifuging. Depending on the quality of the resultant material, this can be followed by blending with base oils and additives to bring the oil back to its original or an equivalent specification. Reprocessed oil is generally returned to its original use.

Reclamation usually involves treatment to separate solids and water from a variety of used oils. The methods used may include heating, filtering, dehydrating, and centrifuging. Reclamation of used oil can give a product of comparable quality to the original, but may contain various contaminants depending on the nature of the process such as heavy metals, by-products of thermal breakdown, and substances associated with specific uses (such as lead, corrosion inhibitors, and polychlorobiphenyl [PCB] derivatives).

Regeneration involves the production of base oils from used oils as a result of processes that remove contaminants, oxidation products, and additives, that is, re-refining involving the production of base oils for the manufacture of lubricating products (Chapter 7). These processes include predistillation, treatment with acids, solvent extraction, contact with activated clay, and hydrotreating. The methods should not be confused with the simpler methods of treating oils, such as those given under reclamation. The regeneration of used oils is widely practiced to obtain the highest degree of contaminant removal leading to the recovery of the oil fraction, which has the maximum viable commercial value. Reclaimed oil is generally used as a fuel or fuel extender.

In fact, the most common method of recycling used oil is to reprocess it into fuel oil. To prepare used oil as a fuel (Chapter 5), the oil is tested for certain types of contaminants, such as excess water, sediments, and PCB derivatives. If the used oil fails this testing, it must be specially treated and managed. In the treatment process, the uncontaminated used oil is slowly heated to cause water to separate and the water is sent to a wastewater treatment plant whereas the oil is filtered and blended with crude oil. This reprocessed oil can be sold as heating fuel oil to many industries, including asphalt plants, cement companies, and steel mills. In many states, reprocessed oil is typically used as a fuel in high-temperature furnaces that melt asphalt for road construction. It is also used in drying ovens for mined clay and for landfill liners.

6.2 REFINERY PROCESSING

Used oils can also be introduced into oil refineries under certain conditions, aiding the manufacture of other refined products. They can also be recycled or reused in a variety of ways. The first option in the waste management hierarchy is to conserve the original properties of the oil allowing for direct reuse. The second option is to recover its heating value—it is important to recycle the hydrocarbon content of used oils. Re-refining could be seen as one of the preferred methods for the disposal of used oil. It has the beneficial effect of reducing the consumption of virgin oils. However, the viability of a re-refining operation needs to be carefully evaluated in each case to ensure a reliable source of feedstock matched to the scale of such a refinery operation, and that the economics of the undertaking are robust enough to withstand the expected price fluctuations that are common in this industry.

However, first and foremost, the composition of used oils is the main element determining its recycle/disposal process options because it could determine the possibility of recycling as a fuel in power plants or as a feedstock for re-refining operations. The main compositional parameters to be considered are (1) water content, (2) heavy metals content, (3) chlorine content, and (4) PCB derivatives content.

6.2.1 General Options

Reprocessing and re-refining involve operations that will separate and remove contaminants in used oil so that this oil becomes suitable for reuse. Contaminants removed in this process will be part of waste streams, which must be disposed of in an environmentally sound manner.

In reprocessing, relatively simple physical/chemical treatments such as settling, dehydration, flash evaporation, filtration, coagulation, and centrifugation are applied to remove the basic contaminants in used oils. The objective is to clean the oil to the extent necessary for less demanding applications, not to produce a product comparable to virgin oil. Direct reprocessing is not feasible for mixed oils; therefore, at source, segregation of used oil stocks is essential. Reprocessed oils are most commonly used in industrial applications.

The other option for used oil is re-refining it to return it for use as an automotive or industrial lubricant. During this process, hazardous materials are separated from the oil and sent to the proper hazardous waste facilities. The main products of re-refining are diesel fuel, high-quality and low-quality lubricants, and heavy fuel oils. Approximately 65% v/v of the used oil can be re-refined to have a composition and properties suitable for use as base oil. The remaining oil and waste by-products can be used as a fuel oil or asphalt extender. It takes 42 US gallons (1 US barrel or 159 L) of crude oil to produce the same amount of lubricating oil that can be obtained by re-refining 1 US gallon (3.785 L) of used oil. In addition, re-refining uses as little as one-third of the energy as refining new oil to produce a base oil, and this process allows the oil to be recycled many times without losing its lubricating quality.

However, re-refining requires modern processes (Chapter 7), which are expensive to operate when all safety and environmental considerations are included in the overall operating system. In most re-refining processes, a continuous feed of used oil is heated and, in stages, it is dewatered and vacuum-distilled into separate grades of distilled oil. These oils may then be hydrotreated to produce a fine, clear product. The by-products that have marginal value include distillation bottoms (used as an asphalt extender or in fuel oil blending) and demetallized filter cakes (used as road base material). The remainder of the material is a residue or waste stream, such as acid tar, spent clay, centrifuge sludge, and process water, which are directed to treatment or disposal.

Apart from economic considerations (Chapter 8), oil regeneration technologies depend to some degree on the quality of the used oil and particularly on the oil not containing significant concentrations of more difficult-to-process oil products such as heavier fuel oils or chlorinated hydrocarbons. The presence of such materials can seriously affect the technical performance of the regeneration process and its ability to produce lubricating oils or similar products of sufficiently good quality.

6.2.2 Conversion to Lubricating Oils

The four most commonly used re-refining technologies, with respect to aiming at ensuring optimal product yield, meeting utility and energy requirements, limiting hazardous chemicals used and waste volumes produced are (1) the acid/clay re-refining process, (2) the vacuum distillation/clay process, (3) the vacuum distillation/hydrotreating processes (hydroprocessing), and (4) the solvent extraction/distillation processes.

The *acid/clay process* has a long operational history and it is appropriate for a wide range of circumstances and is thus readily operable in most countries (Table 6.1). However, a number of studies made on ranking of re-refining by-product waste streams in terms of environmental hazards suggest that the acid/clay process is the least environmentally sound of the four main re-refining processes. The principal reason for this is the large quantity of by-product acid tar produced, which presents difficulties in disposal. It is, therefore, highly recommended not to use such a process in cases where there is no or inadequate capacity or facility to treat and dispose of the acid sludge resulting from the process.

Vacuum distillation involves the distillation of oils under subatmospheric pressure, which lowers the necessary operating temperature and reduces problems of thermal breakdown (Table 6.1). The use of wiped film equipment, which allows material with significant solids content to be more readily processed with reduced thermal breakdown, is increasing. Clay with high adsorptive capacity is used to remove impurities such as heavy metals and breakdown products arising in the use of the oil. The clay is frequently used before distillation to provide a cleaner feed and also to give recovered oil a final polish (Table 6.1).

The distillation process involves the removal of volatile materials produced during service and the final separation of the high-boiling distillate from contaminants (bottoms and residuum). The process results in the recovery of high-quality oil (very low ash and sulfur content) and asphalt (bottoms and residuum) that contains heavy metals and other contaminants. The asphalt may be used as an extender for more conventional refinery asphalt but it is necessary to ensure that heavy metals and any other contaminants do not leach from the asphalt while in service (Speight 2005).

If there was a more economically and environmentally sound way for treating acid tar, the overall process could then be operated in a more efficient way. Technical solutions to many of these problems have been developed and are being increasingly applied; however, the commitment of additional resources is seen as necessary to the rapid further commercialization of existing development and to identify new management methods to overcome the environmental problems of the disposal of acid tar.

Solvent extraction/distillation processes use a solvent that is either miscible with the base oil fraction or some of the impurities in the used oil, enabling a separation of the higher molecular weight contaminants (Table 6.1) from the lubricating oil fraction before it is fractionated into various cuts in the distillation process.

In fact, solvent extraction has replaced acid treatment as the method of choice for improving the oxidative stability and viscosity/temperature characteristics of base oils. The solvent selectively dissolves the undesired aromatic components (the extract), leaving the desirable saturated components, especially alkanes, as a separate phase (the raffinate; Rincon et al. 2005a). For example, a mixture of methyl ethyl ketone and 2-propanol has been used as an extracting material for recycling used engine oils (Martins 1997; Rincon et al. 2005b; Shakirullah et al. 2006). Propane has also been used for solvent refining good used oil (Rincon et al. 2003)—propane is capable of dissolving paraffinic or waxy material and intermediately dissolved oxygenated material. The asphaltene constituents are composed of polynuclear aromatic heteroatomic systems and are insoluble in the liquid propane. These properties make propane ideal for recycling the used engine oil, but there are many other issues that have to be considered (Speight and Ozum 2002; Hsu and Robinson 2006; Gary et al. 2007; Speight 2014). Hydrotreating may also be applied to the product oils, but the economics of the more elaborate treatments need to be carefully evaluated.

Catalytic hydroprocessing of used oils provides a commercially viable alternative to high-temperature incineration or chemical treatment. Selective hydrogenation could be utilized to remove contaminants such as PCB derivatives or heavy metals from used oils. Catalytic hydrogenation of contaminated organic waste streams is carried out at moderate temperatures and pressures (Havemann 1978; Puerto-Ferre and Kajdas 1994; Speight and Ozum 2002; Hsu and Robinson 2006; Gary et al. 2007; Speight 2014). The treated organic phase is generally suitable for reuse as a fuel oil. The use of this technology is primarily constrained by economics, although the disposal costs of organochlorine-contaminated oil could be substantially reduced. Thus, the use of this type of technology could have positive economic benefits.

Membrane technology is another method for the regeneration of used lubricating oils. In this method, three types of polymer hollow fiber membranes—polyethersulfone (PES), polyvinylidene fluoride (PVDF), and polyacrylonitrile (PAN)—are used for recycling the used engine oils. The process is carried out at 40°C (104°F) and at low pressure. The process is a continuous operation because it removes metal particles and dust from used engine oil and improves the recovered oils' liquidity and flash point (Dang 1997). The use of modern high-temperature polymeric membranes has many significant advantages over the conventional methods of oil re-refining and also over expensive ceramic and inorganic filter media.

6.3 COGENERATION

The next option after re-refining is the recovery of oils focusing principally upon its use as a fuel. A large volume of used oil is used solely for its energy content, as a secondary or substitute fuel (under controlled combustion conditions). The inherent high energy content of many used oil streams may encourage their direct use as fuels, without any pretreatment and processing, and without any quality control or product specification. Such direct uses do not constitute good practice, unless it can be demonstrated that combustion of the waste can be undertaken in an environmentally sound manner. The use of used oil as fuel is possible in situations in which any contaminants do not present problems upon combustion, or it can be burnt in an environmentally sound manner without modification of the equipment in which it is being burnt. Normally, used oils for use as fuel need to be subjected to treatments involving some form of settlement to remove sludge and suspended matter. Simple treatment of this type can substantially improve the quality of the material by removing sludge and suspended matter, carbon, and to varying degrees, heavy metals.

Although much of the used oil used as a secondary fuel receives only such basic pretreatment, every encouragement should be given to measures that improve the quality and control of this type of activity. Where fuels are to be marketed broadly, it is certainly desirable that used oils be subjected to both source and quality screening, and that products are supplied to a specification, even if only rudimentary. Where activities of this type are subject to a license, permit, or authorization system, conditions should be specified to ensure that a minimum level of control is established, and that equipment for blending, separation, etc., is provided, used as necessary, and maintained properly.

The partial replacement of fuel with used oils is a technique that is widely applied around the world, in particular under controlled conditions in cement kilns. Used oil is an ideal replacement fuel in the process of cement manufacture (Chapter 5). Cement production involves the heating, calcining, and clinkering of blended and ground raw materials, typically limestone and clay or shale with other materials to form a clinker. The cement industry is energy-intensive and the dominant energy consumed in cement manufacture is fuel for the clinker. Heavy metals present in used oils become incorporated in the clinker and then into the final cement structure. Because the contents of the kiln are alkaline, they trap hydrogen chloride, sulfur oxides, and other acid gases formed in the destruction of chlorine and sulfur-containing compounds.

Other opportunities for the use of used oil as a fuel include: (1) in stone quarries for stone-drying purposes; (2) in asphalt-coating plants; and (3) in smelters handling iron, lead, tin, aluminum, and some precious metals. Used oil has also been used as support fuel in chemical incinerators, in coking plants, in brickworks, electricity-generating stations, and steam-raising boilers.

Most oil-fired domestic, commercial, industrial, or utility boilers can burn used oils. However, combustion of used oils that were not subjected to strict product specifications and under uncontrolled conditions can lead to serious pollution of all environmental media. Also, combustion plants will suffer corrosion when oils with halogen contents are burnt. Large industrial and utility boilers are generally considered to pose relatively low environmental risks because of their combustion efficiency, use of consistent quality fuel, and in some cases, pollution control devices (e.g., electrostatic precipitators, bag filters, high-energy venturi scrubbers), tall dispersive chimneys, and their location (often away from high population densities).

The environmental risks associated with burning used oil in any size boiler could be reduced through the application of comprehensive and integrated management strategies including: (1) pretreatment of used oil to meet established quality specifications, such as settling, centrifugation, vacuum distillation, and solvent extraction; (2) dilution of used oil by blending with virgin fuel oil; and (3) installation of flue gas emission control devices.

It must be noted that used oil can be burned in cement kilns without many of the negative air quality effects normally associated with burning used oil in small-sized to medium-sized boilers. Unless boilers or other combustion processes are equipped with burners of high combustion and contaminant destruction efficiencies or with flue gas treatment devices, used oil burning should be strongly discouraged or prohibited if practicable. This implies that efforts should be made to bring adequate treatment or disposal technologies to remote communities and sparsely populated areas in need of such technologies and implement a program providing the necessary infrastructure for establishing used oil collection/storage/transport/treatment systems for such small-volume oil generators.

Used oils have traditionally been directed to a variety of uses other than re-refining and burning; they are, for instance: road oil, raw material in asphalt production, flotation and forming oil, secondary lubricant, pesticide carrier, weed killer, livestock oil, all-purpose cleaner, and vehicle undercoating.

6.3.1 Diesel Engine Coupled with a Generator

A type of cogeneration plant operates with a diesel engine that drives an electric generator in which the exhaust gases and cooling system for the diesel engine are used to generate hot water or low-pressure steam. The fuel for the diesel engine can be a mixture of used oil, which has been subjected to a rudimentary cleaning process such as settling or centrifugation, and heavy fuel oil. As pointed out by Audibert (2006), there is some danger of precipitating asphaltene constituents in this mixture due to the paraffinic nature of the used oil. Another option would be to use a distillate from the used oil as fuel, but this would increase the cost of the fuel.

Examples of diesel engine generator systems include emergency power systems in critical installations such as hospitals, and oil and gas well drilling operations, which operate in remote areas, far from other sources of power. Particularly, in the case of well drilling operations, these tend to be large engines that generate used oil, but also can consume large quantities of diesel fuel. If the used oil, after a simple cleaning or settling operation, is mixed with other diesel fuel, a large diesel engine can operate with this mixed fuel and these can be significant users of the recycled used oil.

6.3.2 Gas Turbine-Steam Turbine

The fuel fed to a gas turbine has to be essentially free of heavy metals and other impurities that could damage the turbine blades. If a used oil is seen as a potential fuel for a gas turbine unit, it will be necessary to subject the oil to a vacuum distillation to produce a cleaned distillate that can then be gasified in the presence of oxygen to form clean synthesis gas, a mixture of carbon monoxide and hydrogen, which can then be burned in the gas turbine. This gas turbine can drive an electric generator, and the exhaust gases exiting the gas turbine at high temperatures can be used to generate steam, which can also be used to provide mechanical or heat energy for further use in the plant. This cogeneration improves the efficiency of the overall process considerably (Audibert 2006).

6.4 REGENERATION RESIDUES

The early re-refining acid/clay process produced large quantities of toxic acid sludge that was difficult to dispose of in an environmentally acceptable way. Fortunately, this process has been banned in many jurisdictions around the world, and has been replaced by processes in which the waste streams generated are carefully managed to ensure that any discharges to the environment meet the strictest modern requirements. The modern re-refining plants can claim to be environmentally responsible producers of a product that, in many cases, requires less energy to produce than the production of the same product from the original raw materials.

This being the case, however, there are still significant quantities of potentially toxic and environmentally damaging chemicals that are found in the re-refinery or fuel production facilities that have to be handled in an appropriate manner to protect both the personnel in the plant and surrounding areas, as well as the end users of the products.

Probably the most difficult stream to handle in a re-refining operation is the one containing the heavy metal and other contaminant residues that are separated from the lubricating oil fractions. However, the gaseous and aqueous emission streams from such a plant generally also have to be treated and carefully monitored to ensure that potentially harmful or undesirable substances are not emitted in an uncontrolled manner.

Attempts have also been made to dispose of the residue containing the heavy metals by incorporating it in various building materials such as bricks or concrete building blocks. This is a legitimate method, provided the metals are permanently bound in the matrix and do not leach out over time when exposed to water or damp conditions.

6.4.1 Analyses

It is critically important that all incoming used oil to a re-refinery or fuel production facility be analyzed for a number of components. Foremost among these are the PCBs that can still be found in some recycled transformer oils and which could find their way into a used oil collection system. Legislation in most jurisdictions now sets a limit to the PCB level that can be treated in a facility that is not licensed as a hazardous waste treatment site: these facilities generally have to meet certain requirements regarding high-temperature incineration with a sufficiently long residence time to destroy the PCBs. Clearly, a re-refining plant has to avoid accepting used oils that contain PCBs that exceed the regulated levels.

The second important analysis of the incoming used oil is for the halogen levels, which also have to be controlled to avoid the formation of undesirable compounds in the process and the accelerated corrosion caused by chlorides in particular.

The water content of the used oil feed is also of interest, particularly if radiator fluid, which contains glycols, has been mixed in with the used lubricating oil in the waste tank in a service station. Some re-refining operations operate at a slightly higher initial flash temperature to drive off the glycol with the water, and the glycol can then be separated and recycled as a separate stream.

The analyses required for the product lubricating oil streams are the usual base oil tests (Chapter 2), of which the viscosity and viscosity index are the prime distinguishing tests, and the tests required for the fuel products will depend on the particular type of fuel that is produced.

6.4.2 Road Oiling and Asphalt Additive

Used oil has been applied to gravel roads as a dust suppressant for many years. It has been used most commonly in rural areas, which have a high proportion of unpaved roads and are located some distance from other used oil markets (burning and re-refining). Although some road oiling is still common in many areas of the world, its popularity has declined in recent years because of reductions in the proportion of unpaved roadways, problems due to contaminants in used oils (such as PCBs, polychlorinated dibenzodioxins [PCDDs], polychlorinated dibenzofurans [PCDFs]), competition from other used oil end uses (re-refining), availability of alternative dust suppression substitutes (calcium chloride and surfactants), and precluding environmental regulations. In fact, studies suggest that the potential effects generated by road oiling on health and the environment are severe enough to discourage or prohibit such practice. Road oiling with contaminated oil has led to very serious environmental problems, mostly due to leaching and run-off of hazardous constituents of the oil.

Used oils have been used occasionally as cutting stocks and extenders in the manufacture of asphalt. Because used oil constituents are essentially insoluble in water, potential contaminants are coated with viscous asphaltic materials and incorporated into the final product. Leaching of significant contaminant concentrations from finished asphalt roads and roofs is considered unlikely; however, the potential effects of using used oils in asphalt production should be evaluated on a site-specific or region-specific basis. The hot coating of road stones with asphalt has given rise to environmental problems, this leading in certain circumstances to the setting of a 10 parts per million (ppm) limit for PCB derivatives in used oil–based fuels being used for this purpose. As a waste management practice, it should be discouraged. It does not significantly affect the reduction of the volume of wastes that need to be disposed of. A number of countries are, in fact, prohibiting such a use.

6.5 ENVIRONMENTAL MANAGEMENT

Used motor oil contains numerous toxic substances, including polycyclic aromatic hydrocarbons, which are known to cause cancer. In addition, tiny pieces of metal from engine wear and tear, such as lead, zinc, and arsenic, make their way into the oil, further contributing to the polluting potential of used motor oil. Motor oil is exposed to heat and oxygen during engine combustion, which changes its chemical makeup. Because spent motor oil is heavy and sticky, and contains an extensive concentrated cocktail of toxic compounds, it can build up and persist in the environment for years.

Once motor oil leaks from an engine, it has the potential to travel long distances, and most leaked motor oil eventually makes its way into waterways in the form of runoff. Once it reaches waterways, used motor oil is toxic to plants and animals living in the water, and its film can impair natural processes, such as oxygen replenishment and photosynthesis. Used motor oil can also pollute soil and drinking water. According to the U.S. Environmental Protection Agency, 1 gallon of used motor oil can contaminate 1 million gallons of fresh water. If used motor oil reaches sewage treatment plants, even small amounts—50 to 100 ppm—can foul the water treatment process. Soil becomes less productive when exposed to used motor oil.

The environmental effects associated with the other end uses listed earlier vary from one application to another. The nature and extent of concerns for any given application will depend on the volume of oil used, the operational practices of the companies or individuals involved, and the manner in which the oils are ultimately discharged to the environment. Generally, these practices should be avoided unless it can be demonstrated that environmental risks can be effectively controlled on a site-specific basis.

Any decision process used to select a preferred recycling or reuse option must take into account the fact that the matter needs to be dealt with under existing environmental regulations. The contaminants and environmental/health risks associated will ultimately limit the number of acceptable reuse or recycling options of used oils. Furthermore, the availability of waste management resources (collection, storage, transport, and treatment) will restrict the selection of environmentally sound disposal options (including blending, segregation, gravity separation, and strategic storage for the preparation of optimal feedstock blend). Finally, economic viability (which will be affected by transport costs, end uses, pollution abatement investment, etc.), social acceptability considerations as well as regulations (regulations and other economic instruments could be developed and implemented to assist in sustaining the market) would form part of the analytic procedure.

In fact, if used motor oil is contaminated to such an extent with hazardous constituents, it is regulated as hazardous waste in the United States and in many other countries. Composed of a mixture of several hundred organic chemicals, motor oil becomes contaminated with heavy metals and chemical additives as it is used. Recycling used motor oil helps protect the environment while conserving a valuable petroleum resource. As noted previously, used lubricating oil that is not properly disposed of and finds its way into either surface or groundwater can cause significant damage at low concentrations. One part per million (1 ppm) can make water unsafe to drink; 35 ppm can produce a visible oil slick that damages aquatic life, including fish and shellfish; and 50 ppm can foul a wastewater treatment plant.

The harmful effects of used motor oil in an aquatic environment arise from both the characteristics of the oil itself and the contaminants being picked up during use. Aquatic animals can be smothered either directly when covered with oil or indirectly through reduced resistance to infection and disease. Plants can also be killed or their growth stunted. In the long term, toxic substances are released as the oil breaks down, exposing aquatic plants and animals to potentially carcinogenic compounds.

Although some of the reuse and recycling alternatives are technically sound, the costs involved in the re-refining process and combustion of used oils can be very high. Source reduction, in this regard, should be seen as a primary objective in a strategy for hazardous waste management. It is also obvious that the economics of the reuse and recycling of used oils as a preferred option must be examined before considering final disposal. In certain circumstances, however, for instance during re-refining, the preferred process may not provide an adequate return on investment. Viable and ecologically sound alternatives should, in this particular case, be investigated before considering final disposal options. The direct burning of used oils in conventional combustion devices can create serious pollution problems and, although this can be reduced by fitting pollution abatement equipment, this is not, in most cases, very practicable. The burning in specially designed waste incinerators can diminish these problems; however, the process is very expensive, particularly if there is no provision for energy recovery.

To identify suitable and acceptable reuse and recycling options, a number of criteria need to be considered before deciding on which treatment technology to select; these include: (1) the extent to which used oil can be treated to obtain specific products; (2) the potential of harm to human health and the environment; (3) the economic balance and market opportunities; (4) transport requirements/costs; (5) the location of treatment facilities; (6) the processing of the hazardous waste contaminants and by-products of the process itself; and last, but certainly not least; (7) worker safety.

The question as to which treatment technology is most appropriate is de facto related to regulations, availability of facilities and their location, and in most cases to a significant degree on the market mechanism (competitive uses of the products).

Finally, in many situations, a number of supplementary factors need to be considered in the assessment procedure referred to previously to comply with the requirements for environmentally sound and efficient management of hazardous waste. For example, in the case of transboundary movements (interstate transportation) in the United States, consideration of the standards for the environmentally sound management of the recovery operation in the state of export and the state of import should be looked into. Before entering into the authorization (or permitting) process for a particular reuse, recycling, or recovery operation, the following elements should be taken into consideration: (1) site selection, (2) design standards for facilities, (3) training of operators of the facility, (4) environmental assessment, (5) operation/discharge standards, (6) monitoring and control, (7) emergency and contingency plans, (8) records and record-keeping, and (9) decommissioning.

Finally, it should be noted that considerations on crude oil and energy savings can change with time and location, causing the optimal used oil disposal route to change accordingly. An effective collection system is the key step in a responsible used oil management system and is the step that provides the principal environmental benefits. In addition, the performance qualities of some commercial re-refined oils may not match the performance qualities of the equivalent virgin base oils and hydrotreatment will be required for used oil regeneration to produce base oils of similar quality to their virgin counterparts. Also, if the re-refined oils contain toxic compounds that can be hazardous to human health, hydrotreatment of the regenerated base oils will achieve equivalent characteristics to the virgin base oils.

REFERENCES

Ali, M.F., and Hamdan, A.J. 1995. Techno-Economic Evaluation of Waste Lube Oil Re-Refining. *International Journal of Production Economics*, 42: 263–273.

Audibert, F. 2006. *Waste Engine Oil: Rerefining and Energy Recovery*. Elsevier B.V., Amsterdam, The Netherlands.

Dang, C.S. 1997. Rerefining of Used Oils: A Review of Commercial Processes. *Tribotest Journal*, 3: 445–457.

Gary, J.G., Handwerk, G.E., and Kaiser, M.J. 2007. *Petroleum Refining: Technology and Economics*, 5th Edition. CRC Press, Boca Raton, Florida.

Harrison, C. 1994. The Engineering Aspects of a Used Oil Recycling Project. *Waste Management*, 14(3–4): 231–235.

Havemann, R. 1978. The KTI Used Oil Re-Refining Process. Proceedings, 3rd International Conference of Used Oil Recovery & Reuse, Houston, Texas, October 16–18.

Hsu, C.S., and Robinson, P.R. (Editors). 2006. *Practical Advances in Petroleum Processing*, Volume 1 and Volume 2. Springer Science, New York.

Martins, J.P. 1997. The Extraction-Flocculation Re-Refining Lubricating Oil Process Using Ternary Organic Solvents. *Industrial & Engineering Chemistry Research*, 36: 3854–3858.

Puerto-Ferre, E., and Kajdas, C. 1994. Clean Technology for Recycling Waste Lubricating Oils. Proceedings, 9th International Colloquium, Ecological and Economic Aspects of Tribology, Esslingen, Germany, January 14–16.

Rincon, J., Canizares, P., Garcia, M.T., and Garcia, I. 2003. Regeneration of Used Lubricant Oil by Propane Extraction. *Industrial & Engineering Chemistry Research*, 42: 4867–4873.

Rincon, J., Canizares, P., and Garcia, M.T. 2005a. Regeneration of Used Lubricating Oil by Polar Solvent Extraction. *Industrial & Engineering Chemistry Research*, 44: 43–73.

Rincon, J., Canizares, P., and Garcia, M.T. 2005b. Waste Oil Recycling Using Mixtures of Polar Solvents. *Industrial & Engineering Chemistry Research*, 44: 7854–7859.

Shakirullah, M., Ahmed, I., Saeed, M., Khan, M.A., Rehman, H., Ishaq, M., and Shah, A.A. 2006. Environmentally Friendly Recovery and Characterization of Oil from Used Engine Lubricants. *Journal of Chinese Chemical Society*, 53: 335–342.

Speight, J.G. 2005. *Environmental Analysis and Technology for the Refining Industry*. John Wiley & Sons Inc., Hoboken, New Jersey.

Speight, J.G. 2014. *The Chemistry and Technology of Petroleum*, 5th Edition. CRC Press, Boca Raton, Florida.

Speight, J.G., and Ozum, B. 2002. *Petroleum Refining Processes*. Marcel Dekker Inc., New York.

Wolfe, P.R. 1992. Economics of Used Oil Recycling: Still Slippery. *Resource Recycling*, 11(9): 28–35.

Yee, F.C., Yunus, R.M., and Sin, T.S. 2002. Modeling and Simulation of Used Lubricant Oil Re-Refining Process. Proceedings, 2nd World Engineering Congress Sarawak, Malaysia, July 22–26, pp. 1–6.

7 Refining Processes for Used Oil Recycling to Base Oils or Fuels

7.1 INTRODUCTION

The objective of re-refining used lubricating oil is to provide a source of valuable base oil that is suffering from decreased availability. In the re-refining process, used lubricating (mainly automotive) oil undergoes an extensive series of processes to remove contaminants, distill the used oil, and reform the oil molecules to produce new base oils that have the same performance characteristics as virgin base oils made from crude oil (Ott et al. 2010).

It is important to distinguish between re-refined base oils and recycled base oils because recycled base oils using the acid/clay and the clay filtration processes are not the same as re-refined base oils. For example, it is not always possible to produce typical base oil viscosity fractions/grades, and recycled base oils may not have the same performance and oxidation stability because of variable quality, depending on the feedstock.

Furthermore, re-refining technology has advanced from the early processes when used oil was reclaimed by removing water, dirt, sludge, and some volatile compounds (Staengl 2009). Used oil that is re-refined undergoes a similar manufacturing process as new oil made from crude. Typically, key components of the modern re-refining process include vacuum distillation and hydrotreating. The processes used to re-refine used oil are also very different than the methods used to reclaim oil. Reclaimed oil is used oil that has been filtered to remove dirt, fuel, water, and any other heavy particles (Chapter 4). Reclaimed oil cannot pass tests to meet American Petroleum Institute (API) standards or vehicle manufacturer requirements. On the other hand, re-refined oil is used oil that has undergone an extensive process that removes contaminants in addition to water, dirt, and lower-boiling fuel constituents.

Re-refining used oil is complicated by the fact that the oil is difficult to characterize. It is very much dependent on the source of the used oil, the types of used oil, how it was collected, and the intended end-use of the oil. The highly variable quality of used oils drives the complexity of the facilities necessary to treat it for reuse. Used oil is made up of base oils and various additives that impart the desired quality characteristics to the finished product. High-tech motor oil, for example, contains as much as 30% v/v additive components in the finished product.

The product from the re-refining of used oil must meet the same specifications as unused crankcase oil for the re-refined product to be of value. There are a number of standards and specifications that are used to ensure the quality of lubricating oils (American Society for Testing Material [ASTM] Method D4485). In this standard, several other ASTM methods are used to determine the properties of lubricating oil and whether or not the oil meets specifications (Chapter 4).

Thus, the processes that have been developed for treating the used oil may be grouped into the following categories that have common elements:

- The acid/clay processes
- Simple distillation processes
- The distillation/clay and distillation/hydrogenation processes
- The solvent extraction/distillation/hydrogenation processes
- The hydrogenation/distillation processes
- Filtration processes
- Coking processes
- Other processes

In the following sections, some of these processes will be reviewed, together with some of the results that have been obtained with them. The review will not include all the processes that have been proposed in the past because there have been many processes suggested and described over the years, but most of the techniques that have been proposed that have significant variations will be included in this review.

7.2 THE ACID/CLAY PROCESSES

The removal of certain undesirable components from oil by contacting it with sulfuric acid has been known and practised for a long time (Chapter 4; Speight and Ozum 2002; Hsu and Robinson 2006; Gary et al. 2007; Speight 2014). Olefins, aromatics, mercaptans, nitrogen compounds, resins, and asphaltic materials all dissolve to a varying degree in sulfuric acid, depending on the temperature and concentration of the acid, whereas paraffins and naphthenes are only slightly reactive with high concentrations of acid at near atmospheric temperatures. This forms the starting point for the acid/clay process and, which for many years since 1935 (Hartmann 2008), was the most popular process for treating used oil.

Although many plants continue to operate in various parts of the world, this process has fallen into disfavor in many countries due to the environmental problems associated with the waste streams generated (Audibert 2006). Also, in the 1970s, when difficulties were encountered with this process in treating used oils containing some of the modern additives, modifications were made to reduce the amount of acid used (Audibert 2006), and this has been referred to as a modified Meinken process.

There are several variations of the acid/clay process, but a typical application is shown in Figure 7.1, in which the used oil is first filtered to remove coarse particles and then enters a distillation unit to drive off entrained water and some of the light ends at temperatures between 120°C and 180°C (248°F and 356°F). It is then cooled

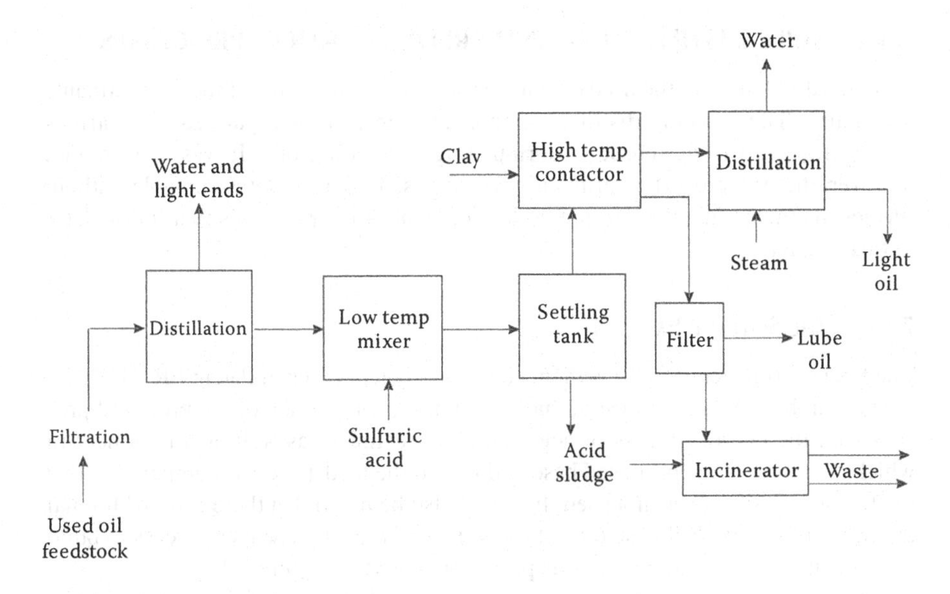

FIGURE 7.1 The acid/clay process.

and contacted with concentrated sulfuric acid at near atmospheric temperature and the acidic sludge that is formed is allowed to settle in settling tanks. Disposal of this sludge has proved to be problematic in many instances: it may be treated with sodium hydroxide or calcium carbonate to neutralize the acid, or it may be taken to an incinerator, which results in an ash and gaseous exhaust emissions, but with significant corrosion levels. One of the improvements made in the Meinken process was to install a falling film distillation unit after the oil dehydration step to produce a distillate that required less acid and, hence, reduced the amount of acid sludge generated (Audibert 2006).

The liquid stream from the top of the separator is led to a reactor vessel where bentonite clay is mixed with it and the temperature is increased to approximately 270°C (518°F; Audibert 2006) to clarify the liquid. The underflow from this reactor is cooled and then filtered to produce a liquid product stream of lube base oil and a filter cake that contains the clay, oil residue, and some of the heavy metal components retained by the clay. Disposal of this filter cake can also be problematic and studies have been reported in Brazil (Teixeira and de A. Santos 2008) where it has been incorporated in ceramic bricks as a disposal method, or it can also be incinerated. The overflow from the reactor can be steam-distilled under vacuum to separate a light oil product from a water stream.

Due to the relatively high losses of oil in the acid sludge and the clay filter cake, the recovery of lube oil is usually lower than in other processes and ranges from 60% to 65% of the water-free oil feedstock (Monier and Labouze 2001). The improvements made in the Meinken process have reduced acid consumption to approximately 3% w/w of the water-free oil feedstock, and the clay consumption to approximately 3.5% w/w of the water-free oil feedstock (Audibert 2006).

7.3 SIMPLE DISTILLATION AND PREDISTILLATION PROCESSES

A simple distillation of the used oil can be used to separate some of the contaminants it contains. The resulting distilled components can then be separated into various boiling fractions to be reused in formulating lubricating oils. It was soon found, however, that the properties of the recovered base oils were not acceptable without further treatment, and this led to the development of the processes that follow later in this chapter.

7.3.1 The Sotulub Process

The Société Tunisienne de Lubrifiants (Sotulub) patent (French Patent FR 93 03275, dated March 3, 1993) is useful in that it includes a comparison of the proposed process with the earlier process of acid and clay treatment, as well as a treatment in which the effect of the proposed base addition to the feedstock was compared with a treatment in which it was not used. It should also be noted that there is no additional treatment of the product base oils, making it a significantly cheaper process to build and operate. An embodiment of this process is shown in Figure 7.2.

The incoming oil is first tested to eliminate oils that have a high content of gasoline, fatty acids, or chlorine. The used oil that passes the tests is preferably filtered and stored in the feedstock tank before passing to the preheater, which heats the oil to a temperature preferably between 140°C and 160°C (284°F and 320°F). A strong base such as sodium or potassium hydroxide, or a mixture of the two, is then mixed with the stream leaving the preheater and enters a flash separator where the water and light hydrocarbons are flashed off. The amount of strong base added is preferably between 1% and 3% w/w of the pure base for the corresponding oil flow rate. The flow from the bottom of the flash separator passes to the gas oil separator, where the gas oil fraction is distilled off. The base oil fractions and the residue contained in the stream from the

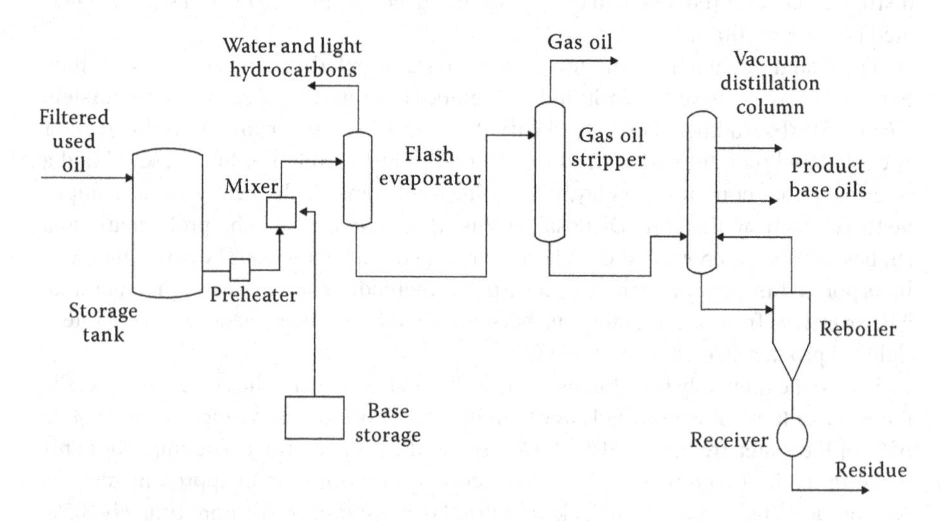

FIGURE 7.2 The Sotulub process. (From Sotulub Patent FR 93 03275.)

bottom of the gas oil column pass to a vacuum distillation column that distills base oil fractions withdrawn as sidestreams from the column, and an underflow that passes to an evaporator that recycles the distillate into the vacuum distillation column, and produces an underflow stream of residue that collects in a receiver vessel before being removed (to be used as tar or bitumen in road-making or as a fuel). It was claimed that good results had been obtained by producing two sidestream base oil cuts, one of 150N and another of 400N to 500N. In these designations, the higher the number, the more viscous the oil and the "N" refers to the word *neutral*, and is a throwback to the early days of refining when residual acidity from the process had to be neutralized.

An alternative arrangement also considered would be to replace the vacuum distillation column and evaporator with a centrifuge to separate the residue from the base oil fractions. A yield or recovery factor of at least 90% of the amounts of oil obtained in the laboratory under ideal conditions, and without consideration of the economic factors, is claimed. The useful products include the base oil fraction, water, residues used in road-making or as a fuel, light hydrocarbons, and gas oil.

Included in the patent document are tables that compare the properties of a feedstock oil, products obtained by distillation in the process without addition of the strong bases, products obtained by the acid/clay process previously used, and products obtained by the new process of this patent. These are shown, respectively, in Tables 7.1, 7.2, 7.3, and 7.4, essentially as they appear in the patent document. The high values of lead in the oils reflect the presence of tetraethyl lead in the gasoline, which was used to increase the octane rating of the fuel at that time.

Audibert (2006) discusses the later Sotulub French patent 96 15380, which modified the previous process by making two or three additions of the strong base at various places in the process. Another variation of the process proposed an oxidation of the oil and adding a strong base to the distillate from the vacuum distillation column before an additional final distillation step, which produces the base oil fractions as sidestreams, and an underflow that is recycled to the feedstock to the vacuum distillation column. The residue, as before, is withdrawn from the bottom of the thin film evaporator fed from the bottom of the vacuum distillation column.

TABLE 7.1
Sotulub Process: Laboratory Distillation of Used Oil
Distillation Yield = 66.45%

Products	Yield (%)	Acidity (TAN) mg KOH/g	Viscosity cSt at 40°C	Color	Appearance
Gas oil	4.06	2.01	2.97	5	
Light cut	23.15	0.75	27.05	<4	Cloudy
Heavy cut	43.30	0.49	74.8	4.5	Cloudy
Residue	19.60	—	—	—	—
Water + gasoline	8.39	—	—	—	—
Losses	1.50	—	—	—	—

Source: Sotulub Patent Fr 93 03275.

TABLE 7.2
Sotulub Process: Processed without Strong Base Pretreatment Distillation Yield = 62.35%

Product	Yield (%)	Acidity (TAN) mg KOH/g	Viscosity cSt at 40°C	Color	Appearance			
Gas oil	6.99	4.70	3.70	5	Cloudy			
Light cut	12.40	0.10	30.57	<4	Cloudy			
Heavy cut	49.95	0.41	75.98	4.5	Cloudy			
Residue	20.27	—	—	—	—			
Water + gasoline + losses (dist.)	10.39	—	—	—	—			
Metals content	**Ba**	**Ca**	**Pb**	**Zn**	**P**	**Cr**	**Fe**	**Si**
Used oil	8	692	1138	694	704	5	84	77
Light cut	<2	<1	6	<1	8	<0.5	<0.5	26
Heavy cut	<2	<1	5	<1	<5	<0.5	<0.5	6
Gas oil	<2	<1	39	2	101	<0.5	8	327
Residue	85	4901	2529	3469	2570	36	652	228

Source: Sotulub Patent Fr 93 03275.

TABLE 7.3
Sotolub Process: Processed Oil with Previous Acid/Clay Process Regeneration Yield = 55.21%

Product	Yield (%)	Acidity (TAN) mg KOH/g	Viscosity cSt at 40°C	Color	Appearance			
Gas oil (without treatment)	6.99	4.70	3.70	5	Cloudy			
Light cut	11.16	0.01	30.04	1.5	C + B			
Heavy cut	44.05	0.05	75.85	3	C + B			
Residue	20.27	—	—	—	—			
Water + gasoline + losses (dist.)	10.39	—	—	—	—			
Losses (chem. treatment)	7.14	—	—	—	—			
Metals content	**Ba**	**Ca**	**Pb**	**Zn**	**P**	**Cr**	**Fe**	**Si**
Light cut	<2	<1	5	<1	<5	<0.5	<0.5	5
Heavy cut	<2	<1	5	<1	<5	<0.5	<0.5	5

Source: Sotulub Patent Fr 93 03275.

TABLE 7.4
Sotulub Process: Used Oil Processed According to Patent FR 93 03275 Regeneration Yield = 62.76%

Product	Yield (%)	Acidity (TAN) mg KOH/g	Viscosity cSt at 40°C	Color	Appearance			
Gas oil	5.41	0.02	6.05	1.5	C + B			
Light cut	12.77	0.01	33.89	2	C + B			
Heavy cut	49.99	0.01	84.91	2.5	C + B			
Residue	20.44	—	—	—	—			
Water + gasoline + losses (dist.)	11.39	—	—	—	—			
Metals content	**Ba**	**Ca**	**Pb**	**Zn**	**P**	**Cr**	**Fe**	**Si**
Light cut	<2	<1	<2	<1	<5	<0.5	<0.5	3
Heavy cut	<2	<1	<2	<1	<5	<0.5	<0.5	3
Gas oil	<2	<1	6	<1	95	<0.5	1	122
Residue	132	7624	2788	3500	2644	40	664	279

Source: Sotulub Patent Fr 93 03275.

7.3.2 The Mohawk Pretreatment Process

Mohawk Lubricants Ltd. of Vancouver, British Columbia constructed a used oil processing plant in North Vancouver based on distillation and solvent extraction in the early 1980s (Canadian Patent CA 2,068,905 dated July 22, 1997; Briggs 2012) before switching to a vacuum distillation and hydrogenation process, which was the first of this type in North America (Newalta 2012; Lube Report 2011). The Mohawk Company was acquired by Newalta Corporation in 2002, which continues to operate one of the two used oil refineries in Canada, the other being in Breslau, Ontario.

In the patent of 1997, a pretreatment of the used oil was proposed to reduce the acidity and plant fouling found in the distillate stream when the used oil was distilled. This pretreatment with an alkali metal component, generally in the form of the hydroxide, but also including the possibility of a salt of any acid weaker than sulfonic such as carbonate, borate, or sulfide, was made while the oil still contained at least 1% of free water before being subjected to the higher temperatures in a distillation process. This could be done with sodium hydroxide, or a mixture of the basic alkali metal chemicals, which are added in a stoichiometric amount to react with the dibasic sulfonates present in the used oil as detergent additives, usually present in the form of zinc, calcium, magnesium, or possibly barium. The treatment chemical required a combination of temperature and reaction time for the reactions to proceed, and an example in a large stirred tank was suggested with a temperature of 82°C (180°F) and a residence time in the tank of 2.5 h.

The mechanism proposed was that the alkali metal replaced, for instance, the dibasic calcium in the calcium sulfonates. When the calcium sulfonates decompose at high temperature in the presence of water, sulfuric acid is produced. However, when sodium sulfonates decompose under similar conditions, sulfuric acid is not one of the products. By testing various treatment chemicals, it was shown that the mechanism of the treatment was not the neutralization of the produced acids, but involved the prevention of acid formation.

7.3.3 The Enprotec Process

The Enprotec process described in the patent of 1998 (US Patent 5,814,207, dated September 29, 1998; Reissued to Avista Resources as US RE38,366, dated December 30, 2003) was based on the use of a cyclonic evaporator to improve on a cyclonic evaporator design, which had been the subject of an earlier patent (WO 91/17804) that contained a series of inverted cones inside the cyclone unit and that claimed to improve the separation of distillate from the liquid. A superheated liquid was injected tangentially into the cylindrical evaporator that operated under vacuum. The 1998 patent described an evaporator, as shown schematically in Figure 7.3, which was essentially empty and had an upper part into which a spray of cooled distillate was injected to condense the vapor stream.

The used oil feeds was combined with a recycled liquid stream from the evaporator to maintain high velocities in the evaporator and this, combined with any grit or particles in the feedstock would tend to scour the internal surface of the evaporator. It was claimed that this configuration and operating mode minimized the deposition of coke on the internal surfaces of the evaporator, an improvement on the units with internal cones, and the spray condenser avoided the condensation of the distillate on a metal surface, thus avoiding the deposition of coke there. To produce a range of base oil products, multiple units of the evaporator as modular entities were proposed, each operating under different conditions of temperature and pressure and arranged in series, as shown in Figure 7.4. The patent did not address the subsequent finishing treatment of the base oil in much detail.

When the cyclonic evaporator and spray condenser is combined with a solvent extraction as a finishing treatment, it is representative of the plants that were built

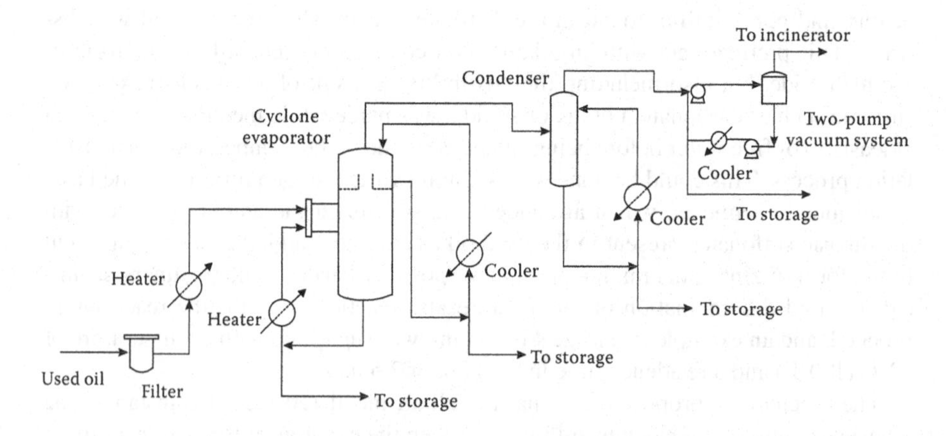

FIGURE 7.3 Simplified Enprotec Process. (From Enprotec US Patent 5,814,207.)

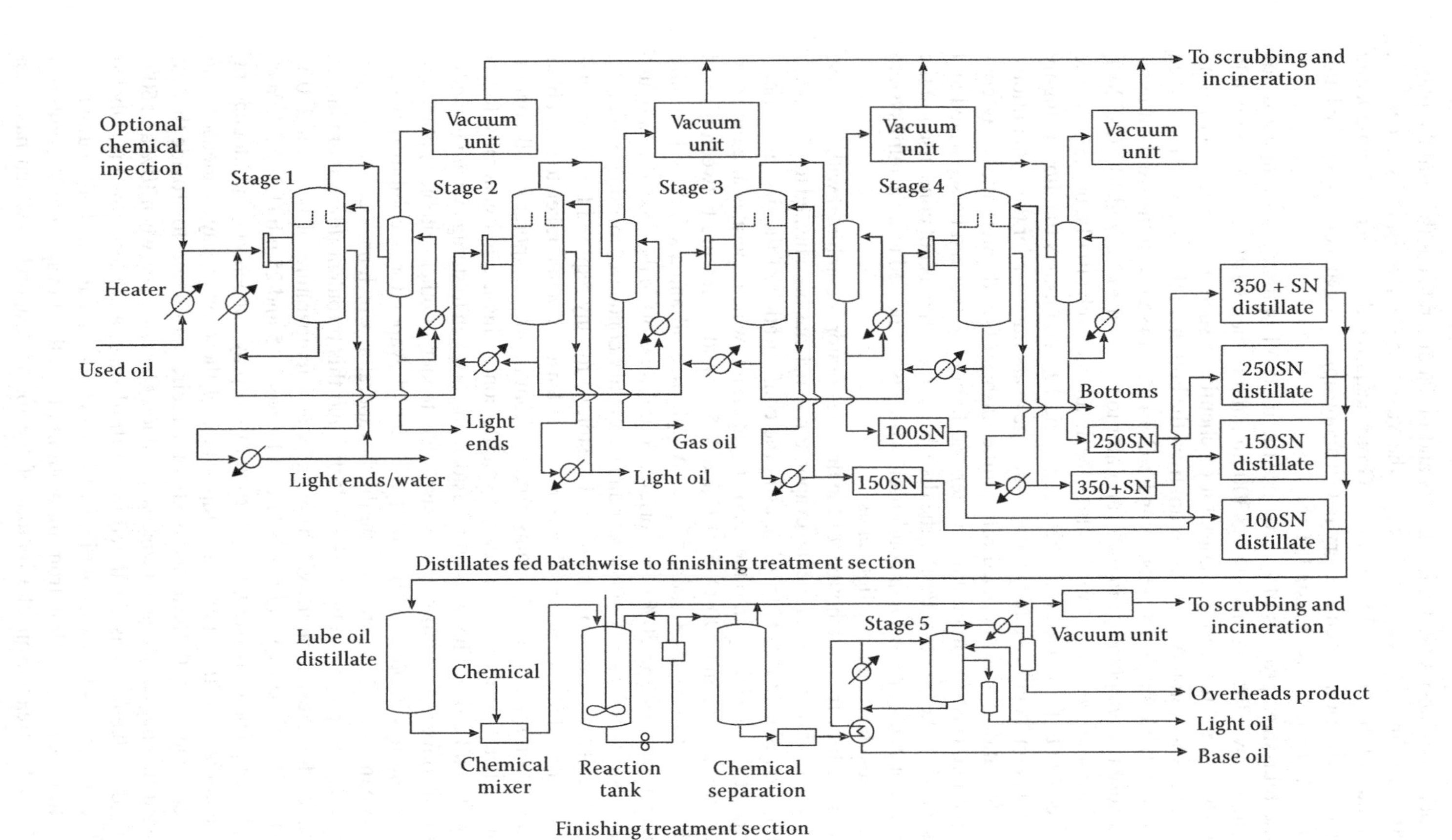

FIGURE 7.4 Four-stage Enprotec Process. (From Enprotec US Patent 5,814,207.)

in Denmark by Dansk Olie Genbrug A/S and in Saudi Arabia at the Unilube plant, where the re-refining process is known as the Vaxon process. Avista Resources Inc., which is part of the Mustad International Group, acquired the plants in Denmark and subsequently the Dollbergen refinery in Germany, upgrading and enlarging the latter to use the Vaxon process (Mustad 2012).

In the four-stage process of Figure 7.4, the used oil is filtered to remove solids with a size above, for instance 100 to 300 μm, and is then heated to approximately 80°C (176°F) with the optional addition of chemicals such as caustic soda before entering the evaporator. A recycle stream from the bottom of the evaporator is heated to between 180°C and 200°C (356°F and 392°F) and mixed with the used oil feedstock to raise the temperature in the evaporator to between 160°C and 180°C (320°F and 356°F) and a pressure between atmospheric and 400 mb (40 kPa or 5.8 psi). The ratio of recycle to fresh used oil is between 5 and 10, and ensures that flow is highly turbulent when it is injected tangentially into the evaporator. Under these conditions, between 5% and 15% of the used oil volume flashes off and enters the spray condenser section, this stream contains water and light ends that can be separated and used as a fuel in the process. Part of the distillate stream that is not condensed in the first spray condenser passes to a condenser, which is connected to a vacuum system and condenses the remaining light ends.

Part of the flow from the bottom of the first evaporator is combined with a recycle stream from the bottom of the second evaporator, which has been heated to a temperature of approximately 280°C (536°F), as in the configuration of the first evaporator, and is injected tangentially into the second evaporator, which operates at a temperature between 260°C and 290°C (500°F and 555°F) and a pressure between 40 and 100 mb (4 and 10 kPa, 0.58 and 1.45 psi). A light fuel oil and a spindle oil, usually forming between 6% and 20% of the original oil, flashes off in the second evaporator and passes through the spray condenser and secondary condenser forming part of the second stage. These components proceed to storage for any required finishing treatment. Any vapor that does not condense in the condenser system is taken to a scrubber and incineration system. The bottoms product from the first evaporator will contain most of the heavy metals and other nonvolatile contaminants, and generally forms between 5% and 15% of the used oil feedstock and is taken to storage where it can be further treated for use as a roofing flux, asphalt extender, or other suitable application.

Part of the flow from the bottom of the second evaporator is combined with a recycle stream from the bottom of the third evaporator, which has been heated to a temperature of approximately 330°C (625°F), and this turbulent flow stream enters the third cyclonic evaporator, which operates at a temperature of between 290°C and 330°C (554°F–626°F) and a pressure of between 15 and 25 mb (1.5–2.5 kPa or 0.22–0.36 psi). The distillate passes through two spray condensers and, depending on the required cuts, the operating temperature of these two spray condensers can be adjusted, together with the adjustment of the recirculation rate, to provide the two base oil cuts having the necessary viscosity values. Generally, between 10% and 50% of the used oil feedstock is separated in the third stage and these streams, labeled as 100SN and 150SN base oils in Figure 7.4, pass to storage for further treatment.

As before, part of the flow from the bottom of the third evaporator is combined with a recycle stream from the bottom of the fourth evaporator, which has been heated to a temperature required to achieve an operating temperature in the fourth evaporator of between 320°C and 345°C and a pressure of between 5 and 15 mb (0.5–1.5 kPa or 0.07–0.22 psi). The base oil fractions flashed off in the fourth stage evaporator, shown as 250SN and 350+SN in Figure 7.4, constitute generally between 10% and 50% of the used oil feedstock and pass to storage units for further treatment.

The further treatment of the base oil products is not specified in the patent document, other than to point to a chemical treatment to remove unstable or other undesirable components.

Tables 7.5, 7.6, and 7.7 from the patent document show, respectively, some properties of the used oil feedstock, the distillate produced using this method, and distillate oil produced by this method, which was subjected to an undefined finishing treatment.

TABLE 7.5
Enprotec Process of US Patent 5,814,207: Used Oil Feedstock Properties

	Value	Method Used
Chlorine, mg/kg	710	IP AK/81
Density, kg/m^3	893.5	NF M 60-172
Metals, mg/kg		ICP
Aluminum	16	
Antimony	9	
Barium	31	
Cadmium	1	
Calcium	2119	
Chromium	3	
Copper	37	
Iron	108	
Lead	214	
Magnesium	274	
Manganese	2	
Molybdenum	4	
Nickel	2	
Silicon	45	
Silver	<1	
Tin	10	
Titanium	2	
Vanadium	1	
Zinc	904	
Phosphorus, mg/kg	842	ICP
Sulfur, % w/w	0.648	ASTM D-2622 (RX)
Total acid number (TAN), mg KOH/g	2.5	NFT 60-112
Viscosity, cSt		
at 40°C	31.07	NFT 60-100
at 100°C	5.349	NFT 60-100
Viscosity index	105	NFT 60-136

Source: Enprotec US Patent 5,814,207.

TABLE 7.6
Enprotec Process of US Patent 5,814,207: Distillate Properties

	Value	Method Used
Chlorine, mg/kg	42	IP AK/81
Color	<7.5	NFT 60-104
Metals, mg/kg		ICP
Aluminum	1	
Antimony	<1	
Barium	<1	
Cadmium	<1	
Calcium	1	
Chromium	<1	
Copper	<1	
Iron	<1	
Lead	1	
Magnesium	<1	
Manganese	<1	
Molybdenum	<1	
Nickel	<1	
Silicon	8	
Silver	<1	
Tin	<1	
Titanium	<1	
Vanadium	<1	
Zinc	<1	
Nitrogen, mg/kg		
Basic	92	
Total	329	
Phosphorus, mg/kg	36	ICP
Sulfur, % w/w	0.419	ASTM D-2622 (RX)
Total acid number (TAN), mg KOH/g	0.15	NFT 60-112
Viscosity, cSt		
at 40°C	31.07	NFT 60-100
at 100°C	5.349	NFT 60-100
Viscosity index	105	NFT 60-136

Source: Enprotec US Patent 5,814,207.

TABLE 7.7
Enprotec Process of US Patent 5,814,207: Chemically Finished Base Oil

	Value	Method Used
Chlorine, mg/kg	3	IP AK/81
Color	<1.5	NFT 60-104
Metals, mg/kg		ICP
Aluminum	<1	
Antimony	<1	
Barium	<1	
Cadmium	<1	
Calcium	<1	
Chromium	<1	
Copper	<1	
Iron	<1	
Lead	<1	
Magnesium	<1	
Manganese	<1	
Molybdenum	<1	
Nickel	<1	
Silicon	<1	
Silver	<1	
Tin	<1	
Titanium	<1	
Vanadium	<1	
Zinc	<1	
Nitrogen, mg/kg		
Basic	10	LPMSA/718
Total	31	LPMSA/718
Phosphorus, mg/kg	<1	ICP
Sulfur, % w/w	0.300	ASTM D-2622 (RX)
Total acid number (TAN), mg KOH/g	<0.05	NFT 60-112
Viscosity, cSt		
at 40°C	29.25	NFT 60-100
at 100°C	5.16	NFT 60-100
Viscosity index	105	NFT 60-136
Cloud point, °C	–7	NFT 60-105
Conradson carbon residue, % w/w	<0.01	ASTM D189
GC distillation, °C		
IBP	299	
5	366	
10	385	
15	396	
20	404	
30	416	
40	426	

(continued)

TABLE 7.7 (Continued)
Enprotec Process of US Patent 5,814,207: Chemically Finished Base Oil

	Value	Method Used
50	434	
60	443	
70	452	
80	463	
85	470	
90	479	
95	490	
FBP	521	
Flash point COC, °C	218	NFT 60-118
Noack volatility, 1 h at 250°C, % w/w	14.3	
Oxidation stability, 2 × 6 h at 200°C	—	IP 48
Viscosity at 40°C, cSt		
Before treatment	29.25	NFT 60-100
After treatment	36.28	NFT 60-100
Conradson carbon residue, % w/w		
Before treatment	<0.01	
After treatment	0.37	
Pour point, °C	−12	

Source: Enprotec US Patent 5,814,207.

7.4 DISTILLATION/CLAY PROCESSES

Clay-based processes are still used in some refineries (Speight and Ozum 2002; Hsu and Robinson 2006; Gary et al. 2007; Speight 2014) and it is not surprising that one of the first alternatives to the acid/clay process was the development of the relatively low capital cost atmospheric and vacuum distillation, followed by treatment of the distillate with an activated clay, which removed some of the color and odor from the product oils, as well as some other undesirable components. An important development in these processes was the regeneration of the spent clay and its recycling for multiple treatments before it eventually has to be replaced.

7.4.1 The Viscolube Italiana S.p.A. Process

Viscolube Italiana S.p.A. patented the process shown in Figure 7.5, which included a pretreatment of the used oil with a basic reagent such as sodium hydroxide to saponify the fatty acids that are present and that helps in their precipitation, but also neutralizes the chlorine that is present in both bound and free forms (Patent WO 94/07798, dated April 14, 1994). The saponification of the fatty acids facilitates their elimination by settling but also destroys their volatility and prevents their being vaporized in the distillation column and contaminating the distilled oils. The elimination of the chlorine has beneficial effects in reducing corrosion in the plant.

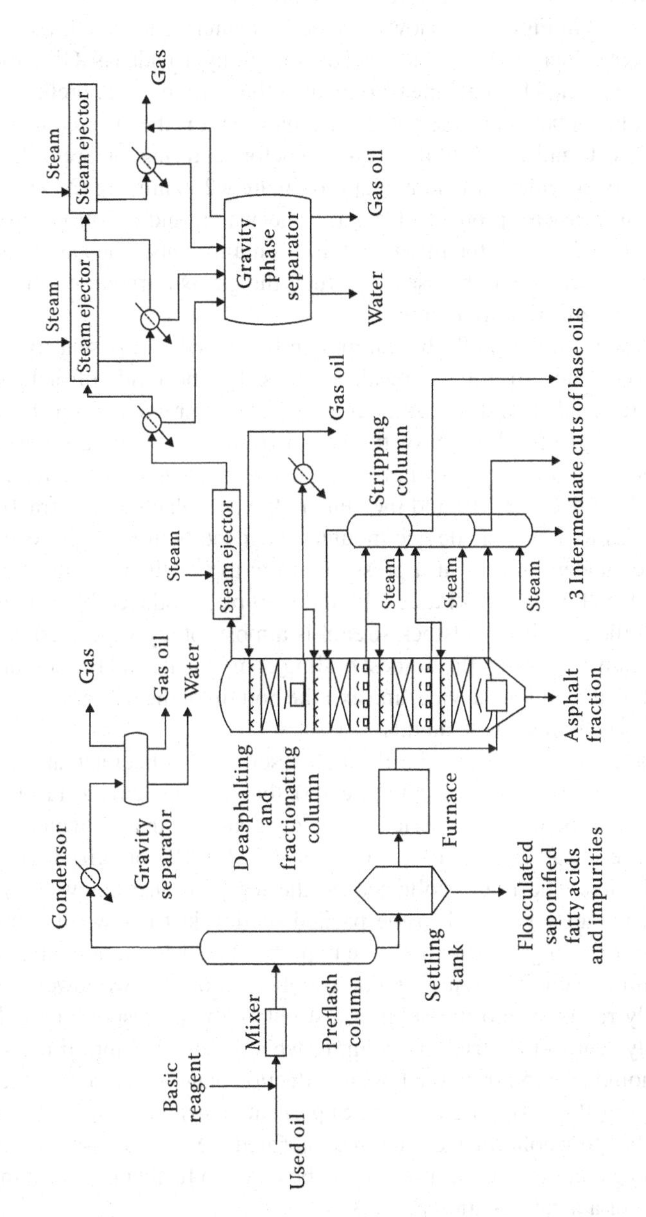

FIGURE 7.5 Viscolube Italiana Process. (From Viscolube Italiana Patent WO 94/07798.)

Viscolube had been working with the propane extraction process under licence from the Institut Français du Pétrole (IFP) since the early 1970s (Audibert 2006) to remove the asphalt fraction, later developing its own thermal deasphalting (TDA) process, and this formed the second step in the process.

The embodiment in Figure 7.5 shows the used oil entering a static mixer together with a basic reagent such as a 30% solution of sodium hydroxide (NaOH), the alkali addition being controlled by a pH measurement at the bottom of the preflash column at a preferred value of between 10.8 and 11.2. This stream is then heated to a temperature between 120°C and 140°C (248°F–284°F) before entering the preflash column. This column operates under a vacuum of approximately 200 mm Hg (27 kPa, 3.9 psi) and produces an overhead product of gas oil and water, and a bottom product of dehydrated oils together with the impurities and contaminants. The gas oil obtained from the overhead stream can be used as a fuel, the gas is burned and the water is taken to a water purification treatment.

The underflow from the preflash column can be cooled and is sent to a settling tank, which is designed to allow a residence time for the fluids of at least 48 h, the saponified fatty acids and the other contaminants entrained in the flocculated components settling to the bottom of the vessel from where they are periodically withdrawn. The overflow stream from the settling vessel is heated to a temperature of approximately 360°C (680°F) and then enters the deasphalting and fractionating column, being fed to a cyclone device mounted near the bottom of the vessel. The column operates under vacuum at a pressure at the top of the column of between 10 and 20 mm Hg (1.3 and 2.7 kPa or 0.18 and 0.39 psi) produced by a steam ejector system, and the entering feedstock stream is almost totally vaporized, leaving a small liquid stream containing the highest boiling components and the contaminants to be separated by the cyclone device, and exit the bottom of the column as an asphalt fraction that also contains heavy metals.

The gas phase rises in the column through a series of collector plates and four packed column sections (as shown in Figure 7.5), the packing being a stainless steel, folded metal sheet type to form a series of zig-zag channels that allow a very small pressure decrease of, at most, 5 to 10 mm Hg (0.7–1.4 kPa or 0.10–0.20 psi) for each packed section. The liquid collected on the top plate just below the topmost packed section is partly recycled to the packed section just below this uppermost plate, partly, after cooling, to just above the uppermost packed section, and partly to a product gas oil stream. The liquid fractions collected on the two lower collection plates are partly refluxed into the column just below the corresponding collection plate, and partly sent to the stripping column, which is divided into three separate sections. The liquid fraction from the lowest collection plate is also sent to the lowest separate section of this stripping column. Superheated stripping steam enters each section of the stripping column and the three intermediate cuts of base oils collected from the stripping column leave as products to be taken to further decolorizing treatment with clays or a hydrogenation process.

The streams from the steam ejector system are taken to a gravity separation vessel where the noncondensing gases, gas oil, and water are separated—the gases being burned in a furnace, the water being recycled after treatment to the steam boilers, and the gas oil being combined with the gas oil fraction from the top of the fractionating column.

The asphalt fraction, which does not contain any solvents and which has a high content of very heavy and viscous oils, can be marketed as a component for bituminous road-making.

The patent claims considerable advantages in costs over the sulfuric acid and propane solvent processes, giving a table of comparative costs at that time, and claiming that the only waste product is the water separated from the used oils in the preflash step, constituting approximately 4% of the feedstock material.

7.4.2 The Pesco-Beam Process

The Roanoke, Virginia company, Pragmatic Environmental Solutions Co. (PESCO), teamed up with their engineering supplier Beam to form the Pesco-Beam Environmental Solutions Company in 1991 and, since then, has designed and built multiple units in various countries based on atmospheric distillation, followed by a two-stage wiped-film vacuum evaporator distillation with the distillate undergoing clay treatment to improve color and odor, and to remove some additional unwanted components. One of the main attractions of this process is the capital cost. Briggs (2012) quoted a construction cost of $3.5 million in 2004 for the ORRCO re-refining plant in Portland, Oregon, with a capacity of 600 barrels of used oil per day (~30 ML/year, ~7,920,000 US gallons/year). Figure 7.6 shows the schematic flow diagram for this process.

The incoming stream of used oil would normally pass through a filter to remove larger solid materials before entering an atmospheric distillation unit where the light ends and water are flashed off, these being recondensed and separated. These light ends can be used for fuel purposes within the process, and the water is treated and disposed of.

The underflow from the atmospheric distillation unit passes to a wiped-film short path evaporator operating under vacuum and which is heated by an externally heated thermal fluid that provides close control of the wall temperature in the evaporator. A light cut of base oil is obtained as distillate from this unit, whereas; the underflow passes to a second wiped film evaporator operating at lower pressure to provide a second heavier cut of base oil. The underflow from the second evaporator contains much of the metal content of the feedstock oil and heaviest hydrocarbons and can be used as an asphalt extender.

The two streams of base oils from the vacuum evaporators pass to banks of columns packed with the activated clay and travel slowly through this clay. After the clay treatment, the oils can be filtered or centrifuged if necessary to remove any remaining suspended clay particles to exit as Group I base oils that can be used for making various lubricants. After treating a certain volume of product oil, the clay is regenerated by a combustion process with air injection in the packed columns before placing the regenerated bank back on stream and switching the flow as necessary. Some success has apparently also been achieved in refining the clay treatment process to produce base oils that meet the requirements for

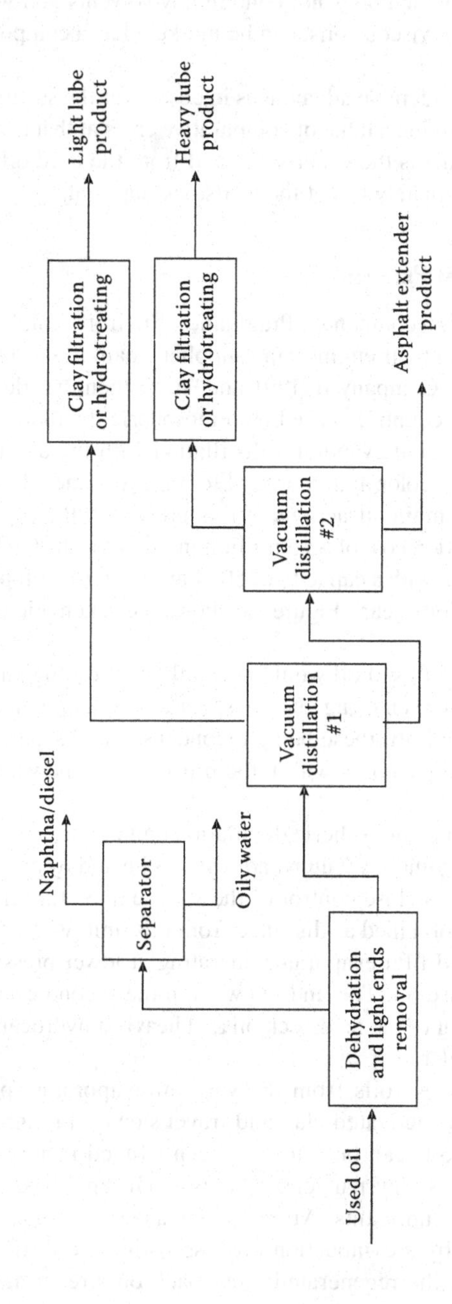

FIGURE 7.6 The Pesco-Beam Process. (Courtesy of Pesco-Beam/ORRCO.)

Group II base oil. Tables 7.8 and 7.9 show some product data as reported by an analytical services laboratory for the ORRCO plant in Portland, Oregon, which was originally designed and built by Pesco in 2003 for the ORRCO-related company, Energy and Material Recovery Inc. (ORRCO 2012), with subsequent modifications by the latter company. In addition, Briggs (2012) reports a product yield for each component as shown in Table 7.10. It should be noted that these are not results from the original plant, but are the results obtained by the operating company in 2010 on the modified plant.

As an option to the clay treatment, Pesco-Beam also offered a process that includes the hydrogenation of the product base oils after the evaporators, but this requires more expensive equipment and operating costs with catalytic hydrogenation at high pressure and temperature and a source of hydrogen.

TABLE 7.8
Product Oil 100N Properties for ORRCO Plant

Parameter	Method	Value
Acid number, mg KOH/g	ASTM D974	0.28
API gravity	D4052	33.1
Ash, %	D482-02	<0.1
BSW, %	ASTM 1796	<0.2
Heating value, BTU/lb	D-240-09	19,875
Carbon residue, %	D-524	0.132
Color	ASTM D1500	0.5 ASTM color units
Copper strip corrosion	ASTM D130	1A
API gravity at −15°C	ASTM D-1250-80	29.2
Flash point, °F	1010	>200
Pour point, °F	ASTM D-97	12
Sulfur, ppm	6010	170
Viscosity at 100°C, cSt	D-445	5.29
Viscosity at 40°C, cSt	D-445	29.38
Viscosity index	D-2270	112
Volatility at 250°C/1 hour, %	In-house	7.43
Volatility at 371°C/1 h, %	In-house	55.49
Cetane index	ASTM D976	53.16
Distillation, °F	D1160	
IBP		443.5 (285.8°C)
90%		693 (367.2°C)
End point		99% at 703.2
Saturates content, % w/w	D-2007-02	93
Aromatics content, % w/w	D-2007-02	<0.1
Polars content, % w/w	D-2007-02	7.0

Source: Reproduced with permission from B. Briggs. ORRCO. Available at http://www.orrco.biz. Accessed on November 7, 2012.

TABLE 7.9
Product Oil 250N Properties for ORRCO Plant

Parameter	Method	Value
Acid number, mg KOH/g	ASTM D974	0.56
API gravity	D4052	32.5
Ash, %	D482-02	<0.1
BSW, %	ASTM 1796	<0.2
Heating value, BTU/lb	D-240-09	19,869
Carbon residue, %	D-524	0.1135
Color	ASTM D1500	1.0 ASTM color units
Copper strip corrosion	ASTM D130	1A
API gravity at −15°C	ASTM D-1250-80	28.7
Flash point, °F	1010	>200
Pour point, °F	ASTM D-97	9
Sulfur, ppm	6010	700
Viscosity at 100°C, cSt	D-445	6.78
Viscosity at 40°C, cSt	D-445	43.06
Viscosity index	D-2270	113
Volatile solids, %	EPA 160.4	3.11
Volatility at 250°C/1 h, %	In-house	3.11
Volatility at 371°C/1 h, %	In-house	45.39
Cetane index	ASTM D976	52.00
Distillation, °F	D1160	
IBP		352.6
90%		690.8
End point		99% at 767.7
Saturates content, % w/w	D-2007-02	93
Aromatics content, % w/w	D-2007-02	1.0
Polars content, % w/w	D-2007-02	6.0

Source: Reproduced with permission from B. Briggs. ORRCO. Available at http://www.orrco.biz. Accessed on November 7, 2012.

TABLE 7.10
ORRCO Product Yields

Product	Yield (%)
Light naphtha	1 to 3
Diesel fuels	7 to 10
Light neutral base oil (VI > 100)	45 to 50
300 neutral base oil (VI > 100)	30
Asphalt flux	~10

Source: Reproduced from B. Briggs, ORRCO in News Articles: What's Happening in Used Oil Recycling in North America and What's New? Available at http://www.orrco.biz. Accessed on November 7, 2012.

7.5 DISTILLATION AND HYDROGENATION PROCESSES

Distillation of petroleum and its products followed by hydrogenation are common refinery options (Speight and Ozum 2002; Hsu and Robinson 2006; Gary et al. 2007; Speight 2014). Thus, the replacement of the clay treatment of the distilled base oil fractions by hydrogenation overcame the problem of disposal of the spent clay and was a more environmentally acceptable process in many cases. The processes suggested generally involve distillation of the used oil followed by processing in a guard reactor to remove most of the heavy metals that would poison the catalyst in a conventional hydrotreatment. The oils are then hydrogenated over a catalyst to remove the remaining oxygen, nitrogen, chlorine, and other undesirable components, and to saturate any unsaturated compounds.

7.5.1 The Mohawk and Chemical Engineering Partners–Evergreen Processes

The Mohawk Oil Company's plant in North Vancouver, British Columbia, now operated by Newalta Corporation, was the first in North America to use a combination of thin-film evaporators with a subsequent hydrotreating step to produce higher grade base oil products. The use of chemical pretreatment of the used oil led to a reliable process that was licensed to Chemical Engineering Partners (CEP) of California in 1989 (Audibert 2006), and a collaboration over the next 5 years on improvement of the hydrotreating catalyst ensued, CEP being the engineering and technology affiliate of Evergreen Holdings of California. Figure 7.7 shows the flow diagram for the early version of the Mohawk process, which was also licensed to the Breslube refinery in Breslau, Ontario, this refinery being acquired by Safety Kleen in 1987.

The Evergreen plant in Newark, California, although still based on a combination of distillation followed by hydrogenation, represents improvements made by CEP, which include a proprietary pretreatment of the used oil and the removal of the diesel fraction from the atmospheric distillation step rather than in the final separation of the base oils, this resulting in a reduced volatility of the base oil products. Figure 7.8 shows the improved CEP–Evergreen process (CEP 2012). Table 7.11 shows typical results for the 110 and 250 base oil products from this refinery (CEP 2012). The CEP web site also indicates that the minimum economic size for a plant using this technology is approximately 15,000 tonnes (33,070,000 pounds) per year, anything smaller would make the process unattractive. The yield of hydrotreated base oil is given as typically 74% of the feedstock rate, with 4% light oil, 4% gas oil, and 13% asphalt flux (CEP 2012). The hydrotreater operates at a pressure of 1300 psig (9 MPa) and a temperature of 315°C (600°F) and an operating cost of between $0.50 and $0.75 per US gallon ($0.13/L and $0.20/L) excluding capital and feedstock cost is given, depending largely on the local cost of fuel, electricity, hydrogen, waste, and waste water.

The Safety Kleen refinery near Chicago uses a process that is similar to the process used in the Safety Kleen plant in Breslau (Kajdas 2000).

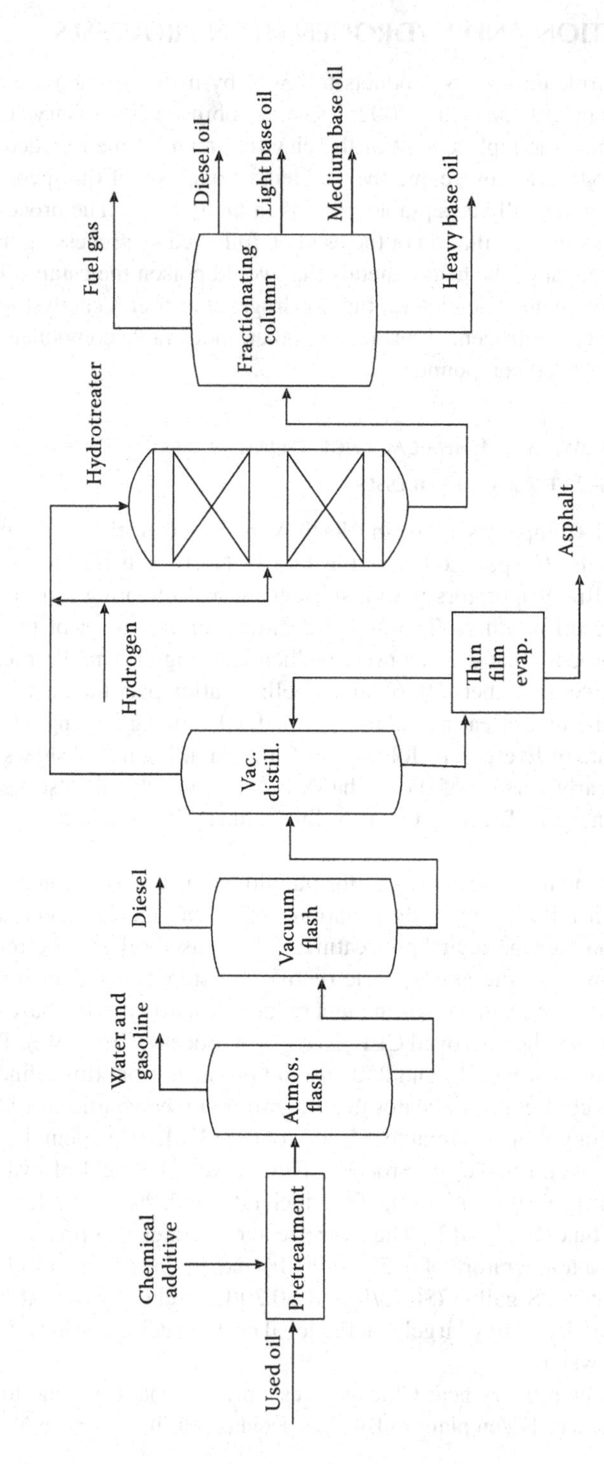

FIGURE 7.7 The Mohawk process. (Courtesy of Mohawk.)

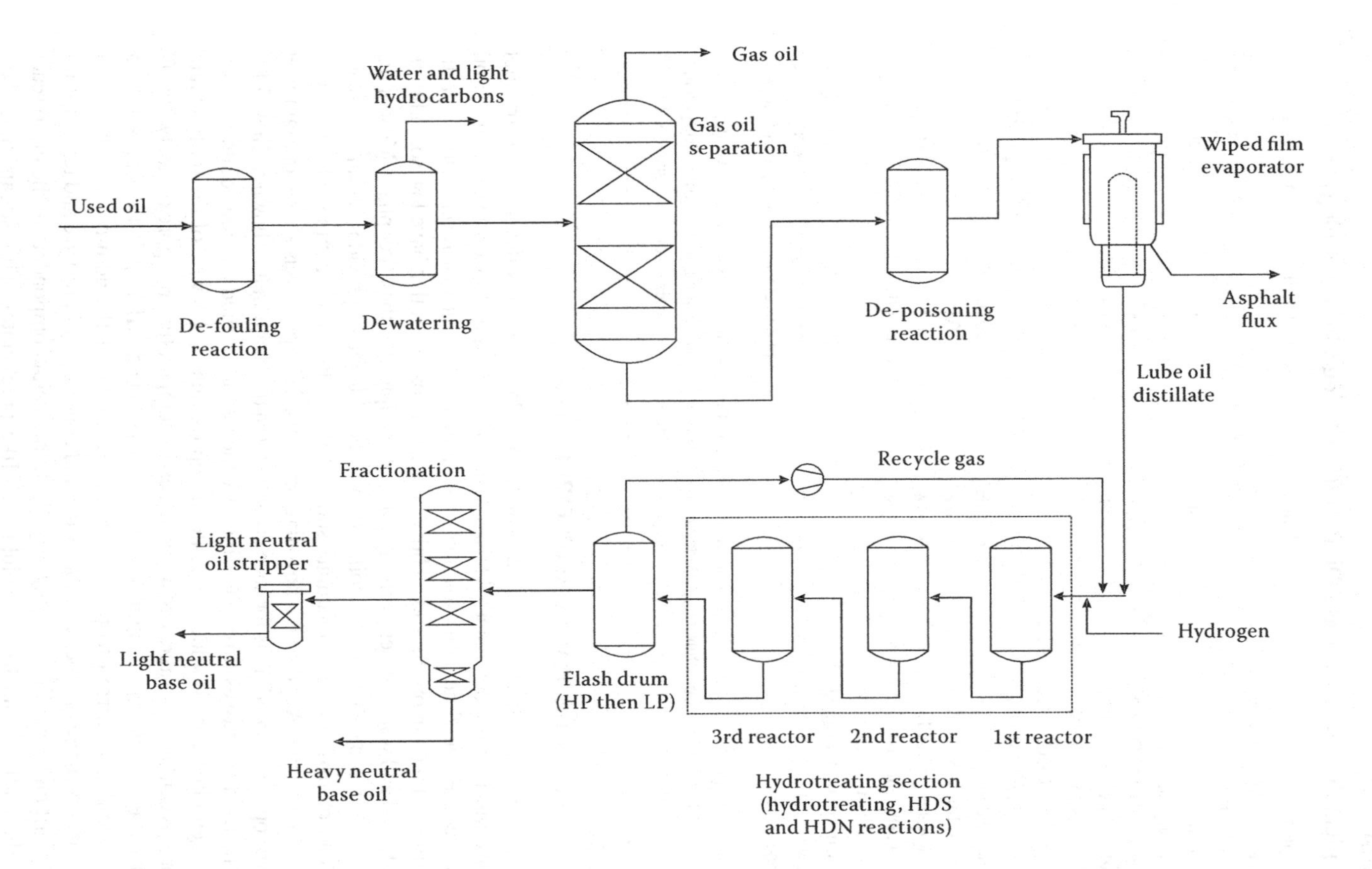

FIGURE 7.8 The CEP–Evergreen Process. (Courtesy of CEP-Evergreen.)

TABLE 7.11
Typical Properties of Base Oil Produced Using CEP Technology

Property	Units	Typical 110	Typical 250
Viscosity, ASTM D 445			
at 100°C	cSt	4.3	7.2
at 40°C	cSt	22	49
Viscosity index, ASTM D 2270	—	102	107
Density, ASTM D 1298	kg/m^3	865	871
API gravity, ASTM D 287	°API	33.9	33
Color, ASTM D 1500	—	<0.5	<1
Flash point, COC, ASTM D92	°F (°C)	390 (199)	470 (243)
Pour point, ASTM D97	°F (°C)	−15 (−26)	−12 (−24)
Total acid number	mg KOH/g	<0.01	<0.01
Total sulfur, ASTM D3120	% w/w	0.02	0.02
% saturates ASTM D 2007	% w/w	90.5	92
Total ash content, ASTM D482	% w/w	<0.001	<0.001
Crackle test—trace water			
Chevron test method 350		Pass	Pass
Odor		None	None
Appearance		Bright and clear	Bright and clear

Source: Reproduced with permission from Chemical Engineering Partners (CEP). Available at http://www.ceptechnology.com/technology.htm. Accessed November 24, 2012. Evergreen, Evergreen Oil Co. Available at http://www.evergreenoil.com. Accessed on November 13, 2012.

7.5.2 The Phillips Petroleum (PROP) Process

This process (US Patent 4,151,072, dated April 24, 1979) was developed to try and remove much of the detergent additives found in used oil and involved treating the oil with a concentrated aqueous solution of an ammonium salt that reacts with the ash-forming metal components and additives to form phosphates that have low solubility in both the oil and water phases, forming a heavy precipitate in the water phase. Most of the water is then separated from the oil, which is then filtered to provide a clean oil stream that can be hydrotreated to saturate some of the hydrocarbon compounds and lower the sulfur content. A schematic flow diagram of the 1979 process is shown in Figure 7.9 and the operating conditions for each component in the diagram are shown in Table 7.12.

The used oil passes through a line in which concentrated aqueous diammonium hydrogen phosphate is added at a rate of approximately 1% w/w of the oil flow rate before passing to a heater, which increases its temperature to approximately 160°C (320°F) at a pressure of approximately 215 psia (1.5 MPa-a). It then passes to a contactor that is stirred to provide intimate mixing of the additive and the metal components in the oil, the residence time and conditions of pressure and temperature being sufficient to allow substantially all of the metal components in the oil to react with the additive and form insoluble ash-like precipitates. This stream may then optionally have a filter aid, such as diatomaceous earth, added to it and then passes

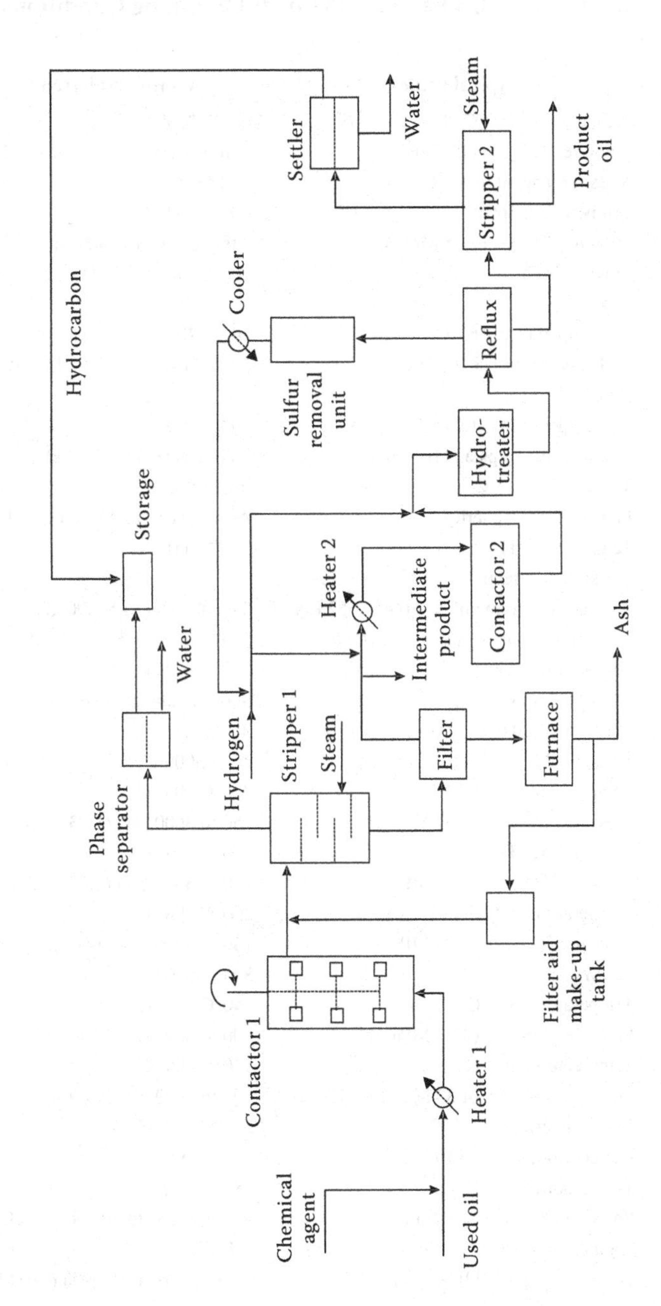

FIGURE 7.9 The Phillips Petroleum Process. (From Phillips US Patent 4,151,072.)

TABLE 7.12
Phillips Petroleum Process: US Patent 4,151,072: Operating Conditions Embodiment 1

Unit	Typical Value	Approx. Preferred Range
Heater 1 for feedstock	Temperature: 160°C	60°C–200°C
	Pressure: 215 psia (1.5 MPa-a)	Atmospheric to 250 psia (1.72 MPa-a)
Treating agent	Mass ratio agent to oil: 0.01:1	0.005:1 to 0.05:1
Contactor 1	Temperature: 160°C	60°C–200°C
	Pressure: 215 psia (1.5 MPa-a)	Atmospheric to 250 psia (1.72 MPa-a)
	Time: 30 min	10 min to 2 h
Stripper 1	Top	
	Temperature: 160°C	60°C–200°C
	Pressure: 16 psia (110 kPa-a)	20 to 2 psia (138–13.8 kPa-a)
	Bottom	
	Temperature: 115°C	60°C–200°C
	Pressure: 16 psia (110 kPa)	20 to 2 psia (138–13.8 kPa-a)
Phase separator	Temperature: 40°C	0°C–80°C
	Pressure: atmospheric	Atmospheric to 45 psia (310 kPa-a)
Filter	Temperature: 115°C	60°C–200°C
	Pressure differential	
	Plate and frame filter: 80 psi (552 kPa)	5 to 100 psi (34.5–690 kPa)
	Contin. rotary drum: 10 psi (69 kPa)	2 to 14 psi (13.8–96.5 kPa)
Furnace	Temperature: 760°C	650°C–870°C
	Pressure: atmospheric	Substantially atmospheric
Filter aid	Mass ratio, aid to oil 0.01:1	0:1 to 0.15:1
Hydrogen charge	111 v/v oil	80 to 3000 v/v oil
Heater 2	Temperature: 370°C	200°C–480°C
	Pressure: 735 psia (5.07 MPa-a)	150 to 3000 psia (1.03–20.7 MPa-a)
Contactor 2	Temperature: 370°C	200°C–480°C
	Pressure: 735 psia (5.07 MPa-a)	150 to 3000 psia (1.03–20.7 MPa-a)
Hydrotreater	Temperature: 360°C	200°C–430°C
	Pressure: 730 psia (5.03 MPa-a)	150 to 3000 psia (1.03–20.7 MPa-a)
Hydrogen charge	222 v/v oil	80 to 3000 v/v oil
Reflux	Temperature: 325°C	290°C–400°C
	Pressure: 705 psia (4.86 MPa-a)	600 to 800 psia (4.14–5.52 MPa-a)
Sulfur removal unit	Temperature: 290°C	150°C–430°C
	Pressure: 700 psia (4.83 MPa-a)	100 to 3000 psia (0.69–20.7 MPa-a)
Cooler	Inlet temperature: 290°C	260°C–370°C
	Outlet temperature: 55°C	40°C–95°C
Stripper 2	Temperature: 370°C	280°C–395°C
	Pressure: 20 psia (138 kPa-a)	Atmospheric to 50 psia (345 kPa-a)
Settler	Temperature: 55°C	0°C–80°C
	Pressure: 16 psia (110 kPa-a)	Atmospheric to 45 psia (310 kPa-a)

Source: Phillips US Patent 4,151,072.

to the stripper vessel where most of the water, gasoline, and other light hydrocarbon fractions flash off, passing to a phase separator where waste water and the light hydrocarbons are separated. The flow from the bottom of the stripper, containing a sulfated ash value of between 0.3% and 10% w/w of the flow excluding the filter aid and excess treating agent, contains the oil, some residual water, the filter aid where applicable, and the precipitated metal compounds. Steam can be injected near the bottom of the stripper to help remove light hydrocarbons and residual water before the bottoms stream passes to a filter to remove the precipitated materials. The filter itself may be coated with a filter aid, or the filter aid may be added to the stream as described previously, and the filter cake is removed to pass to a furnace to recover the filter aid to recycle it as shown in Figure 7.9, or it can be discarded.

The oil, after filtration, has a thermal ash yield between 0.01% and 0.3% w/w, excluding the excess treating agent or filter aid that may have passed through the filter and may be used for various industrial applications without further treatment, being removed from the line as shown in Figure 7.9. However, the preferred method is to pass the oil stream through a second heater to increase its temperature to approximately 370°C (698°F) and pressure of approximately 735 psia (5.07 MPa-a), possibly adding some hydrogen before the heater, and then entering a contactor in which the sulfonates in the oil are decomposed on an adsorbent bed containing bauxite or activated carbon, or other suitable material. The adsorbent promotes the breakdown and decomposition of the ammonium salts of sulfonic acids and ashless detergents contained in the oil and various compositions for this adsorbent are suggested. An alternative treatment avoids the use of a contactor vessel and heats the oil to approximately 380°C (716°F) in the presence of hydrogen and an adsorbent suspended in the oil, and after cooling, the stream is refiltered.

This stream then passes to the hydrotreater where hydrogen is added and the high temperature and pressure in the presence of a catalyst result in the removal of unsaturated compounds, as well as residual sulfur, oxygen, and nitrogen compounds. The oil then passes to a separator-reflux column to remove water and various other materials formed in the previous treatments of the oil. The possible injection of water into this vessel is also suggested to remove most of any HCl that may be present, and part of the H_2S and NH_3, as water-soluble salts. The stream from the top of the reflux unit passes to the sulfur recovery unit to remove H_2S on, for example, a bed of zinc oxide with other possibilities for this H_2S removal also being suggested. The stream then passes through a cooler and can then be water-washed to remove NH_3 before returning the hydrogen as a recycle stream.

The flow from the bottom of the reflux unit passes to the lube-stock stripper where steam is injected to remove the lighter, low-boiling hydrocarbons or, alternatively, hydrogen gas-stripping could be used. The product lube oil stock is cooled and proceeds to storage. The flow from the top of the stripper vessel, containing essentially fuel oil and water, passes to a settling vessel where the water and hydrocarbons separate—with the hydrocarbons being possibly combined with the hydrocarbons from the other phase separator as shown in Figure 7.9.

A table is presented in the patent document, which gives the mass flow rate and composition data for many of the streams in this process, and Table 7.13 shows part of the patent table for the feedstock stream, the intermediate product, and the final product.

TABLE 7.13
Phillips Petroleum Process: US Patent 4,151,072: Mass Flow Rate/Stream Day Embodiment 1

	Feedstock		Intermediate Product		Final Product	
	lbs	kg	lbs	kg	lbs	kg
Oil	6644	3014	6445	2923	6261	2840
5 metals plus P	51	23				
S[a]	15	7	13	5.9	<1	<0.45
O[b]	50	23	44	20	<1	<0.45
N[a]	10	4.5	10	4.5	<0.1	<0.045
H_2O	417	189				
Light Hydrocarbons	300	136	150	68		

Source: Phillips US Patent 4,151,072.
[a] Present in combined form in the used oil.
[b] Present in combined form in the used oil, excluding H_2O.

An alternative process with three contactor vessels is also suggested and is presented in Figure 7.10, whereas Table 7.14 shows the operating conditions in each of the steps in this figure. The used oil feedstock is heated, mixed with an aqueous treating agent containing diammonium hydrogen phosphate, and enters the first contactor where the residence time in the agitated vessel is sufficient to allow the reactions with the ash-forming components in the oil to be substantially complete. The agitation is preferably accomplished by using a heated recycle stream from the bottom of the vessel, which is then reinjected as shown, but a stirrer could also be used.

The mixture then passes to the second contactor, which operates at a temperature between approximately 110°C and 140°C (230°F and 285°F) and a pressure slightly above atmospheric to distill off the majority of the water present in the used oil, together with some light hydrocarbons. This distillate passes to the phase separator which provides a stream of light hydrocarbons which passes to storage, and a waste water stream to be disposed of. The preferred means of agitation in this contactor is again a recycled, heated stream from the bottom of the contactor which is re-injected into the contactor.

The stream from the bottom of the second contactor then passes to the third contactor in which it is mixed with a slurry of diatomaceous earth in a light hydrocarbon carrier fluid, agitation again being provided by reinjection of a heated recycle stream as shown in the diagram. In addition to the heat energy provided by the heaters on any of the recycle streams to the contactors, a heated jacket on any of the contactors is considered as an option. Any residual water and light hydrocarbons in the oil in the third contactor pass overhead to the phase separator as shown.

After mixing with the diatomaceous earth, the stream passes to a filter which may, optionally, be precoated with diatomaceous earth or other material as a filter aid. The filter cake from the filter is taken to a furnace which recovers a dry ash of diatomaceous earth which can be recycled in the process, or be partly disposed of.

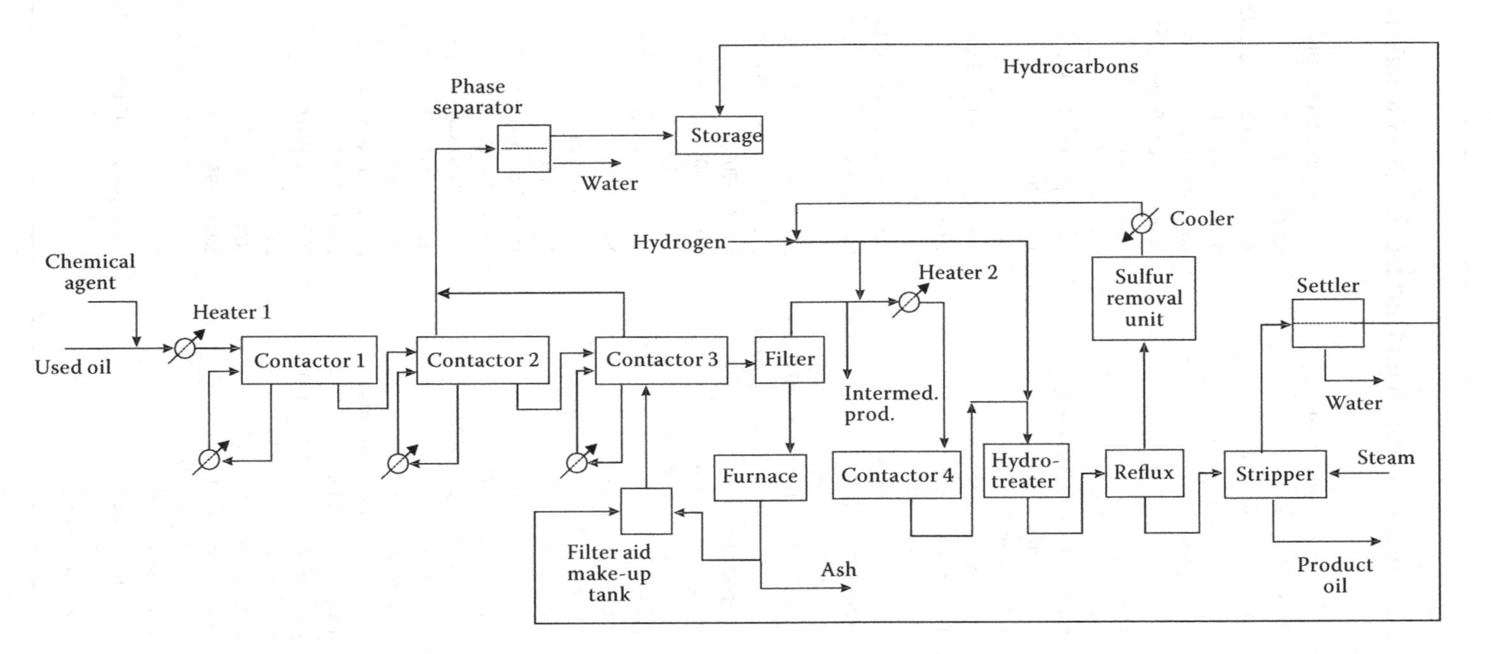

FIGURE 7.10 The Phillips Petroleum Process version 2. (From Phillips US Patent 4,151,072.)

TABLE 7.14
Phillips Petroleum Process: US Patent 4,151,072: Operating Conditions Embodiment 2

Unit	Typical Value	Approx. Preferred Range
Heater 1 for feedstock	Temperature: 95°C	60°C–120°C
	Pressure: 17 psia (117 kPa-a)	Atmospheric to 250 psia (1.72 MPa-a)
Treating agent	Mass ratio agent to oil: 0.01:1	0.005:1 to 0.05:1
Contactor 1	Temperature: 95°C	60°C–120°C
	Pressure: 17 psia (117 kPa-a)	Atmospheric to 50 psia (345 kPa-a)
	Time: 30 min	10 min to 2 h
Contactor 2	Temperature: 125°C	110°C–140°C
	Pressure: 16 psia (110 kPa-a)	5 to 25 psia (34–172 kPa-a)
	Time: 30 min	10 min to 2 h
Contactor 3	Temperature: 95°C	140°C–200°C
	Pressure: 16 psia (117 kPa-a)	5 to 25 psia (34–172 kPa-a)
	Time: 30 min	10 min to 2 h
Phase separator	Temperature: 40°C	0°C–80°C
	Pressure: atmospheric	Atmospheric to 45 psia (310 kPa-a)
Filter	Temperature: 115°C	60°C–200°C
	Pressure differential	
	Plate and frame filter: 80 psi (552 kPa)	5 to 100 psi (34.5–690 kPa)
	Contin. rotary drum: 10 psi (69 kPa)	2 to 14 psi (13.8–96.5 kPa)
Furnace	Temperature: 760°C	650°C–870°C
	Pressure: atmospheric	Substantially atmospheric
Filter aid	Mass ratio, aid to oil 0.01:1	0:1 to 0.15:1
Hydrogen charge	111 v/v oil	80 to 3000 v/v oil
Heater 2	Temperature: 370°C	200°C–480°C
	Pressure: 735 psia (5.07 MPa-a)	150 to 3000 psia (1.03–20.7 MPa-a)
Contactor 2	Temperature: 370°C	200°C–480°C
	Pressure: 735 psia (5.07 MPa-a)	150 to 3000 psia (1.03–20.7 MPa-a)
Hydrotreater	Temperature: 360°C	200°C–430°C
	Pressure: 730 psia (5.03 MPa-a)	150 to 3000 psia (1.03–20.7 MPa-a)
Hydrogen charge	222 v/v oil	80 to 3000 v/v oil
Reflux	Temperature: 325°C	290°C–400°C
	Pressure: 705 psia (4.86 MPa-a)	600 to 800 psia (4.14–5.52 MPaa)
Sulfur removal unit	Temperature: 290°C	150°C–430°C
	Pressure: 700 psia (4.83 MPa-a)	100 to 3000 psia (0.69–20.7 MPa-a)
Cooler	Inlet temperature: 290°C	260°C–370°C
	Outlet temperature: 55°C	40°C–95°C
Stripper 2	Temperature: 370°C	280°C–395°C
	Pressure: 20 psia (138 kPa-a)	Atmospheric to 50 psia (345 kPa-a)
Settler	Temperature: 55°C	0°C–80°C
	Pressure: 16 psia (110 kPa-a)	Atmospheric to 45 psia (310 kPa-a)

Source: Phillips US Patent 4,151,072.

The hot, filtered oil can be withdrawn as an intermediate product which is suitable for various industrial applications, or it can be preferably taken to a hydrogenation process. In this case, it is mixed with hydrogen, heated to a temperature between 200°C and 480°C (390°F and 895°F), and the sulfonates react as in the first embodiment of the patent process. The fourth contactor contains, preferably, bauxite or an activated carbon adsorbent bed, but other adsorbents are also considered. The adsorbent causes the ammonium salts of sulfonic acids and the ashless detergents in the oil to break down and decompose. It also binds some of the resulting products and prevents them from travelling on to the hydrotreater section. Periodic regeneration of these adsorbents is necessary, using conventional means.

An alternative treatment is also contemplated in which the fourth contactor is omitted and the ash components and highly polar constituents are removed by heating the oil, mixing it with hydrogen, cooling it and re-filtering it. Various possibilities for the adsorbent are given, with preferred compositions and combinations of components being discussed.

After treatment, the oil enters the hydrotreater where some of the additives to the original oil are decomposed. As in the first embodiment, the unsaturated materials, residual sulfur, oxygen and nitrogen are removed to yield a product oil for further processing. The catalyst as used in a conventional hydrodesulfurization process would be from a group from the Group VIB and Group VIII metals and combinations thereof on a refractory support.

After hydrotreating, the oil passes to the separator-reflux column which removes water and other treatment by-products. Water can be injected into this column to remove HCl and part of the H_2S and NH_3 as water-soluble salts. The overhead stream from the reflux column contains hydrogen, H_2S, NH_3 and water and passes to the sulfur removal unit in which the H_2S may be removed over a bed of material such as zinc oxide, other materials such as iron oxide and others also being proposed. This stream is then cooled and may then, although not shown in the diagram, be washed with water to remove ammonia before recycling the hydrogen in the process. The stream from the bottom of the reflux unit can be steam stripped, or hydrogen stripped, to remove the light hydrocarbon fractions from the product oil stream which is then cooled and sent to storage. The overhead from the stripper, containing essentially fuel oil and water, is taken to a separator where a hydrocarbon layer can be recycled to the phase separator earlier in the process, as shown in the diagram, or be recycled to the filter aid make-up tank. Any gases in the hydrocarbon stream from the separator can be removed by allowing them to flash off by lowering the pressure.

Typical compositions for some of the streams in this embodiment are the same as given for the first embodiment in Table 7.14. Additional intermediate compositions are given in the patent document, but do not seem to affect the final compositions obtained in either embodiment.

A later patent, US Patent 4,287,049 dated September 1, 1981, had the same flow sheet as the first embodiment of the 1979 patent, but added a polyhydroxy compound to the diammonium hydrogen phosphate treating agent to improve the filterability of the oil, particularly used oil from diesel engines which contain significant amounts of fine soot particles. It was speculated that the polyhydroxy compound facilitated the agglomeration of the soot particles which produced a precipitate which was easier to remove by filtration. Several polyhydroxy compounds were proposed and included glycerol, sugar-alcohols, mono-saccharides and di-saccharides, particularly sucrose, and ethylene glycol, added in a preferable ratio between 0.25 and 0.5% w/w of the total oil. Data was presented to show that the filtration rate of a typical sample of oil was approximately doubled when either sucrose or ethylene glycol were used as the polyhydroxy additive to the diammonium hydrogen phosphate which was used in the demetallization step at the start of the process.

A subsequent patent, US Patent 4,789,460 dated December 6, 1988, proposed the use of a polyethoxyalkylamine together with the ammonium salt treating agent to double or triple the filtration rate of the oil by coagulating soot particles in the oil before the filtration process. The solubility of the polyethoxyalkylamine depends on the chain length of the polymer, shorter chains being soluble in the oil and longer chains being highly soluble in water, and the polyethoxyalkylamine is added together with the ammonium salt at a rate of 0.5% w/w of the final feedstock stream flow. The patent refers to the product marketed by the Rohm and Haas Company under the trademark TRITON RW-SURFACTANTS as being suitable and these are added to the oil and water at a pH of less than seven so that the polyethoxyalkylamine functions as a salt where it is more effective for the required purpose. Figure 7.11 shows a schematic diagram of the process.

The operating conditions in the contactor vessel where the ash-forming reactions take place are a preferred temperature of approximately 100°C (212°F) and a pressure of around 20 psia (138 kPa-a) with a preferred residence time of approximately 30 min. The removal of water in the stripper section is effected by the injection of steam to help remove light hydrocarbons and residual water from the oil stream which then proceeds to a heat soak vessel which results in a lowering of the ash content of the oil after it is filtered. The heat soak involves heating at least a part of the mixture to a preferred temperature between approximately 340°C and 370°C (645°F and 700°F) with a preferred mean residence time between approximately 15 and 60 min. Nitrogen can be injected as a purge to minimize oxidation of the oil which is then cooled and filtered at a temperature between approximately 100°C and 180°C (212°F and 355°F). The treatment of the streams after this parallels the treatments in the earlier patents.

Audibert (2006) provides some data on the Phillips process applied to two used oil samples, the first with a total metals content of 9500 ppm and sulfur content of 0.44% w/w which produced a finished oil with a total metals content of less than 12 ppm, a sulfur content of 0.04% w/w and a viscosity index of 104. The second used oil sample contained a total metals count of 5800 ppm and sulfur content of 0.4% w/w and yielded a finished oil with a metals content of less than 12 ppm, a sulfur content of 0.03% w/w and a viscosity index of 102. Audibert also comments on the relatively high capital costs and operating costs for this process.

7.5.3 The Exxon/Imperial Oil Process

This process (US Patent 4,512,878 to Exxon dated April 23, 1985; Canadian Patent 1209512 to Imperial Oil dated August 12, 1986; German Patent DE 34 05 858 to Exxon dated August 16, 1984) included a heat soaking step to remove much of the phosphorus that remained in the distillate in the following distillation process, the phosphorus being responsible for catalyst deactivation in a subsequent hydrotreating

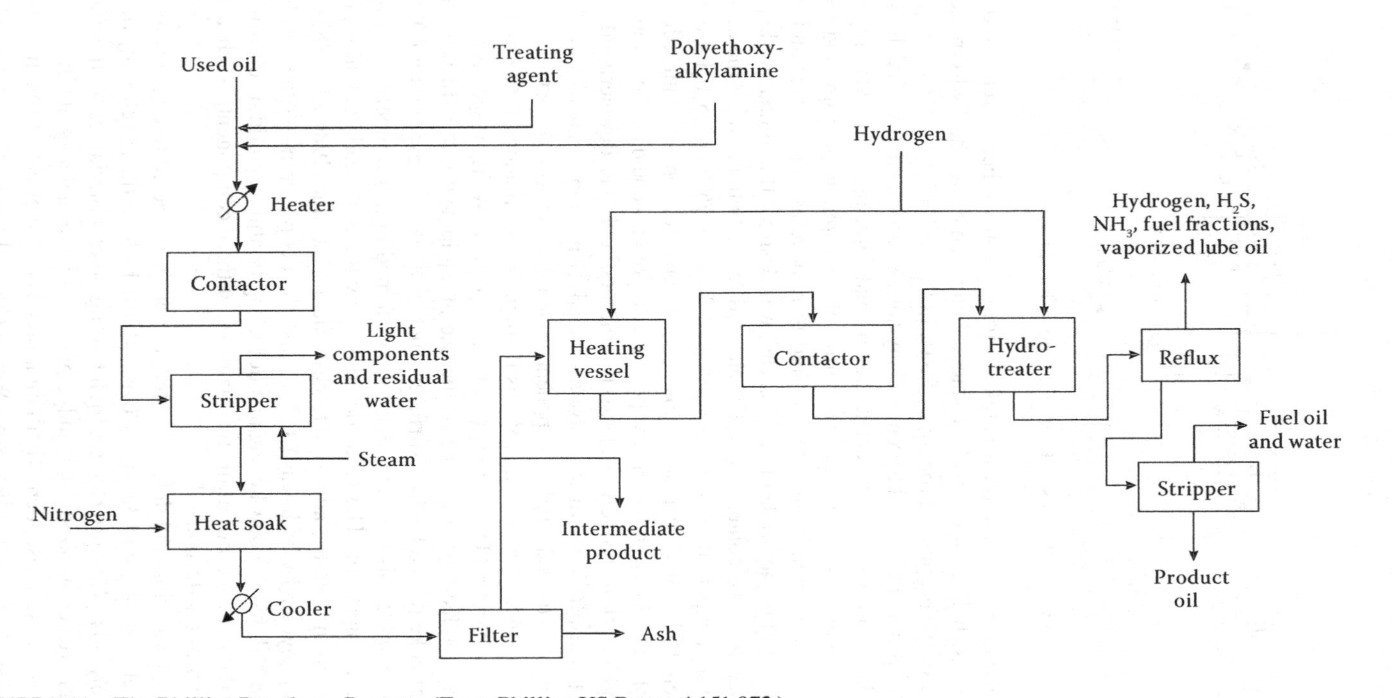

FIGURE 7.11 The Phillips Petroleum Process. (From Phillips US Patent 4,151,072.)

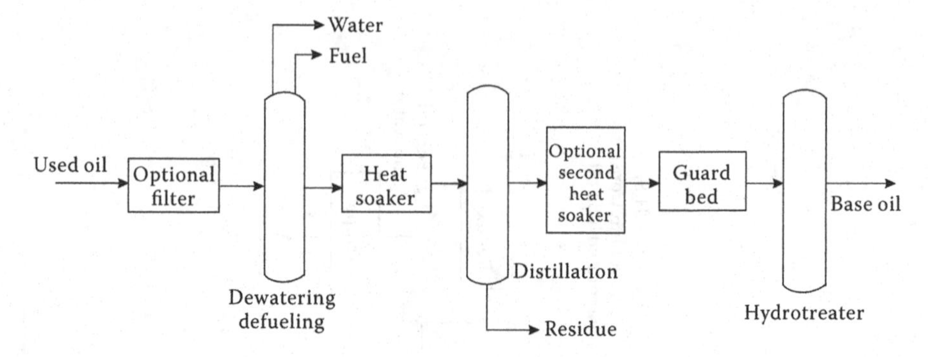

FIGURE 7.12 The Exxon Process. (From Exxon US Patent 4,512,878.)

step. After the heat soak, the oil was taken through a guard bed to remove remaining phosphorus, halides and residual sludge before entering the hydrotreater. An embodiment of the proposed process is shown in Figure 7.12.

The used oil may be filtered to remove any solid matter it contains, such as sand, metal particles or other solids, and enters the optional defueling and dewatering evaporator. This was envisaged to be a combination of atmospheric pressure and vacuum units, and operating conditions were tailored to remove as much water and fuel as possible, considering the grade of base oil that was desired. If a lighter grade oil was desired, more light end content could be tolerated in the bottoms from the defueling step. Operating temperatures between 80°C and 300°C (175°F and 570°F) and pressures between 0.5 kPa (0.07 psi) and ambient conditions are proposed with reflux ratios between 0.5:1 and 5:1, and thin film or wiped film evaporators could be used. Tables 7.15, 7.16, 7.17 and 7.18 show properties of two typical feedstock oils, dewatering/defueling results by batch distillation with an atmospheric equivalent temperature (AET) of 390°C (735°F) required to meet the light component content for an SAE-10 grade oil, continuous atmospheric and vacuum (A and V) distillation, and distillation of the raw used oil in a thin film evaporator (TFE) or a wiped film evaporator (WFE), respectively.

The proposed distillation envisaged the use of equipment that is resistant to the deposition of coke on the heating surfaces, such as a cyclonic evaporator (O'Blasny et al. 1979) or, as in this case and due to the claimed higher efficiency, a thin film or wiped film evaporator. The feedstock oils show the high lead content typical of the contamination from the additives to gasoline at that time (1980 and 1982) in the form of tetraethyl lead which was used to raise the octane rating of the gasoline. It is also suggested that the dewatering and defueling treatment could be eliminated if the used oil did not contain significant quantities of these components, such as might be the case with transformer oil.

After the dewatering and defueling step, the oil is subjected to a heat soak, most preferably between 300 and 320°C (570°F and 610°F) for a time preferably between approximately 30 to 120 min to form high boiling point materials from the majority of the phosphorus compounds and sludge precursors, the majority of which go with the underflow in the subsequent distillation process. The effect of this heat soak is shown in Table 7.19 where a heat soak was performed on the dewatered and defueled oil from Table 7.18 at two temperatures of 280°C and 320°C (535°F and 610°F),

TABLE 7.15
Exxon Process: Properties of Two Used Oil Feedstocks: US Patent 4,512,878

Property	Feedstock A	Feedstock B
Gravity, °API	25.0	25.7
Density, kg/m^3 at 15°C	903.7	899.6
Viscosity, cSt at 40°C	—	50.4
Composition		
Water, liquid % v/v	5	12
Fuels, liquid % v/v	12	15
SAE-10 grade distillate, liquid % v/v	68	57
Residue, liquid % v/v	15	16
Sulfur, % w/w	0.49	0.39
Nitrogen, mass-ppm	NA	550
Halogen (Cl/Br), ppm	770/NA	2500/340
PCB, ppm	<0.1	0.4
Metals, ppm		
Pb	2921	2221
Zn	1229	1025
Ca	1346	1010
Mg	295	166
P	1125	980

Source: Exxon US Patent 4,512,878.

TABLE 7.16
Exxon Process: Batch Distillation Dewatering/Defueling of Used Oil (Feedstock A in Table 7.15)

Property	Dewatering	Defueling
Distillation temperature (vapor), °C	100	228
Distillation pressure, kPa	101 (atmospheric)	0.887
Atmospheric equivalent temperature, °C	100	390
Yield (material removed) from whole used oil, liquid % v/v	i–5	5–17

Source: Exxon US Patent 4,512,878.
Note: The dewatered oil was used as feedstock to the vacuum distillation defueling operation.

and residence times varying between 0.5 and 2 h, and the oil was then distilled in a laboratory scale wiped film evaporator (Pope WFE). It is seen that the heat treatment by passing the oil through a stainless steel coil in a sand bed resulted in a sludge deposition of approximately 75% of the sludge compared with the case with no heat treatment, and a phosphorus content of the distillate of only 3 to 8 ppm compared with the untreated oil when the treatment was at 320°C (610°F) for half an hour.

TABLE 7.17
Exxon Process: Continuous A&V Distillation Dewatering/Defueling of Used Oil (Feedstock B in Table 7.15)

Property	Dewatering	Defueling
Throughput, m^3/day	2.2	1.1
Temperature (bottom), °C	250	270
Pressure, kPa	Atmospheric	3.3
Reflux ratio	None	3:1
Overhead yield, % w/w of feedstock	13.8	12.4

Source: Exxon US Patent 4,512,878.
Note: The dewatering distillate containing 83% w/w water, 17% w/w fuels.

TABLE 7.18
Exxon Process: Thin Film and Wiped Film Evaporator Distillation of Raw Used Oil: Feedstock A (Luwa and Pfaudler are Manufacturers of the Evaporators)

Distillation Equipment	Luwa TFE	Pfaudler WFE
Feedstock rate, kg/m^2-h	344–402	180–288
Hot oil distillation temperature, °C	276	264
Distillation pressure, kPa	1.33–2.0[a]	4
Overhead yield on feedstoc, liquid % v/v	17.0	16.9

Source: Exxon US Patent 4,512,878.
[a] Measured at the exit from an external condenser.

TABLE 7.19
Exxon Process: Effect on Distillate Quality of Pilot Heat Soaking Before WFE (Pope Pilot Plant) Distillation of Dewatered/Defueled Oil from Table 7.17

	Luwa Dewatered/Defueled Oil					
Heat soaking	None	Yes (pilot plant)				
Temperature, °C	—	280	320			
Residence time, h	—	1.0	1.0	0.5	1.0	2.0
WFE (Pope) Distillation						
Temperature, °C	289	292	289	289	292	298
Pressure, kPa	0.5	0.4	0.4	0.5	0.4	0.4
Phosphorus, ppm	240	25	6/3	8	6/3	5
Sludge, ppm at 310°C	2030	450	490	460	490	390

Source: Exxon US Patent 4,512,878.

An additional test used a Pfaudler WFE with an internal de-entrainment device to minimize carry-over to the distillate produced the results represented in Table 7.20 and show a much lower sludge content in this distillate at comparable phosphorus levels compared with Table 7.19, both giving values much less than in the distillate obtained without heat treatment of the oil.

Another test result presented in the patent document was carried out in a large scale pilot plant heat soaker with a throughput of 0.62 m^3 (164 US gals) per day of Feedstock B oil as in Table 7.21 that had been dewatered and defueled as in Table 7.17. After the heat soak, the oil passed to the WFE where it was separated into a distillate and a residue. For comparison purposes, the results from the laboratory scale WFE performed on the non-heat soaked material fed to the heat soaker are also included in this table. It is seen that the phosphorus contained in the distillate decreases with increasing heat soaker temperature. At soak temperatures of less than 280°C (535°F), the phosphorus content of the distillate was high, a condition that would be seriously deleterious to the catalyst life in a subsequent hydrotreatment process. At the other extreme, at temperatures above 340°C (645°F), the dispersant type additives in the oil decomposed and sludge which had been suspended in the oil started to settle out. As this sludge formed approximately 2% v/v of the oil, problems were encountered with plugging in the heat soaker. Also at temperatures above 340°C (645°F), larger amounts of light ends were distilled off and it proved to be impossible to meet the target specification in this case for the viscosity of the lube oil fraction of 29 to 31 cSt at 40°C (104°F). For this reason, an optimal heat soak temperature between 300°C and 320°C (570°F and 610°F) was inferred.

Data is provided in the patent document regarding the blending of the residue from the Pfaudler WFE operated with heat soaked oil with 85/100 penetration paving asphalt and show that up to approximately 8 liquid volume percent of the blend could be made up of the residue material.

TABLE 7.20
Exxon Process: Effect on Distillate Quality of Pilot Heat Soaking Before WFE (Pfaudler WFE) Distillation of Dewatered/Defueled Oil from Table 7.18

Heat soaking	No	Yes
Temperature, °C	—	320
Residence time, h	—	0.5
WFE distillation		
Temperature, °C	278	288
Pressure, kPa	0.19	0.28
Phosphorus, ppm	337	3
Sludge, ppm at 310°C	2860	113

Source: Exxon US Patent 4,512,878.

Note: The oil not heat treated (Feedstock A in Table 7.14) was dewatered and defueled in a Pfaudler WFE as in Table 7.17. The heat treated oil feedstock was dewatered and defueled in a batch distillation unit at approximately 390°C atmospheric equivalent temperature (AET) and the other conditions shown in Table 7.15.

TABLE 7.21
Exxon Process: Pilot Plant Case for Heat Soaking/WFE Distillation of Dewatered/Defueled Oil (Feedstock B in Table 7.17)

Heat soaking	None	Yes (in pilot plant)					
Temperature, °C	—	250	280	300	320	340	
Residence time, h	—	0.5	0.5	0.5	0.5	0.5	
WFE distillation	—						
Temperature, °C	305	300	300	300	300	300	
Pressure, kPa	0.27	0.27	0.27	0.27	0.2	0.2	
WFE distillate							Target
Phosphorus, ppm	72	33	24	18	14	12	
Sludge, ppm	—	—	213	—	116		
Toluene insolubles, ppm	—		55	61	60		
TAN, mg KOH/g	—	—	—	—	—	0.53	—
Viscosity at 40°C, cSt	31.64	29.60	29.40	28.57	29.39	26.50	29–31

Source: Exxon US Patent 4,512,878.

Note: The oil not heat soaked was distilled using the Pope laboratory scale WFE on the heat soaker feedstock.The heat soaking temperature was the heating coil skin temperature.

After the distillation of the heat soaked oil, the distillate still contains halides, trace levels of phosphorus, and carry-over sludge. Removal of the chlorine and bromine is particularly desirable before hydrotreating to avoid the formation of corrosive hydrogen chloride and hydrogen bromide in the hydrotreater. Batch tests were carried out to determine the effectiveness of various adsorbents in removing these three contaminants from the oil and included Fuller's earth, charcoal, lime and activated alumina and it was found that the activated alumina showed the best reduction in phosphorus and halogens. This material was therefore proposed for a guard bed to treat the oil before the hydrotreater, the guard bed to operate at a temperature preferably between approximately 180°C and 340°C (355°F and 645°F), a pressure between about atmospheric and approximately 5 MPa (725 psi), a space velocity preferably between approximately 0.5 and approximately 1 v/v/h, and the preferred use of hydrogen as a gas phase to prevent any coke formation in the guard bed and be compatible with the following hydrotreater. This was studied in the pilot plant facility where a typical distillate oil (Norsk Esso 130N) that had been vacuum-fractionated and filtered through a 3-μm filter was mixed with hydrogen and flowed over a fixed bed of activated alumina at a temperature between 180°C and 320°C (356°F and 608°F), hydrogen at a pressure of 3.4 MPa (493 psi) and a flow rate of 1.5 kmol/m^3 (0.0425 kmol/ft^3), and a space velocity of 1. Table 7.22 shows the results of this test where the phosphorus and halide contents were reduced with increasing temperature to levels of 1 ppm and 17 ppm, respectively.

A pilot plant scale test was also reported in the patent document where activated alumina was used to reduce the phosphorus and halide content of a distillate obtained from Feedstock B that had been dewatered and defueled as in Tables 7.17 and 7.21. The results in this case are shown in Table 7.23 and show a large decrease

TABLE 7.22
Exxon Process: Pilot Plant Treatment of Typical Luwa Distillate with Alumina

		Luwa TFE Distillate Treatment Temperature, °C			
	Feedstock	180	280	300	320
Phosphorus, ppm	77	55	6	3	1
Halogens, ppm	90	80	60	30	17

Source: Exxon US Patent 4,512,878.

in the content of phosphorus, halides and trace metals. This stream is then suitable as feed to the hydrotreatment process.

The hydrotreatment proposed was carried out over a conventional catalyst of nickel/molybdenum, although cobalt/molybdenum catalyst could also be used, at temperatures preferably between 260°C and 320°C (500°F and 610°F), a preferred hydrogen

TABLE 7.23
Exxon Process: Pilot Plant Test of Activated Alumina Treatment of Distillate from WFE Fed with Heat Soaked Oil from Feedstock B Dewatered and Defueled as in Tables 7.17 and 7.21

	Feedstock	Treated Oil
Conditions		
Liquid hourly space velocity (LHSV)	—	1.0
Gas rate, kmol/m^3	—	7.5
Temperature, °C	—	320
Pressure, MPa hydrogen	—	3.5
Inspections		
Density, kg/m^3 at 15°C	870.3	870.3
Viscosity, cSt at 40°C	30.46	30.71
Viscosity, cSt at 100°C	5.17	5.20
Viscosity index	98	98
Total nitrogen, ppm by mass	150	160
Basic nitrogen, ppm by mass	36	41
Sulfur, % w/w	0.25	0.23
Color by ASTM	D8	D8
Phosphorus, ppm by mass	20	<3
Halogens, ppm by mass	140	61
Metals		
Pb	12	1
Si	11	1
Ca	7	2
Zn	5	2

Source: Exxon US Patent 4,512,878.

pressure between approximately 3 and 5 MPa (435 and 725 psi), a preferred space velocity of between approximately 0.5 and 2 v/v/h, and a gas flow rate of preferably between 1.5 and 5.0 kmol/m^3 (0.0425 and 0.142 kmol/ft^3). The properties of a target product—SAE-10 grade oil—were compared with SAE-10 oil made from Western Canadian crude and the results are shown in Table 7.24. Various possible catalysts and carriers are suggested with a preferred catalyst being cobalt/molybdenum on an alumina carrier, or nickel/molybdenum on alumina, where the cobalt, nickel and molybdenum are in the elemental, oxide or sulfide form, but preferably as a sulfide.

It was found that the alumina guard bed was effective in removing the phosphorus and halide compounds from the TFE or WFE distillates when the feedstock oil had been heat soaked, but oil which had not been heat soaked contained too many of the sludge precursors and these caused plugging of the guard bed after approximately 200 h of operation. An optional second heat soak and settling step was also

TABLE 7.24
Exxon Process: Hydrofining Process Applied to Alumina Treated Oil

	Feedstock	Hydrofining Product		New Oil SAE-10 Grade
Conditions				
Liquid hourly space velocity (LHSV)			1.0	
Gas rate, kmol/m^3			7.5	
Temperature, °C		290	284	
Pressure, MPa hydrogen		3.5	5.6	
Inspections				
Density, kg/m^3 at 15°C	870.3	868.2	868.2	874.1
Viscosity, cSt at 40°C	30.71	29.46	29.55	29.50
Viscosity, cSt at 100°C	5.20	5.09	5.10	4.97
Viscosity index	98	99	99	90
Total nitrogen, mass ppm	160	43	47	
Basic nitrogen, mass ppm	41	23	24	
Sulfur, % w/w	0.23	0.09	0.10	0.07–0.10
Color, ASTM	D8	<1.0	<1.0	<1.0
Halogens, ppm	61	<2	<2	—
Phosphorus, ppm	<3	<1	<1	—
TAN mg KOH/g	—	0.02	0.01	Typ. 0.02
Aromatics, % w/w	—	14.7	14.8	
Saturates, % w/w	—	83.3	83.2	Min. 80
Polars, % w/w	—	1.7	1.8	
GC distillation, ASTM D2887				
IBP, °C		307	308	
5%, °C		359	359	
10%, °C		371	372	
Volatility, % off at 368°C		8	8.5	10 max

Source: Exxon US Patent 4,512,878.

TABLE 7.25
Exxon Process: Effect of Optional Second Heat Soak/Settling on Toluene Insolubles Contained in WFE Distillate

	Toluene Insolubles, ppm Second Heat Soak Residence Time, h				
Temperature, °C	**0 (Fresh Distillate)**	**24**	**40**	**72**	**120**
25	230	112	110	—	100
150	298	92	44	15	7

Source: Exxon US Patent 4,512,878.

proposed for the distillate coming from the TFE or WFE as shown in Figure 7.12 when the sludge content, as measured as toluene insoluble constituents, was higher than approximately 100 to 300 ppm. It was found that the first heat soak resulted in a distillate from the WFE that still contained approximately 150 to 400 ppm of residual sludge at 310°C (590°F). This sludge was made up of 100 to 300 ppm of toluene insoluble constituents at room temperature, and approximately 50 to 100 ppm of sludge that can form at the high temperatures of 300°C to 310°C (570°F to 590°F) from the sludge precursors. The second heat soak was aimed at reducing these sludge components by letting the oil stand at a temperature between 125°C and 150°C (255°F and 300°F) for 48 to 72 h. The results obtained from a distillate made from oil that had been heat soaked at 320°C (610°F) for half an hour and then distilled in the laboratory in the Pope WFE evaporator are shown in Table 7.25 where the yield of toluene insoluble constituents decreased from approximately 200 ppm to approximately 15 ppm and the sludge precursors were reduced from approximately 300 ppm to less than 150 ppm.

Comparing the final composition and properties of the oil with the present requirements for lube oil base stocks, it is seen that the viscosity index is higher than the minimum required for a Group I or a Group II oil, but the sulfur and saturates requirements meet only the requirements for a Group I oil. If a Group II base oil is required, it is possible that a more vigorous hydrotreatment or alternative catalysts might lower the sulfur content and increase the proportion of saturated constituents.

7.5.4 The Kinetics Technology International B.V. (KTI) Process

This process (US Patent 4,941,967 dated July 17, 1990) takes a feedstock of used oil that has been filtered and dewatered, passes it through a predistillation column operating under vacuum and with a short residence time to remove a lighter gas oil component, the underflow passing to either a single wiped film evaporator, or two wiped film evaporators to produce an overhead product that is taken to a settling tank before joining the gas oil stream and passing to hydrogenation and distillation columns to produce diesel oil and base oil products. The underflow from the last wiped film evaporator constitutes the residue from the process. This is shown in Figure 7.13

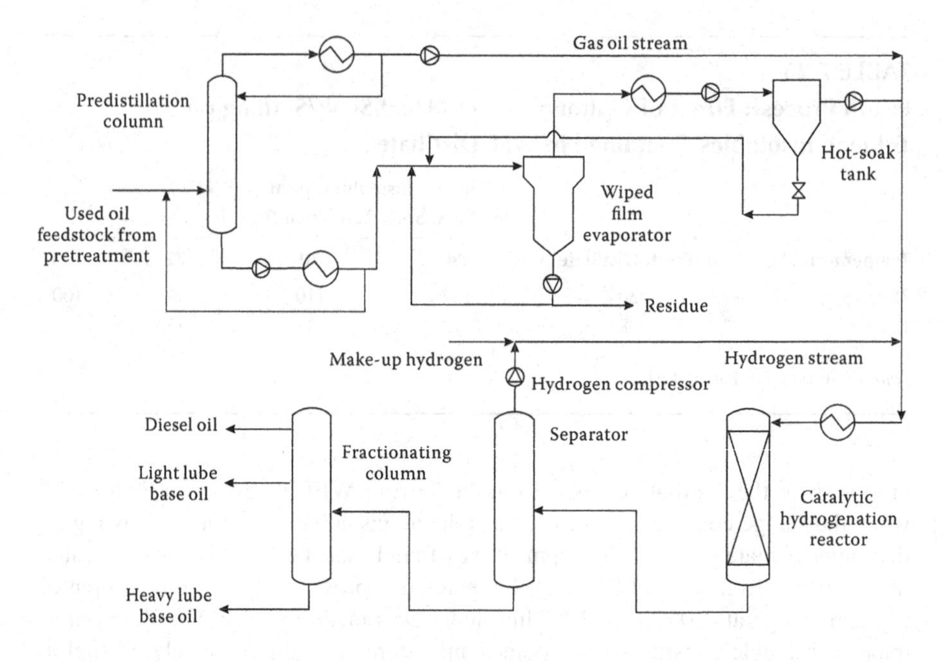

FIGURE 7.13 The KTI process with single wiped film evaporator (example 1 version). (From KTI US Patent 4,941,967.)

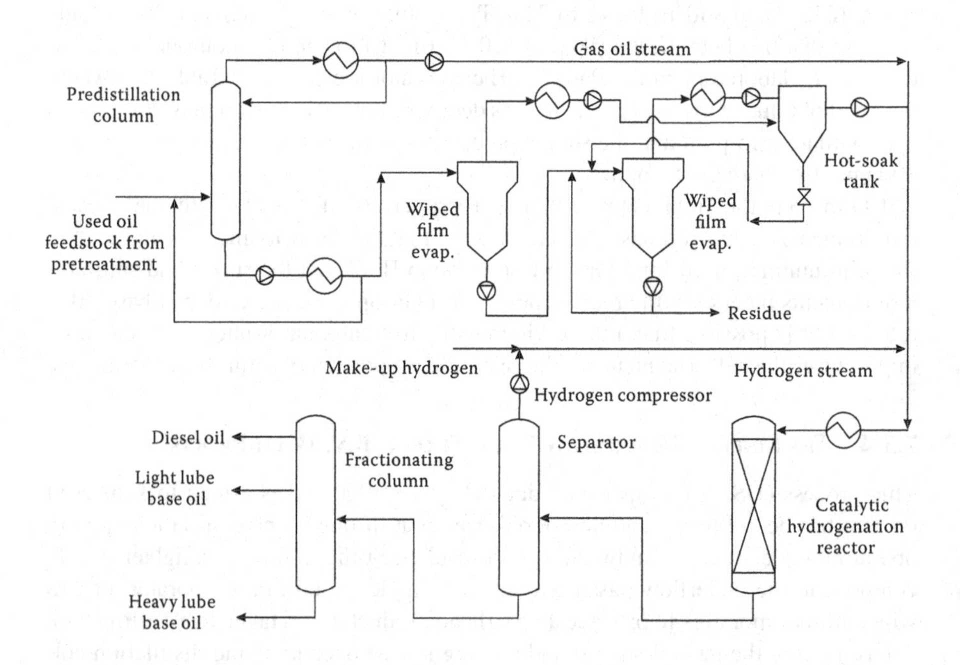

FIGURE 7.14 The KTI process with two wiped film evaporators (example 2 version). (From KTI US Patent 4,941,967.)

for the single wiped film evaporator case (Example 1), and in Figure 7.14 for the two wiped film evaporator case (Example 2).

7.5.4.1 Version 1 with a Single Wiped Film Evaporator

Before entering the process, the used oil is filtered and distilled to remove sludge forming impurities, water and light components such as gasoline and this is the pre-treated oil feedstock to the process in Figures 7.13 and 7.14. This enters the predistillation column which operates under a low pressure of 2 kPa (0.29 psi) and at a temperature of 220°C (430°F). The gas oil fraction evaporated is condensed and partially refluxed into the predistillation column. The underflow stream with the lube oil fractions is heated and splits into two streams, one being recycled into the incoming feedstock stream to the predistillation column, and the other going on to a wiped film evaporator operating at a pressure of 0.2 kPa (0.029 psi) and a temperature of 345°C (653°F). The underflow from the wiped film evaporator splits into two streams, one being recycled into the feedstock stream to the wiped film evaporator and the other being the residue from the process. The overflow from the wiped film evaporator, the light lube oil fractions, are condensed and enter a hot-soak vessel where heavy impurities settle out, the underflow from the hot-soak vessel containing these impurities is returned to join the feedstock stream into the wiped film evaporator. The materials in the hot-soak vessel are maintained at a temperature between 150°C and 250°C (300°F and 480°F) for a residence time of between 11 and 30 h.

The condensate from the top of the hot-soak vessel is combined with the gas oil stream from the predistillation column and forms the feedstock to the second part of the process. This feedstock is combined with a stream of hydrogen and enters the catalytic hydrogenation vessel operating at 320°C (610°F) and 6 MPa (870 psi) to reduce the sulfur, nitrogen, chlorine and oxygen content of this stream. The hydrogenated stream passes to a separation vessel where the free hydrogen is removed to be recompressed and recycled, the hydrocarbon stream passing to a fractionation column which separates it into a diesel oil fraction, a light lube oil base stock and a heavy lube oil base stock.

7.5.4.2 Version 2 with Two Wiped Film Evaporators

This process is very similar to the first but with a second wiped film evaporator being used as shown in Figure 7.14. The light lube oil fraction from the first evaporator is taken to the hot-soak vessel whereas the underflow is taken to the second evaporator. The overflow from the second evaporator is taken to the soak vessel whereas the underflow is again split into two streams, the one being recycled into the feedstock stream to the second evaporator and the other forming the residue stream from the process. The two evaporators operate at different conditions which are tabulated in Table 7.26, together with some of the results obtained under these operating conditions for the two versions of the process.

Audibert (2006) provides information on the location of plants built according to this process, the KTI relube process, as being in Greece for LPC in 1982 and in Tunisia in 1983, and that KTI subsequently licensed new units or provided upgrading to plants in Europe (Haberland), Africa (Alexandria), the Middle East (Syria) and the USA (Evergreen Oil).

TABLE 7.26
KTI Process Data: US Patent 4,941,967

	Units	Example 1	Example 2
Temperature in predistillation column	°C	220	220
Pressure in predistillation column	kPa	2	2
Temperature in first wiped film evaporator	°C	345	320
Pressure in first wiped film evaporator	kPa	0.2	1.5
Temperature in second wiped film evaporator	°C	—	345
Pressure in second wiped film evaporator	kPa	—	0.15
Temperature in hot-soak tank	°C	180	?
Residence time in hot-soak tank	h	24	26
Temperature in hydrotreater	°C	320	320
Pressure in hydrotreater	kPa	6000	6000
Temperature in fractionation column	°C	200	200
Pressure in fractionation column	kPa	3	3
Feedstock rate of dry used lubricating oil	kg/h	5000	5000
Gas oil fraction from predistillation column	kg/h	410	120
Condensate (free from impurities) from hot-soak tank	kg/h	4180	2560
Residue product from wiped film evaporator	kg/h	310	280
Residue recycling rate of bottoms from evaporator	kg/h	800	200
Diesel fuel obtained as product	kg/h	520	190
Total lubricating base oil product	kg/h	4020	2460

Source: KTI US Patent 4,941,967.
Note: Example 1 for single wiped film evaporator version. Example 2 for two wiped film evaporators version.

7.5.5 The Viscolube/Axens Revivoil Process

The Viscolube plants in Italy (Fehrenbach 2005) are now based on the so-called Revivoil process developed jointly with the French Axens company (a group company of the Institut Français du Pétrole or IFP Energies Nouvelles) to replace the treatment with clay with a hydrogenation process. Essentially, the TDA distillation step technology contributed by Viscolube was combined with the catalytic hydrogenation expertise of IFP to define the new Revivoil process to produce Group II base oil products. Audibert (2006) suggests that for small capacity plants of less than 10,000 tonnes per year (11,000 short tons per year; 22,050,000 pounds), a clay adsorption treatment is suitable for the refining stage, whereas hydrotreatment is used for plants with capacities greater than 30,000 tonnes per year (33,000 short tons per year; 66,150,000 pounds).

The technology has been applied at a plant in Ceccano, Italy (Fehrenbach 2005) and the plant at Pieve Fissiraga, Italy has been using TDA and hydrotreatment since June 2003 (Audibert 2006), and reference is also made to the plants in Indonesia, Poland and two plants in Spain. The Group II base oil products have also been tested for mutagenicity and polynuclear aromatic or polycyclic aromatic compounds, showing acceptable low levels in the product streams.

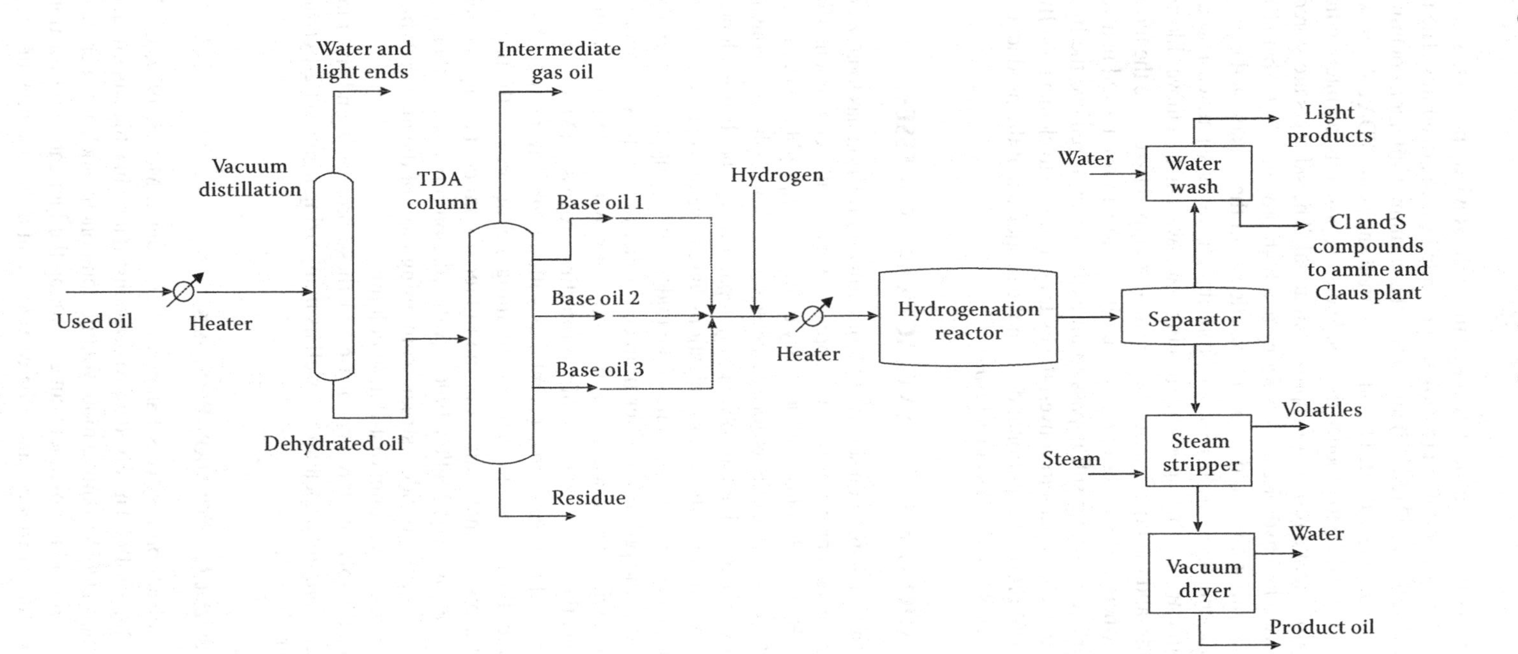

FIGURE 7.15 The Viscolube/Axens Revivoil Process. (From description by Fehrenbach 2005.)

Figure 7.15 shows the schematic flow diagram for the Revivoil process as described by Fehrenbach (2005) and Audibert (2006) has presented similar diagrams. The incoming used oil is heated to 140°C (285°F) and enters the vacuum distillation column where water and light ends are flashed off. The oil stream then enters the TDA column where it is distilled at 360°C (680°F) to produce three side cuts of base oils, an intermediate gas oil overhead product, and a residue of asphaltic material. The side cuts are stored to be separately processed in a hydrotreater operating at a high pressure of 100 bar (10 MPa; 1450 psi) with a catalyst chosen to favor the reaction with the unsaturated compounds of sulfur and nitrogen. The product stream from the reactor is separated into a vapor stream and a liquid stream, the vapor stream being water-washed to remove chlorine and sulfur compounds, and the liquid stream being steam-stripped to remove the most volatile compounds, whereas the water remaining in the oil stream is removed in a vacuum dryer. The streams containing sulfur pass to an amine plant to remove the hydrogen sulfide and then to a Claus plant to convert the hydrogen sulfide to elemental sulfur.

Audibert (2006) presents additional data on the properties of the product oils and the pertinent process economics at that time.

7.6 DISTILLATION/SOLVENT EXTRACTION PROCESSES

Solvent extraction of a distillation separated lubricating oil fraction has long been used as a means of testing the lubricating oil for the presence of asphaltic constituents (Speight and Ozum 2002; Hsu and Robinson 2006; Gary et al. 2007; Speight 2014). Indeed, one of the commonly-used test methods for the determination of asphaltene constituents (ASTM D2007) was a combination solvent-clay separation on the distilled lubricating oil fraction. Afterward, a number of organizations have developed processes that have a solvent extraction step to separate some of the undesirable components from the lube oil fraction. Some of the processes differ from each other in that the solvent or diluent used may combine with the lubricating oil fraction or it may be immiscible with it.

As an example, the first step in one of the Interline processes (Mellen 1996) is to mix the stream with the solvent which combines with the lube oil fraction. A settling tank then separates the heavier underflow containing the majority of the heavy metal compounds and asphaltic materials from the lighter stream containing the lube oil fraction and the solvent (e.g., Mellen 1996). A distillation process can then be used to recycle the solvent and provide a stream containing most of the lube oil which can then be fractionated in a vacuum distillation column.

Various aliphatic solvents have been used in these processes and the majority have used propane, butane or pentane as well as many using *n*-methyl-2-pyrrolidone (NM2P) or other chemicals.

7.6.1 THE LUBRIZOL CORPORATION PROCESSES

This immiscible solvent process (US Patent 4,021,333 dated May 3, 1977; US Patent 4,154,670 dated May 15, 1979) involved a two-stage distillation in which an initial distillate fraction is evaporated containing more volatile components such as water, gasoline, kerosene and fuel oil, and a subsequent cut containing the lubricating oil fraction. It is stated that the distillation would usually be carried out at a reduced pressure of 1.5 to

10 torr (mm Hg), the first distillate fraction being conducted with a vapor temperature of up to approximately 250°C (480°F), and the second cut up to approximately 290°C (555°F) at these pressures. The liquid residue would be used as an extender for asphalt or rubber whereas the lubricating oil fraction is taken to a liquid–liquid extraction column where an organic liquid that is immiscible with the oil fraction is used to remove impurities that are soluble in the extractant at the 20°C to 50°C (68°F and 122°F) and atmospheric pressure conditions. The next step would be to allow the mixture to separate into two layers, one being the lubricating oil fraction and the other the extractant with impurities, and hence separate the lubricating oil which could be further distilled to separate any small amount of residual extractant from the lubricating oil. The extractant stream with the majority of the impurities can be separated from the impurities by distillation and the extractant is recycled for reuse. A range of extractant organic liquids is proposed, generally with a boiling point between 120°C and 225°C (250°F and 435°F) at atmospheric pressure and a specific gravity with respect to water at 4°C (39°F) between 0.90 and 1.15, with particular preference being expressed for ethylene glycol monomethyl ether, dimethylformamide and *n*-methyl-2-pyrrolidone.

It is also proposed that a preliminary step could involve dilution of the feedstock material with an organic material that would allow solid impurities to be removed by filtration or centrifugation. An additional option involved heating the incoming feedstock with an aqueous solution of a strongly alkaline material such as sodium hydroxide to concentrate metallic constituents, including metal-containing additives, of the used oil in a sludge which can be readily removed in the subsequent dilution step to provide a cleaner feedstock to the distillation process.

Little information is provided on the impurities that are expected to be removed by the extractant liquid in the liquid-liquid extraction process, apart from a reference to degradation products derived from the oil itself or from additives therein being contained in the recycled oil.

The process does contemplate the treatment of petroleum-based mineral oils as well as synthetic oils of various types.

A subsequent patent to the same company, US Patent 4,154,670 dated May 15, 1979, proposes a similar process without the initial distillation process. The feedstock is mixed with a diluent and insoluble impurities are removed by decantation, filtration or centrifugation. The liquid-liquid extraction is then carried out with a range of suggested extractants with the same preferred extractants as in the first patent. Vacuum distillation of the oil then removes the diluent at temperatures up to approximately 125°C (257°F) and also other volatiles such as fuel components that may be dissolved in the oil. It is suggested that the product lubricating oil fraction may be subjected to further refining steps such as hydrogenation, solvent extraction, treatment with clay or similar processes, and would be suitable for use as a lubricant, bunker fuel, petrochemical intermediate or similar application.

7.6.2 The Bartlesville Energy Research Center (BERC)/ US Dept. of Energy (DOE) Process

This process (US Patent 4,073,719 dated February 14, 1978; US Patent 4,073,720 dated February 14, 1978) was developed in an effort to move away from the use of strong acids and caustics in the treatment of the used oil which were said to remove higher molecular weight diaromatic and polynuclear aromatic polar compounds that were associated with natural lubricity of the base oil with an adverse effect on this parameter of the lubricant product. It was claimed that the sludge produced in the process was chemically neutral and could be used as a road surfacing agent or as a source of heavy metals, that the amount of waste generated was less than in other processes, that between 60% and 75% of the used oil feedstock was recovered for recycling as a base oil stock, and that all of the purification steps were mild, thus maintaining the natural lubricity and anti-oxidation properties of the recycled petroleum. The process is also of interest in the mixture of solvents that were proposed in this patent. Figure 7.16 shows part of the flow diagram for the process described in this patent document, the rest of the diagram showed the blending and additive mixing steps and these have been omitted from Figure 7.16.

The used oil enters a vacuum distillation unit to strip water and volatile hydrocarbons to prevent the formation of azeotropes with the solvent which would interfere with later solvent recovery. This unit operates at a temperature between approximately 150°C and 175°C (300°F and 350°F) and a pressure between approximately 2 and 10 mm Hg (0.27 and 1.33 kPa). The stripped oil is then mixed with a solvent in a ratio of approximately 1 part oil to 3 parts solvent, the solvent consisting of a mixture of approximately 1 part 2-propanol with 1 part methylethyl ketone and 2 parts 1-butanol, this resulting in the oil dissolving in the solvent and precipitating a sludge of metal compounds, oxidation products and other impurities. This mixing takes place preferably at ambient temperature, temperatures down to approximately 10°C increasing the effectiveness of the solvent in precipitating more of the metal compounds and other undesirable components whereas temperatures up to approximately 30°C to 40°C (86°F–104°F) reduce the effectiveness. Generally, approximately 10% w/w of the oil is precipitated in the sludge.

The solvent and oil mixture is then separated from the sludge, either by allowing the sludge to settle in a tank overnight and then decanting the solvent and oil mixture, or a centrifuge could be used immediately after mixing in either a continuous or a batch operation.

The solvent and oil mixture then enters a solvent recovery unit consisting of an evaporator/stripper system which, in a pilot scale study consisted of a continuous-feed distillation column operating at a pressure of 150 mm Hg absolute (20 kPa-a) at 175°C (345°F). As the residual solvent content of the oil at these conditions was found to be approximately 0.1% of the solvent, a second pass through the column at a pressure of 1 mm Hg absolute (0.13 kPa-a) was made to reduce this residual value.

The oil then passes to a vacuum distillation unit to remove additional impurities such as volatile hydrocarbons and the remaining asphaltene constituents and metals. Either a single overhead cut or multiple cuts, as shown in Figure 7.16, may be taken where the operating temperature in the column is between 300°F and 600°F (149°C and 316°C) at a pressure between 100 and 200 mm Hg (13.3 and 26.6 kPa) so that the temperature of the oil is kept below the point at which cracking could occur. If multiple cuts are taken, the streams would be stored and taken separately to the next stage of treatment.

The side streams of base oil boiling fractions are then taken either to a clay treatment or a mild hydrogenation treatment to decolorize and deodorize the oil before taking it

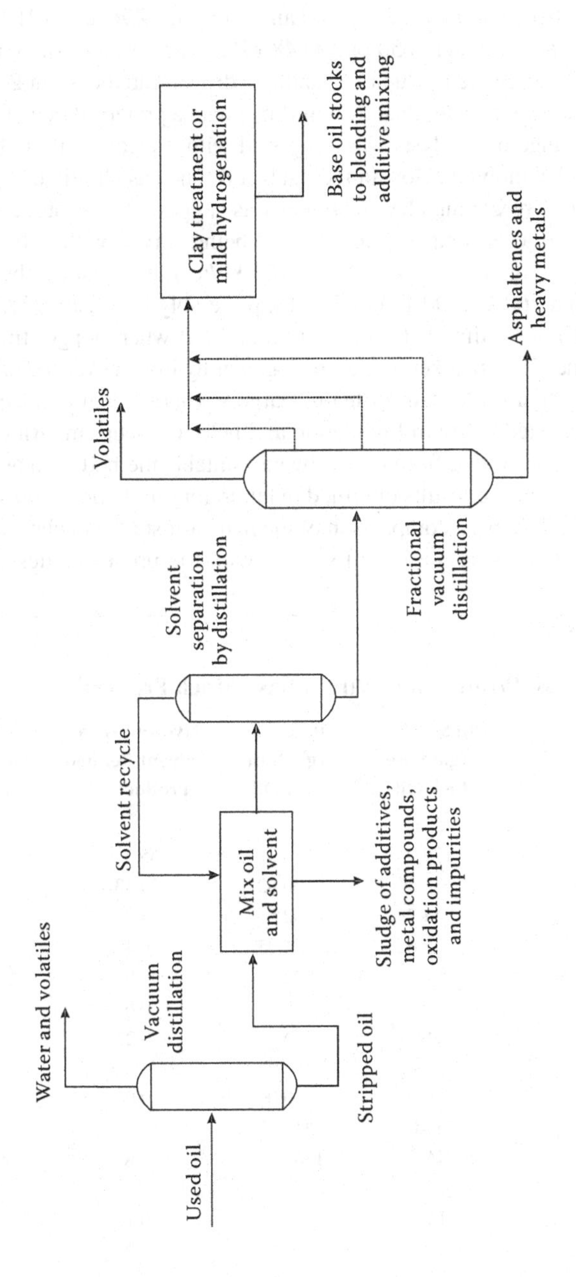

FIGURE 7.16 The BERC/DOE process for base oil recovery. (From BERC/DOE US Patent 4,073,719.)

to a blending and additive mixing process. The preferred treatment was the mild hydrogenation alternative, which operates at a temperature of approximately 260°C to 371°C (500°F–700°F) with a preferred temperature in the range of 316°C (600°F), a hydrogen partial pressure between approximately 400 and 900 psig (2.76 and 6.21 MPa-g) with a preferred value approximately 650 psig (4.48 MPa-g), a space velocity between 0.5 and 2.5 v/v/h with a preferred value of 1, and a hydrogen rate between 250 and 2000 standard cubic feet per barrel (44.5 and 35 nm^3/m^3) with a preferred rate of 1500 SCFB (267 nm^3/m^3). A range of catalysts was suggested with typical catalysts being cobalt molybdate and nickel molybdate on an inert substrate such as alumina.

An acid-activated bleaching clay treatment was suggested if a source of hydrogen was not available at a reasonable price, the oil being mixed with between 0.2 and 1 pounds of clay per gallon of oil (24 and 120 kg/m^3) and heating the mixture to between 150°C and 260°C (300°F and 500°F), preferably between 195°C and 215°C (380°F and 420°F), for a time between 30 min and 3 h where longer times increase the oxidation of the oil. A blanket of inert nitrogen or hydrogen over the oil in the tank could reduce oxidation of the oil. A steam sparge—where steam is injected into the oil—was also suggested to control oxidation and help to sweep impurities out of the oil. The clay is then separated from the oil by any suitable method such as filtration.

Several examples of the results obtained in laboratory and pilot plant studies were given and Table 7.27 shows a comparison of the hydrofinished and clay-treated products obtained by processing a used oil sample with the processes described in the

TABLE 7.27
BERC/DOE Process: Product and Virgin Base Stock Properties

Property	190 SUS SN Virgin Base Stock Oil	150 SUS SN Virgin Base Stock Oil	Hydrofinished Solvent-Refined Product Oil	Clay-Treated Solvent-Refined Product Oil
Viscosity				
SUS at 100°F	179.0	144.3	165.5	182.9
cSt at 100°F	38.30	30.62	35.33	39.15
SUS at 210°F	44.7	42.5	44.2	45.6
cSt at 210°F	5.62	4.95	5.49	5.91
Viscosity index	91	92	99.8	1.3
Acid number	0.0	0.0	0.0	0.0
Carbon residue, Ramsbottom %	NA	NA	0.23	0.23
Ash, %	0.00	0.00	0.00	0.00
Aniline point, °F	217.0	217.7	218.1	220.0
Oxidation stability, ASTM D943, h	NA	1364[a]	NA	1340[a]
Copper corrosion, ASTM D130	1a	1a	1a	1a

Source: BERC/DOE US Patent.

[a] Base stock contained 0.3% w/w BHT oxidation inhibitor and 0.05% corrosion inhibitor for D943 test only.

patent document, compared with two virgin base stock oil results. Based on the properties in the table, it was claimed that the re-refined oils were indistinguishable from the virgin base stock oils. No information was provided on the sulfur content or polynuclear aromatic hydrocarbon (PAN) levels in the product oils.

7.6.3 The Snamprogetti Processes

The first patent from 1979 (German patent DE 29 01 090 dated July 19, 1979; US Patent 4,406,778 dated September 27, 1983) was widely patented in Europe at a time when the move away from acid and clay treatments for the re-refining process was becoming more popular. This patent proposed a predistillation step, followed by a solvent extraction and then vacuum distillation to obtain the lighter base oil components which were separated from the heavy base oils. The heavy base oils were then subjected to a heat treatment to modify the structure of the contaminants, mainly the sulfonates and phenates of calcium, barium and magnesium added to the lube oils as detergents and surfactants, to make them less soluble in the lubricating oil, followed by a second solvent extraction of this heavy base oil stream. All the base oil product streams were stored in tanks and were then taken to a catalytic hydrofinishing process which operated on each fraction separately. It was claimed that the heat treatment of only the heavy base oil fraction resulted in better heat efficiencies for the process, that the second extraction step could be better optimized and adjusted in treating only a part of the recycled oil, and that longer catalyst life was obtained in the hydrofinishing step due to the efficient removal of contaminants in the foregoing steps. This first process is shown in Figure 7.17 where the dashed lines represent the operation on the heavy base oil fraction which is carried out in an alternating fashion with the treatment of the full feed stream.

The used oil feedstock is preheated in the first furnace to a temperature between 180°C and 230°C (355°F and 445°F) and passes to the predistillation column where water and light hydrocarbons are taken off overhead whereas the base oils and

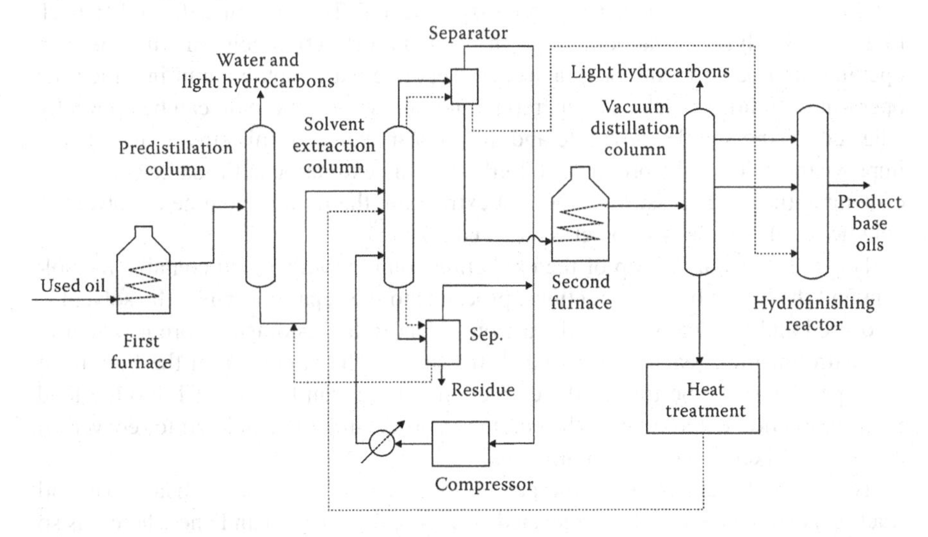

FIGURE 7.17 The Snamprogetti Process. (From Snamprogetti Patent DE 29 01 090.)

contaminants leave the bottom of the column. The solvent, which is a normal paraffin of low molecular mass and where propane is particularly suitable, enters the column near the base whereas the underflow from the predistillation enters the column near the top, the ratio of the volume flow of solvent to volume flow of oil being comparatively low at between 3 and 10 because maximum contaminant removal is not required in this step.

The extraction is carried out at a temperature between 30°C (86°F) and the critical temperature of the solvent and a pressure between 24.5 and 49 bar (2.45 and 4.9 MPa; 355 and 710 psi). The oil and most of the solvent leave from the top of the extraction column whereas some solvent and the contaminants leave from the bottom of the column, both these streams going to separators where the solvent is recovered and recycled through a compressor. The underflow from the bottom of the lower separator contains the contaminants. The overhead product, containing the base oils, exits the bottom of the overhead separator and proceeds through a second furnace where it is heated and led to a vacuum distillation column where a sump temperature of over 300°C (570°F) is maintained to separate it, for example, into the remaining light hydrocarbons from the top of the column, two sidestreams of base oil components and a stream from the bottom of the column of the heavy base oil components and the remaining contaminants. The sidestreams are taken to tank storage before the next step of hydrofinishing which operates separately on these to produce final base oil products of different viscosities.

The hydrofinishing operation with hydrogen is carried out over catalysts consisting of the sulfides of metals of the Group VI and VIII periodic table of elements on an aluminum oxide carrier, operating at a temperature between 250°C and 420°C (480°F and 790°F), a space velocity between 0.1 and 5 volumes/volume per hour, and the recycled hydrogen rate of between 15 and 850 normal-liters per liter (2 and 113.3 scf/US gal).

The heavy base oil stream from the bottom of the vacuum distillation column enters a holding tank where it undergoes an adiabatic heat soak between 300°C and 450°C (570°F and 840°F) for between 1 and 120 min for the reactions with the sulfonates and phenates to proceed as previously indicated. This stream is then taken back to a second solvent extraction, either in the same extraction column which is then operated in a batch mode, or in a second, separate extraction column in which the operating conditions such as temperature and ratio of solvent to oil can be optimally adjusted for the smaller flow rate and different make-up of this stream and also to improve the color of the product oil. In this second extraction at the same conditions of temperature and pressure as the first extraction, the ratio of volume of solvent to volume of oil is increased to between 5 and 20.

The stream from the top of the extraction column, which now contains the solvent and the heavy base oil fractions, proceeds to the separator where the solvent is recovered and the heavy base oil fraction is sent to tank storage before it is sent to the hydrofinishing reactor for its batch treatment. The stream from the bottom of the separator at the bottom of the extraction column can be recycled into the feed to the extraction column when the column is processing the whole oil to recover oil remaining mixed with the contaminants.

Two examples are given in the patent document, one where the heavy base oil fraction is not heat-soaked and recycled to a second extraction and one where it is so treated. Tables 7.28 and 7.29 show some of the properties of the streams in each case.

TABLE 7.28
Snamprogetti Process Patent DE 29 01 090: Single Extraction Process for Heavy Base Oil Fraction

	Process Yield %	ASTM Color	Viscosity cSt at 210°F	Neutr. No. mg KOH/g	Metals (ppm)						
					Ca	Ba	Zn	Pb	P	Cl	Br
1. Feed	100.0	>8	14.8	8.6							
2. Predistillation and heat treatment											
Water	6.9			10.52							
Light hydrocarbons	6.1			3.56							
Residue	87.0	>8	14.08	7.7	1500	420	770	2350	980	250	320
3. Extraction with propane											
Refined material	93.66	>8	10.07	0.34	10	<5	<5	<5	35	49	—
Residue	6.34										
4. Vacuum fractionation											
Gas oil	2.84										
Light base oils	26.95	3.5	5.23	0.48	<5	<5	<5	<5	45	60	43
Middle base oils	44.47	4.5	9.92	0.30	<5	<5	<5	<5	10	17	10
Heavy base oils	25.74	>8	32.39	0.38	36	<5	10	16	70	25	<5
5. Hydrofinishing											
Light base oils	98.0	1.5	5.1	<0.03	<5	<5	<5	<5	10	<5	<5
Middle base oils	98.0	2	9.5	<0.03	<5	<5	<5	<5	10	<5	<5
Heavy base oils	98.0	>8	30.0	<0.03	<5	<5	<5	<5	10	<5	<5

Source: Snamprogetti Patent DE 29 01 090.

TABLE 7.29
Snamprogetti Process Patent DE 29 01 090: Heat Treatment and Recycling of Heavy Base Oil Fraction

	Process Yield %	ASTM Color	Viscosity cSt at 210°F	Neutr. No. mg KOH/g	Metals (ppm)						
					Ca	Ba	Zn	Pb	P	Cl	Br
1. Feed	100.0	>8	14.68								
2. Predistillation and heat treatment											
Water	6.9			4.51							
Light hydrocarbons	3.4			2.19							
Residue	89.7		13.27	3.16	1560	400	900	2600	850	650	580
3. Extraction with propane											
Refined material	94.2		9.94	0.81	230	<5	125	120	470	140	50
Residue	5.8	>8									
4. Vacuum fractionation											
Light base oils	28.0	4	5.15	0.44	<5	<5	<5	<5	180	95	20
Middle base oils	47.8	7	9.58	0.28	<5	<5	<5	<5	18	20	<5
Heavy base oils	23.3	>8	31.32	1.35	1000	10	530	510	1490	30	18
5. Heat treatment of heavy	100.0		30.5	1.33	980	10	500	570	1460	25	12
6. Extraction of heavy with C_3											
Refined material	93.0	<7.5	28.4	0.13	<5	<5	<5	<5	10	15	<5
Residue	7.0										
7. Hydrofinishing											
Light base oils	98.0	<2	4.98	<0.03	<5	<5	<5	<5	<10	<5	<5
Middle base oils	98.0	<2.5	9.3	<0.03	<5	<5	<5	<5	<10	<5	<5
Heavy base oils	98.0	<4	28.0	<0.03	<5	<5	<5	<5	<10	<5	<5

Source: Snamprogetti Patent DE 29 01 090.

The later US Patent to Snamprogetti in 1983 proposed treating the whole feed stream in a thermal treatment and heat soak before taking it to an extraction column which operates with solvent gas such as propane under conditions of temperature and pressure that ensure the solvent is in a supercritical condition. The overhead stream from the extraction column contains the base oil fractions and this goes to a series of separation columns where the operating conditions are adjusted to permit a series of cuts of base oil fractions to be separated and allowing the solvent to be recycled. Figure 7.18 shows an embodiment of the process.

In this process the used oil enters a predistillation column where the water and light hydrocarbons are distilled off overhead, the stream from the bottom of the column is then subjected to a heat treatment that promotes the dissolution of the additives in the next stage of the process. The heat treatment is shown as a separate step in Figure 7.18, but the patent allows for a heat treatment before the predistillation column, after the same column, or in the predistillation column itself, the oil being heated to a temperature between 200°C and 420°C (390°F and 790°F) for a time between 1 min and 2 h. A light paraffinic hydrocarbon gas or carbon dioxide gas under supercritical conditions, and preferably propane, is introduced together with the heated oil into the extraction column, the operating conditions of temperature and pressure being chosen to give the best separation of the oil from the additives and contaminants, avoiding supercritical conditions that would also result in dissolving the additives.

The temperatures in the extraction column and in the later separation columns is between the critical temperature of the gas used and a temperature that is 100°C (180°F) greater than the critical temperature, and a pressure that is greater than the critical pressure. The oil dissolved in the supercritical gas is removed from the top of the column and fed to a first separation column operating at a higher temperature or lower pressure (or both) than the conditions in the extraction column. The undissolved hydrocarbon components, together with the undissolved impurities, the additives, and part of the solvent gas are taken from the bottom of the extraction column

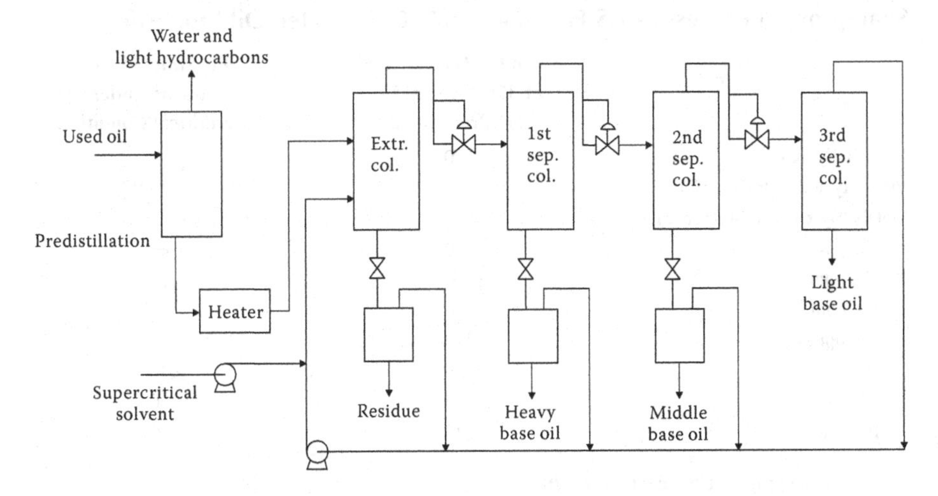

FIGURE 7.18 The Snamprogetti Process. (From Snamprogetti US Patent 4,406,778.)

and expanded through a valve into a container that separates the solvent gas for recycle from the residue stream.

The fluid phase leaving the top of the first separation column is taken to a second separation column, again operating within the limits earlier discussed, but in any case at a higher temperature or lower pressure (or both) than the first separation column. The stream leaving the bottom of the first separation column is again expanded through a valve into a vessel that allows separation of the solvent gas for recycle and a heavy base oil component.

Again, the fluid leaving the top of the second separation column contains the oil components dissolved in the solvent gas and this is fed to a third separation column operating at a higher temperature or lower pressure (or both) than the second separation column. The flow from the bottom of the second separation column is expanded through another valve and is taken to the separation vessel which yields a recycle solvent gas stream and a middle base oil liquid fraction.

The fluid leaving the top of the third separation column is essentially solvent gas and is recycled whereas the underflow from this third column forms the light base oil stream.

All the expansions though the valves after the separation columns are controlled to give the same final pressure for all the recycled gas streams which are combined and fed to a compressor which recycles the solvent gas to the extraction column. Several alternatives are presented to improve the separation, including heating various stages, increasing the temperature at the top of some of the columns to provide an internal reflux, or injecting part of the recycled and recompressed solvent gas into the bottom of the separation columns.

Two examples are given in the patent document, the second of which involved treating a spent oil at 350°C (660°F) for 215 s, predistilling it and then extracting it continuously with propane under supercritical conditions at a temperature of 140°C

TABLE 7.30
Snamprogetti Process of US Patent 4,406,778: Extracted Oil Properties

	Spent Oil Feed Treated at 350°C for 215 s and Predistilled	Oil Extracted with Propane under Supercritical Conditions
Yield % by mass	100	93
Viscosity at 210°F (98.9°C), cSt	13.2	9.9
Metals (by x-ray fluorescence)		
Ba, ppm	420	<5
Ca, ppm	1500	10
Pb, ppm	2350	<5
Zn, ppm	770	<5
P, ppm	980	<40
Cl, ppm	250	60
Br, ppm	320	30

Source: Snamprogetti US Patent 4,406,778.

(285°F) and a pressure of 120 kg/cm^2 (11.77 MPa, 1707 psi), a propane to oil ratio of 10:1 measured as liquid volumes at 15°C (59°F). This overhead stream was flashed at atmospheric conditions to give a product oil with the properties shown in Table 7.30.

7.6.4 The Texaco Inc. Process

This process (US Patent 4,328,092 dated May 4, 1982) refers to the separation of the aromatic and unsaturated components from the saturated components in a lubricating oil feedstock by solvent extraction with *n*-methyl-2-pyrrolidone. The incoming feedstock is contacted in a countercurrent flow with *n*-methyl-2-pyrrolidone at a temperature of between 50°C and 120°C (122°F and 250°F) and a solvent to oil ratio between 1:1 and 3:1. The two streams from this contact tower are the primary raffinate stream from the top and the primary extract stream from the bottom. The primary raffinate stream goes to a raffinate recovery unit in which the solvent is separated from the lubricating oil stream to yield a solvent refined product and a solvent stream that is recycled for use in the process.

The primary extract stream is cooled to a temperature approximately 10°C (18°F) or more below the temperature at which it leaves the contactor, and this stream separates into two liquid phases that can be separated in a decanter vessel. The two streams from the decanter vessel are the secondary raffinate stream from the top and the secondary extract stream from the bottom.

The secondary raffinate stream, which is relatively poorer in aromatic hydrocarbons than the stream to the decanter vessel, is returned to the extraction tower as a separate stream or by mixing it with the feed stream.

The secondary extract stream, relatively richer in aromatics than the stream fed to the decanter vessel, passes through heat exchangers and then to a flash tower operating at 170 to 205 kPa (25 to 30 psi) to recover the solvent from the extract.

The solvent is recycled in the process and the extract from the bottom of the tower is taken to a high pressure flash tower operating at 375 to 415 kPa (54 to 60 psi). The solvent from the top of this tower is cooled and passes to the low pressure flash tower from which the recovered solvent is taken to an accumulator for recycle in the process. The stream from the bottom of the high pressure tower is taken to an extract recovery system which may consist of a vacuum flash tower and stripper or other suitable process to remove the remaining solvent, the extract stream forming a product stream from the process. Suitable heat exchange between various streams results in energy efficiencies for the process and several optional arrangements are given for some of the streams. Figure 7.19 represents a simplified flow diagram of the Texaco process in which some heat exchangers and pumps have been omitted in this diagram for the sake of clarity.

7.6.5 The Delta Central Refining Inc. Process

This process (US Patent 4,360,420 dated November 23, 1982) envisages a seven-stage treatment for used oil in which a multistage distillation process precedes an extraction process using tetrahydrofurfuryl alcohol (tetrahydrofuryl carbinol), which reportedly has a greater selectivity and higher yields of raffinate while having a

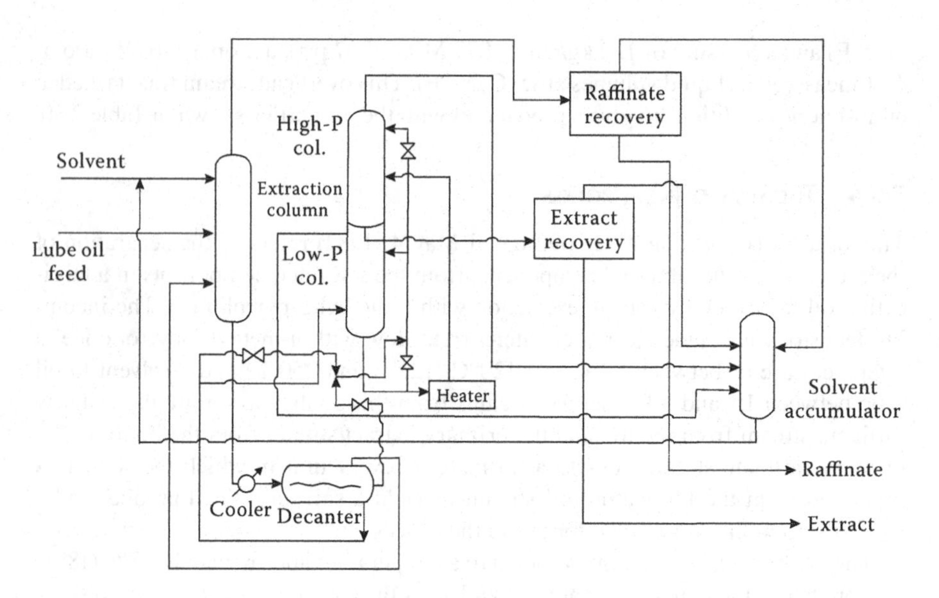

FIGURE 7.19 Simplified flow diagram of the Texaco process. (From Texaco US Patent 4,328,092.)

greater affinity for the contaminants found in used oils. This solvent is miscible with water, has a specific gravity with respect to water of 1.054 at 20°C (68°F), is colorless and is claimed to have low toxicity. An embodiment of the process is shown in Figure 7.20.

The used oil feed is heated to between 38°C and 93°C (100°F–200°F) and enters the first stage evaporator which operates at essentially atmospheric pressure and further heats the oil to a temperature between 104°C and 204°C (220°F and 400°F). This separates the water, gasoline and light volatiles from the fraction containing the lubricating oils which then passes to a heat exchanger to raise the temperature to between 150°C and 205°C (300°F and 400°F). The water, gasoline and light volatiles may be taken for further processing to separate these components.

After entering the second stage evaporator which operates at a pressure between 20 and 150 mm Hg absolute (2.7 and 20 kPa-a), the lubricating oil fraction is heated to a temperature between 121°C and 250°C (250°F and 500°F) and this separates an underflow of fuel oil containing heavier components than the light volatiles in Stage 1, and an overflow which passes to another heat exchanger which raises the temperature to between 180°C and 235°C (350°F and 450°F) before it enters the third stage evaporator. This unit is either a thin film or wiped film evaporator that operates at a pressure between 2 and 5 mm Hg absolute (0.27 and 0.67 kPa-a) and produces a light lube fraction and the feed to the fourth stage evaporator.

One of the features of the patent is that the light lube fraction from the top of the third stage evaporator is split into a feed to the extraction part of the process and a stream that is recirculated, as opposed to a conventional reflux stream in a distillation, to combine with the incoming feed from the second stage. It is claimed that this lowers the vaporization temperature of the distillate in the evaporator, reduces

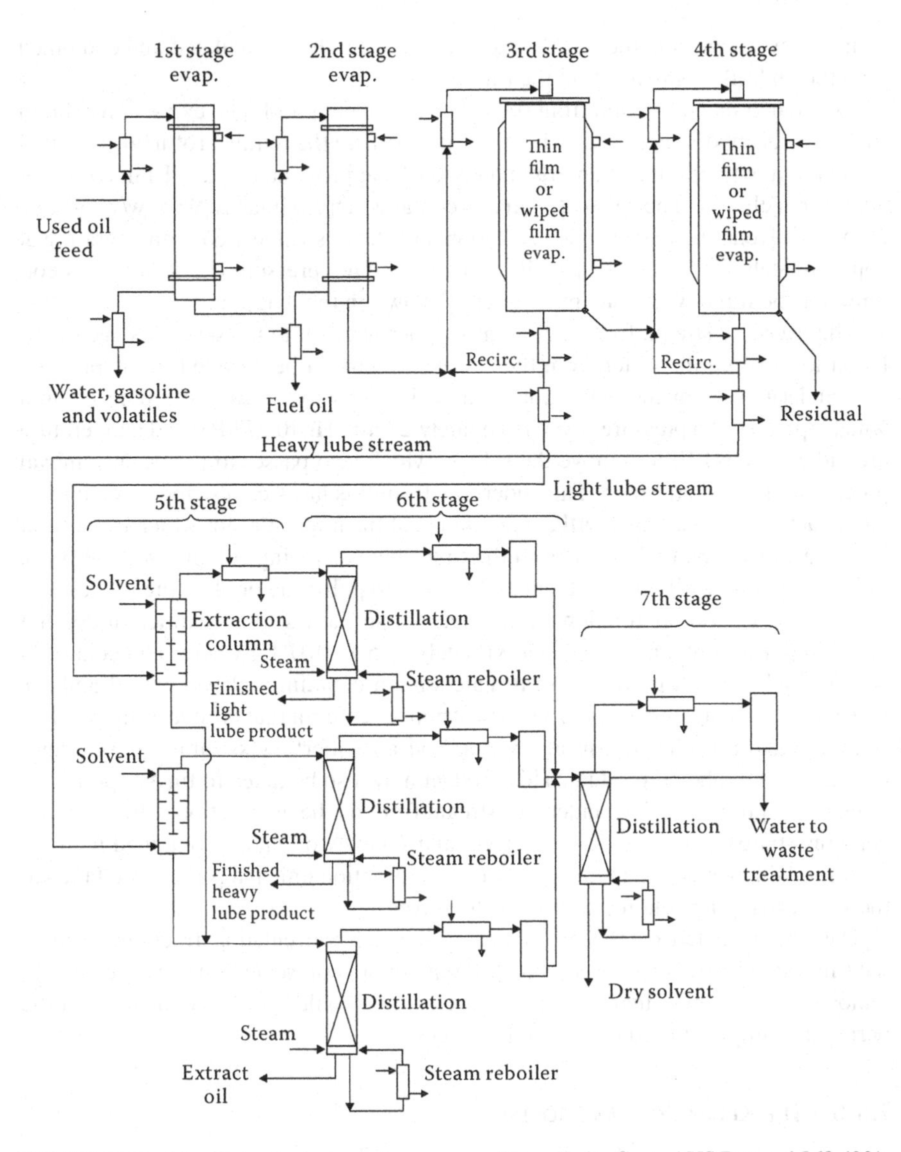

FIGURE 7.20 The Delta Central refining process. (From Delta Central US Patent 4,360,420.)

fouling and reduces cracking and coking tendencies. This recirculation is controlled to be between 0 and 300% w/w of the flow rate of the incoming feed stream and can be adjusted to give different results in the evaporator. The underflow from the third stage passes to a heat exchanger that heats it to between 205°C and 255°C (400°F and 490°F) before entering the fourth stage evaporator which operates at a pressure between 0.5 and 3 mm Hg absolute (0.07 and 0.4 kPa-a) and a temperature between 157°C and 316°C (315°F and 600°F), and produces a heavy lube fraction stream and a very heavy residual stream. A recirculation stream is used in a similar fashion to the previous stage and between 0 and 300% w/w of the feed stream flow rate is taken from the overflow from the fourth stage evaporator to be recirculated and combined with the underflow stream from the third stage.

The light lube oil fraction from stage 4 passes to a liquid-liquid extraction column such as a rotary disc contactor where it is mixed with the tetrahydrofurfuryl alcohol solvent at a temperature of approximately 65°C (approximately 150°F) in a ratio of approximately 1:1. The raffinate stream contains approximately 95% w/w oil and 5% w/w solvent and passes to a distillation and steam stripping column operating at approximately 60°C (140°F) and 10 mm Hg absolute pressure (1.3 kPa-a) this column may be fitted with a steam reboiler as shown in the diagram.

The overflow stream from the steam stripper is taken to the seventh stage distillation to separate the water from the solvent which can be recycled in the process. The underflow from the extraction column is taken to a steam stripping column which operates at a pressure of approximately 20 mm Hg (0.27 kPa) and temperature around 71°C (160°F) to remove the solvent which also passes to the water removal process in the seventh stage. The underflow from this last steam stripping column is an extracted oil stream that will contain much of the impurities for which the solvent has an affinity. The underflow from the first steam stripping column in stage 6 is a finished light lube oil product that can be taken to polishing or blending processes.

The heavy lube oil fraction from stage 4 also passes to an extraction column operating at a temperature of approximately 225°F (107°C) where it is contacted with the solvent and produces a raffinate stream containing 95% w/w oil and 5% w/w solvent that again passes to a steam stripping column that produces an overflow stream containing the solvent and water, and an underflow stream representing a finished heavy lube oil product. This product may also be taken to further polishing or blending processes. The underflow stream from the heavy fraction extraction unit contains approximately 95% w/w solvent and 5% w/w oil and is combined with the corresponding stream from the light fraction extraction unit and is then also taken to the steam stripping column to recover the solvent.

Data is presented on the effect of varying the recirculation fractions used in the third and fourth stage evaporators, but no data is presented on the efficiency of removal of the contaminants in the used oil. This would, of course, depend on the particular sample of used oil fed to the process.

7.6.6 The Krupp Koppers Process

The Krupp Koppers GmbH company developed a process (German Patent DE 36 02 586 dated July 30, 1987) which used a supercritical solvent that is a gas at normal conditions to extract the base oil from the impurities in the used oil, the oil then being separated from the solvent by depressurizing the stream and taking the oil fraction to a hydrogenation process with the optional use of bleaching clay to remove other impurities from the final products. A gasifier is used to dispose of some of the waste streams and the process was claimed to be more environmentally friendly than many previous processes. Figure 7.21 shows the flow diagram for the process.

The used oil is filtered to remove solid particles and passes to the distillation column where it is distilled at atmospheric pressure and a temperature between 120°C and 250°C (250°F and 480°F). The overhead stream which contains a mixture of gas oil

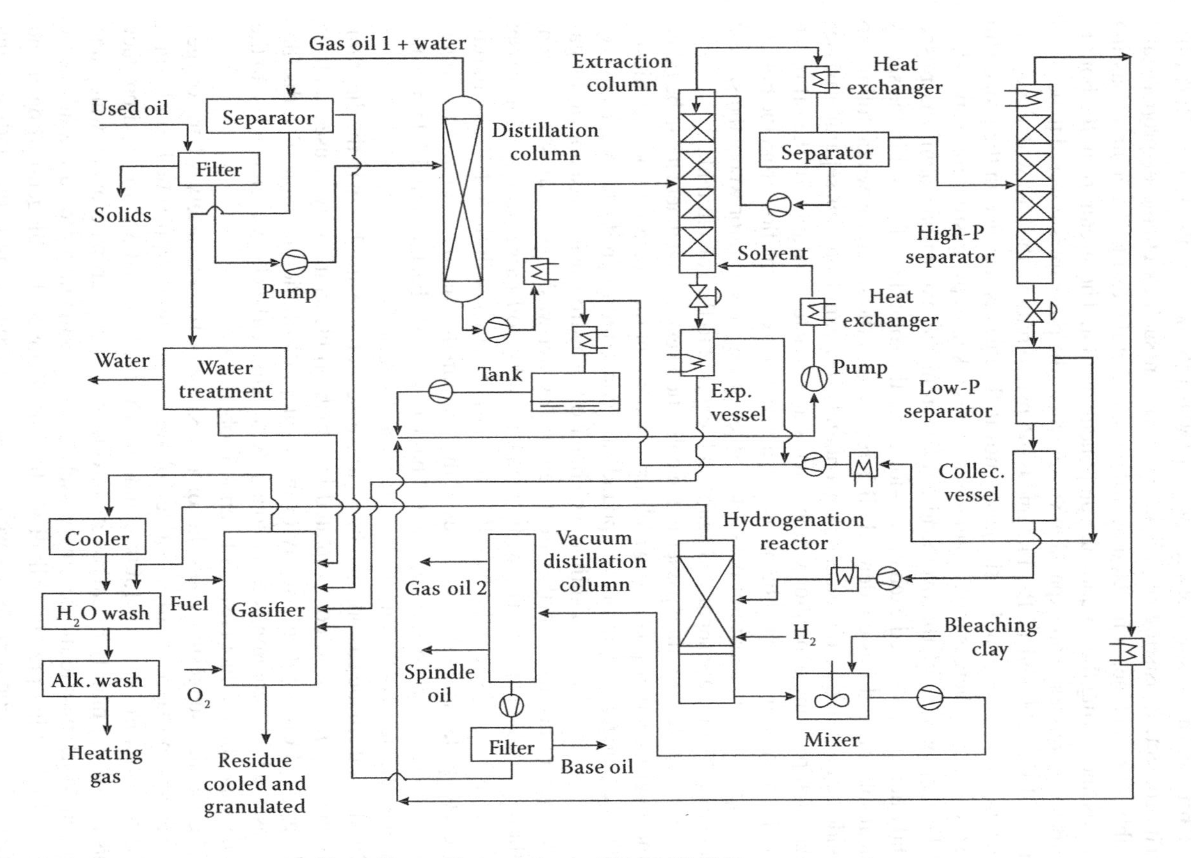

FIGURE 7.21 The Krupp Koppers process. (From Krupp Koppers Patent DE 36 02 586.)

and water passes to a separator, the gas oil being taken to a gasifier and the water being taken to a water treatment plant. The pressure and temperature of the stream from the bottom of the distillation column is adjusted before being injected to the middle part of the extraction column, which operates at a temperature between 20 and 80°C (68°F and 176°F) and a pressure between 50 and 150 bar (5 and 15 MPa; 725 and 2175 psi).

The solvent, consisting of a light hydrocarbon such as ethane or propane and possibly some butane, is injected near the bottom of the extraction column, passing upwards and taking the soluble oil components with it. The stream from the bottom of this vessel passes to an expansion vessel where the pressure is lowered to between 1 and 0.01 bar (100 and 1 kPa, 14.5 and 0.145 psi) to remove the gas, with the option of heating the vessel to aid in the gas removal. The gas is recycled in the process to the solvent tank whereas the insoluble residue is taken to the gasifier for disposal.

The temperature of the overhead stream from the extraction column is raised slightly in a steam-heated exchanger—lowering the solubility of the oil in the supercritical solvent, and this condensed oil is removed in the separator which follows to provide a reflux to the column. The solvent and oil stream from this separator passes to the high pressure separator which effects a complete separation of the oil phase from the solvent, the solvent being recycled in the process. The flow from the bottom of the high pressure separator passes to the low pressure separator where the solvent remnants in the oil are separated again and recycled to the solvent tank.

The solvent-free oil then passes to the collection vessel from where it can be taken to the hydrogenation process, which uses a conventional catalyst such as nickel, where the halogen and sulfur compounds are converted to products such as hydrogen chloride and hydrogen sulfide which leave the hydrogenation reactor in the gas phase and are taken for further treatment in a water-wash and alkali-wash system as shown in the diagram. The hydrogenated product oil passes to a mixing vessel where bleaching clay can be mixed in the stirred vessel with the oil. This mixture of bleaching clay and oil then passes to the vacuum distillation column which operates at a pressure between 0.002 and 0.1 bar (0.2 and 10 kPa, 0.029 and 1.45 psi) and separates the oil into various boiling fractions.

Product oils in the gas oil, spindle oil and base oil range are shown in the diagram. An embodiment is also contemplated in which no bleaching clay is used and the hydrogenated oil is taken directly to the vacuum distillation column. The distillation residue in the column is taken to a filter to remove the clay which is taken to the gasifier, and the product base oil fraction is obtained.

The gasifier suggested could be of the Koppers-Totzek variety where oxygen and air, and possibly small quantities of steam, are used with a fuel to obtain a flame temperature between 1300°C and 2000°C (2370°F and 3630°F) and produce a gas consisting mainly of carbon monoxide, hydrogen, carbon dioxide, water, and nitrogen which pass to the cooling vessel and then to the high pressure water wash, followed by a high pressure alkali wash. This removes the halogen-hydrogen and sulfur-hydrogen compounds, as well as the metal-containing compounds coming from the oil additives, and leaves a stream which can be used as fuel gas. The gasifier residue is taken in a molten liquid state to a water bath for granulation and cooling.

Several other embodiments of the process were also considered in the patent document, including a multi-step separation instead of the single high pressure and low pressure separators shown in Figure 7.21. An example process using ethane as the solvent was also described, the ethane being used at a temperature of 43°C and a pressure of 100 bar (10 MPa, 1450 psi). Table 7.31 shows the flow rates and properties of some of the streams from this example case.

The capital and operating costs for this process could be higher than the costs for a simpler process, requiring a high pressure vessel for several of the components and

TABLE 7.31
Krupp Koppers Process: German Patent DE 36 02 586: Example Case

Parameter	Units	Value
Used oil (with H_2O and light hydrocarbons)		
Feed rate	kg/h	175
Total chlorine content	ppm	1000
PCB/PCT content[a]	ppm	50
Total metals content	ppm	2588
Extraction column		
Temperature	°C	43
Pressure	bar (MPa)	100 (10)
Packing	—	Raschig rings
High-P separator		
Temperature	°C	150
Pressure	bar (MPa)	100 (10)
Flow rate from bottom	kg/h	126
Bleaching clay used	kg/h	2.8
Distillation column/filter		
Base oil produced	kg/h	74
Total chlorine content	ppm	11
PCB/PCT content	ppm	0.5
Total metals content	ppm	—
Spindle oil produced	kg/h	35
Total chlorine content	ppm	11
PCB/PCT content	ppm	0.5
Total metals content	ppm	—
Gas oil 2 produced	kg/h	14
Total chlorine content	ppm	11
PCB/PCT content	ppm	0.5
Total metals content	ppm	—
Oily clay	kg/h	5.6
Gasifier		
Operating temperature	°C	>1500
Oxygen used	kg O_2/kg mixture	0.95

Source: Krupp Koppers Patent DE 36 02 586.

[a] PCB = polychlorinated biphenyls; PCT = polychlorinated terphenyls.

several compression stages, but it does represent an elegant method of treatment for the recycling of the used oil.

7.6.7 The Interline Hydrocarbon Inc. Process

An embodiment of the Mellen patent from 1996 assigned to Interline Hydrocarbon Inc. (US Patent 5,556,548 dated September 17, 1996; US Patent 6,174,431 dated January 16, 2001) is shown in Figure 7.22. This was proposed as a batch process where the used oil was mixed with a solvent, the resulting heavier dirt and sludge allowed to settle to the bottom of a container, and the oil and solvent then passed through an activated charcoal filter to remove lead and other metallic components. This stream was then heated to evaporate the solvent which passed through a cooled region on its way back to the liquid solvent storage tank. Several possible solvents were suggested and included acetone, isopropyl alcohol or a hydrocarbon from the methane series, and a ratio of solvent to oil of 10 to 1 was claimed to give the best results. No means of removing the settled dirt and sludge from the settling stage was addressed as this was not proposed as a continuous process.

A second patent from 2001 also assigned to Interline Hydrocarbon Inc., proposed the use of moderate heat and one re-refining step without the preremoval of any water. The process consists of a pretreatment with an aqueous base, such as a 45% solution of potassium hydroxide together with a phase transfer catalyst which aids in the transfer of the base from the aqueous phase to the organic phase and some of the chemical reactions that involve some of the impurities in the oil, at a moderate temperature of less than 200°F (93°C), which helps to reduce acidity and fouling in the subsequent steps, cooling and carefully mixing this with propane while avoiding

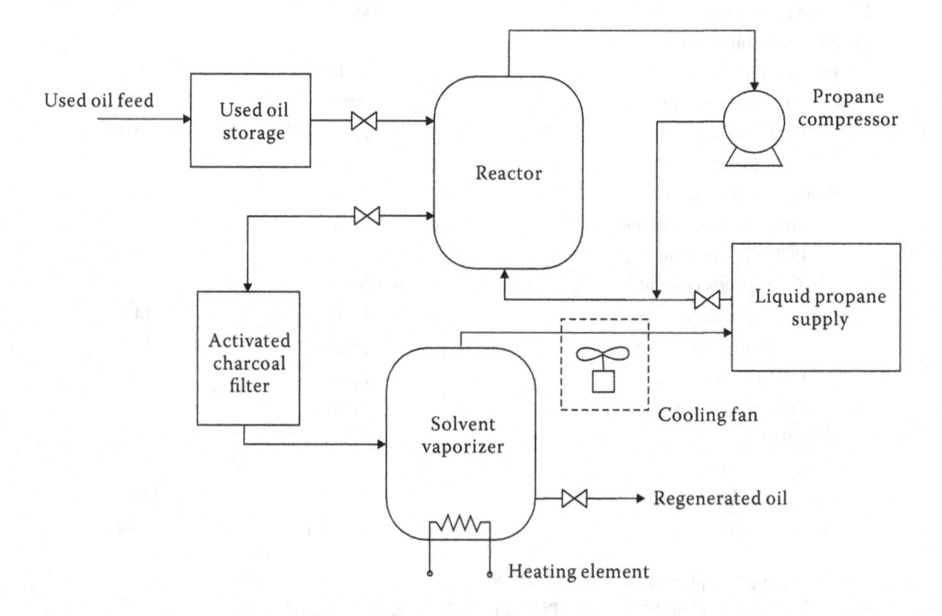

FIGURE 7.22 The Interline Process. (From Interline US Patent 5,556,548.)

the formation of emulsions. The impurities are then allowed to settle out in a settling vessel or extraction vessel, the heavy fraction from the bottom of these vessels goes to a stripper and then constitutes the asphalt residuum product. The stream from the top of the settling or extraction vessels goes to a stripper and then can be distilled to form a base oil product. The propane is recycled and reused in the process.

This process is represented in Figure 7.22. The used oil feed, aqueous basic solution and optional phase transfer catalyst at a temperature of less than 93°C (200°F) are fed to the first mixing vessel for gentle mixing that avoids turbulent conditions that would result in the formation of emulsions, the mixture then passing to the second mixing vessel for final mixing. The base solution used could be any of a number envisaged, or a mixture of these, all of which should be effective in precipitating impurities from the used oil. The phase transfer catalyst could also be any of a number of chemicals, most preferably a quaternary ammonium salt such as tricaprylyl methyl ammonium chloride (trioctylmethylammonium chloride) which has high solubility in both aqueous and organic phases, is relatively cheap and innocuous and has high reactivity in phase transfer catalysis-OH reactions.

Both batch processing or continuous operations were envisaged and water content can be adjusted to be in the 5% to 10% v/v range, mixing continuing for between 30 and 90 min. This stream is then cooled to approximately 32°C to 43°C (90°F–110°F) and is then mixed with liquid propane, again avoiding turbulence which would lead to the formation of emulsions. The ratio of used oil to propane in some embodiments could be in the range of 3:1 to 10:1 (see comment below). This mixture then enters an extraction vessel where the impurities flocculate and settle gently to the bottom of the vessel from where they can be withdrawn as a stream containing asphaltic materials, water and a high percentage of the metals and phosphorous compounds, polymers and other impurities in the used oil. This heavy stream can then pass to a stripper where the propane and water are removed to yield an asphalt residuum which can be used as an asphalt extender.

A lighter fraction consisting of the oil and propane mixture is withdrawn from near the top of the extraction vessel and passes to a second settling vessel to allow further settling of the impurities to the bottom of the vessel. The impurities are again removed from the bottom of the settling vessel whereas the oil and propane mixture is removed from near the top of the vessel and passes to a stripper to remove the propane from the base oil, the propane then being condensed and recycled in the process. The base oil stream could either be used without further processing, or could be distilled to provide a number of light oil and base oil fractions with different properties.

An alternative embodiment of the process is shown by the dashed flow line in Figure 7.23 in which liquid propane is injected near the bottom of the extraction vessel to provide a gentle, non-turbulent countercurrent flow of propane and the settling impurities where the ratio of oil to liquid propane is maintained preferably around 1:5 with one part liquid propane being in the countercurrent flow. This ratio would seem to indicate that the ratio of oil to propane in the range of 3:1 to 5:1 given above should be reversed.

The removal of the zinc compounds, said to be mainly responsible for fouling in the refining stages, was highlighted and data was presented to compare the

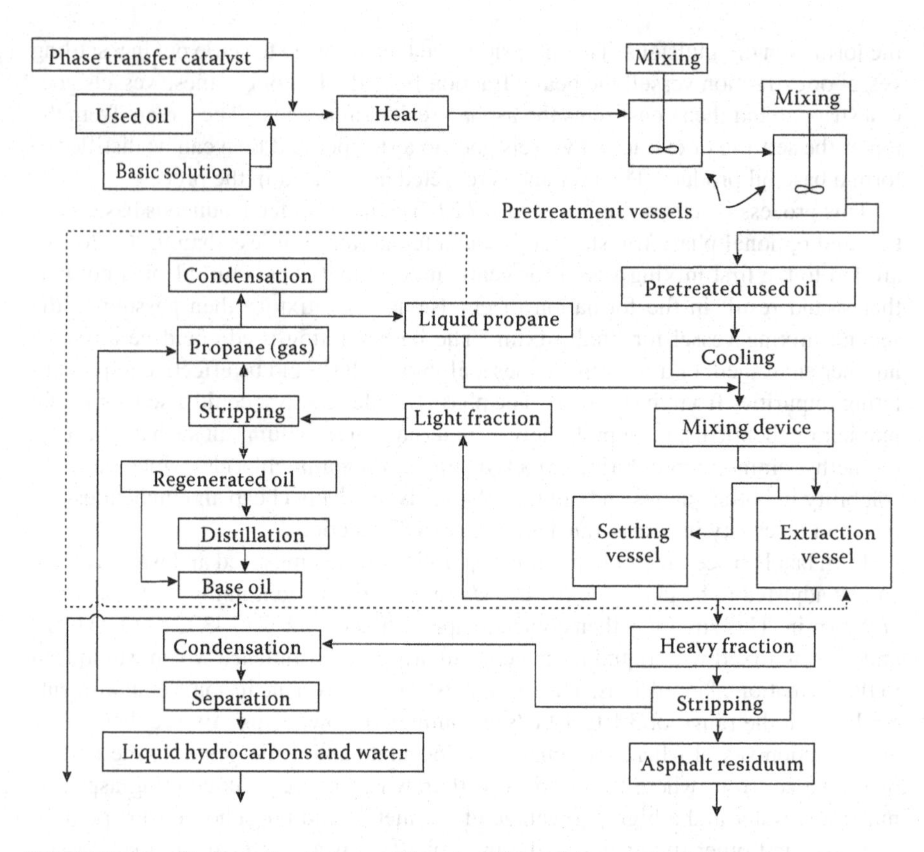

FIGURE 7.23 The Interline Process. (From Interline US Patent 6,174,431.)

effectiveness of three methods of mixing the propane with the used oil that had not been exposed to a pretreatment bubbling the propane through the oil, mixing the propane with the oil by flowing through a venturi, and by mixing the components in a globe valve. The globe valve mixing was presented as the best option for reducing the level of all of the impurities considered.

Data was also provided to show that treating used oil by the preferred methods presented in this patent could reduce the level of lead in the used oil by 52% w/w, sodium by 88% w/w, magnesium by 98% w/w, calcium by 88% w/w, phosphorus by 73% w/w and zinc by 85% w/w. These reductions were also accompanied by a significant reduction in the viscosity of the base oil obtained by the invented method compared with the used oil viscosity. Of course, the actual values obtained for these reductions will depend to some extent on the particular compounds present in the used oil, and would have to be confirmed by measurements with any new sample of used oil.

Several plants based on the Interline technology have been built in various parts of the world, namely in the USA, Korea, the U.K., Australia, Spain, and the United Arab Emirates. In any particular economic environment the economic viability of the plant would depend on a number of factors such as collection costs for the used oil, components and compounds present in the used oil, local utilities costs, waste disposal costs, tax and other incentives that might be available from government agencies wanting to promote a more environmentally friendly recycling of used oil, the market and future prices for the base oils produced, and all of these would have to be a part of a comprehensive evaluation for any potential installation.

7.6.8 The Shell Upgrading Process

The Shell patent of 2003 (Patent WO 03/033630 dated April 24, 2003; US Patent 7,261,808 dated August 28, 2007) was designed to take the product from a preprocessed used oil, processed by various techniques or, for example, by the Interline Hydrocarbon Inc. patent process of US Patent 6,174,431 of 2001, and further upgrade the product base oils to improve the pour point and Health, Safety and Environment characteristics of the oil to comply with industry standards to formulate new lubricants. It was claimed that this was achieved in the following four-step process:

1. The preprocessed oil is contacted with a hydrodemetallization catalyst in the presence of hydrogen at a preferred temperature between 330°C and 420°C (625°F and 790°F) and a pressure between 20 and 150 bar (2 and 15 MPa, 290 and 2175 psi). This can be achieved in a two-step contact where the first hydrodemetallization catalyst has a high uptake capacity for metals such as nickel, vanadium and molybdenum as well as reducing the content of halogens such as chlorine and fluorine as well as phosphorus, calcium, zinc and silicon, and then the second hydrotreating catalyst has a relatively higher activity for desulfurization and denitrification reactions, the second treatment taking place at a temperature preferably between 350°C and 400°C (660°F and 750°F), and a pressure preferably between 20 and 150 bar (2 and 15 MPa, 290 and 2175 psi).
2. After a gas rich in hydrogen is used to lower the temperature of the oil by 10°C to 40°C (18°F–72°F), the oil is then contacted with a presulfided dewaxing catalyst such as nickel in the presence of hydrogen at a temperature preferably between 350°C and 400°C (660°F and 750°F), and a pressure preferably between 20 and 150 bar (2 and 15 MPa, 290 and 2175 psi), the operating conditions and catalyst being chosen to lower the pour point to a value preferably between –12°C and –20°C (+10°F and –4°F).
3. Contacting the oil from the dewaxing treatment with a hydrotreating catalyst to saturate any unsaturated compounds, improve the color and stabilize the oil. This step is carried out at a temperature preferably between 340°C and 400°C (645°F and 750°F), and a pressure preferably between 20 and 150 bar (2 and 15 MPa, 290 and 2175 psi).

Various configurations are suggested in the patent document, such as co-current or counter-current flow of the oil and hydrogen, and various flow rates are suggested together with suggestions for suitable catalyst options for each step. One preferred configuration is shown as a cocurrent flow system in Figure 7.24, and Table 7.32 shows the properties of a test sample from the Interline reclamation process that was treated according to this patent process as embodied in Figure 7.24. The operating

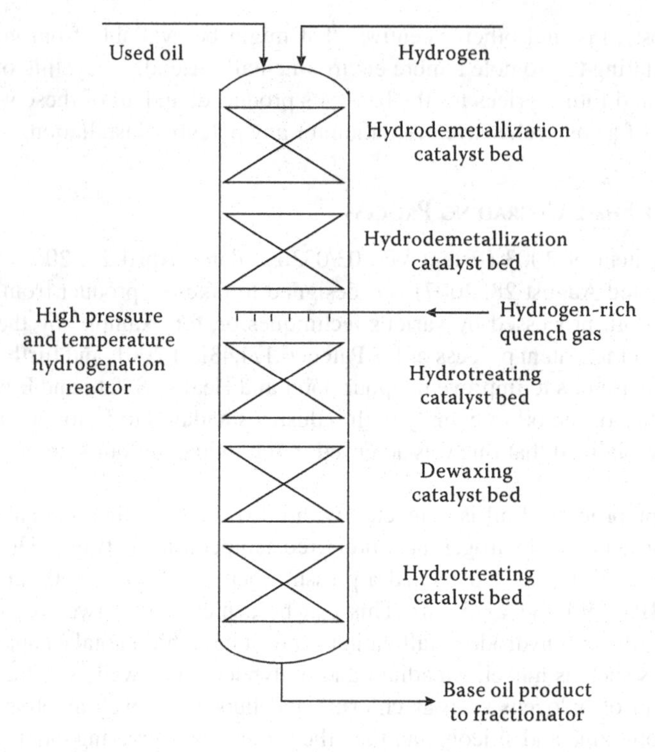

FIGURE 7.24 The Shell upgrading process. (From Shell US Patent 7,261,808.)

TABLE 7.32
Shell Upgrading Process of Patent WO 03/033630: Feed Oil Properties

Property	Value
Sulfur, ppm	5600
Nitrogen, ppm	228
Total metals, ppm by mass	97
Phosphorus, ppm by mass	31
Calcium, ppm by mass	32
Zinc, ppm by mass	10
Silicon, ppm by mass	10
Chlorine, ppm by mass	50
Kinematic viscosity at 100°C, cSt	6.7
Viscosity index	105
Pour point, °C	–7
Boiling range, °C	
IBP	355
50% v/v	450
95% v/v	535

Source: Shell Patent WO 03/033630.

pressure was 51.6 bar (5.16 MPa; 748 psi) and gas rate 500 normal liters per kilogram (8 standard cubic feet per pound) of feed. Other operating conditions and the catalyst types for each step were given in the patent document, the resulting product oil being distilled into three fractions, a light fraction boiling below 370°C (700°F) and comprising approximately 3% w/w of the feed material, and two product oils with the properties shown in Table 7.33. This showed an improved pour point, near-zero metal and chlorine contents and a reduced sulfur and nitrogen content without much change in the viscosity index, these fractions of the base oil products meeting the requirements for a Group I base oil.

A second example with the same feed oil was carried out at a pressure of 121 bar (12.1 MPa, 1755 psi) and gas flow rate of 1000 normal liters (264 US gallons) per kilogram (16 standard cubic feet per pound) of feed to obtain product base oils meeting the requirements for a Group II oil as shown in Table 7.34.

TABLE 7.33
Shell Upgrading Process of Patent WO 03/033630 (Example 1 Product Oil Properties)

Property	Product Fraction 1	Product Fraction 2
Kinematic viscosity at 100°C, cSt	4.7	9.35
Viscosity index	96	102
Pour point, °C	–20	–11
Sulfur content, mg/kg	259	535
Saturates content, % w/w (ASTM 2007)	77	73
Metals content, ppm by mass	<1	<1
Chlorine content, ppm by mass	Not detectable, below detection limit	Not detectable, below detection limit

Source: Shell Patent WO 03/033630.

TABLE 7.34
Shell Upgrading Process of Patent WO 03/033630 (Example 2 Product Oil Properties)

Property	Product Fraction 1	Product Fraction 2
Kinematic viscosity at 100°C, cSt	4.4	8.6
Viscosity index	105	109
Pour point, °C	–11	–7
Sulfur content, mg/kg	9	20
Saturates content, % w/w (ASTM 2007)	93	91
Metals content, ppm by mass	<1	<1
Chlorine content, ppm by mass	Not detectable	Not detectable

Source: Shell Patent WO 03/033630.

7.6.9 The Sener Ingenieria y Sistemas S.A./Interline Process: The Ecolube Plant

In 1997 (Angulo Aramburu 2003), the Spanish company Sener Ingeniería y Sistemas S.A. entered into a cooperative agreement with Interline to improve the process using propane extraction of the used oils. The main aims were to operate a continuous process and further improve the process to produce a product meeting the specifications of virgin base oils without using clay or hydrofinishing treatments, and to be competitive for smaller plants of less than 30,000 tonnes per year (33,000 short tons per year; 66,150,000 pounds) while being environmentally friendly with no emissions of odors, waste or liquids. The improvements using a pretreatment of the used oils with chemicals from the 2001 Interline patent were incorporated, and the Sener team further improved the process to reduce fouling and improve the quality of the product base oils, as well as including systems to avoid the emission of odors and other environmental contaminants. Figure 7.25 represents an embodiment of the process. The Ecolube plant at Fuenlabrada near Madrid in Spain was built in 1999/2000 by Sener and started preproduction in mid-2000, being officially opened in January 2002.

At that time Sener was the majority shareholder in the Ecolube plant (51%) with other participants being the Spanish oil recycling company Tracemar (39%), and the Spanish Environment Ministry (10%). Data from this first year was reported to be at the level of 30,000 tonnes per year (33,000 short tons per year; 67,200,000 pounds) with operations for 8000 h/year, the yield of SN-150 and SN-350 being between 67 and 70% w/w on a water-free basis, with between 6 and 8% w/w of SN-80 base oil. The asphalt product constituted between 20 and 24% w/w on a dry basis of the feed and had a low (0.2% w/w) content of polynuclear aromatic constituents (PNAs).

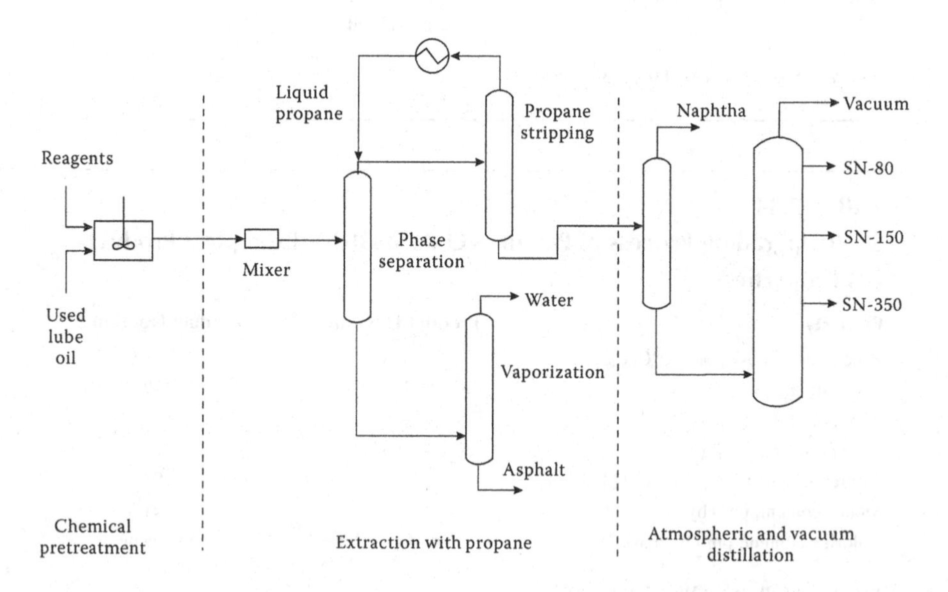

FIGURE 7.25 The Sener process. (Adapted from Angulo Aramburu 2003.)

Light low-boiling products made up between 3 and 4% w/w of the fed material on a dry basis and were burned in a thermal destruction unit with heat recovery from the flue gases. The water content of the used oil feed, which also contained glycols, was usually in the 4% to 6% w/w range and was recovered by distillation from the asphaltic fraction from the solvent extraction step of the process, this water being sent to a waste water treatment plant in tank cars.

Typical analyses of the SN-150 and SN-350 oils from the Sener plant (Angulo Aramburu 2003) are presented in a table which show a paraffin content for the SN-150 and SN-350 products of 69.5 and 71.8%, respectively, and a viscosity index of greater than 100 in both cases.

The Sener publication states that the SN-150 and SN-350 base oils would be classified as Group I oils according to API 1509, Appendix E specifications although some of the properties meet the requirements for Group II oils with their saturates content of between 91 and 92% and a viscosity index of greater than 100: the sulfur content of the re-refined oil was not given and may not meet the value of less than 0.03% w/w required for Group II oils. The content of polynuclear aromatics (PNAs, also known as polynuclear aromatic hydrocarbons, PAHs) in the products was given as less than the limit allowed for non-toxic materials.

The 2003 article also gives information on the asphalt product, operating costs and utilities, and investment costs at that time at this plant as follows:

a. Asphalt product. The product contained some of the additives that make it suitable for use with commercial asphalts. By adding 5% of this product to a B40/50 asphalt, an asphalt was obtained which met the specifications for a B60/70 asphalt having good ductility and low fragility which could also be used to make waterproof materials for the construction industry. The typical properties of the blended asphalt were compared in a table with the B60/B70 specifications.
b. Operational data and utilities. The plant operated with a staff of 28, made up of 20 operations and maintenance personnel and 8 management, commercial, administration and laboratory personnel. Two operators were required per shift. The required utilities and consumables were as shown in a table which showed 90 kWh of electrical energy was consumed per tonne (2205 pounds) of used oil processed, and the propane consumption was 3.5 kg per tonne (7.0 pounds per short ton; 7.8 pounds per 2240 pounds) of used oil.
c. Investment and re-refining costs. These were presented in two graphs in the report. This data may be generalized in terms of arbitrary currency units and compared with models of the form:

$$\text{Capital cost 2} = \text{Capital cost 1} \times \left[\frac{\text{Plant capacity 2}}{\text{Plant capacity 1}}\right]^y$$

$$\text{Operating cost 2} = \text{Operating cost 1} \times \left[\frac{\text{Plant capacity 2}}{\text{Plant capacity 1}}\right]^z$$

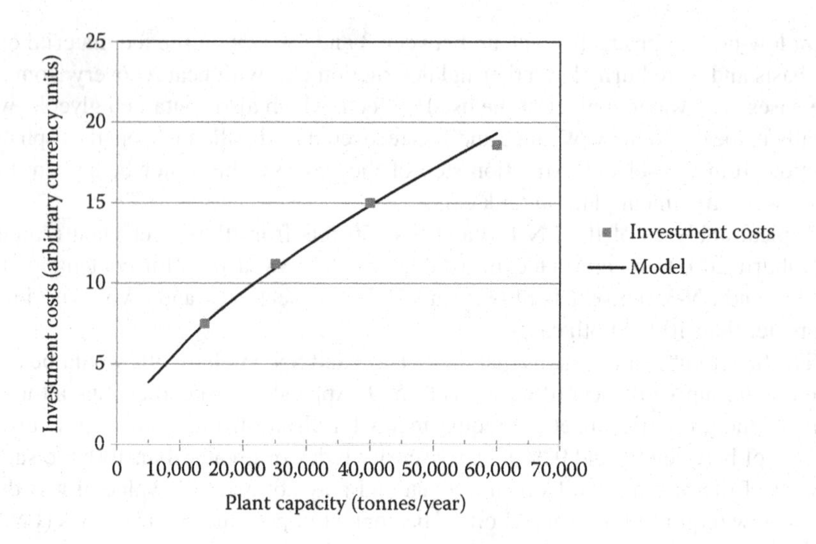

FIGURE 7.26 Capital cost and model for a Sener Ecolube-type plant. (Adapted from Angulo Aramburu 2003.)

The base case is taken as the 14,000 tonne per year (15,432 short tons; 30,864,000 pounds) case. For the capital or investment cost case, it is seen in Figure 7.26 that an exponent of $y = 0.65$ provides a good fit to this data. For the operating costs per tonne (2205 pounds) shown in Figure 7.27, a negative exponent of $z = -0.45$ gives a good representation of the data over this range. Included in the investment costs are the process units and off-site costs, the tank farm, buildings for offices and laboratories,

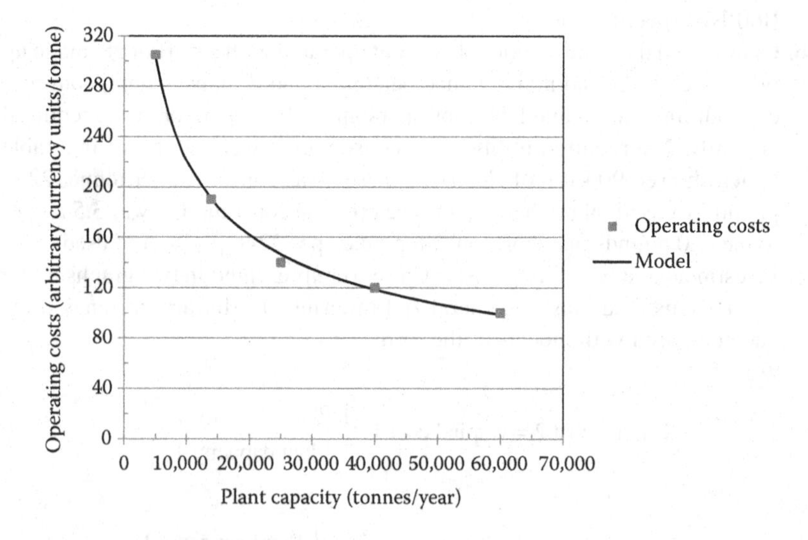

FIGURE 7.27 Operating costs and model for a Sener Ecolube-type plant. (Adapted from Angulo Aramburu 2003.)

engineering costs and start-up costs. It may be noted that the fuel gas costs were given as 650 therms per tonne (1 therm = 100,000 British thermal units [BTUs]; 2205 pounds) of used oil, translating to 69 GJ per tonne (2205 pounds) of used oil, and this would be a significant cost item.

The investment cost of the process units was stated as representing approximately 70% of the total investment cost. It was also reported that the process was competitive at that time at nominal capacities around 30,000 tonnes per year (33,000 short tons per year; 66,150,000 pounds). Although the actual costs have been obscured in these figures, it has been reported (ARI 2002) that the Fuenlabrada plant was constructed at a cost of 9.91 million euros (approximately 13.8 million US dollars) in 1998. No further information is available on the accuracy of this number, or on what was included in it.

In reviewing the results from this plant, it should be noted that the above data represent the results obtained with the used oil fed to the plant at that time and the particular operating conditions: this will always be a problem when the composition of the feed material to a process does not have a fixed composition but changes with time. With the increasingly demanding specifications required by the car manufacturers for engine lubricating oils, and the increasing amount of synthetic lubricating oil being used, it may be expected that the quality of the re-refined oil should also be increasing as the synthetic components should pass through this process without being degraded or destroyed.

7.6.10 The Modified Sener Process

One of the modifications made to the flow sheet was covered in the 2007 patent (US Patent 7,226,533 dated June 5, 2007) in which it was claimed that less fouling in the process equipment and less cracking of the lubricating oil fractions were achieved by keeping the process temperatures for the oil under 350°C (660°F). Figure 7.28 shows an embodiment of the modifications to the original plant that were covered in the later patent.

In this version, the incoming used oil passes through a propane deasphalting process according to previous specifications and then a concentrated aqueous solution of alkaline hydroxide is introduced into the stream which is then combined with the recycle from the bottom of the flash evaporation vessel, passing then to a heat exchanger heated by a thermal fluid on the shell side. The thermal fluid temperature is between 250°C and 320°C (480°F and 610°F) and the oil stream is heated to a temperature between 180°C and 260°C (355°F and 500°F), the mixture of vapor and liquid forming inside the tubes and keeping flow velocities high. This mixture passes then into a flash evaporation vessel operating at near atmospheric pressure and in which light hydrocarbons, solvents and water are evaporated. This comparatively small flow as an overhead stream, after being condensed, is separated into three streams of non-condensing gases, hydrocarbons which may be refluxed into the flash evaporator, if required, to prevent the heavier fractions from being entrained with the overhead stream vapors, and the water stream.

The evaporation of the water from the aqueous solution of the alkaline hydroxide results in an anhydrous form of this chemical having greater reactivity, and also

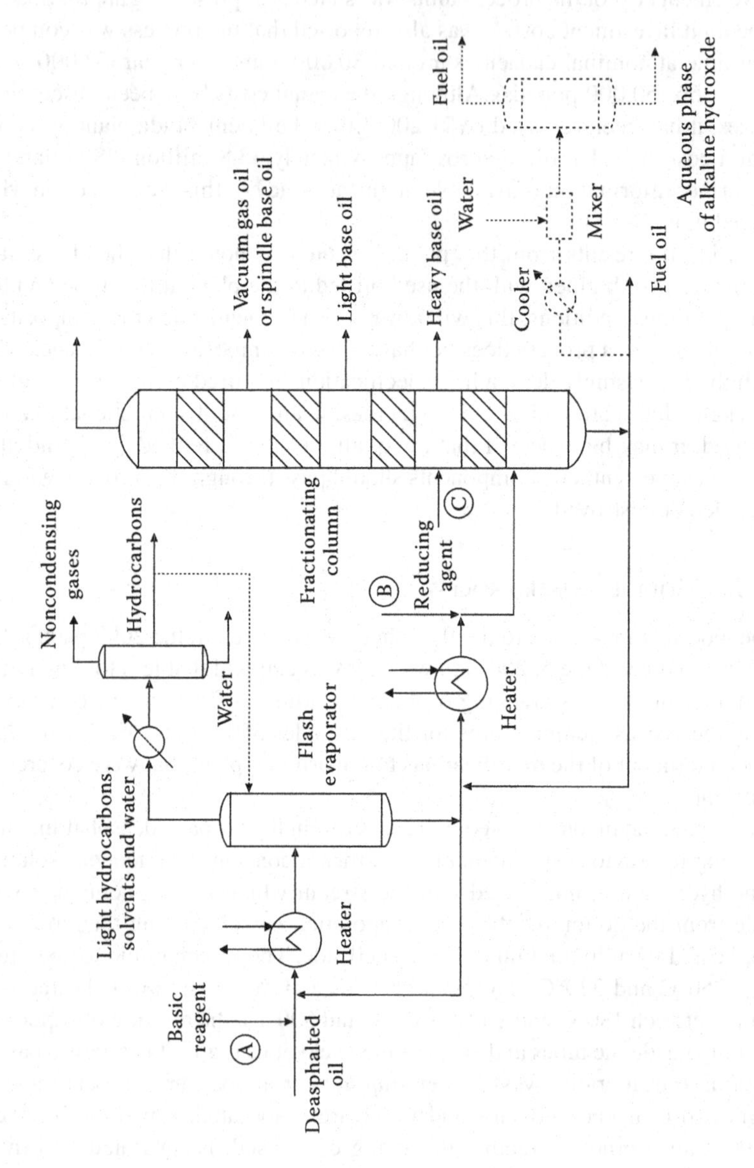

FIGURE 7.28 The modified Sener process. (From Sener US Patent 7,226,533.)

makes it unnecessary to use high pressure equipment to contain the vapor pressure of water in the 200°C to 300°C (390°F–570°F) range.

The flow from the bottom of the flash evaporator is split into two streams, one being recycled into the feed of the deasphalted oil stream before the first heater and serving to keep the flow velocities in the heater tubes high to avoid deposits and contaminant fouling. The ratio of flows of the recycle stream to the deasphalted oil stream is usually in the range of 1:1 to 5:1. The rest of the underflow from the flash separator is then combined with a stream coming from the bottom of the fractionating column as shown in Figure 7.28. This latter combined stream then passes through another heat exchanger operating with a thermal heating fluid on the shell side at a temperature between 350°C and 390°C (660°F and 735°F), heating the oil stream in the tubes to a temperature preferably between 315°C and 335°C (600°F and 635°F). This stream then enters the flash vaporization zone of the fractionating column which operates at a pressure generally between 2 and 10 mbar (0.1 to 0.5 kPa, 0.0145 and 0.0725 psi) at the top, and which is packed with a low pressure drop packing so that the pressure at the base of the column is generally between 10 and 20 mbar (0.5–1 kPa, 0.0725–0.145 psi).

This fractionation column can be operated with between two and five side cuts, Figure 7.28 showing three streams of a light cut of vacuum gas oil or spindle base oil, a medium cut of light base oil, and heavy base oil. It was found that the addition of small amounts of a reducing or hydrogenating agent such as hydrazine to the feed to the fractionating column provided an improvement in the quality of the base oils obtained, and that this reducing agent could be added at any of the points labeled A, B or C.

The flow from the bottom of the fractionating column is split into two streams, one being recycled into the feed to the fractionating column just before the second heater, reducing the fraction that is vaporized in the tubes and keeping the flow velocities high to minimize fouling, and the other stream forming a product of asphaltic type material that can be used as an asphalt extender or thinner material. An optional treatment of the product stream containing the asphaltic materials and the residual alkaline hydroxide is shown in dashed line form in which the stream is first cooled and is then mixed with water before passing to a separator to obtain a fuel oil product stream and an aqueous stream containing the residual alkaline hydroxide which can be recycled in the process.

Table 7.35 shows example data included in this patent document for the treatment of a sample of used oil having the properties shown in Table 7.36. A continuous flow process was used to treat a stream of 1000 kg/h (2205 lbs/h) of this oil with 2500 kg/h (5513 lbs/h) of liquid propane in a standard propane deasphalting process with the mixture passing to a phase separator, producing an overhead flow of propane and an underflow of 890 kg/h (1962 lbs/h) of extract. The basic reagent of 4 g of potassium hydroxide per kilogram (4 lbs/1000 lbs) of this extract were added to the feed to the flash atmospheric vaporization and 0.2 grams of hydrazine per kilogram (0.2 pounds per 1000 pounds) of extract were added to the recycle stream of heavy oil from the vacuum distillation column.

The 890 kilograms per h (1962 pounds per h) of extract was combined with a flow of 900 kilograms per h (1985 pounds per h) of recirculation liquid per hour to a

TABLE 7.35
Product Oil Properties for Sener Process: US Patent 7,226,533

	Light Base Oil	Heavy Base Oil
Color (ASTM D500)	<1.5	<2.0
Water (Karl-Fisher), %	<0.01	<0.01
Viscosity (ASTM D445), cSt at 100°C	5.3	8.0
Acidity, mg KOH/g	0.03	0.02
Corrosion of copper strip (ASTM D130)	1a	1a
Viscosity index (ASTM D2270)	>100	>100
Ramsbottom carbon (ASTM D524), %	<0.05	<0.05
Aniline point, °C	102	106
Aromatic carbon, %	8.9	8.1
Paraffinic carbon, %	69.5	71.8
Naphthenic carbon, %	21.6	20.1

Source: Sener US Patent 7,226,533.

TABLE 7.36
Properties of the Feed Oil for the Sener Process Products in Table 7.35

Property	Value
Color	Dark
Flash point (C.O.C.), °C	165
Viscosity (ASTM D445), cSt at 100°C (1 cSt ≡ 10^{-6} m^2/s)	12.6
Water (ASTM D95), %	4.5
Metals, ppm	3500
Distillation (ASTM D1160)	
Initial boiling point, °C	224.5
Final boiling point, °C	527.7
Total distilled, %	89.0

Source: Sener US Patent 7,226,533.

heat exchanger heated with thermal fluid at 275°C (525°F) to a temperature of 225°C (435°F). The atmospheric flash evaporation produced 30 kilograms (66 pounds) of light fractions per hour from the top of the separator and 1760 kilograms per h (3881 pounds per h) from the bottom, 900 kilograms per h (1985 pounds per h) of the latter stream were recirculated to the feed stream. The remaining 860 kilograms per h (1896 pounds per h) of bottoms from the flash separator were mixed with 3500 kilograms per h (7718 pounds per h) of bottoms from the vacuum column and heated in the second heat exchanger with a thermal oil at 370°C (700°F) to a temperature of 325°C (615°F). This stream was fed to the vacuum column operating at a pressure at the top of 5 mbar (0.5 kPa, 0.07 psi) and at the bottom of 12 mbar (1.2 kPa, 0.17 psi), and packed with a low-pressure drop packing. This produced the following products:

- 30 kg/h (66 pounds/h) of spindle oil
- 370 kg/h (816 pounds/h) of light base oil
- 310 kg/h (684 pounds/h) of heavy base oil
- 140 kg/h (309 pounds/h) of fuel oil

It is claimed that the addition of hydrazine in the distillation carried out in the presence of a basic compound, as described in the patent document, contributes to achieving the same quality as a first refined base oil. No information was provided on the sulfur content of the oils.

7.6.11 Further Modification to the Sener Process

In an effort to design a plant that would be suitable for moderate capacities of 15,000 to 30,000 tonnes (16,538 to 33,078 pounds; 33,070,000 to 66,150,000 pounds) per year, be environmentally friendly and operate at low temperatures to avoid cracking of the lubricating oil components, and does not include a costly hydrogenation step, nor treatment with acid and clay, the process proposed in this patent (US Patent 7,431,829 dated October 7, 2008) was described. The flow diagram in Figure 7.29 represents this version in which a chemical reagent is added to demetallize the used oil at a comparatively low temperature and this is followed by atmospheric and vacuum distillation, both with the addition of alkaline hydroxide solutions to displace the ammonia from the demetallized oil and produce base oils with superior qualities in terms of odor, color, acidity and the copper corrosion test.

The chemical reagent that is added to the incoming used oil can be a single or combination of components such as monoammonium phosphate, di-ammonium phosphate (DAP), or the sulfates, or others according to the metals present in the used oil, and are added in an aqueous solution in a preferred range of 0.5% to 5% w/w of the incoming oil. These reagents form low solubility salts with the metals, mainly calcium, zinc and magnesium, which leads to most of them precipitating out and these can then be separated from the oil. Using ammonium phosphates, a residual metals content of approximately 100 ppm is quoted. The reactions take place in a reactor section, which can take various forms, operating at a temperature of between 120°C and 180°C (250°F and 355°F) and pressures between 3 and 11 bar (300 and 1100 kPa; 44 and 160 psi) with a residence time between 10 and 120 min.

The mixture then enters an adiabatic decompression section in which some of the lighter components flash off, are condensed and separated into a light hydrocarbon stream which contains solvents, gasoline and kerosene fractions. The liquids from the bottom of the decompression section are cooled and taken to a separation vessel which yields a stream of demetallized oil, an oily sludge stream which contains the metal salts, and an aqueous stream that contains the remaining reagent. A continuous centrifugation method is the preferred method to separate the demetallized oil from the oily sludge of metal salts.

The demetallized oil then has an alkaline hydroxide such as sodium hydroxide or potassium hydroxide added at a preferred level of 0.5% to 3% w/w of the oil flow, and is then heated to a temperature of less than 300°C (570°F) in a tubular heat exchanger with a thermal fluid on the shell side and designed to maintain high oil

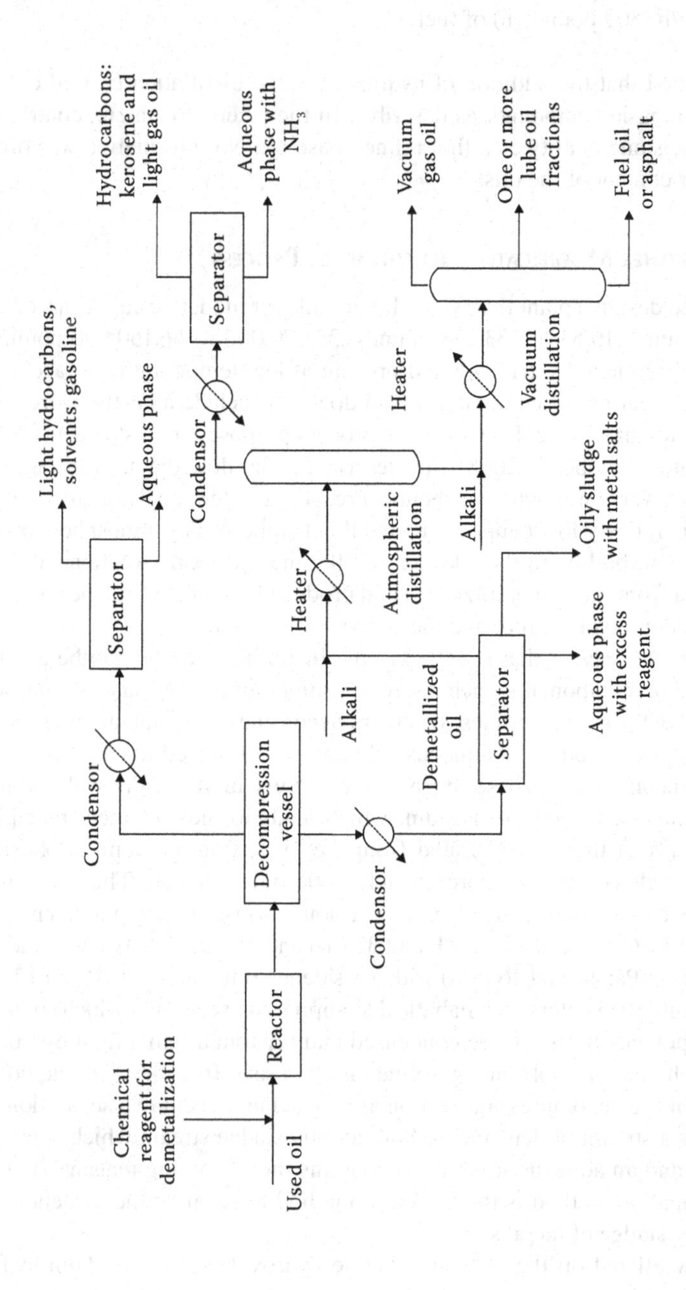

FIGURE 7.29 The modified Sener process. (From Sener US Patent 7,431,829.)

velocities through the tubes. Although the distillation could be carried out in a conventional column, another embodiment of the process has a preferred configuration for the distillation unit as a tubular heat exchanger where the evaporation takes place inside the tubes, the lighter components being evaporated at a temperature between approximately 200°C and 300°C (392°F and 570°F). It should be noted that Figure 7.29 shows the conventional distillation column following the heater so that the alternate embodiment could be visualized by considering the column as being a separator unit. The vaporized stream from the atmospheric distillation contains the remaining water, light hydrocarbons, and solvents, together with the ammonia released by the effect of the alkaline hydroxides. This stream is condensed and is then separated into a stream of organic materials containing kerosene and light gas oil components, and an aqueous stream containing the ammonia.

Additional alkaline hydroxide is then added to the liquid stream from the atmospheric distillation and it is again heated in a tubular heat exchanger system with a thermal fluid on the shell side at a temperature of less than 385°C (725°F), and enters the vacuum distillation unit which operates at a pressure between 2 and 10 mbar (0.2 and 1 kPa; 0.029 and 0.145 psi) at the top of the column and with a column feed temperature between 310°F and 335°C (590°F and 635°F). Again, the preferred embodiment could be to have the distillation take place inside the tubes of the tubular heat exchanger. A packed column is preferred giving a low pressure drop through the packing.

An example is given in the patent document in which a sample of a particular used oil was treated according to these methods and showed good color, viscosity and acidity properties with a very low residual water content, and an overall recovery of the base oil fractions of 67% w/w. Starting with a metals content in the 1 kilogram (2.205 pounds) of used oil of 1750 ppm of calcium and 854 ppm of zinc, treatment with 25 g (0.055 pounds) of DAP in an aqueous solution for 60 min at 150°C (300°F) in an autoclave at 6 bar (600 kPa; 87 psi) pressure with mechanical agitation resulted in 0.910 kilograms (2.01 pounds) of demetallized oil with a calcium content of 39 ppm and a zinc content of 30 ppm, 0.03 kilograms (0.066 pounds) of sludge containing the metal phosphates, and an aqueous phase containing the excess ammonium phosphate.

7.6.12 Institut Français du Pétrole (IFP) Process

This process (US Patent 5,759,385 dated June 2, 1998) provides for a dehydration of the used oil feed which also produces a light component stream, followed by a vacuum distillation to produce a residue and at least one overhead distilled gas oil fraction and several possible side streams. The residue goes to a liquid-liquid extraction column operating with a paraffinic hydrocarbon solvent with 3 to 6 carbon atoms, preferably propane, to produce an overhead clarified hydrocarbon stream which, after solvent removal, goes to a hydrotreater for catalytic hydrogenation. The underflow from the extraction unit also has solvent stripped from it and forms a residue product for sale as a fuel or bitumen extender. The light fractions from the dehydrator, the gas oil and sidestreams from the vacuum distillation unit, and the clarified oil stream from the extraction unit are stored and treated separately in a

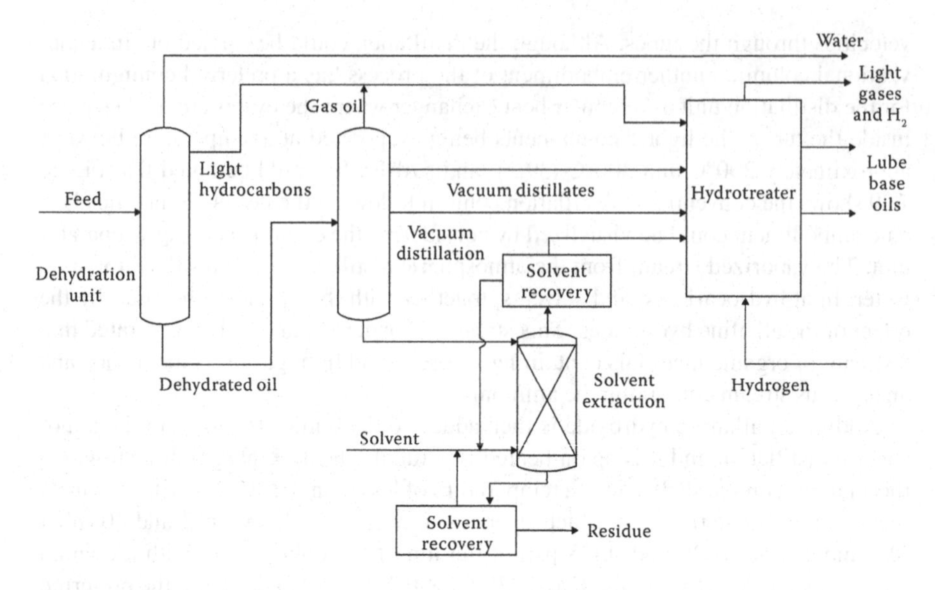

FIGURE 7.30 The IFP process. (From IFP US Patent 5,759,385.)

hydrotreater unit to form the base oil products. An embodiment of the process is shown in Figure 7.30.

After being filtered if it contains any particulate matter, the used oil enters the dehydration vessel to remove the 2% to 4% of water normally contained in the oil. The evaporator operates at atmospheric pressure or a slight vacuum at an example temperature between 120°C and 180°C (250°F and 355°F), in any event less than 240°C (465°F) and preferably less than 200°C (390°F), distilling off the water and some of the light fractions such as gasoline which makes up 1% to 2% of the feed, solvents, glycol and some of the additives. The dehydrated oil passes then to a vacuum distillation unit which produces an overhead product of gas oil which is very rich in chlorine and contains some metals such as silicon, having an end point boiling point between 280°C and 370°C (535°F and 700°F). One or more vacuum distillate streams are also produced containing, as an example, spindle oil and base oils suitable for engines. The residue from the bottom of the column contains the majority of the metals and metalloids, as an example in the 6000 to 25,000 ppm range, and mainly precipitated polymers, and has an initial boiling point between 450°C and 500°C (840°F and 930°F).

The vacuum column residue is generally cooled to a temperature of less than 70°C (158°F), preferably less than 60°C (140°F), and passes to a solvent extraction column which uses a light paraffinic hydrocarbon solvent with between 3 and 6 carbon atoms, preferably propane, at a preferred ratio of solvent to oil between 5:1 and 15:1, to produce a clarified oil from the top of the column and a concentrated residue from the bottom from which solvent can be recovered and recycled. The solvent used in the extraction, which could also be a mixture of hydrocarbons, can be split into two fractions, one of which is used to dilute the feed and regulate the temperature of the feed into the column, and the other is injected directly into the bottom part of the column to adjust the bottom temperature of the column and extract the oil still in the residue. The extraction in this column is carried out at a temperature between 40°C (104°F) and the critical temperature of the hydrocarbonat a pressure that is high enough to keep the hydrocarbon in the liquid phase with the extraction zone having the highest possible temperature gradient, preferably greater than 25°C (45°F). When propane is used, the preferred temperature is between 45°C (113°F) and the critical temperature of the hydrocarbon, and the pressure can be in the range of 30 to 40 bar (3–4 MPa, 435–580 psi).

An additional embodiment foresees the extraction operation taking place under supercritical pressure conditions of between 40 and 70 bar (between 4 and 7 MPa, 580 and 1015 psi), and phase separation is achieved by heating with no vaporization or condensation so the solvent is recycled at a supercritical pressure. This eliminates the evaporation and re-condensation of the solvent required in the first, lower pressure, embodiment.

The residue stream from the extraction column contains a high percentage of solvent acting as a carrier for the precipitated residues. To maintain the viscosity in a manageable range, a viscosity reducer can be added to this stream before the solvent is reheated and vapor stripped to provide a solvent that is compressed and condensed to be recycled. The residue, containing no solvent, can be sold as a fuel or be used as a bitumen extender.

The light hydrocarbons from the dehydration column, the vacuum distillates from the vacuum distillation column, and the clarified oil from the extraction column are best processed separately or in various combinations in a hydrotreater with hydrogen using at least one catalyst. This catalyst contains at least one oxide or sulfide of at least one group VI metal or at least one group VIII metal, such as molybdenum, tungsten, nickel, or cobalt on a support such as alumina, silica-alumina, or a zeolite, a preferred catalyst being based on nickel and molybdenum sulfides supported on alumina. The hydrotreater operating conditions are

- Space velocity of 0.1 to 10 volumes of liquid feed per volume of catalyst per hour.
- Reactor inlet temperature between 250°C and 400°C (480°F and 750°F), preferably between 280°C and 370°C (535°F and 700°F).
- Reactor pressure in the range 5 to 150 bar (0.5–15 MPa, 72.5–2175 psi), preferably between 15 and 100 bar (1.5–10 MPa, 218–1450 psi).
- Advantageously with pure hydrogen recycling in the range between 100 and 2000 Nm^3 per m^3 (100 and 2000 scf/ft^3) of feedstock.

It is claimed that the hydrotreatment is of high quality with the catalysts retaining a high degree of activity and having a longer lifetime as the preceding treatments produce a very pure vacuum distillate and a bright stock cut from the clarified oil with a low residual metal content of less than 5 and 20 ppm respectively. The hydrotreatment of the gas oil fraction from the vacuum distillation unit eliminates the chlorine and reduces the sulfur content in this stream.

A final distillation step for these products can also be carried out to adjust the cut-points, if this is found to be necessary. The final products obtained have the following properties, as also shown for an example case in Tables 7.37, 7.38, 7.39 and 7.40:

- The oils from the corresponding distilled oil fractions.
- Bright stock oil from the clarified oil stream.
- A mixture of gas and light hydrocarbons containing purge hydrogen.
- Optionally, a gasoline-gas oil cut from the gas oil cut and the light fractions containing gasoline.
- Oils that comply with specifications (that are not defined), and have very satisfactory thermal stability and stability to light exposure.
- A slight change in viscosity with respect to the used oil feed and, in some cases, a slightly altered pour point.
- A metal content of less than 5 ppm and a chlorine content less than 5 ppm, usually undetectable.
- A polynuclear aromatic compound content that is usually of the same order as that of the base oils obtained by hydrorefining, 0.2% to 0.5% w/w, and can equal that of oils obtained by solvent refining, for example furfural, of approximately 1.5% w/w.

The example case in Tables 7.37 through 7.40 involved, as a first step, removal of the 4% w/w of water in the used oil feed by atmospheric distillation, together with 2.4% w/w of light fraction. The dehydrated oil with the properties shown in Table 7.37 was sent to the vacuum distillation unit and the two side stream vacuum distillates (VD) were combined, this representing a boiling range between 280°C and 565°C (535°F and 1050°F), and this was sent to the hydrotreatment unit. Table 7.38 represents the properties of the combined side stream cut and the vacuum residue (VR) cut.

The vacuum residue cut was sent to the solvent extraction unit which operated with propane at a propane/oil ratio of 8:1, a pressure of 39 bar (3.9 MPa; 566 psi), a temperature at the top of the extraction column of 85°C (185°F) and at the bottom of the column of 55°C (131°F). The propane was separated from the residue by vaporization, and the residue was mixed with dehydrated oil or a viscosity-reducing agent to be used as a fuel or sold as a blending agent in asphalt cements. The clarified oil was obtained by vaporizing the propane from the overhead stream from the extraction column, yielding a bright stock cut (BS). Table 7.38 shows the corresponding properties for the vacuum residue and bright stock after clarification with propane.

The mixture of vacuum distillates (1 + 2) and the bright stock oil were sent separately to the hydrotreater which used a catalyst of nickel sulfide and molybdenum sulfide on an alumina support, operating at a temperature of 300°C to 280°C (570°F–535°F), a partial pressure of hydrogen of 50 bar (5 MPa; 725 psi), a residence time of one hour and a hydrogen recycle rate of 380 Nm^3/m^3 (380 scf/ft^3) of feed. Table 7.40 shows the properties of the products compared with the properties of the respective feeds. These data shows that the product oils would meet the specifications for a Group I lubricating oil base stock, but would need additional sulfur removal to meet the requirements for a higher classification.

TABLE 7.37
IFP Process: US Patent 5,759,385: Dehydrated Oil Properties

Property	Units	Dehydrated Oil
Specific gravity at 15°C	—	0.892
Color ASTM D1500	—	8+
Pour point	°C	−18
Viscosity at 40°C	cSt (10^{-6} m^2/s)	102.11
Viscosity at 100°C	cSt	11.7
Viscosity index	—	102
Total nitrogen	ppm	587
Sulfur	% w/w	0.63
Chlorine	ppm	280
Conradson carbon	% w/w	1.56
Sulfated ash	% w/w	0.9
Phosphorus	ppm	530
Flash temp. open cup	°C	230
Neutralization number	mg KOH/g	0.92
Metals (total)	ppm	3.445
Ba	ppm	10
Ca	ppm	1114
Mg	ppm	324
B	ppm	16
Zn	ppm	739
P	ppm	603
Fe	ppm	110
Cr	ppm	5
Al	ppm	20
Cu	ppm	18
Sn	ppm	1
Pb	ppm	319
V	ppm	1
Mo	ppm	3
Si	ppm	31
Na	ppm	129
Ni	ppm	1
Ti	ppm	1

Source: IFP US Patent 5,759,385.

7.6.13 IFP/Snamprogetti Process

The further development of the IFP propane extraction process by Snamprogetti involved a second propane extraction step after the vacuum distillation with a hydrofinishing step as shown in Figure 7.31 (Audibert 2006; Kajdas 2000). After atmospheric distillation to remove water and light ends, the additives and partly degraded polymers, together with other impurities, are separated in a propane extraction step.

TABLE 7.38
IFP Process: US Patent 5,759,385: Vacuum Distillation Stream Properties

	Units	VD Cut (1 + 2)	VR Cut
Specific gravity at 15°C	—	0.8768	0.9302
Color ASTM D1500	—	8	Black
Pour point	°C	–9	–15
Viscosity at 40°C	cSt (10^{-6} m^2/s)	49.39	959.5
Viscosity at 100°C	cSt	7.12	55.96
Viscosity index	—	101	111
Total nitrogen	ppm	180	1535
Sulfur	% w/w	0.47	1.00
Chlorine	ppm	45	830
Phosphorus	ppm	15	1740
Conradson carbon	% w/w	0.08	5
Flash point open cup	°C	231	283
Sulfated ash	% w/w	0.005	3
Sediment	ppm	0.05	0.6
Neutralization number			
Total acid	mg KOH/g	0.14	
Strong acid	mg KOH/g	0	
Base	mg KOH/g	0.24	
Metals (total)	ppm	≈11	11,444
Ba	ppm	<1	30
Ca	ppm	<1	3711
Mg	ppm	<1	1077
B	ppm	<1	51
Zn	ppm	<1	2462
P	ppm	6	1995
Fe	ppm	<1	365
Cr	ppm	<1	15
Al	ppm	<1	64
Cu	ppm	<1	59
Sn	ppm	<1	22
Pb	ppm	<1	1060
V	ppm	<1	2
Mo	ppm	<1	7
Si	ppm	3	95
Na	ppm	2	425
Ni	ppm	<1	2
Ti	ppm	<1	2

Source: IFP US Patent 5,759,385.

TABLE 7.39
IFP Process: US Patent 5,759,385: VR and Bright Stock Clarified with Propane

	Units	VR	BS Clarified with C_3
Specific gravity at 15°C	—	0.9302	0.895
Color ASTM D1500	—	Black	8+
Pour point	°C	−15	-9
Viscosity at 40°C	cSt (10^{-6} m^2/s)	959.5	377
Viscosity at 100°C	cSt	55.96	25.4
Viscosity at 150°C	cSt		
Viscosity index	—	111	89
Total nitrogen	ppm	1535	375
Sulfur	% w/w	1.00	0.786
Chlorine	ppm	830	20
Phosphorus	ppm	1740	15
Conradson carbon	% w/w	5	0.60
Flash point open cup	°C	283	332
Sulfated ash	% w/w	3	<0.005
Sediment	ppm	0.6	<0.05
Neutralization number			
Total acid	mg KOH/g		0.3
Strong acid	mg KOH/g		0.0
Base	mg KOH/g		0.55
Metals (total)	ppm	11,444	≈19
Ba	ppm	30	<1
Ca	ppm	3711	1
Mg	ppm	1077	<1
B	ppm	51	1
Zn	ppm	2462	1
P	ppm	1995	<1
Fe	ppm	365	<1
Cr	ppm	15	<1
Al	ppm	64	<1
Cu	ppm	59	<1
Sn	ppm	22	6
Pb	ppm	1060	<1
V	ppm	2	<1
Mo	ppm	7	<1
Si	ppm	95	7
Na	ppm	425	3
Ni	ppm	2	<1
Ti	ppm	2	<1

Source: IFP US Patent 5,759,385.

TABLE 7.40
IFP Process: US Patent 5,759,385: Properties After Hydrotreatment

	Units	VD Cut (1 + 2)	Hydrogenated VD Cut (1 + 2)	BS Cut	Hydrogenated BS Cut
Specific gravity at 15°C	—	0.8768	0.872	0.895	0.893
Color ASTM D1500	—	8–	1–	8+	2.5
Pour point	°C	–9	–6	–9	–6
Viscosity at 40°C	cSt (10^{-6} m^2/s)	49.39	47.39	377	373.48
Viscosity at 100°C	cSt	7.12	7.00	25.40	25.10
Viscosity index	—	101	104	89	88
Total nitrogen	ppm	180	65	375	217
Sulfur	% w/w	0.47	0.182	0.786	0.443
Chlorine	ppm	45	0	20	0
Phosphorus	ppm	15	0	15	0
Conradson carbon	% w/w	0.08	0.014	0.60	0.39
Flash point open cup	°C	231	220	332	309
Neutralization number					
Total acid	mg KOH/g	0.14	0.06	0.3	0.02
Strong acid	mg KOH/g	0.0	0.0	0.0	0.0
Base	mg KOH/g	0.24	0.13	0.55	0.36
Polycyclic aromatics	% w/w		<0.5		<0.5
Metals (total)	ppm	11	1	19	1
Ba	ppm	0	0	0	0
Ca	ppm	0	0	1	0
Mg	ppm	0	0	0	0
B	ppm	0	0	1	0
Zn	ppm	0	0	1	0
P	ppm	6	0	0	0
Fe	ppm	0	0	0	0
Cr	ppm	0	0	0	0
Al	ppm	0	0	0	0
Cu	ppm	0	0	0	0
Sn	ppm	0	0	6	0
Pb	ppm	0	0	0	0
V	ppm	0	0	0	0
Mo	ppm	0	0	0	0
Si	ppm	3	0	7	1
Na	ppm	2	1	3	0
Ni	ppm	0	0	0	0
Ti	ppm	0	0	0	0

Source: IFP US Patent 5,759,385.

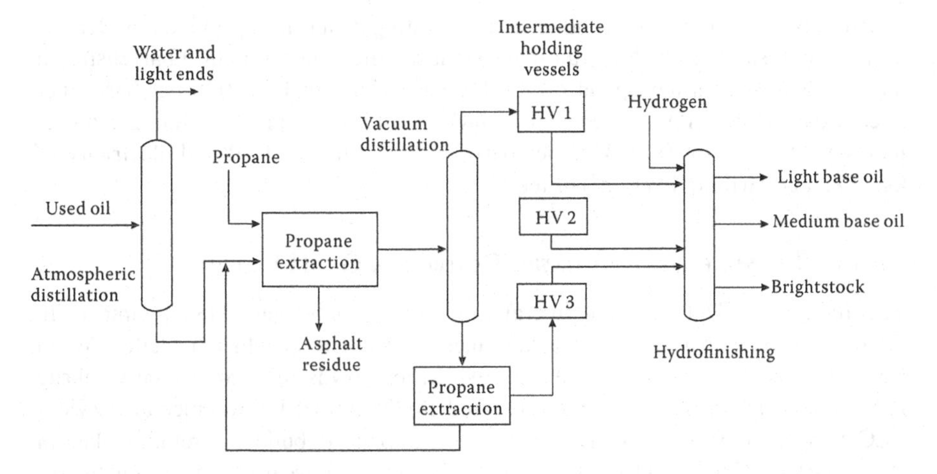

FIGURE 7.31 The Snamprogetti modification of the IFP process. (Adapted from Audibert 2006 and Kajdas 2000.)

The base oil fractions then enter a vacuum distillation column to obtain various base oil fractions, and a residue stream which then enters a second propane extraction process to lower the metals content and reduce the resins and asphaltic content of this stream. The final processing stage involves hydrogenation of the base oil and bright stock fractions in separate treatments from the intermediate holding tanks to produce the product base oil fractions.

Further development by IFP produced several processing variations which generally include the following steps:

- Distillation at atmospheric pressure to separate water and light ends
- Vacuum distillation to obtain light and medium base oil cuts
- Treatment of the vacuum distillates in a hydrofinishing step to obtain finished base oils
- Propane deasphalting of the vacuum distillation residue to obtain a bright stock fraction
- Hydrogenation of the bright stock fraction in a reactor which contains two catalyst beds, one for demetallization and another for hydrofinishing of the bright stock

7.6.14 The Bechtel Process

The Bechtel process (Bechtel 2012; Audibert 2006), known as the MP Refining^{SM} process, uses *n*-methyl-2-pyrrolidone in a solvent extraction of crude oil distillates and residual stocks to produce paraffinic raffinate or naphthenic raffinate that can be processed further into lube oil base stocks. It is claimed that the process uses less energy than the older phenol or furfuryl extraction processes. To meet Group II specifications for lube oil, Bechtel also offers the Hy-Raff^{SM} process which treats raffinate from an NM2P or furfuryl-based solvent extraction process to produce a product that meets Group II lube oil specifications.

Additional economic data and utilities operating data is also provided by Bechtel on their web site, and includes an investment requirement for a 7000 barrel/stream day (1113 m^3/stream day) plant on the US Gulf Coast of US$4000 per barrel per stream day (US$25,159/$m^3$ per stream day) in 2012 and typical operating requirements for fuel of 130,000 BTU per barrel (863 MJ/m^3) of feed and electricity of 0.8 kWh per barrel (5 kWh/m^3) of feed.

7.6.15 The Mineraloel-Raffinerie Dollbergen GmbH Process

The refinery at Dollbergen (Mineralöl-raffinerie Dollbergen [MRD] GmbH) in Germany is now part of the Mustad International Group which includes Avista Resources Inc. and Avista Oil AG and which also owns the Dansk Olie Genbrug A/S plant in Denmark. They also acquired the Universal Environmental Services LLC company in Georgia, USA in late 2011 with plans to build a re-refining plant in Peachtree City, Georgia. The processes described in the above patents represent the situation before possible process modifications that may have been made since that time.

The 2004 patent (German Patent DE 198 52 007 A1 dated May 18, 2000; US Patent 6,712,954 dated March 30, 2004), which was preceded by the German patent, describes a process in which "the waste oil is treated by means of distillation, thin-film evaporation in a high vacuum, optional fractionation for separation into layers of different viscosities and subsequent extraction with *N*-methyl-2-pyrrolidone and/or *N*-formylmorpholine." It is also claimed that the toxic components such as polycyclic aromatic hydrocarbons (PAHs) and polychlorinated biphenyl derivatives (PCBs) are removed in an almost quantitative manner. The process also enables the recycling of the desirable and expensive synthetic oil components like polyalpha-olefins (PAOs) which might be destroyed in more stringent treatment processes such as high temperature hydrogenation. In doing this, the company claims to produce base oil which is in some respects superior to base oils made conventionally from crude oil.

Figure 7.32 represents a simplified version of the main steps in an embodiment of the MRD process. Tables 7.41, 7.42, and 7.43 represent properties of the reclaimed base oil resulting from the process and which were included in the patent document.

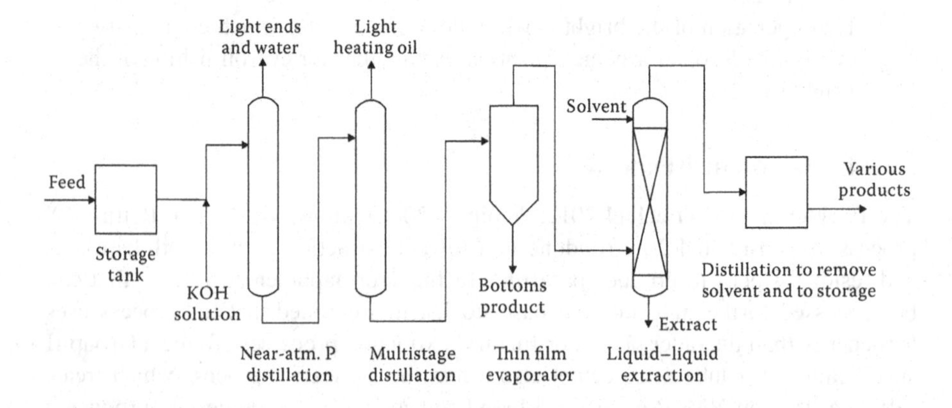

FIGURE 7.32 The MRD GmbH process. (From MRD US Patent 6,712,954.)

TABLE 7.41
Reclaimed Oil Properties for MRD Process: US Patent 6,712,954 (NMP/Oil Ratio = 1.5)

	Units	Raffinate
Temperature	°C	80 isotherm
Return oil phase from extract		Yes
NMP/oil ratio	v/v	1.5
Yield	% w/w	84
Color ASTM		0.5
Neutralization number	mg KOH/g	0.01
Viscosity 40°C	mm^2/s	20.93
Viscosity 100°C	mm^2/s	4.23
Viscosity index		106
Aromatics percentage CA (IR)	%	3.5
Poly aromatic hydrocarbons (PAH) Sum n Grimmer	mg/kg	0.257
Benzopyrene	mg/kg	0.0034

Source: MRD US Patent 6,712,954.

In the example given in the patent document, the feed of used oil is fed together with approximately 0.5% of an aqueous solution of potassium hydroxide of between approximately 5% and 50% concentration to a distillation column which removes the light ends and water. This column operates at near-atmospheric pressure or slight vacuum of up to approximately 60 kPa (8.7 psi) and a temperature of approximately

TABLE 7.42
Reclaimed Oil Properties for MRD Process: US Patent 6,712,954 (NMP/Oil Ratio = 1.8)

	Units	Raffinate
Temperature	°C	80 isothermal
Return oil phase from extract		Yes
NMP/oil ratio	v/v	1.8
Yield	% w/w	85
Color ASTM		L 1.5
Neutralization number	mg KOH/g	<0.03
Viscosity 40°C	mm^2/s	36.05
Viscosity 100°C	mm^2/s	6.07
Viscosity index		114
Aromatics percentage CA (IR)	%	3.9
Poly aromatic hydrocarbons (PAH) Sum n Grimmer	mg/kg	<1
Benzopyrene	mg/kg	—

Source: MRD US Patent 6,712,954.

TABLE 7.43
Reclaimed Oil Properties for MRD Process: US Patent 6,712,954 (NMP/Oil Ratio = 1.5)

	Units		Raffinate	
Temperature	°C	80 isothermal	80 isothermal	80/25 gradient
Return oil phase from extract		Yes	Yes	No
NMP/oil ratio	v/v	2.0	1.1	1.1
Yield	% w/w	84	92	92
Color ASTM		1.0	L 2.0	2.0
Neutralization number	mg KOH/g	<0.01	0.03	0.04
Viscosity 40°C	mm^2/s	36.00	36.44	37.03
Viscosity 100°C	mm^2/s	6.08	6.07	6.10
Viscosity index		116	112	110
Aromatics percentage CA (IR)	%	3.2	4.7	4.6
Poly aromatic hydrocarbons (PAH) Sum n Grimmer	mg/kg	0.024	0.553	0.078
Benzopyrene	mg/kg	0.002	0.020	0.005

Source: MRD US Patent 6,712,954.

140°C (285°F). The alkali is claimed to be a highly effective agent in binding acid components in the feed oil, including extensive demetallization of the feed. The hydroxide is also claimed to form "soaps" that help form a free-flowing and homogeneous residue in the thin film evaporator residue, which follows later in the process. The potassium form is claimed to be better than the sodium form used in other patented processes and which leads to undesirable precipitations and agglomerations, making it possible to avoid the step of mechanical separation of solid precipitations. The use of potassium hydroxide is also claimed to lead to other undefined benefits in process mode and chemical effects in the subsequent steps of the process.

The bottoms from the water and light ends removal column then pass to a multi-stage distillation step in which fuel oil and diesel fractions are removed, the fuel oil corresponding to a boiling cut of 170°C (340°F) and the diesel fraction representing a boiling cut of 385°C (725°F). This column operates under vacuum at a pressure of approximately 6 kPa-a (0.87 psia) and a temperature at the bottom of the column of 260°C (500°F), and separates the lube oil components in the bottoms from a middle distillate in the tops which has an end boiling point of 380°C (715°F). The bottoms from this column pass to a thin film evaporator which operates under vacuum at a pressure of 0.3 kPa-a (0.044 psia) and a thermal oil carrier temperature of 385°C (725°F), producing a lube oil fraction distillate and a bottoms product.

The distillate fraction from the thin film evaporator is taken to a fractionating process operating at a pressure of 8 kPa-a (1.16 psi) and temperature of 280°C (535°F) to obtain various boiling cuts which are then treated separately in a liquid-liquid extraction column with the solvent, which could be either *n*-methyl-2-pyrrolidone (NMP) or *n*-formylmorpholine (NFM). The solvent to oil ratio of 1.5:1 is used with an isothermal temperature over the whole column of 80°C (176°F), and the column is a conventional extraction column with a screen at the bottom and a packing above this screen. Undesirable components such as the polycyclic aromatic hydrocarbons (PAH) pass into the solvent phase to be removed in the bottoms product. The raffinate or overflow from the liquid-liquid extraction column is taken to a solvent recovery step to obtain solvent to be recycled in the process, the product oils being used to formulate new lubricating oils. The extract or bottoms stream from the extraction column also passes through a solvent recovery process with the remaining extract being used as a heating fuel oil or fuel oil diluent.

The bottoms stream from the thin film evaporator (TFE1) is taken to a further thin film evaporator stage (TFE2), not shown in Figure 7.32, which operates at a higher vacuum of 0.01 kPa-a (0.00145 psia) and a temperature of 410°C (770°F) which produces a high viscosity lubricating oil fraction which is then extracted with the solvent at a solvent to oil ratio of 2:1 and an isothermal column temperature of 90°C (194°F). The raffinate from this extraction is a high grade, high viscosity lubricating oil component whereas the extract can be used as an extender for heating oil components or as a heating oil itself. The bottoms product from the thin film evaporator in this section (TFE2) can be used as a heating oil extender as used in the steel industry.

The properties of the base oils produced are shown for three examples with different operating conditions in Tables 7.41, 7.42, and 7.43 and they are claimed to be of good color, low neutralization number, and high viscosity index, with the polynuclear aromatic hydrocarbons (PNAs) content far below 1 mg/kg (1 ppm) and levels of benzopyrene reduced to a range of less than 0.1 mg/kg (0.1 ppm), the levels of polychlorinated biphenyl derivatives (PCBs) being below the detection limit of the analytical process. It is claimed that no other re-refining process for used oils to base oils known at that time was capable of removing PAH components to the same extent as this process, and that hydrogenation processes, although giving better yields with lower viscosity indices, did not remove the PAH components without extreme hydrogenation conditions and the use of rare metal catalysts, which was not customary in the practice of lubricating oil/base oil production. The safety of the non-toxic solvent is also stressed compared with the difficulties of working with hydrogen and the toxic and corrosive products of hydrogenation such as H_2S and HCl.

The last example in Table 7.43 shows the effect of a temperature gradient in the extraction column of 80°C (176°F) at the top of the column to 25°C (77°F) at the bottom, rather than the isothermal operation in the other cases. The presence of up to 5% vegetable oils in the used oil feed is also claimed not to affect the process.

7.7 HYDROGENATION/DISTILLATION PROCESSES

Catalytic hydrogen treatment (hydroprocessing) of used oils provides a commercially viable alternative to high temperature incineration or chemical treatment (Chapter 4). Selective hydrogenation could be utilized to remove contaminants such as polychlorobiphenyl derivatives (PCBs) or heavy metals from used oils. The use of this technology has been primarily constrained by economics, most of all because the product was often designated *fuel oil* and used as such.

The modern objective of re-refining used lubricating oil is to provide a source of valuable base oil that is suffering from decreased availability. With this emphasis on the production of high-value base oil (rather than a lower-value fuel that is still in plentiful supply), the use of this type of technology could have positive economic benefits (Chapters 8 and 9).

7.7.1 The Universal Oil Products Hylube Process

Universal Oil Products (UOP) developed a process (US Patent 4,818,368 dated April 4, 1989; US Patent 4,882,037 dated November 21, 1989; US Patent 4,923,590 dated May 8, 1990; US Patent 5,176,816 dated January 5, 1993; US Patent 5,302,282 dated April 12, 1994; US Patent 5,904,838 dated May 18, 1999) represented in Figure 7.33 in which the incoming feed material containing temperature-sensitive hydrocarbons is mixed with a hot hydrogen-rich stream as it enters the process and passes immediately to a flash zone in which part of the mixture is flashed off as a vapor stream. This vapor stream is partially condensed to provide a vapor stream which can be hydrogenated over a catalyst. This ensures the following:

- The temperature-sensitive material in the feed is heated quickly to a temperature at which it can be distilled off, limiting its time at a higher temperature
- Avoiding heating the feed by indirect means through a heating surface, for instance, which could cause coking and coke deposits on the heating surface
- Providing a source of hydrogen to minimize the formation of hydrocarbon polymers at elevated temperatures
- The hydrogen also acting to reduce the partial pressure and residence time of the hydrocarbon material during vaporization in the flash zone
- The partial condensation of the vaporized stream removes undesirable heavier components from the stream to be hydrogenated over a catalyst
- Utility costs are minimized due to the integrated nature of the hydrogenation processes while producing a distillable heavy hydrocarbon stream that does not require hydrogenation

The process described in the second patent of 1989 involves contacting the feed stream with a hot stream containing preferably more than 90 mole-% hydrogen, keeping the hydrocarbon feed stream temperature preferably lower than approximately 250°C (480°F) before it enters the flash zone which is maintained at a temperature between approximately 380°C and 460°C (715°F and 860°F), a pressure between about atmospheric and 2000 psig (13.8 MPa-g), a hydrogen circulation rate between approximately 168 and 5056 normal m^3 per cubic meter (or scf/ft^3), and a preferred average residence time in the flash zone between approximately 1 and 10 s. The vapor stream from the flash separator is cooled to condense the undesirable high molecular weight components that are then separated from the hydrocarbon vapor stream that passes to the hydrogenation reaction system, the underflow from the separator forming a low-ash fuel oil product which would still require the removal of dissolved hydrogen and light hydrocarbon gases. The hydrogenation over a catalyst takes place in a suitable type of contactor, various types being considered, with various catalysts and support possibilities being suggested. The operating conditions are suggested as being at a preferred pressure between approximately 100 and 1800 psig

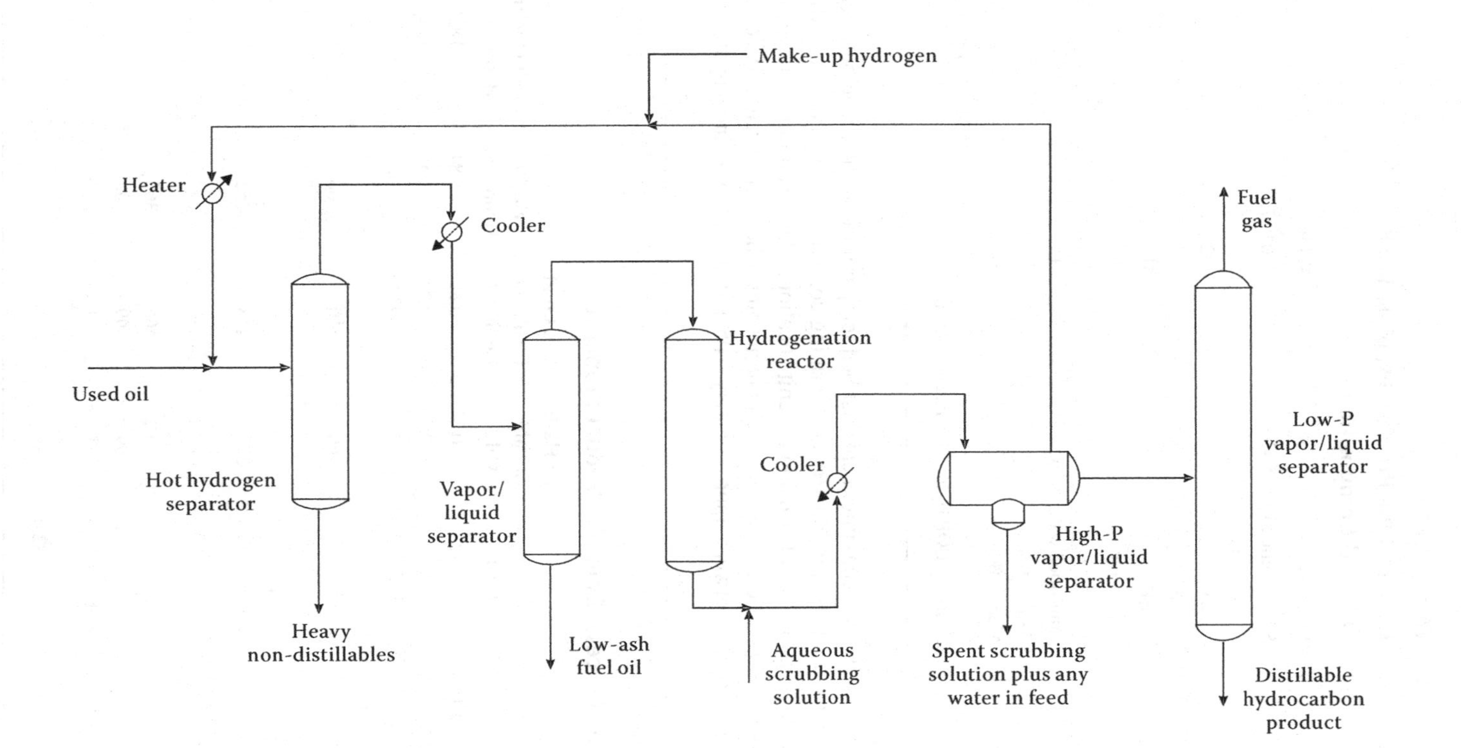

FIGURE 7.33 The UOP Hylube process. (From UOP US Patent 4,882,037.)

TABLE 7.44
UOP Hylube Process: Example Used Oil Feed Composition

Property	Value
Specific gravity at 15°C	0.907
Distillation (D-1160), °C	
IBP	92
50%	394
EP	514
% over	88
% residue	12
Emulsified water, % w/w	19
Ash, % w/w	1.15
Metals, % w/w	0.41

Source: UOP Hylube US Patent 4,882,037.

(690 and 12,411 kPa-g), a maximum catalyst bed temperature in the range between 350°C and 454°C (660°F and 850°F), the operating conditions being chosen so the desired dehalogenation, desulfurization, denitrification, olefin saturation, oxygenate conversion and hydrocracking reactions are accomplished. The preferred liquid hourly space velocities between approximately 0.05 and 20 per hour, and a hydrogen

TABLE 7.45
UOP Hylube Process: Example Product Properties

Property	Feed	First Flash Distillable Oil	Part. Cond. Distillable Liquid HC	HC to Hydrogen	Product from Hydrogen
Specific gravity at 15°C	0.9072	0.87	0.88	0.84	0.825
Sulfur, % w/w	0.23	0.25	0.22	0.5	<0.01
Chloride, % w/w	0.2	0.22	<0.05	1.3	<0.01
Distillation % D-1160, °C					
IBP	97	60	310	60	60
10	102				
30	344				
50	394	387	418	252	249
70	429				
90					
EP	514	565	565	360	363
% over	88	99	99	98	99
% residue	12	1	1	2	

Source: UOP Hylube US Patent 4,882,037.

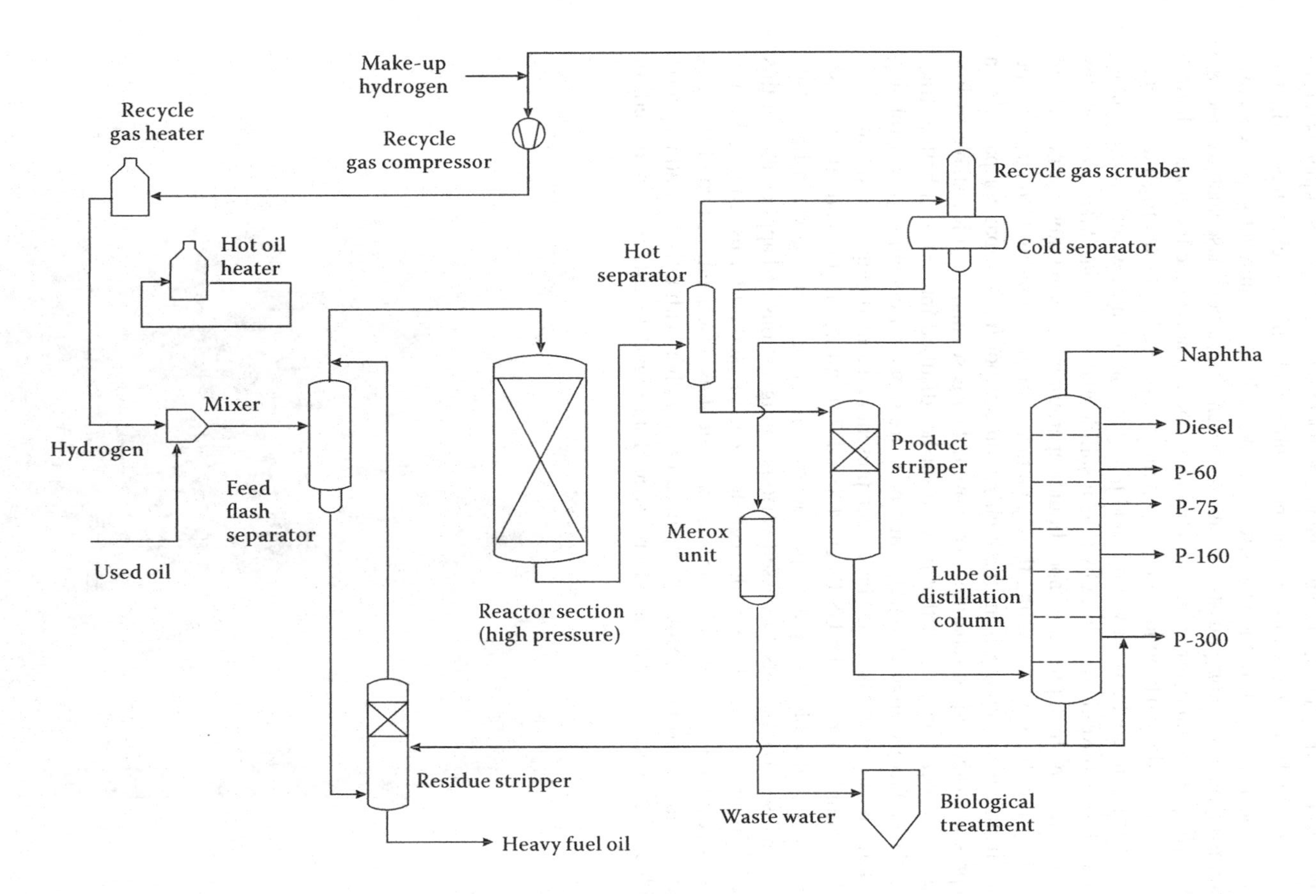

FIGURE 7.34 The Hylube process as implemented by Puralube. (Courtesy of Puralube, Zeitz.)

circulation rate preferably between approximately 300 and 20,000 standard cubic feet per barrel (50.6 and 3371 normal m^3 per cubic meter) are also stated.

The product stream from the hydrogenation reactor is mixed with an aqueous scrubbing solution where the composition of the scrubbing chemical is chosen according to the components present in the feed stream to the hydrogenation reactor: if the stream contains halogenated compounds, the agent chosen would be a base such as calcium hydroxide, potassium hydroxide or sodium hydroxide to neutralize the acids of hydrogen chloride, hydrogen bromide or hydrogen fluoride formed in the hydrogenation reactor. If the stream contains only sulfur and nitrogen compounds, water alone might be sufficient to remove the resulting hydrogen sulfide and ammonia. After cooling, this stream enters the high pressure vapor-liquid separator which separates a vapor stream of hydrogen-rich gas for recycling through a heater to the incoming feed stream. The liquid hydrocarbon stream from the high pressure vapor-liquid separator passes to a low pressure vapor-liquid separator to remove lighter gases and provide a stable, hydrogenated hydrocarbon product that may be further distilled into various boiling fractions. The aqueous stream from the high pressure separator contains spent scrubbing solution plus any water that may have been present in the incoming feed stream.

An example given in the US Patent 4,882,037 document had a feed oil composition as shown in Table 7.44 and the product stream compositions shown in Table 7.45.

The subsequent patents addressed various configurations and applications of this type of process, all of which could include a feed of used oil for re-refining, but the basic concept was contained in the original patent as described. In two of the subsequent patents, co-processing of the used oil together with either an atmospheric residue oil stream or pyrolysis oil produced from organic waste was envisaged. In

FIGURE 7.35 The Puralube Plant at Tröglitz, Germany. (Courtesy of Puralube, Zeitz.)

1995, Puralube Inc. of Wayne, Indiana, acquired the exclusive rights to the Hylube™ process and in 1997 set up Puralube GmbH in Germany (Puralube 2012).

A refinery to produce 45,000 tonnes (49,600 short tons; 99,225,000 pounds) per year of Group II base oils and 25,000 tonnes (27,562 short tons; 55,125,000 pounds) per year of by-products was completed in 2004 and a second train with the same capacity was added and started production in late 2008. Kalnes et al. (Kalnes et al. 2008) and Hartmann (2008) report Group II and Group II+ base oil products in the SN-75, SN-160 and SN-300 range, marketed as PUR-75, PUR-160 and PUR-300, respectively, all having excellent properties. In addition, the Puralube web site (Puralube 2012) shows a PUR-60 product also being currently produced, together with naphtha, gas oil or diesel, heavy fuel oil and asphalt extender products. The yield of base oil products is given as 90,000 tonnes (99,000 short tons, 198,000,000 pounds) per year from the 150,000 tonnes (165,000 short tons, 330,000,000 pounds) per year of used oil feed, quoted on a wet basis, and 50,000 tonnes (55,000 short tons, 110,000,000 pounds) per year of by-products. Figure 7.34 shows the configuration of the process as implemented at Zeitz, near Leipzig in Germany, and the photographs in Figures 7.35 and 7.36 show this plant.

FIGURE 7.36 Overview of the Puralube Plant at Tröglitz, Germany. (Courtesy of Puralube, Zeitz.)

As the quality of the used oil being recycled continues to improve with an increasing fraction being fully synthetic components, it may be expected that the quality of the re-refined lubricating oil base stock can potentially continue to improve as well. As many of the synthetic oils have viscosity indices greater than 120, it is possible that some of the Group II+ base oils produced by this process may have a viscosity index that matches the requirement for Group III base oils. The table presented by Kalnes et al. (Kalnes et al. 2008) shows the properties of the PUR-160 base oil product compared with the requirements for the various groups of base oil. Wadle (2009) has also presented data on various tests for carcinogenic potential of the product oils, showing them to be well below the level that is currently acceptable.

7.8 DISTILLATION/FILTRATION/HYDROGENATION PROCESSES

Although filtration processes are in common refinery and industrial use, attempts to filter the lube oil fractions at ambient temperatures and using organic type filter membranes were not very encouraging, due to the high viscosity of these oils at low temperatures and the inability of the membrane materials to tolerate higher temperatures. Dilution of the oil with a solvent such as *n*-hexane reduces the viscosity of the oil and hence improves the filterability of the oil, but the subsequent separation of the solvent by distillation adds to the complexity and cost of the recovery process. The use of ceramic membranes allows the filter temperature to be increased and avoid the use of a solvent, the process described by Maier in US Patent 5,250,184 dated October 5, 1993, and assigned to Studiengesellschaft Kohle mbH now resulting in a microporous ceramic membrane capable of operating at temperatures up to 500°C (930°F).

7.8.1 The Commissariat a l'Energie Atomique et Compagnie Française de Raffinage Process

This process (US Patent 4,411,790 dated October 25, 1983) was aimed at developing a high temperature ultrafiltration method using a mineral filter medium for application to both used oil and the reduction of the asphaltene content of a hydrocarbon stream coming from the bottom of a vacuum distillation column. The filter medium has a metal or ceramic support which is coated with a layer of at least one metal oxide chosen from the group including titanium dioxide, magnesium oxide, aluminum oxide and mixed oxides based on alumina, and silica. The pore sizes envisaged ranged from 50 to 250 Å (1 Å = 10^{-10} m) and the operation could be carried out at temperatures between 100°C and 350°C (212°F and 660°F) with a differential pressure between 1 and 20 bar (100 and 2000 kPa; 14.5 and 290 psi).

The filter is arranged in various configurations, one being as a cylindrical arrangement with an internal diameter of the ceramic support of 15 mm (0.59 inches), an external diameter of 19 mm (0.75 inches), and a length of 800 mm (31.5 inches) with a pore radius of the support being approximately 1 μm, and the inside of the ceramic being coated with the ultrafiltration layer of the metal oxides. The flow of the oil passing through the filter is normally from the inside to the outside chamber of the filter mounting vessel, the active surface of each filter element being 360 cm^2 (55.8 square inches) and a filtration rate of up to 500 L/day/m^2 (132 US gallons/day/m^2, 0.292 barrels per day per square foot) being achieved for a used motor oil with 200 Å filters, an operating temperature of 200°C (392°F), a differential pressure of 5 bar (500 kPa; 72.5 psi) and a concentration factor of 3.

Periodic cleaning of the filters is achieved by applying a higher pressure on the outer filtrate side of the filter to reverse the flow through the filter and displace the retained materials, the differential applied pressure being between 1 and 30 bar (100 and 3000 kPa; 14.5 and 435 psi). The cleaning cycle is relatively short and is applied at fixed time intervals. One of the embodiments of the patent process is shown in Figure 7.37.

As applied to the recovery of used oil, the incoming oil is first heated in a heater using a heat transfer fluid to a temperature around 90°C (194°F). Most of the water and solids suspended in the oil are removed in the first centrifuge before the oil is heated to around 180°C (355°F) before entering the near-atmospheric pressure distillation column where the remaining water, gasoline and chlorinated solvents in the oil are removed overhead. The underflow from this column enters the next distillation column which operates under reduced pressure to produce a top product of gas oil, together with various sidecuts of oil that enter various intermediate storage vessels, the contents of which are each treated separately in the subsequent processing.

The flow from the bottom of this column is split, part going to a heater to be recycled into the column and this providing the heat required in the column, but also providing a thermal shock which breaks down the additive molecules in the oil and facilitating the removal of heavy metals in its subsequent treatment.

The other part of the underflow stream from the column goes to a centrifuge operating at a temperature around 180°C (355°F), which separates the sludge containing the heavy metals. Each of the storage vessel contents is separately passed through the ultrafiltration unit before passing to a catalytic hydrogenation unit to decolorize the product oil. The oil not passing through the filter is recycled to the heater unit before the vacuum distillation column.

Another embodiment of the process described in the patent document has two atmospheric distillation columns with the first operating at a temperature around 180°C (355°F) and the second at a temperature around 360°C (680°F). The underflow from the second column passes through a centrifuge operating at a temperature of around 180°C (355°F) to an ultrafiltration unit, the ultrafiltrate then passing to a vacuum distillation column for fractionation and subsequent hydrogenation. Figure 7.38 shows this version of the process.

Table 7.46 shows the results of the ultrafiltration of a used oil sample carried out at 200°C (392°F) with an ultrafiltration module with 7 barriers coated with a sensitive layer of a mixed oxide of aluminum, and magnesium mixed with silica, with a pore radius of 200 Å, a differential pressure of 5 bar (500 kPa; 72.5 psi), a linear velocity in the filter of 3.23 m/s (10.6 ft/s), and a corresponding rate of filtration of 650 liters/day/m^2 (172 US gallons per day per square meter; 0.378 barrels/day/ft^2).

No information was provided on the catalytic hydrogenation step, but the application of known techniques should result in the production of a base oil product.

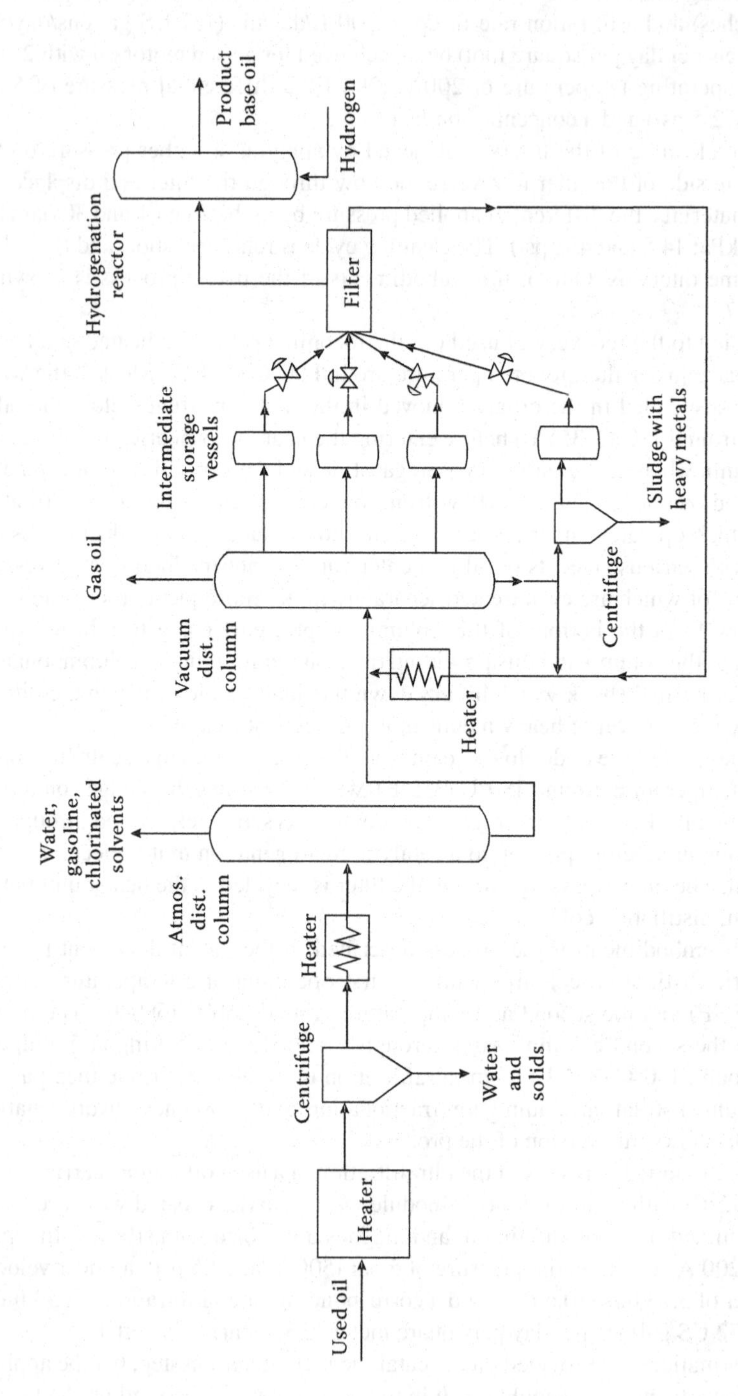

FIGURE 7.37 The CEA/CFR process. (From CEA/CFR US Patent 4,411,790.)

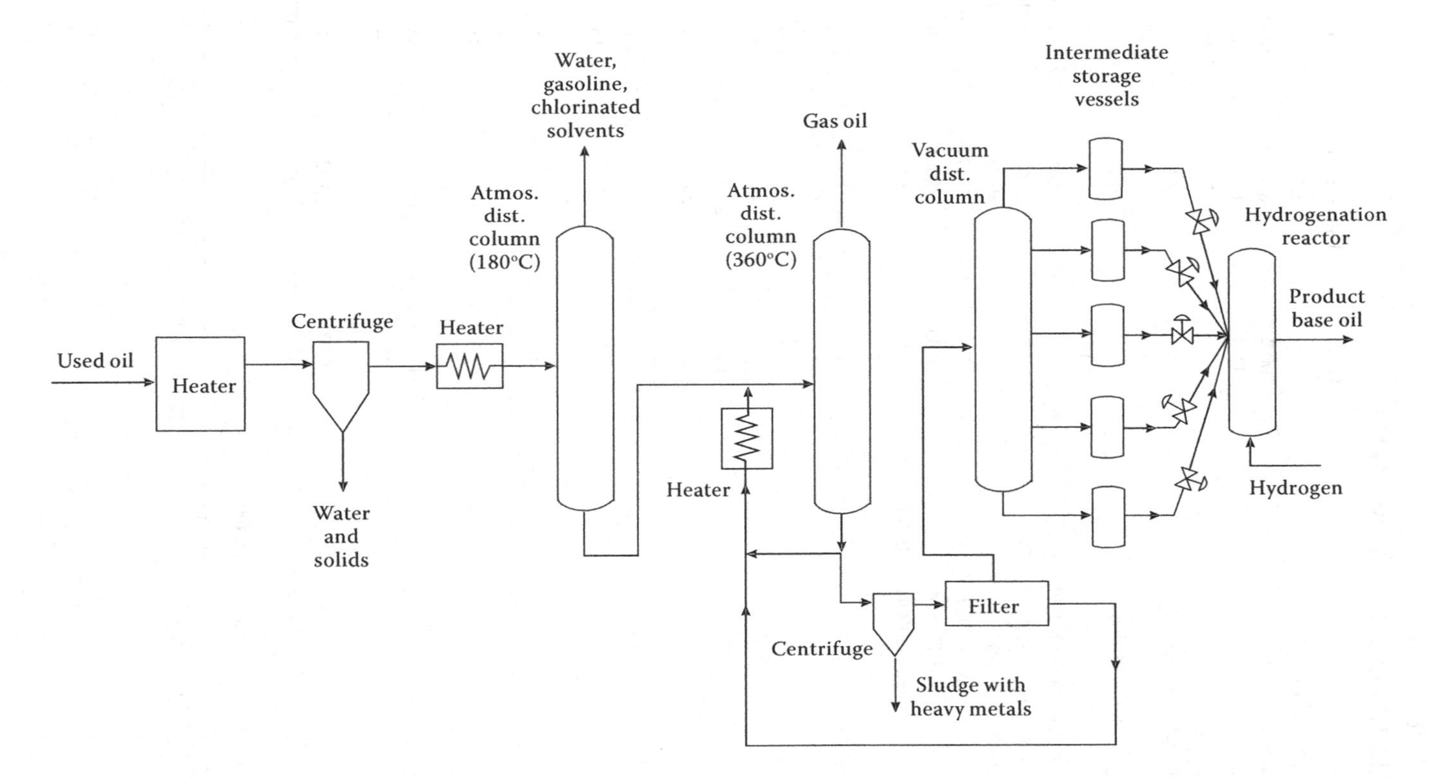

FIGURE 7.38 The CEA/CFR process. (From CEA/CFR US Patent 4,411,790.)

TABLE 7.46
CEA/CFR Ultrafiltration of Used Oil

Property	Feed	Ultrafiltrate	Concentrate
Viscosity, NFT 60-100			
cSt at 40°C	100.2	87.2	127.92
cSt at 100°C	11.8	10.2	14.05
Viscosity index	107	97	108
TAN, mg KOH/g, ASTM664		0.1	
pH	0.6		1
TBN, mg KOH/g, ASTM664		0.1	
pH	0.1		0.1
TBN, mg KOH/g, ASTM12896	2.4	0.41	5.20
Pentane insolubles, m-%, ASTM893	0.78	<0.05	0.91
Toluene insolubles, m-%	0.31	<0.05	0.51
Conradson carbon, m-%, NFT 60-116	2.1	0.47	2.9
Sulfated ash, m-%, NFT 60-143	0.35	<0.005	0.68
Saponification index, mg KOH/g, NFT 60-110	5.0	2.1	6.0
ppm, mass			
Ba	85	<10	176
Ca	602	<10	1241
Mg	96	<10	199
B	<10	<10	<10
Zn	15	<10	31
P	520	29	1079
Fe	152	6	315
Cr	3	<1	6
Al	6	<2	13
Cu	2	<1	4
Sn	3	<2	5
Pb	241	8	502
Ag	<1	<1	<1
Si	15	3	31
Na	41	<5	85
Mo	<2	<2	<2
% w/w			
S	0.97	0.92	1.01
N	≤0.05	≤0.05	≤0.05

Source: CEA/CFR US Patent 4,411,790.

7.9 PROCESSES PRODUCING PRIMARILY A FUEL PRODUCT AND POSSIBLY A BASE OIL PRODUCT

The original processing of used lubricating oil involved *reclamation processes* which usually involve treatment to separate solids and water from a variety of used oils (Chapter 4). The methods used may include heating, filtering, dehydrating and centrifuging and the product was used as a fuel or fuel extender.

Ciora and Liu (2000) quote a level of 70% of the used motor and other oils that were generated in the United States at the end of the 1990s being recycled as fuel. These processes can vary from practically no treatment to simple filtration and more elaborate treatments which include distillation and solvent extraction to produce a burner fuel, to the cracking processes which produce a coke product together with fractions of fuel gas, gasoline and diesel fuel as the main products.

Regulatory bodies in an increasing number of jurisdictions in Europe and the Americas are moving to discourage with suitable legislation the recycling of used oil to a lower level of use on a hierarchical structure which requires recycling to the same level of use to be preferred if this is technically, economically and organizationally feasible. In general, this hierarchical structure, which has been developed in the European Union (Hartmann 2008) with similar legislation in some US states and Canadian provinces, defines the following steps for disposal of materials such as used oil, each subsequent step being defined as a lower level process:

- Prevention of waste generation
- Recycling of the material back to its original utility or new products, materials or substances, but does not include energy recovery. For used oils, this would exclude the processes that produce burner fuels as a product
- Recovery, including energy recovery from a burner fuel, for example
- Disposal of the material in a permanent storage facility or other disposal facility

It is for this reason that many operators of recycling plants that do not produce a product that is acceptable to the lubricant manufacturers have made great efforts to upgrade the recycling process they had been using in the past. In many parts of the world, however, the recycling of used oil to use as a burner fuel or other energy recovery system is still an acceptable process.

7.9.1 Filtration Processes

The simple filtration of used oil is aimed at removing particulate matter with a process that uses much less energy than the processes using distillation or solvent extraction, but results in a product oil that is usually aimed at the burner fuel market. Such a process is found in the studies carried out by the Ceramem company in the 1990s (Audibert 2006), but which do not seem to have been further developed since then. This burner fuel will still contain the dissolved impurities and much of the sulfur and carcinogenic materials that, if not removed in a hydrogenation process, result in a market price for the product that is lower than the price obtainable for a base oil product. Treatment with a clay adsorbent after the filtration results in some improvement of the properties of the product oil, as in the following process, and the second patent claims a base oil product that can be used to reformulate lubricating oil.

7.9.1.1 The Ciora and Liu Processes

This process (US Patent 6,024,880 dated February 15, 2000; US Patent 6,117,327 dated September 12, 2000 and assigned to Media and Process Technology Inc.) represents a combination of a filtration of the used oil, followed by a clay treatment to remove color and odor from the filtrate as the filtered oil still has a color and odor that is very similar to that of the used oil feed material. The regeneration of the clay and its reuse is also a feature of the process. Various types of ceramic filter media were examined and included alumina (Al_2O_3), zirconia (ZrO_2), titania TiO_2), silica (SiO_2), and mixtures of these, with nominal pore sizes varying from approximately 50 Å to 10 μm. Various types of solvent cleaning processes for the filter membrane were examined and chloroform was found to be an effective solvent for the materials retained on the filter surface and which have to be periodically removed by flushing of the filter with the solvent.

Various types of clay minerals were investigated to determine their efficacy in decolorizing and deodorizing the filtered oil, and included activated alumina, zeolite, silica gel, anionic and cationic clay. Regeneration of the clay with a high temperature process in the presence of air, and a process of regeneration with various organic solvents were shown to be effective in restoring the adsorptive properties of the clay which could be used for multiple cycles before having to be replaced. Figure 7.39 shows the flow diagram for this type of process. The dotted lines in the diagram show the flow lines for one variation of the membrane cleaning process and the adsorber regeneration process.

The used oil is pumped through a prefilter with a pore size of 40 μm to remove the larger solids and impurities before entering the holding tank where it can be heated to the temperature required for the filtration step. A pump that has vibration isolators on both sides of it to prevent vibrations being transmitted to the filter passes the oil to the inside of the ceramic filter units which in one test consisted of pore sizes between 0.05 μm and 0.2 μm at a temperature between 130°C and 150°C (265°F and 300°F) with a differential pressure of 15 to 20 psi (103–138 kPa; 14.9 to 20 psi). A significant reduction in the ash content of the permeate was observed. In another test a used oil with a kinematic viscosity of 27.1 cSt at 40°C (104°F) and an ash content of 0.87% w/w was treated with an alumina ceramic membrane filter with nominal pore sizes of 0.1 and 0.2 μm at 140°C to 160°C (285°F–320°F) and differential pressures of 12 to 54 psi (83–372 kPa).

The throughput of the filters was found to be 4 to 6 L (1.1–1.6 US gallons) per hour per bar (4 to 6 L/h/100 kPa; 0.0974 to 0.0146 ft³/h/psi) for the 0.1 μm filter and between 6 and 15 liters (1.6 and 4 US gallons) per hour per bar (0.0146 and 0.0365 ft³/h/psi) for the 0.2 μm filter. The ash content of the oil after filtration through the 0.1 μm filter was found to be 0.157 and 0.109% w/w after the second and fourth day, respectively, and through the 0.2 μm filter it was approximately 0.197% and 0.109% w/w, also after 2 and 4 days of operation. The oil not passing through the filter medium was recycled to the holding tank as shown in the diagram.

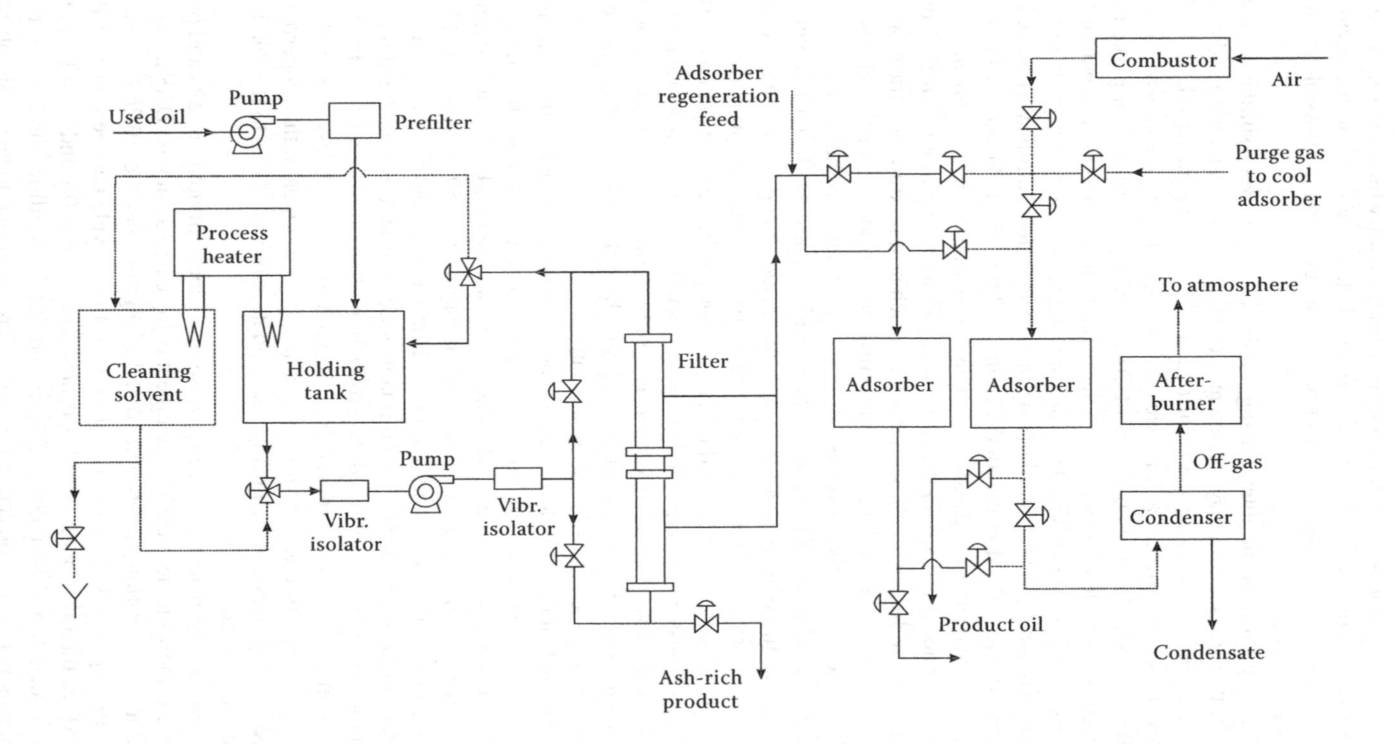

FIGURE 7.39 The Ciora/Liu filtration and clay treatment process. (From Ciora & Liu US Patent 6,024,880/6,117,327.)

When the flow rate of oil through the filter had decreased to 70% to 80% of the initial flow rate, a solvent cleaning cycle was initiated in which the solvent flows through the filter unit and returns to the solvent holding tank while the permeate lines from the filter are isolated. No information was given on the possible recovery of the solvent or its disposal.

The permeate through the filter, having a color and smell very similar to that of the used oil, then passes to a clay treatment section. After testing various adsorbent clays for their ability to remove the color and smell of the permeate oil, as well as their ability to be regenerated, an anionic clay powder of hydrotalcite was found to be effective, operating at a temperature of 70°C (158°F). The test in a packed column of 46 cm (1.81 ft) length and outside diameter 12.5 mm (0.5 inches) pumped oil with a viscosity of 28 cSt at 40°C (104°F) at a rate of approximately 2 cc per hour (7.1×10^{-8} ft^3/h) through the clay bed. Similarly, a test with activated alumina was carried out at room temperature and a flow rate of 0.2 cc/min (42.6×10^{-8} ft^3/h) and was found to be effective in both decolorizing and deodorizing the permeate oil. In addition, the clay could be regenerated thermally in the presence of air to restore more than 80% of its adsorption capacity. Regeneration with a solvent such as a mixture of tetrachloroethylene and methanol, or a 50/50 mixture of methylene chloride and methanol, restored between 50% and 70% of the adsorption capacity of the activated alumina clay.

A variation was also described in which the incoming oil is diluted with a solvent such as kerosene, diesel fuel or jet fuel, but no further information was provided on this option.

No further information was provided in the first patent on the properties of the deashed, decolorized and deodorized oil and without additional analyses and information it is not possible to see whether the oil could be used as a lubricating oil blending stock without further treatment. It should, however, be a superior burner fuel when compared with the untreated used oil where the removal of the ash containing much of the metals would be an advantage.

The second patent of 2000 described a process in which a chemical treatment of the feed oil was proposed to condition the metals to be removed in the membrane filtration process and the use of a clay treatment was omitted. This was claimed to be necessary as membranes with as small a pore size as 50 Å were not able to achieve a significant removal of certain of the heavy metals and filtration alone was not enough to remove completely contaminants such as iron. It was claimed that the process of the second patent allowed the production of an oil that had had almost all of the heavy metals and ash removed from it while using much less energy than a distillation process, while also being less sensitive to variations in the properties of the oil feed to the process, and showing less oxidation of the oil due to the relatively low temperatures used.

The chemical treatment considered the use of various ammonium salts and polyalkoxyalkylamine with the preferred treatment being carried out with DAP dissolved in water at a concentration of between 10 and 500 grams per liter (0.264 gallon) of water, and the aqueous solution is added to the oil to provide between 0.1 and 10% w/w of DAP in the oil. This mixture is heated for between 0.1 and 24 h to a temperature between 100°C and 180°C (212°F and 355°F) to allow the DAP to react with the metals in the oil. The actual reactions were suggested to be the dissolved or dispersed contaminants reacting to form insoluble products which could also clump together as aggregates, and which could be removed by a subsequent filtration step through a 40 μm filter. As the viscosity of the oil after the removal of these solids was found to remain at a lower level than the untreated oil viscosity, an improvement in the throughput rate of the oil through the filter was also observed, as well as increased stability of the posttreatment oil to oxidation.

It was also suggested that a variation of the process could involve the chemical treatment of the retentate oil that did not pass through the membrane filter and which still contained a concentrated ash and metals component. In this embodiment the used oil would not be chemically treated before the ultrafiltration, and the metals and ash would be removed in the post-ultrafiltration step, the chemically treated oil being filtered through a 40 μm filter before being returned to the feed tank or being discharged as a product. In this case, the permeate through the ultrafiltration membrane might be expected to have higher levels of the metals than the embodiment with pre-ultrafiltration chemical treatment. These processes are shown in Figure 7.40.

Table 7.47 shows the metals content of the feed oil (the column "As-received" in the table) and the effectiveness of the DAP and membrane filtration processes in removing these metals from a used oil with the properties shown in Table 7.48. In this case,

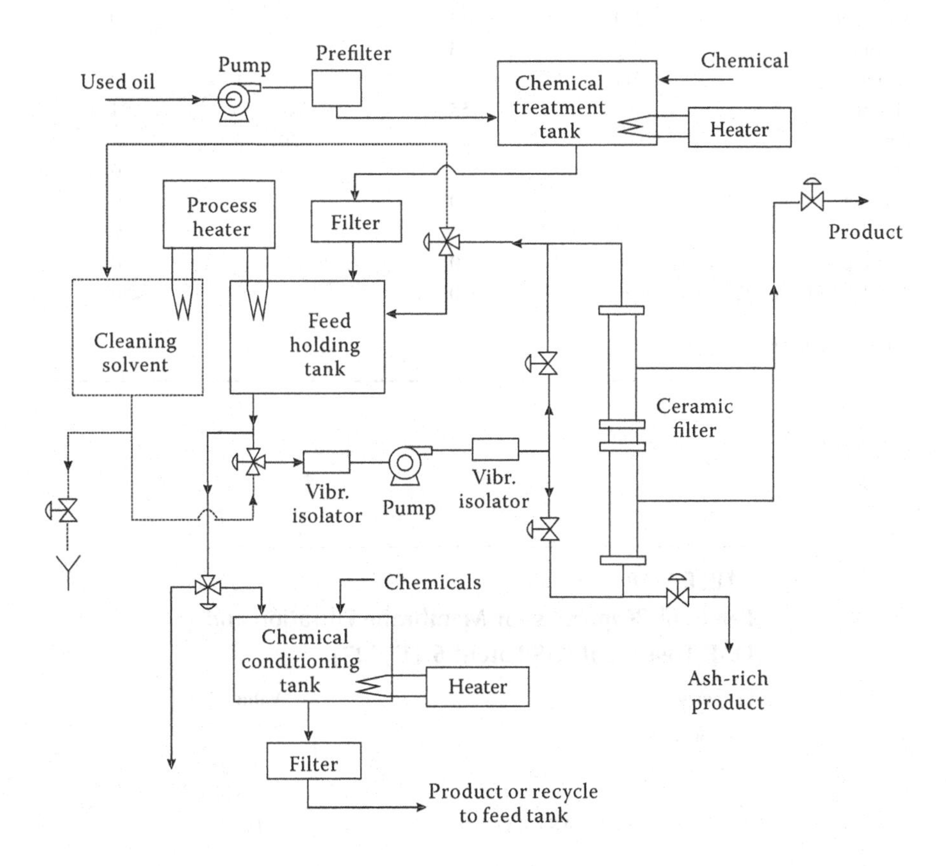

FIGURE 7.40 The filtration and DAP treatment process. (From Ciora & Liu US Patent 6,024,880/6,117,327.)

TABLE 7.47
Membrane Filtration and DAP Treatment of US Patent 6,117,327

	Contaminant Concentration Levels (ppm) in Oil Sample			
Contaminant	**As-Received**	**Membrane Only**	**DAP Only**	**Membrane + DAP**
Iron (by ICP)	205	39	35	1
Chromium	5	2	3	0
Lead	67	12	0	0
Copper	202	20	193	4
Tin	8	1	6	0
Aluminum	45	10	5	0
Nickel	3	0	2	0
Silver	1	0	1	0
Manganese	5	0	0	0
Silicon	112	62	52	83
Boron	30	8	23	8
Sodium	103	4	18	1
Magnesium	244	13	6	0
Calcium	726	15	15	0
Barium	17	0	0	0
Phosphorus	495	155	381	154
Zinc	860	165	20	8
Molybdenum	3	3	0	0
Titanium	2	0	0	0
Vanadium	1	0	1	0
Potassium	0	0	0	0
Ash (by ASTM D482)	6020	970	1520	<50

Source: Membrane Filtr. US Patent 6,117,327.

TABLE 7.48
Feed Oil Properties for Membrane Filtration and DAP Treatment (US Patent 6,117,327)

Property	Value
Viscosity, cSt	
at 40°C	63.3
at 100°C	11.5
Ash content, % w/w (ASTM D482)	0.602

Source: Membrane Filtr. US Patent 6,117,327.

a 2 L (0.53 US gallons; 0.071 ft^3) sample was mixed with an aqueous mixture of 100 cc (0.00353 ft^3) of water containing 36.12 grams (0.078 pounds) of DAP, the stirred mixture was then heated to 160°C (320°F) and water was evaporated from the oil over a two-h period. This oil was then kept at 160°C (320°F) for a 24-h period with 50 cc (0.00177 ft^3) samples being withdrawn at 2, 4, and 24 h: these samples were filtered through a –325 mesh screen and the ash contents were determined to be 0.179, 0.180, and 0.171% w/w, respectively. The metals analysis, using inductively coupled plasma (ICP) techniques, on the 24-h sample is shown in the table column as "DAP only".

The remaining sample was cooled to approximately 50°C (122°F) and filtered through a –325 mesh screen before passing to a membrane filter system with a nominal pore size of 500 Å and operating at a temperature of 120°C to 135°C (248°F–275°F) with a differential pressure of 10 to 20 psi (69 to 138 kPa; 10–20 psi). The flux rate through the filter was measured as 3.8 L (1 US gallon) per square meter per hour per bar (8.6×10^{-4} ft^3/h/psi) at 125°C (255°F). The permeate oil was analyzed for ash content (~0.0012% w/w) and metals content as shown in the table as the column "DAP + Membrane". A sample of the used oil was also passed through the membrane filter under the same operating conditions and the analysis of this permeate is shown in the table as the column "Membrane only".

On the basis of these results, it was claimed that the product oil could be recycled for use in engine lubricant formulation.

7.9.2 Cracking and Coking Processes

When the used oil is heated to temperatures approaching 400°C (750°F) the long hydrocarbon chains start to crack into shorter chain products. If no additional hydrogen is available to the cracked products, hydrocarbons ranging from light gases, through a gasoline fraction and a diesel oil fraction, together with heavier fractions and a coke residue are obtained. If the heating of the oil takes place through the wall of a vessel, the coke will deposit on the heated wall and the process will have to be shut down periodically to remove the coke deposits as these reduce the rate of heat transfer through the wall and reduce the efficiency of the heating process. This has meant that this type of process has generally been confined to limited size batch treatments and the coke removal has been done by mechanical means, in some cases by having personnel enter the vessel to manually scrape and chip the coke off the heated surface. This has made this type of process less suitable for larger throughput plants which aim for continuous operation. A process developed by Koch for his Alphakat company has a novel method of heating the oil without using a heated wall surface and has been widely patented throughout the world.

7.9.2.1 The Alphakat Process

This process (German Patent DE 103 56 245 dated January 1, 2007; US Patent 7,473,348 dated January 6, 2009) was a development of an earlier German patent (DE 10049377 dated October 31, 2002) in which catalysts of sodium-doped aluminum silicate were used to crack hydrocarbon-containing feed materials into diesel and gasoline products. Feed materials included used oil, plastics, grease, wood, and halogenated compounds, some being mixed with carrier oil.

The disadvantage of the earlier process was the heating of the reactants through a wall which led to coke deposits forming on the wall and which interfered with the heat transfer, resulting in higher temperatures at the wall and unwanted reactions with the catalyst materials with additional deposits on the walls. This process, known as the KDV process (Katalytische Drucklose Verölung, or catalytic pressure-less conversion to oil), proposes mixing a suitable catalyst of fully crystallized Y-molecules doped with sodium, potassium, calcium or magnesium-aluminum silicates (depending on whether the feed contains mineral hydrocarbons, highly-halogenated compounds, biological materials or wood, respectively) with the feed. This mixture is pumped around in a flow circuit to generate the required heat energy with the energy input from the pumps and counter-rotating agitators until, at a pressure of around 0.9 bar (90 kPa; 13.1 psi), the temperature is high enough at between 290°C and 350°C (555°F and 660°F) (Alphakat 2012) to crack the oil molecules and evaporate the liquids and diesel fraction from the mixture. The high degree of agitation serves to clean the internal surfaces of the vessels and components in the circuit. The liquid products then pass to a distillation column and the overhead condensed diesel fraction forms the product.

No additional heat is supplied to the reactants, and it was claimed that approximately 10% of the product fuels is required to power a generator system to provide the energy required to operate the process. The small amounts of gases formed in the process are combined with the air fed to the generator system, so that no light products are produced. A conversion rate of approximately 80% of the hydrocarbons in the feed to a diesel fraction is quoted in the information on the Alphakat web site (Alphakat 2012).

A number of plants based on this technology have been built in different countries around the world, some concentrating on waste plastics as feed material, but the economic data and product marketability for its application to used oil recycling would have to be evaluated in each specific environment.

7.10 OTHER PROCESSES

A number of other processes have been proposed, many of which have not reached the implementation stage or, for various reasons, have not been pursued. In addition, there have been patents issued for processes which contain small variations or different combinations of the previous processes. The following represents an example of a recent patent application with some variations in the process equipment used.

7.10.1 The CleanOil Ltd. Process

This process (US Patent 8,088,276 B2 dated January 3, 2012) represents a combination of techniques with some variations that distinguish it. A multilayer filter is used to filter the incoming oil which then passes to a heater and then to a cyclone separator which separates much of the water in the feed. The oil is then heated and separated in a flash tower into an overhead stream of water and light ends and a bottoms stream that is then mixed intimately with propane before entering a series of extraction towers which separate an asphalt product from the stream. The extraction towers, in one embodiment, have an inverted cone-shape to facilitate the removal of

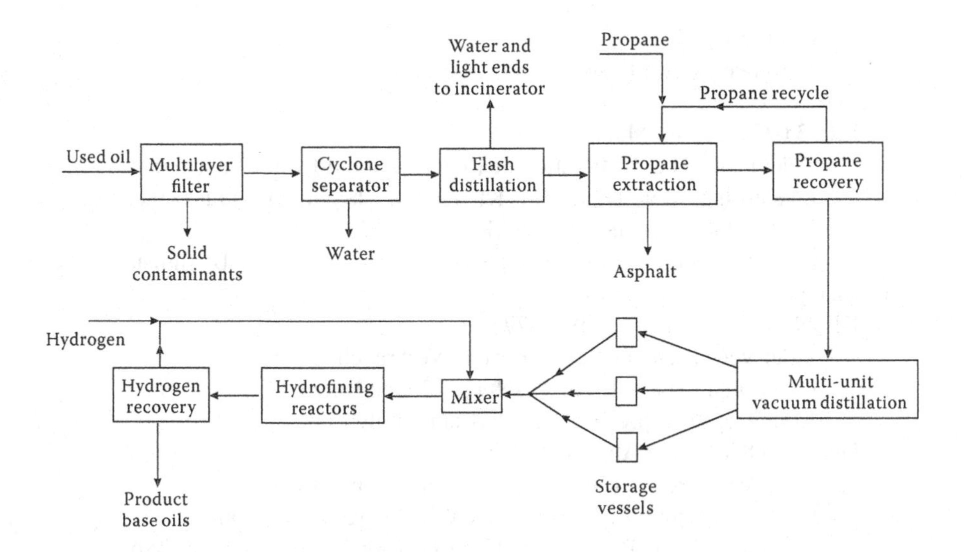

FIGURE 7.41 The CleanOil process. (From CleanOil US Patent 8,088,276.)

resins and asphalts from the bottom of the tower. The overhead stream from these extraction towers then proceeds to a propane recovery section which recovers the propane for recycling and includes a gas-stripping section to remove residual propane and water. The oil stream then passes to a multi-unit vacuum distillation system which removes light ends and the base oil fractions continue to a hydrofining section where the stream is mixed with hydrogen using an in-line static mixer and passes through multiple units which contain suitable catalysts and operating conditions to remove sulfur, nitrogen, chlorine compounds and other oxidizing components, and to saturate the olefins and stabilize the oil. A simplified flow diagram of this process is shown in Figure 7.41.

No information was provided in the patent document on the properties of the product oils that would be expected with this process.

PATENTS

Canada

CA 1 209 512, dated August 12, 1986.
Title: Used Oil Re-refining.
Inventor(s): Reid, L.E., Yao, K.C., Wittenberg S. and Ryan D.G.
Assignee: Imperial Oil Limited, Canada.

CA 2 068 905, dated July 22, 1997.
Title: Waste Lubricating Oil Pretreatment Process.
Inventor(s): Wilson, T.A.
Assignee: Mohawk Canada Ltd., Canada.

CA 2 178 381, dated November 13, 2007.
Title: Methode de raffinage d'Huile et Appareil Servant a Mettre en Oeuvre Ladite Methode/Oil Re-refining Method and Apparatus.

Inventor(s): Kenton, K.J.
Assignee: Avista Resources Inc., USA.

France

FR 93 03275, dated March 3, 1993.
Title: Procédé et Installation de Regeneration d'Huiles Lubrifiantes.
Inventor(s): Merchaoui, H., Khalef, N., Jaafar, A., Ouazzane, A., Boufahja, A. and Meziou, S.
Assignee: Societe Tunisienne de Lubrifiants—SOTULUB, Tunisia.

Germany

DE 29 01 090, dated July 19, 1979.
Title: Verfahren zur Regenerierung Verbrauchter Öle.
Inventor(s): Antonelli, S., Peschiera, B. and Borza, M.
Assignee: Snamprogetti S.p.A, Milano, Italy.

DE 34 05 858, dated August 16, 1984.
Title: Verfahren zur Wiederaufbereitung von Altölen.
Inventor(s): Reid, L.E., Yao, K.C., Wittenberg S. and Ryan D.C.
Assignee: Exxon Research and Engineering Co., New Jersey, USA.

DE 36 02 586 A1, dated July 30, 1987.
Title: Verfahren zur Aufarbeitung von Altöl.
Inventor(s): Wetzel, R., Coenen, H. and Kreuch, W.
Assignee: Krupp Koppers GmbH, Essen, Germany.

DE 100 49 377, dated October 31, 2002.
Title: Katalytische Erzeugung von Dieselöl und Benzinen aus kohlenwasserstoffhaltigen Abfällen und Ölen.
Inventor(s): Oberländer, I., Koch, C. and Gruhnert, W.
Assignee: EVK Dr. Oberländer GmbH & Co., KG.

DE 103 56 245 B4, dated January 1, 2007.
Title: Verfahren zur Erzeugung von Dieselöl aus kohlenwasserstoffhaltigen Reststoffen sowie eine Vorrichtung zur Durchführung dieses Verfahrens.
Inventor(s): Koch, C.
Assignee: Alphakat GmbH

DE 198 52 007, dated May 18, 2000.
Title: Verfahren zur Wiederaufarbeitung von Altölen, die mit dem Verfahren erhältlichen Grundöle und deren Verwendung.
Inventor(s): Pöhler, J., Mödler, M., Bruhnke, D. and Hindenberg, H.
Assignee: Mineralöl-Raffinerie Dollbergen GmbH, Uetze, Germany.

United States

US 3,930,988, dated January 6, 1976.
Title: Reclaiming Used Motor Oil.
Inventor(s): Johnson, M.M.
Assignee: Phillips Petroleum Co., Bartlesville, Oklahoma.

US 4,021,333, dated May 3, 1977.
Title: Method of Rerefining Oil by Distillation and Extraction.
Inventor(s): Habiby, E.N. and Jahnke, R.W.
Assignee: The Lubrizol Corporation, Cleveland, Ohio.

US 4,073,719, dated February 14, 1978.
Title: Process for Preparing Lubricating Oil from Waste Lubricating Oil.
Inventor(s): Whisman, M.L., Reynolds, J.W., Goetzinger, J.W. and Cotton, F.O.
Assignee: United States Department of Energy.

US 4,073,020, dated February 14, 1978.
Title: Method for Reclaiming Waste Lubricating Oils.
Inventor(s): Whisman M.L., Goetzinger, J.W. and Cotton, F.O.
Assignee: United States Department of Energy.

US 4,140,212, dated February 20, 1979.
Title: Cyclonic Distillation Tower for Waste Oil Rerefining Process.
Inventor(s): O'Blasny, R.H., Sparks, T.F., Tierney T.J. and Hunter, J.S.
Assignee: Vacsol Corp., Kansas.

US 4,151,072, dated April 24, 1979.
Title: Reclaiming Used Lubricating Oils.
Inventor(s): Nowack, G.P., Tabler, D.C. and Johnson, M.M.
Assignee: Phillips Petroleum Co., Oklahoma.

US 4,154,670, dated May 15, 1979.
Title: Method of Rerefining Oil by Dilution, Clarification and Extraction.
Inventor(s): Forsberg, J.W.
Assignee: The Lubrizol Corporation, Wickliffe, Ohio.

US 4,287,049, dated September 1, 1981.
Title: Reclaiming Used Lubricating Oils with Ammonium Salts and Polyhydroxy Compounds.
Inventor(s): Tabler, D.C. and Johnson, M.M.
Assignee: Phillips Petroleum Co., Bartlesville, Oklahoma.

US 4,328,092, dated May 4, 1982.
Title: Solvent Extraction of Hydrocarbon Oils.
Inventor(s): Sequeira Jr., A.
Assignee: Texaco Inc. White Plains, Texas.

US 4,360,420, dated November 23, 1982.
Title: Distillation and Solvent Extraction Process for Rerefining Used Lubricating Oil.
Inventor(s): Fletcher, L.C., Beard, H.J. and O'Blasny, R.
Assignee: Delta Central Refining Inc., Natchitoches, Louisiana.

US 4,399,025, dated August 16, 1983.
Title: Solvent Extraction Process for Rerefining Used Lubricating Oil.
Inventor(s): Fletcher, L.C. and O'Blasny, R.H.
Assignee: Delta Central Refining Inc. Louisiana.

US 4,406,778, dated September 27, 1983.
Title: Spent Oil Recovery Process.
Inventor(s): Borza, M., Leoncini, S. and Modenesi, A.
Assignee: Snamprogetti S.p.A., Milan, Italy.

US 4,411,790, dated October 25, 1983.
Title: Process for the Treatment of a Hydrocarbon Charge by High Temperature Ultrafiltration.

Inventor(s): Arod, J., Bartoli, B., Bergez, P., Biedermann, J., Caminade, P., Martinet, J.-M., Maurin, J. and Rossarie, J.
Assignee: Commissariat a l'Energie Atomique; Compagnie Française de Raffinage, France.

US 4,512,878, dated April 23, 1985.
Title: Used Oil Re-refining.
Inventor(s): Reid, L.E., Yao, C. and Ryan D.G.
Assignee: Exxon Research and Engineering Co., New Jersey.

US 4,789,460, dated December 6, 1988.
Title: Process for Facilitating Filtration of Used Lubricating Oil.
Inventor(s): Tabler, D.C. and Johnson, M.M.
Assignee: Phillips Petroleum Co., Oklahoma.

US 4,818,368, dated April 4, 1989.
Title: Process for Treating a Temperature-sensitive Hydrocarbonaceous Stream Containing a Non-distillable Component to Produce a Hydrogenated Distillable Hydrocarbonaceous Product.
Inventor(s): Kalnes, T.N., James Jr., R.B. and Staggs, D.W.
Assignee: UOP, Illinois.

US 4,882,037, dated November 21, 1989.
Title: Process for Treating a Temperature-sensitive Hydrocarbonaceous Stream Containing a Non-distillable Component to Produce a Selected Hydrogenated Distillable Light Hydrocarbonaceous Product.
Inventor(s): Kalnes, T.N. and James Jr., R.B.
Assignee: UOP, Illinois.

US 4,923,590, dated May 8, 1990.
Title: Process for Treating a Temperature-sensitive Hydrocarbonaceous Stream Containing a Non-distillable Component to Produce a Hydrogenated Distillable Hydrocarbonaceous Product.
Inventor(s): Kalnes, T.N. and James Jr., R.B.
Assignee: UOP, Illinois.

US 4,941,967, dated July 17, 1990.
Title: Process for Re-refining Spent Lubeoils.
Inventor(s): Mannetje, L.M.M. and Laghate, A.S.
Assignee: Kinetics Technology Intenational B.V., Zoetermeer, Netherlands.

US 5,176,816, dated January 5, 1993.
Title: Process to Produce a Hydrogenated Distillable Hydrocarbonaceous Product.
Inventor(s): Lankton, S.P. and James Jr., R.B.
Assignee: UOP, Illinois.

US 5,250,184, dated October 5, 1993.
Title: Procedure for the Preparation of Microporous Ceramic Membranes for the Separation of Gas and Liquid Mixtures.
Inventor(s): Maier, W.F.
Assignee: Studiengesellschaft Kohle mbH, Germany.

US 5,302,282, dated April 12, 1994.
Title: Integrated Process for the Production of High Quality Lube Oil Blending Stock.
Inventor(s): Kalnes, T.N., Lankton, S.P. and James Jr., R.B.
Assignee: UOP, Illinois.
US 5,556,548, dated September 17, 1996.
Title: Process for Contaminated Oil Reclamation.
Inventor(s): Mellen, C.R.
Assignee: Interline Hydrocarbon Inc., Casper, Wyoming, USA.
US 5,759,385, dated June 2, 1998.
Title: Process and Plant for Purifying Spent Oil.
Inventor(s): Aussillious, M., Briot, P., Bigeard, P.-H. and Billon, A.
Assignee: Institut Français du Pétrole, France.
US 5,814,207, dated September 29, 1998.
Re-issued as US RE38,366 E, dated December 30, 2003.
Title: Oil Re-refining Method and Apparatus.
Inventor(s): Kenton, K.J.
Assignee: Enprotec International Group N.V., Switzerland.
Re-issued to Avista Resources Inc., Houston, Texas.
US 5,904,838, dated May 18, 1999.
Title: Process for the Simultaneous Conversion of Waste Lubricating Oil and Pyrolysis Oil Derived from Organic Waste to Produce a Synthetic Crude Oil.
Inventor(s): Kalnes, T.N. and James Jr., R.B.
Assignee: UOP, Illinois.
US 6,024,880, dated February 15, 2000.
Title: Refining of Used Oils using Membrane- and Adsorption-based Processes.
Inventor(s): Ciora, R.J. and Liu, P.K.T.
Assignee: —
US 6,117,327, dated September 12, 2000.
Title: Deashing and Demetallization of Used Oil Using a Membrane Process.
Inventor(s): Ciora, R.J. and Liu, P.K.T.
Assignee: Media and Process Technology Inc.
US 6,174,431, dated January 16, 2001.
Title: Method for Obtaining Base Oil and Removing Impurities and Additives from Used Oil Products.
Inventor(s): Williams, M.R. and Krzykawski, J.
Assignee: Interline Hydrocarbon Inc., Alpine, Utah.
US 6,712,954 B1, dated March 30, 2004.
Title: Method for Reprocessing Waste Oils, Base Oils Obtained According to Said Method and Use Thereof.
Inventor(s): Pöhler, J., Mödler, M., Bruhnke, D. and Hindenberg, H.

Assignee: Mineralöl-Raffinerie Dollbergen GmbH, Uetze-Dollbergen, Germany.

US 7,226,533, dated June 5, 2007.

Title: Process for Re-refining Used Oils by Solvent Extraction.

Inventor(s): Angulo Aramburu, J.

Assignee: Sener Grupo de Ingenieria, S.A., Madrid, Spain.

US 7,261,808, dated August 28, 2007.

Title: Upgrading of Pre-processed Used Oils.

Inventor(s): Grandvallet, P., Hagan, A.P. and Huve L.G.

Assignee: Shell Oil Company, Houston, Texas.

US 7,431,829, dated October 7, 2008.

Title: Method for Regenerating Used Oils by Demetallization and Distillation.

Inventor(s): Angulo Aramburu, J.

Assignee: Sener Grupo de Ingenieria S.A., Madrid, Spain.

US 7,473,348 B2, dated January 6, 2009.

Title: Diesel Oil from Residues by Catalytic Depolymerization with Energy Input from a Pump-Agitator System.

Inventor(s): Koch, C.

Assignee: Alphakat GmbH

US 8,088,272 B2, dated January 3, 2012.

Title: Oil Re-refining System and Method.

Inventor(s): Marden, A.L.

Assignee: CleanOil Limited, Hong Kong.

World Intellectual Property Organization

WO 91/17804, dated November 28, 1991.

Title: Enhanced Vacuum Cyclone.

Inventor(s): Kenton, K.J.

Assignee: Enprotec Inc. N.V., GB.

WO 94/07798, dated April 14, 1994.

Title: Process to Re-refine Used Oils.

Inventor(s): Miñana, J., Schieppati, R. and Dalla Giovanna, F.

Assignee: Viscolube Italiana S.P.A., Pieve Fissiraga, Italy.

WO 94/21761, dated September 29, 1994.

Title: Process and Plant for the Regeneration of Lubricating Oils.

Inventor(s): Merchaoui, H., Khalef, N., Jafaar, A., Ouazzane, A., Boufahja, A. and Meziou S.

Assignee: Societe Tunisienne de Lubrificants—Sotulub, Tunisia.

WO 03/033630 A1, dated April 24, 2003.

Title: Upgrading of Pre-processed Used Oils.

Inventor(s): Grandvallet, P., Hagan, A.P. and Huve, L.G.

Assignee: Shell Internationale Research Maatschappij B.V., The Hague, Netherlands.

REFERENCES

Alphakat. 2012. Available at http://www.alphakat.de. Accessed on November 30, 2012.

Angulo Aramburu, J. 2003. Rerefining Used Oils by Propane Extraction: A Proven Technology, Ingenieria Quimica, Rev. No. 400, pp. 55–61, April 2003. Available at http://www.ingenieriaquimica.es/files/pdf/iq/400/02ARTICULOAB.pdf. Accessed on November 17, 2012.

ARI. 2002. Aggregate Research Latest News, September 2, 2002. Sener Pioneering Oil Recycling Technology Could Be Exported Abroad. Available at http://www.aggregateresearch.com/articles/1152/Sener-pioneering-oil-recycling-technology-could-be-exported-abroad.aspx. Accessed on November 29, 2012.

Audibert, F. 2006. *Waste Engine Oils, Rerefining and Energy Recovery*, 1st Edition. Elsevier, Amsterdam, Netherlands.

Bechtel. 2012. Bechtel Corp (Bechtel Group). Available at http://www.bechtel.com. Accessed on November 29, 2012.

Briggs, B. 2012. ORRCO in News Articles: What's Happening in Used Oil Recycling in North America and What's New? Available at http://www.orrco.biz. Accessed on October 30, 2012.

CEP. 2012. Chemical Engineering Partners (CEP). Available at http://www.ceptechnology.com/technology.htm. Accessed on October 20, 2012, and http://www.ceptechnology.com/technology_faq.htm. Accessed on November 24, 2012.

Ciora, R.J., and Liu, P.K. 2000. US Patent 6,117,327, September 12, 2000.

Evergreen. 2012. Evergreen Oil Co. Available at http://www.evergreenoil.com. Accessed on November 13, 2012.

Fehrenbach, H. 2005. IFEU Report (Institut für Energie- und Umweltforschung GmbH) Ecological and Energetic Assessment of Re-Refining Used Oils to Base Oils: Substitution of Primarily Produced Base Oils Including Semi-synthetic and Synthetic Compounds. Study commissioned by the Groupement Européen de l'Industrie de la Régénération (GEIR). Available at http://www.geir-news.com/index2.html. Accessed on November 22, 2012.

Gary, J.G., Handwerk, G.E., and Kaiser, M.J. 2007. *Petroleum Refining: Technology and Economics*, 5th Edition. CRC Press, Boca Raton, Florida.

Hartmann, C. 2008. The European Re-refining Industry—Overview and Evolution. Presented at NORA Conference, Palm Springs, November 5–7, 2008.

Hsu, C.S., and Robinson, P.R. (Editors). 2006. *Practical Advances in Petroleum Processing*, Volume 1 and Volume 2. Springer Science, New York.

Kajdas, C. 2000. Major Pathways for Used Oil Disposal and Recycling: Parts 1 and 2. *Tribotest Journal*, 7-7, 61, September 2000 and 7-2, 137, December 2000. Available at http://www.tribologia.org/ptt/kaj/kaj10.htm and http://www.tribologia.org/ptt/kaj/kaj11.htm. Accessed on November 17, 2012.

Kalnes, T.N., VanWees, M., and Schuppel, A. 2008. High Quality Base Oil via the Hylube™ Process. Available at http://www.aiche-fpd.org/listing/7.pdf. Accessed on November 2, 2012.

Lube Report. 2011. Newalta to Double Rerefinery; Vol. 11 Issue 43, October 26, 2011. Available at http://www.imakenews.com/lng/e_article002250553.cfm?x=b11,O,w. Accessed on November 22, 2012.

Mellen, C.R. 1996. US Patent 5556548. Process for Contaminated Oil Reclamation.

Monier, V., and Labouze, E. 2001. Critical Review of Existing Studies and Life Cycle Analysis on the Regeneration and Incineration of Waste Oils, Taylor Nelson Sofres S.A. Consulting Report to European Commission DG Environment, A2-Sustainable Resources—Consumption and Waste. Available at http://ec.europa.eu/environment/waste/studies/oil/waste_oil.pdf. Accessed on November 22, 2012.

Mustad. 2012. Mustad. Available at http://www.mustadinternational.com. Accessed on October 17, 2012.

Newalta. 2012. Newalta Corporation. Available at http://www.newalta.com. Accessed on October 30, 2012.

ORRCO. 2012. ORRCO. Available at http://www.orrco.biz. Accessed November 7, 2012.

O'Blasny, R.H., Sparks, T.F., Tierney, T.J., and Hunter, J.S. 1979. Cyclonic Distillation Tower for Waste Oil Rerefining Process. US Patent 4,140,212 to Vacsol Corp., February 20, 1979.

Ott, L.S., Smith, B.L., and Bruno, T.J. 2010. Composition-explicit distillation curves of waste lubricant oils and resourced crude oil: A diagnostic for re-refining and evaluation. *American Journal of Environmental Sciences* 6 (6), 523–534.

Puralube. 2012. Puralube. Available at http://www.puralube.de. Accessed on November 2, 2012.

Speight, J.G., and Ozum, B. 2002. *Petroleum Refining Processes*. Marcel Dekker Inc., New York.

Speight, J.G. 2014. *The Chemistry and Technology of Petroleum*, 5th Edition. CRC Press, Taylor and Francis Group, Boca Raton, Florida.

Staengl, L. 2009. Re-refining Today, Paper presented at NORA Technical Meeting, Denver, June 24–26, 2009.

Teixeira, S.R., and de A. Santos, G.T. 2008. Incorporation of Waste from Used Oil Re-refining Industry in Ceramic Body: Characterization and Properties, Revista Ciências Exatas, Universidad de Taubauté (Unitau), Brazil, Vol. 14, No. 2. Available at http://periodicos.unitau.br/ojs-2.2/index.php/exatas/article/viewFile/751/585. Accessed on November 20, 2012.

Wadle, H. 2009. The 2nd Hylube™-Refinery in Operation, presented at LAB Benelux Technical Seminar, Turnhout, Belgium, September 24, 2009. Available at http://www.lubsbelgium.be/nl/documents/Puralube.pdf. Accessed on November 7, 2012.

8 Economics and Other Evaluation Criteria

8.1 INTRODUCTION

Reuse of used lubricating oil has been known and practiced for several decades, especially with the use of these oils as burner fuel (Chapter 5). Until recently, the reliability of the re-refined product was an issue. However, the profusion of re-refining processes—each of which has something to add—has produced product oils that are able to meet base oil specifications and are used (when mixed with the necessary additives) for the desired lubricating purposes. As a result, the re-refined products have moved from being a less predictable mix of components to a highly refined high-value product (base oil) that is of sufficient quality to compare equally with the virgin base oil derived from petroleum sources and which can be used without further purification. This can also be translated into savings to the consumer buying a quart of recycled oil insofar as re-refining replaces the need for more of a valuable and decreasing virgin resource—which would lead to higher process costs for lubricating oil.

The economic potential for re-refining and the effect of the re-refined products on the consumer market can be assessed from the contents of this chapter and from the examples used in Chapter 9.

Thus, one of the essential pieces of information in considering any project to recycle used lubricating oil is the analysis of the economics. There may be other criteria that affect the decision to implement a recycling project, and these can include the benefits of eliminating potential pollutants from the environment or lowering the greenhouse gas emissions in producing a product oil. Such additional considerations have been reported in several life cycle assessment (LCA) studies (Fehrenbach 2005; Monier and Labouze 2001; Kalnes et al. 2006), which have investigated the energy requirements and environmental effects of the recycling process and the production of lubricating oil from crude oil, and these will also be discussed here. However, to assess the economic attractiveness of the recycling project, it is essential to obtain reliable data on the capital and operating costs for the project and to be able to investigate the sensitivity of the project economics to variations in these costs.

In determining the costs of the project, one of the first considerations is the scale of the recycling plant. How big should it be? This will be determined to a large extent by the availability of recycled oil in the vicinity of the plant. There will be a cost associated with the collection of this oil from the industrial sources, this cost being made up of a potential payment to the supplier for each liter (0.264 gal.) or gallon of the oil, and the transportation and storage costs to get the oil to the recycling plant. As the source of the recycled oil gets further and further away from the plant, the costs associated with its collection will increase and there will be a point in which it no longer makes sense to expand the collection area beyond a certain limit. That being said, there are of course different methods of collection that can be investigated, ranging from trucks with their high mobility and flexibility, to shipping by rail in tank cars, each with a different transportation cost. To a large extent, these costs will be determined by the population density or industrial density in a particular area. In large urban areas of the United States or in continental Europe or Asia, the density of recycled oil sources will be much higher than in less densely populated areas such as parts of Canada or the Midwestern United States. It therefore becomes very important to assess the costs of the feed material to the plant in each case, and to realize that these costs can vary strongly in different environments.

8.2 RECYCLED OIL COLLECTION COSTS

Most of the collection of used oil will most likely be carried out with tanker trucks that will make a collection run to a particular area on a regular basis. Depending on the volumes of oil to be collected, there will be some scope to work with trucks with various capacities, a variable number of trucks, and a variable frequency of collection. This information can be built into a model that allows the collection runs to be planned so that the capital and operating costs of the collection process can be used in a manner that is as optimal as possible. Having such a model will enable the prediction of the average collection cost for the collected oil to be used as feed to the plant. This will also make it possible to examine the potential effect on the feed costs of expanding the collection zone.

One of the more difficult costs to estimate for the collection process will be the cost of fuel for the collection trucks. Because the trucks will be covering a lot of ground, and the cost of diesel fuel is determined primarily by the cost of crude oil, which has shown great volatility in the recent past, it is worth constructing a model that allows the effect of variation of this parameter to be examined in some detail, and hence, the sensitivity of the economics to this significant variable.

If the collection process can be built around local collection using trucks that then take the oil back to a storage facility, which then feeds into a rail transport system by tanker railcars that take the oil to a single processing plant, it may be possible to consider a larger plant that could have associated economies of scale. Where significant waterway routes are available, barge or even tanker transportation could also be considered.

Another consideration for the collection process is whether to try and collect automobile antifreeze separately from the lubricating oil. Although many auto service stations will find it convenient to dump brake fluid, antifreeze, and lube oil into the same collection tank, a better method may be considered. If a separate collection tank is installed at each collection point for the antifreeze and lube oil, tanker trucks can be constructed with separate tanks for these components and this would make it easier to recycle the antifreeze for reuse, also simplifying the oil recycling process.

8.3 CAPITAL COSTS OF THE PLANT

All the fixed costs of constructing a recycling plant need to be considered in some detail:

- Land acquisition costs in an area zoned for this type of plant
- Road or rail access costs, potential pipeline links to other facilities
- Power, water, waste disposal, and communications costs
- Site preparation, building, and office costs
- Permit and environmental study costs
- Tank storage costs for feed and product materials
- Processing equipment capital cost
- Instrumentation and process control costs
- Erection and installation costs
- Safety provision costs

The choice of a site for the plant will depend on suitable land being available that is zoned for an operation of this type. In most cases, an application will have to be made for a construction permit that will address all aspects of the operation, but will include a plan for handling the environmental effects of the plant. Potential emissions to the atmosphere must be within permitted values, disposal plans for water and other effluent streams have to be addressed, and spill containment within the plant will be part of the capital cost of the plant.

Refining plants are usually constructed without any enclosures or protection from the weather in the form of a roof or walls, and this has generally been seen as desirable when dealing with highly combustible materials such as oil and gas. As the re-refinery is, however, quite simple, plants designed to operate in very cold climates such as Northern Canada could benefit from at least partial enclosure or protection from the elements, and this would be an additional construction cost.

If rail tank car delivery of the recycled oil or transport of the products to market is foreseen as desirable, either from the start or at some later date, a site would be needed where such a rail link to the plant either exists or can be constructed in the future. Similar considerations would apply to a pipeline link to the plant, and these costs may be included in the development plan.

Instrumentation and process control equipment will be included in the capital cost for the plant and will also form an important part of the safety provisions. In dealing with highly combustible materials like oil and gas, it is essential to monitor all critical parameters in the plant operation and to have an automated response capability if any situation is detected that could lead to damage to the equipment or a dangerous situation. An emergency response plan to handle any foreseeable operational problems should be developed that includes training of operational and management personnel.

The actual cost of the processing equipment will vary, depending on where the equipment is built. Because labor and material costs vary considerably in different countries, in some cases, it may advantageous to have some of the equipment built in a large supplier's foreign facilities and have it shipped to the plant site. If the planned plant site is close to an established petroleum processing industry fabrication facility, differences in fabrication costs may be offset against the transportation costs and a local fabricator may have an advantage for a smaller scale plant, which may be the case for many re-refining operations.

8.4 OPERATING COST CONSIDERATIONS FOR THE PLANT

The variable or operating costs of the plant will include the following:

- Interest cost for a bank loan
- The cost or gate price of the feed oil
- Personnel and training costs, benefits, general, and administrative expenses
- Supply costs for utilities such as heating, electricity, natural gas and water, sewer, and waste disposal costs
- Chemical costs, hydrogen cost if used, and clay cost if used
- Quality control and analytical laboratory costs
- Plant maintenance costs
- Site maintenance and security
- Municipal and other local taxes
- Marketing costs for products
- Depreciation of the plant
- Insurance costs
- Income tax regime and potential available credits for recycled materials

There may be some system of local or general subsidies in place that are meant to promote the re-refining of lubricating oils, and these can be used to offset some of the operating costs or, in some instances, the capital costs of the plant.

Marketing costs can include certification of the base oils produced by an organization such as the American Petroleum Institute, and this can be a significant cost item that has to be taken into account. Before this certification is obtained, the product oils may have to be sold at a steep discount for other uses and this would have to be built into any model for a process that produces the types of base oils requiring certification.

8.5 ECONOMIC INDICATORS

Once the economic data has been assembled and projected forward into the future for some years, there are several economic indicators that may be used to indicate how good an investment the project will be. We shall briefly review two of these indicators, the net present value (NPV) and the internal rate of return (IRR).

8.5.1 Net Present Value

In comparing the profitability of various projects in which capital may be invested, it is important to look at the return earned on the investment over the life of the project. The NPV for a project takes into account that a dollar earned in the future is not the same as a dollar in hand today (Peters et al. 2003; Ulrich 1984). A discount rate is assumed for future earnings, for example, 15% per year, in which this discount rate could represent the average rate of return on the investment, and a NPV may be calculated for the project for any period of operation from the present into the future. In most cases, a term will be chosen for the project, 10, 15, or 20 years, or longer, and the NPV will be calculated for this chosen term. In comparing projects that are competing for the same capital investment, the one with the higher NPV should represent the more attractive investment.

$$\text{NPV} = \sum_{n=0}^{N} \frac{C_n}{(1+r)^n}$$

C_n = net cash flow for year n
r = discount rate or rate of return that could be earned for the capital investment in the market with a similar amount of risk
n = time in years

8.5.2 Internal Rate of Return

The IRR is a complementary criterion that gives the discount rate that makes the NPV of all costs (negative) equal to the NPV of all earnings (positive), or in other words, where the NPV is reduced to zero (Peters et al. 2003; Ulrich 1984). In most cases, the future earnings should be after tax values, particularly if the tax treatment for different projects might be different, but a before tax treatment is also informative. The IRR is calculated by finding a discount rate or yield value by trial and error or other means that, when applied to the future net cash flows, reduces the NPV to zero.

Care should be taken when investments are being compared where different amounts of capital are to be invested or different lengths of operation of the projects are anticipated. For example, if two projects are being compared where the first project requires a larger capital investment than the second, the first may have a lower IRR value than the second whereas the NPV value for the first is higher than the second. Which project is more attractive will be decided by comparing the rate of return (IRR) values or the absolute increase in the capital invested (NPV) values. Most companies tend to give more weight to the NPV value for a project evaluation.

In most cases, where various projects are being compared, both NPV and IRR should be examined, and the advantages and weaknesses of each method should be understood. In deciding what type of refining plant to use for the recycled oils, these parameters will provide information to be used in the decision, which is based on whether the aim is to maximize the rate of return in the project or to maximize the magnitude of the generated funds.

8.6 EXCEL SPREADSHEET FOR A RECYCLING PLANT

As the actual values for the capital and operating variables will probably be different for each particular application, it is useful to set up a spreadsheet on which any variable may be changed. After this change, the spreadsheet will recalculate and the effect of each variable may be easily seen—in this way, the sensitivity of the project economics to assumed values of the variables can be quickly evaluated.

Table 8.1 shows such a spreadsheet set up in Microsoft Excel. This, and the spreadsheet for Table 8.2, may be downloaded from the web site of the publisher of this

TABLE 8.1
Pro-Forma Income Statement for Example Plant 1

Assumed Inflation Rate			Year	Year 1	Year 2	Year 3	Year 4	Year 5	Year 6	Year 7	Year 8	Year 9	Year 10
									Lube Oil Base Stock Price 700$/t				
			%/y		2.5	2.5	2.5	2.5	2.5	2.5	2.5	2.5	2.5
Item	**Vol%**	**Amount**	**Unit Value (Year 1)**	**Year 1**	**Year 2**	**Year 3**	**Year 4**	**Year 5**	**Year 6**	**Year 7**	**Year 8**	**Year 9**	**Year 10**
Revenue								**K$**					
Product sales	100.0	16,400,000 L											
Light lube base	42.5	6,970,000 L	0.62 $/L	4294	4401	4511	4624	4739	4858	4979	5104	5231	5362
Heavy lube base	42.5	6,970,000 L	0.62 $/L	4294	4401	4511	4624	4739	4858	4979	5104	5231	5362
Asphalt extender	15.0	2,460,000 L	0.25 $/L	606	621	637	653	669	686	703	721	739	757
Total Revenue				**9193**	**9423**	**9659**	**9900**	**10,148**	**10,401**	**10,661**	**10,928**	**11,201**	**11,481**
Expenses								**K$**					
Chemicals		1	20 K$/y	20	21	21	22	22	23	23	24	24	25
Salaries & Ben. - oper.		10	90 K$/y	900	923	946	969	993	1018	1044	1070	1097	1124
Salaries & Ben. - mgt.		2	140 K$/y	280	287	294	302	309	317	325	333	341	350
Salaries & Ben. - off.		2	80 K$/y	160	164	168	172	177	181	186	190	195	200
Other G&A		1	100 K$/y	100	103	105	108	110	113	116	119	122	125
Clay		2 chg/y	25 K$/chg	50	51	53	54	55	57	58	59	61	62
Electrical power		1	50 K$/y	50	51	53	54	55	57	58	59	61	62
Natural gas		1	40 K$/y	40	41	42	43	44	45	46	48	49	50

Water supply	1	20 K$/y	20	21	21	22	22	23	23	24	24	25
Feed cost	20,000,000 L/y	0.27 $/L	5400	5535	5673	5815	5961	6110	6262	6419	6579	6744
Storage cost	1	10 K$/y	10	10	11	11	11	11	12	12	12	12
Transportation	1	10 K$/y	10	10	11	11	11	11	12	12	12	12
Water disposal	1	10 K$/y	10	10	11	11	11	11	12	12	12	12
Waste disposal	1	10 K$/y	10	10	11	11	11	11	12	12	12	12
Insurance	6000 K$	2 %/y	120	123	126	129	132	136	139	143	146	150
Municipal taxes	1	20 K$/y	20	21	21	22	22	23	23	24	24	25
Interest on loan												
Permits			100	0	0	0	0	0	0	0	0	0
Total Expenses			**7300**	**7380**	**7565**	**7754**	**7947**	**8146**	**8350**	**8559**	**8773**	**8992**
EBITDA			**1893**	**2043**	**2094**	**2146**	**2200**	**2255**	**2311**	**2369**	**2429**	**2489**
Interest on loan												
Depreciation	See below	10 %/y	315	645	677	708	741	775	809	844	880	917
Earnings Before Tax			**1578**	**1398**	**1418**	**1438**	**1459**	**1480**	**1502**	**1525**	**1548**	**1572**
Capital Expenditures												
Plant	6000	K$	6000									
Maintenance	6000	5 %/y	300	308	315	323	331	339	348	357	366	375
Total Capital			**6300**	**308**	**315**	**323**	**331**	**339**	**348**	**357**	**366**	**375**

(continued)

TABLE 8.1 (Continued)
Pro-Forma Income Statement for Example Plant 1

Assumed Inflation Rate			Year	Year 1	Year 2	Year 3	Year 4	Year 5	Year 6	Year 7	Year 8	Year 9	Year 10
									Lube Oil Base Stock Price 700$/t				
			%/y		2.5	2.5	2.5	2.5	2.5	2.5	2.5	2.5	2.5
Item	**Vol%**	**Amount**	**Unit Value (Year 1)**	**Year 1**	**Year 2**	**Year 3**	**Year 4**	**Year 5**	**Year 6**	**Year 7**	**Year 8**	**Year 9**	**Year 10**
Depreciation													
Plant		6000	10 %/y	300	600	600	600	600	600	600	600	600	600
Maintenance			10 %/y	15	45	77	108	141	175	209	244	280	317
Total Depreciation				**315**	**645**	**677**	**708**	**741**	**775**	**809**	**844**	**880**	**917**
IRR and NPV Calculations			**Today**	**Year 1**	**Year 2**	**Year 3**	**Year 4**	**Year 5**	**Year 6**	**Year 7**	**Year 8**	**Year 9**	**Year 10**
Earnings before tax			K$	1578	1398	1418	1438	1459	1480	1502	1525	1548	1572
Plus: Depreciation			K$	315	645	677	708	741	775	809	844	880	917
Operating cash flow			K$	1893	2043	2094	2146	2200	2255	2311	2369	2429	2489
Less: Capital			−6000 K$	−300	−308	−315	−323	−331	−339	−348	−357	−366	−375
Free cash flow			−6000 K$	1593	1736	1779	1823	1869	1916	1964	2013	2063	2115
IRR (Before Tax)				**−73%**	**−31%**	**−8%**	**6%**	**14%**	**19%**	**22%**	**24%**	**26%**	**27%**
NPV (Before Tax)				**−4615**	**−3302**	**−2133**	**−1090**	**−161**	**667**	**1405**	**2063**	**2650**	**3172**

TABLE 8.2
Pro-Forma Income Statement for Example Plant 2

										Lube Oil Base Stock Price 610$/t			
Assumed Inflation Rate		**%/y**		**Year 1**	**Year 2**	**Year 3**	**Year 4**	**Year 5**	**Year 6**	**Year 7**	**Year 8**	**Year 9**	**Year 10**
					2.5	**2.5**	**2.5**	**2.5**	**2.5**	**2.5**	**2.5**	**2.5**	**2.5**
Item	**Vol%**	**Amount**	**Unit Value (Year 1)**	**Year 1**	**Year 2**	**Year 3**	**Year 4**	**Year 5**	**Year 6**	**Year 7**	**Year 8**	**Year 9**	**Year 10**
Revenue									**K$**				
Product sales	100	46,500,000 L/y											
Water	5	2,500,000 L/y											
Naphtha	2	1,000,000 L/y	0.65 $/L	650	666	683	700	717	735	754	773	792	812
Diesel oil	10	5,000,000 L/y	0.60 $/L	3000	3075	3152	3231	3311	3394	3479	3566	3655	3747
Light lube base	34	17,000,000 L/y	0.80 $/L	13,600	13,940	14,289	14,646	15,012	15,387	15,772	16,166	16,570	16,985
Heavy lube base	34	17,000,000 L/y	0.80 $/L	13,600	13,940	14,289	14,646	15,012	15,387	15,772	16,166	16,570	16,985
Heavy fuel oil	11	5,500,000 L/y	0.35 $/L	1925	1973	2022	2073	2125	2178	2232	2288	2345	2404
Asphalt extender	2	1,000,000L/y	0.25 $/L	250	256	263	269	276	283	290	297	305	312
Losses	2	1,000,000 L/y											
Total Revenue				**33,025**	**33,851**	**34,697**	**35,564**	**36,453**	**37,365**	**38,299**	**39,256**	**40,238**	**41,244**
Expenses									**K$**				
Chemicals		1	25,000 $/y	25	26	26	27	28	28	29	30	30	31
Salaries & Ben. - oper.		12	90,000 $/y	1080	1107	1135	1163	1192	1222	1252	1284	1316	1349
Salaries & Ben. - office		2	80,000 $/y	160	164	168	172	177	181	186	190	195	200
Salaries & Ben. - mgt		2	140,000 $/y	280	287	294	302	309	317	325	333	341	350
Other G&A		1	100,000 $/y	100	103	105	108	110	113	116	119	122	125

(continued)

TABLE 8.2 (Continued)
Pro-Forma Income Statement for Example Plant 2

										Lube Oil Base Stock Price 610$/t			
Assumed Inflation Rate		%/y		Year 1	Year 2	Year 3	Year 4	Year 5	Year 6	Year 7	Year 8	Year 9	Year 10
					2.5	2.5	2.5	2.5	2.5	2.5	2.5	2.5	2.5
Item	Vol%	Amount	Unit Value (Year 1)	Year 1	Year 2	Year 3	Year 4	Year 5	Year 6	Year 7	Year 8	Year 9	Year 10
Catalysts		1 chg/y	50,000 $/chg	50	51	53	54	55	57	58	59	61	62
Electrical power		1	200,000 $/y	200	205	210	215	221	226	232	238	244	250
Natural gas		1	200,000 $/y	200	205	210	215	221	226	232	238	244	250
Water supply		1	100,000 $/y	100	103	105	108	110	113	116	119	122	125
Feed cost		50,000,000 L/y	0.27 $/L	13,500	13,838	14,183	14,538	14,901	15,274	15,656	16,047	16,448	16,860
Hydrogen		25,000 m^3/y	0.50 $/$m^3$	13	13	13	13	14	14	14	15	15	16
Storage cost		1	20,000 $/y	20	21	21	22	22	23	23	24	24	25
Transportation		1	35,000 $/y	35	36	37	38	39	40	41	42	43	44
Water disposal		1	20,000 $/y	20	21	21	22	22	23	23	24	24	25
Waste disposal		1	20,000 $/y	20	21	21	22	22	23	23	24	24	25
Insurance		45,000,000 $	2 %/y	900	923	946	969	993	1018	1044	1070	1097	1124
Municipal taxes		1	20,000 $/y	20	21	21	22	22	23	23	24	24	25
Interest on loan													
Permits				100	0	0	0	0	0	0	0	0	0
Total Expenses				**16,823**	**17,141**	**17,569**	**18,008**	**18,459**	**18,920**	**19,393**	**19,878**	**20,375**	**20,884**
EBITDA				**16,203**	**16,710**	**17,128**	**17,556**	**17,995**	**18,445**	**18,906**	**19,379**	**19,863**	**20,360**

Interest on loan												
Depreciation	See below	10 %/y	2250	4772	4959	5151	5347	5548	5754	5966	6182	6404
Earnings Before Tax			**13,952**	**11,938**	**12,169**	**12,405**	**12,648**	**12,897**	**13,152**	**13,413**	**13,681**	**13,955**
Capital Expenditures												
Plant	45,000,000	$										
Maintenance	45,000,000	4 %/y	1800	1845	1891	1938	1987	2037	2087	2140	2193	2248
Total Capital			**46,800**	**1845**	**1891**	**1938**	**1987**	**2037**	**2087**	**2140**	**2193**	**2248**
Depreciation												
Plant	45,000,000	10 %/y	2250	4500	4500	4500	4500	4500	4500	4500	4500	4500
Maintenance		10 %/y	0	272	459	651	847	1048	1254	1466	1682	1904
Total Depreciation			**2250**	**4772**	**4959**	**5151**	**5347**	**5548**	**5754**	**5966**	**6182**	**6404**
IRR Calculation (in K$)		**Today**	**Year 1**	**Year 2**	**Year 3**	**Year 4**	**Year 5**	**Year 6**	**Year 7**	**Year 8**	**Year 9**	**Year 10**
Earnings before tax		K$	13,952	11,938	12,169	12,405	12,648	12,897	13,152	13,413	13,681	13,955
Plus: Depreciation		K$	2250	4772	4959	5151	5347	5548	5754	5966	6182	6404
Operating cash flow		K$	16,203	16,710	17,128	17,556	17,995	18,445	18,906	19,379	19,863	20,360
Less: Capital		–45,000 K$	–1800	–1845	–1891	–1938	–1987	–2037	–2087	–2140	–2193	–2248
Free cash flow		–45,000 K$	14,403	14,865	15,237	15,618	16,008	16,408	16,818	17,239	17,670	18,112
IRR (Before-Tax)			**–68%**	**–24%**	**–1%**	**12%**	**20%**	**25%**	**28%**	**30%**	**31%**	**32%**
NPV (Before-Tax)			**–30,598**	**–15,732**	**–496**	**15,122**	**31,130**	**47,538**	**64,357**	**81,596**	**99,265**	**117,377**

book. In using these spreadsheets, it is recommended that the files be copied onto a computer and the working copy of the spreadsheet then be given a new name before it is saved on the computer. In this way, if a formula or cell entry is accidentally overwritten or corrupted, the working copy can be deleted and the master copy can be reloaded.

The spreadsheet has been set up to allow the user to alter critical input parameters and the cells in which these inputs are allowed are shaded in green. Care should be taken not to enter data or other inputs in cells that contain formulas as this will corrupt the working of the spreadsheet, making it necessary to reload the master copy.

Some of the input parameters, such as the NPV discount rate assumed, the price of a plant of a certain size, and the price of lube base stock oil are entered in the area to the right of the actual evaluation part of the spreadsheet. There are also conversion options for various units in this area of the sheet.

Some additional features of these spreadsheets, which may not initially be apparent, will now be reviewed.

8.6.1 Capital Cost Varies as the Feedrate Varies

If the size of a plant is doubled, or in other words, the feedrate to the plant is doubled, the capital cost of the plant required would not double. There are economies of scale that benefit a larger plant and it has been found that, in typical refining operations, the relationship between the capital cost for the plant and the size of the plant can be represented by the rule of thumb expression (Peters et al. 2003; Ulrich 1984):

$$\left(\frac{\text{Capital cost of large plant}}{\text{Capital cost of small plant}}\right) = \left(\frac{\text{Feedrate of large plant}}{\text{Feedrate of small plant}}\right)^{0.6}$$

There are some obvious limitations to the use of this approximation in calculating the capital cost of a plant of a different size, and it will only apply over a range that does not differ too much from the size of the plant of known cost. It is, however, useful to obtain some idea of the variation of plant cost with variation of the size. This relationship has been built into the spreadsheet so that any change in the feedrate to the plant will result in a recalculated capital cost that reflects this situation. As the size of a plant is varied, it is expected that the operating costs, except perhaps for personnel costs and some taxes, will in general vary almost in direct proportion to the feedrate, so this factor may not be automatically applied to the operating costs. Chemical costs and utility costs will certainly be almost proportional to the feedrate to the plant.

8.6.2 Operating Costs Vary as the Feedrate Varies

It may be expected that the operating costs for electricity, chemicals, and similar elements will vary almost in proportion to the feedrate to the plant or, in other words, the exponent on the ratio term for the feedrates will not be 0.6 as for the capital costs but will more nearly approach 1. An exponent value of 0.95 has been built into the spreadsheets to account for changes in this type of parameter. This may be modified to suit particular circumstances.

8.6.3 Assumed Inflation Rate

In the spreadsheets, a constant rate of inflation may be assumed. This may be viewed as an average rate over the period of the evaluation. If a variable rate is desired and it is felt that the rates can be predicted with some reliability, a different rate for each future year may then be incorporated into the calculation.

8.6.4 Earnings Before Interest, Taxes, Depreciation, and Amortization

A parameter included in the spreadsheet is the earnings before interest, taxes, depreciation, and amortization (EBITDA). It is used to examine the effect of canceling out the effect of different tax regimes, capital cost effects, financing decisions, and varying interest requirements. As such, it is an intermediate value in the spreadsheet calculation that can be useful in evaluating the project.

8.6.5 NPV and IRR Functions in Excel

The NPV and IRR functions are built into Microsoft Excel and are used in these spreadsheets. For short operating times, it may be found that the IRR function encounters difficulty in converging to a solution and an error message may appear on the sheet. If this occurs, it will be necessary to become more familiar with the IRR function through the Help files, and to modify the starting parameters to achieve convergence. In this case, the formula entered in the IRR line of the spreadsheet may have to be changed to provide a better first value or guess for the IRR that is closer to the final value.

8.7 ECONOMIC EVALUATION OF A SMALL AND A LARGE PROJECT

The two examples that follow will deal with the recycling of used oil to produce a lube oil base stock of different qualities. A recycling plant that cracks the used oil to produce a diesel-like product without hydrogenation may also be evaluated on these spreadsheets, but would be a simpler case for the primary treatment. The costs for a secondary stabilization treatment or further processing of the product could be added to the spreadsheet if required.

8.7.1 Example 1: Low Vacuum Distillation with Subsequent Clay Treatment

The first example of the spreadsheet evaluation of the economics of a process deals with a simple plant that might be used for recycled oil volumes of up to 20,000 or 25,000 tonnes (44,100,000 or 55,125,000 lbs.) per year. This would include an initial filtration to remove any solids, a dehydration step, and then a two-stage low vacuum distillation in a wiped-film evaporator to produce two grades of distillate product. These distillates would be passed through a clay bed to absorb some of the sulfur and heavy metals that might be in these streams, the clay being regenerated periodically by high temperature in the packed bed. The products would generally be a Group I base stock, or perhaps Group II if the clay treatment was extensive, together with the heavy residue from the bottom of the second evaporator.

As shown in Table 8.1, a project life of 10 years has been assumed for this example and the NPV and IRR values relate to this period. Clearly, any other project life may be assumed and the sheet may be expanded by adding additional columns for the later years, and copying the formulas from the last column in this example into the added columns.

The capital costs for such a plant will vary, depending on where the plant is constructed, transportation costs to the site, and labor costs for construction of the plant. For this example, a cost of approximately $6 million is assumed but the effect of varying this cost may be easily examined with the spreadsheet.

The prices received for the products may also vary over the life of the evaluation and the sheet could be modified to include a variable price if this could be predicted with some degree of confidence.

An alternative to the clay treatment step would be to incorporate a hydrogenation plant to remove much of the sulfur compounds and saturate the more reactive unsaturated compounds. This requires hydrogen generation as well as the additional capital costs for the high-pressure and temperature contactors, often with a catalyst fixed bed, ebullated bed, or fluidized bed. These additional capital and operating costs may be built into the spreadsheet. The products from such a plant will produce a higher value base stock oil that can be offset against the higher costs, and the spreadsheet allows a rapid initial screening of these options.

8.7.2 Example 2: Re-Refining in a Hydrogen-Rich Atmosphere

The second example of the use of the economics evaluation spreadsheet is for a plant that handles larger quantities of recycled oil of approximately more than 40,000 tonnes (88,200,000 lbs.) per year, and the larger scale operation and economies of scale justify the higher capital and operating costs of a hydrocracking/hydroforming treatment of the entering feedstock material. This more vigorous treatment and cracking reactions will result in the production of some products in the gasoline and diesel range as well as the base stock oils, and these are included in this spreadsheet. These base stock oils should easily meet the requirements of a Group II+ oil, which has a higher value than the products from the smaller plant in the first example.

This spreadsheet in Table 8.2 also includes a column to allow the percentage of each of the products to be varied if this information is available. The values that have been included in the spreadsheet represent estimates of the product distributions that might be expected from a plant of this type, but are not based on any particular operating plant. This spreadsheet could, of course, also be used for an evaluation of the case in Example 1.

If the recycling plant is located near other petrochemical operations, it may be possible to purchase hydrogen from another operator and avoid the costs of a hydrogen generation plant. This will, of course, only be possible if there is excess capacity in hydrogen generation nearby.

8.8 DISCUSSION OF SOME RESULTS OF THE ECONOMIC EVALUATION

One of the great advantages of an economic evaluation spreadsheet is that it allows a quick assessment of the sensitivity of the economic return to a change in some of the input parameters. Because the sales price of a lube oil base stock varies over the years, particularly when the price of crude oil varies, it is important to investigate the change in return on investment if these prices change. Similarly, but perhaps to a lesser extent, the price of the recycled oil may change with time as collection costs vary.

In deciding which type of plant to use for the recycled oil, it may be said that, in general, for smaller throughputs of up to approximately 20,000 tonnes (44,100,000 lbs.) per year, the simpler type of plant in Example 1 will provide better economic returns, and larger throughputs may justify the use of a hydrogenation step. For volumes of more than approximately 40,000 tonnes (88,200,000 lbs.) per year, the full hydrocracking type of operation will provide good returns, but the return for this type of process drops off quickly when volumes drop much lower than this level. For volumes between approximately 20,000 and 40,000 tonnes (44,100,000–88,200,000 lbs.) per year, a very careful evaluation needs to be made and the sensitivity of the return to variation of assumed input parameters should be determined. Of course, consideration can always be given to increasing the feedrate to the plant by expanding the collection area for the recycled oil and, hence, the cost of the feed delivery to the plant, and these options can be investigated in conjunction with the spreadsheet.

8.9 COMPARISON OF RE-REFINING WITH PRODUCTION FROM CRUDE OIL

Re-refining used oil is particularly important because of (1) the large quantities generated globally; (2) the potential for direct reuse; (3) the potential for reprocessing, reclamation, and regeneration; and (4) because these oils may have detrimental effects on the environment if not properly handled, treated, or disposed of. Used lubricating oil and other oils represent a significant portion of the volume of organic waste liquids generated worldwide. The three most important aspects of used oils in this context are (1) contaminant content, (2) energy value, and (3) hydrocarbon properties. Removal of these contaminants is an important (if not, the major) issue and plays a major role in any LCA of oil recycling.

8.9.1 Life Cycle Assessments

The LCA method, although still applied in various modes and under continuous development, is becoming increasingly popular for comparing the environmental cost of producing a product from a recycled feedstock with the environmental cost of producing the same product from the original source, or the same substitution effect when a recycled product substitutes or replaces the original product. The standard defined by ISO 14040 is

> LCA considers the entire life cycle of a product, from raw material extraction and acquisition, through energy and material production and manufacturing, to use and end of life treatment and final disposal. Through such a systematic overview and perspective, the shifting of a potential environmental burden between life cycle stages or individual processes can be identified and possibly avoided. (ISO 14040:2006)

For instance, if used oil is recycled as feedstock to a refinery, it may have to undergo additional steps to remove some contaminants, but the energy and feedstock costs and resources consumed in bringing the original crude oil to the refinery will be avoided. If the products from the refinery are directly comparable, and care is taken to compare products that meet the same strict performance criteria such as the American Petroleum Institute's criteria for lube oil base stocks, a comparison may be made between the two processes in terms of the energy consumed and, hence, in the greenhouse gases that are emitted, per unit of product in each case.

A similar comparison may be made for cases in which recycled oil is used to replace cement kiln fuel or road tar heaters in industrial burners. In this case, the resources used for the production of the natural gas or heavy fuel oil, its transportation from the wellhead, and subsequent refining is compared with the resources to collect and recycle the used oil to a comparable fuel product.

In making these comparisons, the LCA method as defined by a standard such as ISO 14040 and discussed in more detail elsewhere (Fehrenbach 2005; Monier and Labouze 2001) considers several categories of impact assessment such as resource depletion, global warming, acidification, terrestrial nitrification, and human toxicity, which includes carcinogenic risk potential and fine particulates. In addition, wastewater data may also be considered. A characterization factor is assigned to each component in each of these categories, for instance in the Global Warming category, each kilogram of methane (CH_4) can be assigned a characterization factor of 21, whereas each kilogram of carbon dioxide (CO_2) has a characterization factor of 1 and a kilogram of nitrous oxide (N_2O) has a factor of 310 associated with it. These factors, which reflect the different environmental effects of each component, are used to obtain an equivalent mass of CO_2 released to or taken from the environment per unit mass of product. This would then reflect the effect on global warming of the particular production process. Because the weighting and other assumptions in the process are at times quite subjective, it is important that the same assumptions are made for each of the models for the processes being compared. ISO 14040 also stresses the need for transparency in these assumptions for the impact assessment process and that "there is no scientific basis for reducing LCA results to a single overall score or number, since weighting requires value choices."

In the study by Fehrenbach (2005), a more detailed discussion of the results and assumptions regarding the type of fuel displaced by burning used oil may be found. In general, however, he reported a considerable benefit in re-refining used oil to a base oil product in terms of the preservation of resources and the decrease in the burden on the environment when the re-refining process was compared with the original large-scale refining of crude oil. For the comparison of re-refining used oil and burning it as a fuel, he reported LCA environmental conclusions that favor the re-refining to a base oil rather than using it as a burner fuel.

Similar conclusions of an LCA analysis have also been reported by Kalnes et al. (2006), who compared the products of the UOP Hylube process as implemented by Puralube in Zeitz, Germany, with the production from virgin sources and displaced products. They also considered the increasing fraction of synthetic base oils in marketed lube oils, and hence also in the recycled oils, and the increase in the overall life cycle burden of the lube oil products. They reported that, in almost every category of the LCA analysis, re-refining of the used oil by the Hylube process is more environmentally acceptable than burning the used oil as a fuel in cement kilns.

Boughton and Horvath (2004) investigated the release of harmful emissions to the atmosphere, which was dominated by the heavy metals zinc and lead, after the burning of recycled lube oils as a burner fuel. They reported that the "zinc and lead emissions were the primary contributors to the terrestrial and human toxicity impact potentials that were calculated to be 150 and 5 times higher, respectively, for used oil combusted as fuel than for rerefining or distillation (into a marine diesel oil fuel)." These emissions were also some 50 to 100 times the emissions of lead and zinc from burning a fuel oil derived from crude oil.

An LCA study on used oil recycling organized by CalRecycle (McFall 2011) is expected to be published in late 2013, and this was described as an independent study with a peer review by a group from a different organization. A careful study of this type with transparent and public criteria and assumptions should provide a strong basis for an unbiased assessment of the relative merits of the competing recycling processes under the assumed conditions.

8.9.2 Performance Criteria

An additional criterion in evaluating the recycled lube oil base stock product is the question of whether the product is better or worse in its performance as a lube oil base stock. Although some customers may have an initial aversion to using a lube oil that has some content of recycled oil, the basis for this judgment needs to be examined carefully. This aversion is founded on the assumption that the recycled lube oil product is in some way inferior to the original lube oil produced from crude oil or the synthetic lube oil process. This has led, in some cases, to a concentration of the marketing effort for the products of re-refined lube oils in the gear and industrial lubricating oil sectors. However, some major automobile manufacturers in Europe and North America are now much more favorably inclined toward the recycled products as engine lubricants, provided they meet the required performance criteria.

Obviously, the quality of the lube oil base stock product will depend on the re-refining process used and there can be a variation ranging from Group I to Group III products. In 2010, at least one large manufacturer of engine oils in North America introduced a new product, which contained 50% re-refined oil, claiming that the new product meets all the same performance criteria as the previous formulations that were based on production from crude oil alone.

With the great variability in quality of the produced products, depending to a large extent on the particular process used, but also to a lesser extent on the variability in the feed material composition, it is therefore imperative to have a stringent quality control on any re-refining process to ensure the products are of the required quality and meet the criteria established for the target market, and are also consistent in this quality. This is, of course, no different from the usual vigilance and flexibility required in a refinery that processes crude oil from different sources and with different compositions. If all the harmful components in the recycled oil have been removed, and the product meets the established performance criteria, there should be no reason not to accept them as a premium lube oil base stock having the additional advantage of environmental benefits. In addition, there will be an increasing fraction of recycled oils that are from high-value synthetic oils with their corresponding high-performance properties, and these will contribute to improving the quality of the recycled oil and giving it premium qualities when compared with the nonsynthetic lube oil base stocks made from crude oil.

8.9.3 The United Nations Environmental Program's Sustainability Assessment of Technologies Evaluation

Where the economic performance of a project used to be of paramount importance in the past, environmental aspects have assumed greater weight and LCAs and similar analyses have been used to justify the preference for re-refining over the use of recycled lubricating oils as a fuel in many cases. Recognizing that sociocultural issues may also be of importance, particularly in underdeveloped or developing countries, the International Environmental Technology Center of the United Nations Environmental Program (IETC/UNEP) developed the Sustainability Assessment of Technologies (SAT) method to aid in the selection of a process that is cognizant of each of the following four areas: the technological suitability of the process, the environmental considerations, the economic or financial aspects, and potential sociocultural considerations.

The method (UNEP 2012) starts by defining the problem and obtaining baseline data, and then sets up criteria and indicators in each of the four areas under consideration, identifying problems or issues and target solutions. Consultative meetings with stakeholders can influence these choices. A screening process can eliminate some choices at this stage if they do not meet critical requirements. The criteria for attaining the desired goals are then assigned weighting factors, again in a consultative or repeated estimate manner, and the options are assigned scores related to their performance in each category. The evaluation can then proceed by one of several options in making a choice of the most suitable process for the local conditions.

This systematic process will obviously only be as good as the estimates and criteria used in the evaluation process. In the volatile world of oil price fluctuations and changing environmental regulations, a sensitivity analysis to investigate the stability of the outcome to variations of the parameters used in the analysis would be advisable, and this would be one of the strengths of the SAT method.

REFERENCES

Boughton, B., and Horvath, A. 2004. Environmental assessment of used oil management methods. *Environmental Science & Technology* 38(2): 353–358. Available at http://www.pubs.acs.org/doi/pdf/10.1021/es034236p. Accessed on January 14, 2013.

Fehrenbach, H. 2005. Ecological and Energetic Assessment of Re-refining Used Oils to Base Oils: Substitution of Primarily Produced Base Oils Including Semi-synthetic and Synthetic Compounds. Final report by IFEU (Institut für Energie- und Umweltforschung GmbH, Heidelberg) to the GEIR (Groupement Européen de l'Industrie de la Régénération), 29 p. Available at http://www.noranews.org/attachments/files/106/IFEUReportLongVersion.pdf or http://www.geir-rerefining.org/documents/LCA_en_short_version.pdf. Accessed on January 29, 2013.

ISO 14040. 2006. International Organization for Standardization (ISO). Environmental Management—Life Cycle Assessment—Principles and Framework. International Standard ISO 14040:2006(E), 2nd Edition. Available at http://www.iso.org/iso/catalogue_detail?csnumber=37456. Accessed on January 29, 2013.

Kalnes, T.N., Shonnard, D.R., and Schuppel, A. 2006. LCA of a Spent Lube Oil Re-refining Process. Paper presented at 16th European Symposium on Computer Aided Process Engineering and 9th International Symposium on Process Systems Engineering, W. Marquardt, and C. Pantelides (Editors). Elsevier, Amsterdam, Netherlands. Available at http://www.nt.ntnu.no/users/skoge/prost/proceedings/escape16-pse2006/Part%20A/Volume%2021A/N52257-Topic2/Topic2-%20Oral/1073.pdf. Accessed on January 28, 2013.

McFall, D. 2011. California Targets Used Oil. *Lubes 'n' Greases,* 17(1): 36–40. Available at http://www.digital.olivesoftware.com/Olive/ODE/LNG/.

Monier, V., and Labouze, E. 2001. Critical Review of Existing Studies and Life Cycle Analysis on the Regeneration and Incineration of Waste Oils. Report by Taylor Nelson Sofres Consulting S.A. to the European Commission, DG Environment, A2—Sustainable Resources—Consumption and Waste. Available at http://www.ec.europe.eu/environment/waste/studies/oil/waste_oil.pdf. Accessed on December 22, 2012.

Peters, M.S., Timmerhaus, K.D., and West, R. 2003. *Plant Design and Economics for Chemical Engineers*, 5th Edition. McGraw Inc., New York.

Ulrich, G.D. 1984. *A Guide to Chemical Engineering Process Design and Economics*. John Wiley & Sons Inc., New York.

UNEP Compendium. 2012. Compendium of Recycling and Destruction Technologies for Waste Oils. Available at http://www.unep.org/ietc/Portals/136/Publications/Waste%20Management/IETC%20Waste_Oils_Compendium-Full%20Doc-for%20web_Nov.2012.pdf. Accessed on January 25, 2013.

9 Choice of a Process Technology

9.1 INTRODUCTION

In choosing which of the available technologies would be most suitable for any particular application, there are a number of factors to consider and the solution for one application may very well not be optimal for another. In particular, it will be of primary importance to consider the following factors:

- The availability of a reliable supply of used oil as feed to the plant in sufficient quantities throughout the operating year, with sufficient flexibility to handle temporary fluctuations in supply, both upward and downward.
- The availability of a suitably zoned site for the plant with required operating permits including environmental impact or life cycle assessments, if required.
- Suitable utility supply to the site for electricity, water, natural gas, possibly hydrogen and other chemicals, waste streams or incinerator requirements, truck or rail access, loading and off-loading facilities, and adequate storage potential for feed and products. A sampling and laboratory analysis facility should be available to sample and test every load of used oil that enters the facility for polychlorinated biphenyls (PCBs), chloride, and other constituents that are potentially a problem in the recovery process, or are outside the limits prescribed by legislation.
- The economic viability of the process is obviously critical, taking into account possible subsidies or tax advantages from governmental organizations, normally anticipated fluctuations in the market price for the products as the price of crude oil fluctuates, changing requirements for lubricating oil properties as engine manufacturers require increasingly better-performing oils, and currency fluctuations where these can have an effect on the operation.
- The existence of a market for the proposed products, which is sustainable for the life of the project. For instance, the market for Group I base oils for use in automotive lubricants has been shrinking whereas the demand for Group II and Group III base oils has been expanding.
- Safety considerations and requirements according to local regulations.
- The availability of a pool of suitably qualified operating personnel to staff the plant.

Because these considerations may differ markedly from one country to another, it is instructive to review some of the studies that have been carried out in different parts of the world and some of the publications influencing the choice of a process, and attempts to increase the acceptability of re-refined oil products. A selection of reports and study results is listed in the References section of this chapter and these will be investigated in the following sections. It will be seen that some of the studies will come to a conclusion regarding the most suitable process among those considered. In many cases, the selection or comparison criteria will not be clear, process technology will have advanced or at least changed, new processes may now be available, the properties of the used oil feed material may have changed, market conditions and requirements for the products may be different, and legislative and environmental conditions may have changed. Caution would therefore be advisable in extrapolating conclusions and recommendations contained in these reviews and reports to present-day conditions in any particular situation and location.

The following sections will provide an overview of each of the publications, ordered by year of publication and going back to the early 1990s, but more details can usually easily be obtained from the online source if desired.

9.2 RELEVANT REPORTS

There is a collection of relevant published reports that give descriptions and indications of the performance of some of the re-refining technologies. These reports are listed here to introduce the reader to the sources of information on which parts of this book are based.

9.2.1 Process Screening for a Plant in India

This report was prepared for the US–Asia Environmental Partnership and World Environmental Center and was sponsored by United States Agency for International Development (USAID) (Tolia 1994). The chairman of a company in Bombay, India, visited several organizations in the United States with the aim of finding the most suitable process, partners, and funding sources to build a recycling plant in Bombay.

The three processes that were considered were:

1. A distillation process or distillation with hydrofinishing as proposed by Texaco and Kinetics Technology International (KTI)
2. The Evergreen or Mohawk/Chemical Engineering Partners (CEP) process
3. The Interline process

After discussions with a number of environmental service companies—Universal Oil Products (UOP), Interline Resources, Bechtel, Evergreen, KTI, and Texaco—the Mohawk process (with developments by KTI and CEP) was identified as the most favorable and further interactions were planned. No discussion of the selection criteria used in this evaluation was included in this report.

9.2.2 Basel Convention Guidelines on Waste Management

These guidelines were developed by the Technical Working Group of the Basel Convention on the Control of Transboundary Movements of Hazardous Wastes and their Disposal and were adopted at the third meeting of the Conference of the Parties to the Basel Convention in September 1995 in Geneva. They cover the environmentally sound management of wastes subject to the Basel Convention, including used oil, and include waste generated nationally and disposed of within the national territory as well as waste imported through a transboundary movement, or produced as a result of the treatment of imported wastes (Basel Convention Technical Guidelines [BCTG] 1995).

The guidelines start with a discussion of what constitutes a used oil and the sources of this material before a general discussion of recycling, which includes reprocessing and regeneration, both of which will produce a product that can be returned to its original use, and reclamation which will generally result in a product that can be used as a fuel or fuel extender. The three re-refining technologies considered to be the most commonly used at that time are then discussed in general terms:

1. The acid–clay process, identified as the least environmentally sound of the three
2. The vacuum distillation/clay process
3. The vacuum distillation/hydrotreating or hydroprocessing process

The production of a specification and a license, permit, or authorization system for reclamation processes leading to a fuel product was recommended as a quality control and environmental protection measure. Although the burning of used oil in cement kilns results in less negative air quality effects due to the high combustion temperatures in the kilns, "the use of used oil as a fuel in boilers or other combustion processes that are not equipped with burners with a high combustion and contaminant destruction efficiency, or with flue gas treatment devices, should be strongly discouraged or prohibited if practicable."

Other reuses were then considered and included road oiling and asphalt production. Reference was made to studies that suggest that "the potential impact generated by road oiling on health and the environment are severe enough to discourage or prohibit such practice, as road oiling with contaminated oil has led to very serious environmental problems." Using used oil products in the manufacture of asphalts was recommended to be examined on a site-specific or region-specific basis, although the leaching of contaminants from finished roads or roofs was considered to be unlikely. The hot coating of road stones with asphalt "should be discouraged as a waste management practice" due to the environmental problems encountered, despite the limits on the PCB content in used oil–based fuels used in this manner, and some countries have prohibited this practice.

The guidelines end with a discussion of the "Elements to be Considered for Environmentally Sound Management," including the economic aspects and health and safety considerations. The basic criteria for a selection process for the environmentally sound reuse or recycling of used oils were listed, including end-of-life decommissioning of the plant. Examples of specifications for a used oil–derived fuel and maximum contaminant content for used oils were also included.

9.2.3 Process Screening for a Plant in Saudi Arabia

This report by members of the Department of Chemistry at the King Fahd University of Petroleum and Minerals in Dharan (Ali et al. 1995) initially considered three processes:

1. The Meinken process
2. The Mohawk–CEP process
3. The KTI process

The intention was to compare the economic viability of a 50,000 tonnes (110,300,000 lbs.) per year plant for each of the processes because there were existing Meinken plants in Jeddah of 10,000 and 80,000 tonnes (22,100,000 and 176,400,000 lbs.) per year at that time. Unfortunately, at that time, insufficient data was available on the economics of the KTI process, so the comparison was basically between the Meinken and Mohawk processes.

The actual cost data from that time may no longer be relevant today, but show the need to carry out an analysis for every planned plant location. For instance, it was assumed that the Mohawk plant would be located next to a refinery so that a separate hydrogen plant would not be necessary with the hydrogen being purchased, eliminating a large capital cost component. It was found that the raw materials cost was the largest component of the production cost. Chemicals such as lime, activated clay, and ammonia having to be imported from Europe, and the cost of process water was high because it came from desalination plants.

Annual sales value for the products were stated as $16.41 million for the Meinken process and $15.38 million for the Mohawk process, which is surprising as the Meinken process with no hydrotreatment might be assumed to produce a lower value product than the Mohawk process, which did include hydrotreatment. The fixed capital costs have value today probably only in a relative sense at $28.75 million for the Meinken process and $17.71 million for the Mohawk process, and the internal rates of return for the processes were 11.24% for the Meinken process and 45.36% for the Mohawk process, making the Mohawk process, in this case, the more profitable operation. No analysis was made for the Mohawk process if a hydrogen plant capital cost had to be included.

9.2.4 Manual on Re-Refined Lubricants Use in California

This resource manual, written by J.E. Zachary, was developed by the Community Environmental Council of the Gildea Resource Center in Santa Barbara, California, using a grant from the California Integrated Waste Management Board (CIWMB) "to develop educational materials and conduct workshops on the technology, application, quality and market for re-refined oil with the project goal to overcome quality perception barriers and increase market demand for re-refined oil in California."

The purpose of the manual was "to provide fleet administrators, automotive technicians, purchasing agents, and others with information they seek out when evaluating whether to purchase re-refined lubricants" (Zachary 1996).

The manual addresses the following items:

- Myths about re-refined oil
- Re-refining technology—the Mohawk process
- Standards for lubricating oil
- Car manufacture warranties
- Evaluation and test cases of re-refined oil
- Procurement, including examples of government directives
- Mandates, guidelines, and request for proposal suggestions

The myths that were seen as needing clarification were the following: (1) nobody uses re-refined oil, (2) re-refined oil will make engines fail, (3) re-refined oil is too expensive, (4) re-refined oil is contaminated, and (5) re-refined oil will void manufacturer warranties.

The manual addresses the standards, testing methods, and limits that are set by vehicle and engine manufacturers in conjunction with the International Lubrication Standardization and Approval Committee (ILSAC), the American Society for Testing and Materials (ASTM), the Society of Automotive Engineers (SAE), and the American Petroleum Institute (API), and that the laboratory and field tests by the National Bureau of Standards (NBS), U.S. Army, Department of Energy (DOE), U.S. Postal Service, and the Environmental Protection Agency (EPA) have all established that re-refined oil that meets the established standards is essentially indistinguishable from the oils made from virgin crude oil and, in some cases, actually exceeds the performance of new oils. In addition, there is a significant reduction in the energy required to produce a re-refined oil product when compared with the energy required to produce a lubricating oil from crude oil. Statements from several of the leading car manufacturers, including Chrysler, Ford, General Motors, and Mercedes Benz, confirm that the source of the lubricating oil to be used in their engines is not significant as long as the oil meets their performance specifications and is certified by a body such as API, and that the oil meets these specifications on a continuing basis.

The manual refers to the growing number of federal agencies and local government agencies that are increasingly promoting the use of re-refined oil. The U.S. EPA in a 1988 guideline, requires "all federal agencies, all state and local government agencies and contractors that use federal funds to purchase such products to implement a preference program that favors the purchase of re-refined oil to the maximum extent practicable." As the re-refined products are subject to a reduced tax rate, suggestions were made for the level of re-refined product in the marketed oil, which should qualify for the "re-refined" tax rate, and a value of 70% was common. Many of the government agencies promoting the use of the re-refined products also allowed for a preference for this choice when the price was up to 5% higher than oil made from virgin crude oil.

Also of interest in this manual was the brief discussion of the Mohawk re-refining process, which at that time, was used at four locations in North America: the Safety-Kleen refineries in East Chicago, Illinois and Breslau, Ontario, Canada, the Evergreen operation in Newark, California, and the Mohawk plant in Vancouver, British Columbia, Canada.

9.2.5 Used Oil Management in New Zealand

This report, prepared by the Ministry for the Environment, reviews the used oil situation in a relatively small country where approximately 30 ML (7,920,000 US gallons) per year of used oil were estimated to have been generated from approximately 60 ML (15,840,000 gallons) of lubricating oils consumed at that time. Of this, some 21 ML (5,544,000 US gallons) were accounted for, leaving approximately 9 ML (2,376,000 US gallons) per year that were apparently being improperly disposed of. As such, the 30 ML (7,920,000 US gallons) per year represented the largest source of nonwatery liquid waste streams in the country and made this a priority for the Pollution and Waste Group of the Ministry (New Zealand 2000).

The majority of the used oil was being recycled as burner fuel or used in road oiling for dust suppression. A re-refining operation by Dominion Oil in Auckland, processing 7 ML (1,848,000 US gallons) of used oil per year, closed in mid-1998 due to poor economics and ongoing environmental problems such as odor emissions. The report concludes that due to the small volumes and increasing specialization of the oil markets, "the re-refining of used oil in New Zealand is unlikely to be viable under the current conditions." Some niche markets did exist, for instance, in producing specialty hydraulic oils or in the re-refining of transformer oils.

The main contaminants expected in the used oils were reviewed and the key issues in re-refining were identified as controlling the emissions and destroying the highly toxic residues. Burning of the fuel produced by reprocessing with filtration or gravity separation of some of the contaminants was judged to not be a problem when used in a high-temperature burner such as a cement kiln with its long residence times, but to be of more concern in low-temperature burners in which the potentially carcinogenic contaminants were not destroyed. This conclusion was supported by some modeling of air pollution caused by burner emissions. The practice in some other countries was reviewed, including the European Union (EU), United States, Australia, Belgium, and the Rose Foundation operation in South Africa. This latter foundation, which is a voluntary undertaking funded by the suppliers of lubricating oil products, was set up in 1994 and in 1998/1999 collected some 36 ML (9,504,000 US gallons) for recycling, representing a collection rate of approximately 70% of the available oil. The work by National Oil Recyclers Association (NORA) in the United States in getting ASTM specification in 1995 for industrial grade reprocessed fuel oil in four grades as a national standard was also highlighted as a possible step that could be emulated.

The process of road oiling was discussed in that the control of these practices in New Zealand rested at the regional council level and was permitted in some jurisdictions although it was prohibited in others. Although the practice was declining, it still persisted in rural and remote areas. Much of the same distributed control measures also applied to the use of burners of various types in different parts of the country, and the need for controls, policies, or guidelines was addressed in looking for solutions to some of the problems that had been identified. Among the various options for managing the effects of the used oil, different policy tools such as input specifications for the burner fuels, certification of the burners, accreditation of the fuel suppliers, establishment of rules governing where certain things such as road oiling could be done, output controls such as pollution control equipment or ambient air monitoring, were all discussed. It was also suggested that national guidelines, which could be used by regional councils to set local codes of practice, could be set up and would also be an option. The report concluded with a solicitation of feedback on the questions that had been raised and the issues that had been discussed.

9.2.6 Overview of Re-Refining from Poland

Part 1 of these two publications by C. Kajdas of the Warsaw University of Technology dealt with the general problem of used oil, citing the consumption of approximately 5 million tonnes (11,025,000,000 lbs.) per year of lubricating oil in the EU, with approximately 50%, or 2.5 million tonnes (5,512,500,000 lbs.), being generated per year as used oil (Kajdas 2000). Of this, approximately 60% or 1.5 million tonnes (3,307,500,000 lbs.) per year was being collected, leaving approximately 1 million tonnes (2,205,000,000 lbs.) per year of used oil as unaccounted for and which probably ended up being dumped in landfills and other undesirable disposal methods. The processes of recycling, which included reclaiming, reprocessing, and re-refining—with increasing complexity in that order—were discussed and the method of re-refining was identified as being the safest method of dealing with the used oil. Although the most common use of used oils was as a fuel for industrial burners and furnaces, unless certain checks and controls are in place to govern this practice, the danger remains that the contaminants and carcinogenic components in the oil could be released to the environment in undesirable quantities. Where small quantities of used oil are generated in a large geographical area, it was recognized that combustion might be the only reasonable alternative, but the destruction of the valuable base oil resource was, in general, not the best choice.

Simple recycling processes were considered, including the mobile units by CHEM-ECOL, which are truck-mounted or trailer-mounted and clean the oil using a vacuum dehydration process followed by passage through a series of filters down to micron size.

A general discussion of the major re-refining processes followed, including the hydrogenation processes of KTI, Mohawk, BERC/NIPER, and the Phillips Petroleum PROP process. It was identified that the KTI process involved treatment at temperatures not exceeding 250°C (482°F), and that an 82% yield was attained. Several plants were in operation after the first industrial plant was completed in Greece in 1982.

The discussion of the Mohawk technology, a development of the KTI process, included the following: "the proprietary feature of the Mohawk technology that distinguishes it from other vacuum distillation/hydrogenation approaches is understanding of the chemistry of the lubricating oil additives found in the waste oil under the influence of time and temperature. This has led to processing techniques and operating conditions that provide superior performance in terms of on-stream time, catalyst life, and corrosion resistance." It was noted that the Mohawk process had been licensed to Evergreen in Newark, California and Breslube in Ontario, Canada.

The BERC/NIPER (the Bartlesville Energy Research Center, now the National Institute of Petroleum and Energy Research) process was similar to the Mohawk process, but with the addition of a solvent treatment to reduce fouling and coking precursors.

The discussion of the Phillips Petroleum PROP process noted the use of diammonium phosphate to remove metals, but with the spent catalyst forming a hazardous waste product.

Part 2 of the publication dealt with the other re-refining processes, starting with the Safety-Kleen plant in East Chicago, which was processing 250,000 tons per year of used oil, making it the largest in the world. It was noted that the product base oils met API standards for these materials.

The comments on the Institut Français du Pétrole (IFP)/Snamprogetti process noted that Snamprogetti extended the IFP Selectopropane propane extraction process to include a propane extraction step before and after the vacuum distillation, and that IFP had then developed the process further to produce a new marketed technology.

In considering the UOP direct contact hydrogenation (DCH) process, otherwise known as the Hylube process, it was noted that no environmentally undesirable by-products were produced, and that the process could be integrated with other refinery operations.

With the Viscolube process, yields of 75% base oil products could be achieved with hydrofinishing, and the process had been improved by working with IFP to produce the IFP/Viscolube or Revivoil process, which produced high-quality products comparable with virgin oil products.

The Resource Technology process, which used cyclonic vacuum distillation towers without any trays or packings, together with clay treatment and filtration was included, as was the Interline process, which was said to make smaller volume processing economic. The 27,000 tonnes (54,000,000 lbs.) per year Interline plants in Salt Lake City and Stoke-on-Trent in the United Kingdom were mentioned. The Rose/Kellogg or Residuum Oil Supercritical Extraction process, developed by Kerr-McGee and sold to the Kellogg Company was said to enable energy and capital cost savings over other solvent extraction processes.

Comments on the other vacuum distillation and clay treatment processes included the identification of problems with coking and carryover of resinous materials in the distillation, moderated by the use of wiped-film evaporators in plants in Germany, Australia, and New Zealand. The Entra process involved a linear tubular reactor with high temperatures combined with very short residence times in the order of milliseconds, and dechlorination with metallic sodium and the use of Fuller's earth to improve the color of the product oils. Finally, the Degussa process for transformer oils and the handling of PCBs was capable of producing product oils that met the ASTM specifications for new transformer oils.

The final advantages for re-refining of used oil quoted the IFP data on the feed and energy requirements in tonnes of fuel oil equivalent (FOE) to produce 1 tonne (2205 lbs.) of mineral base oil being between 1.4 and 1.6 tonnes (3087 and 3528 lbs.) FOE for production from conventional processing of Middle East crude oil, and between 1.1 and 1.2 tonnes (2426 and 2646 lbs.) FOE for production from re-refined oil. The preferred technologies at that time seemed to be converging on a two-step procedure of vacuum distillation and hydrotreating, and the Mohawk, Safety-Kleen, and Revivoil processes produced high-quality base oils and concentrated the contaminants in the distillation residue.

9.2.7 Overview of Re-Refining from Spain

The Regional Activity Centre for Cleaner Production (RAC/CP) was set up in 1996 under an agreement between the Spanish Ministry of the Environment and Rural and Marine Affairs and the Government of Catalonia. It is located in Barcelona and operates under the Mediterranean Action Program, which was set up in 1975 under the United Nations Environmental Program (UNEP) for the production and development of the Mediterranean basin and promotes environmentally friendly projects and developments in all the countries bordering the Mediterranean. This report was generated to review the main processes for the reuse of both mineral oils from industry and vegetable oils from catering companies and restaurants (RAC/CP 2000).

This study identified two groups of countries that each had a different interest in this technology. The first group consisted of countries that did not belong to the EU or fall within its sphere of influence but were still interested in the results of actions that had an effect on the environment. The second group of countries was made up of the countries belonging to the EU or its sphere of influence and whose actions were subject to EU-wide directives or regulations, which were aimed at a consistent approach by all members of the community.

After reviewing the possible contaminants in used oil, the three options available for recycling the oil were discussed, namely, the reprocessing of materials such as hydraulic oils for reuse, the re-refining of base oils to return them to their original use, and combustion to recover the energy in the oil. In discussing the collection of the oil, it was noted that the separation of the used oils at their source made subsequent processing easier and that this was a desirable goal. The collection in Italy, Spain, and France was tabulated for several years preceding this report, and the collection system was described.

The re-refining processes were then discussed, including the acid–clay Meinken and modified Meinken processes, the vacuum distillation and hydrogenation processes, and the processes that included solvent extraction. This was followed by a discussion of energy recovery processes or the use of the oil, after cleaning, as a fuel in various industries, and a discussion of the economics of the recycling processes.

Three practical cases for the reuse of used oil were then discussed, the Catalan Waste Oil Treatment Company (Cator S.A.), operating a treatment plant near Tarragona that processed 30,000 tonnes (66,150,000 lbs.) per year of used oil using the Vaxon process to produce a yield of 50% to 60% of the used oil as a base oil product, 17% to 20% as a natural bitumen, 5% as potassium salts that could be used as fertilizers, 12% to 13% as light hydrocarbons used as fuel in the plant, and 7% to 15% water. The re-refined base oils had achieved both API and Association des Constructeurs Europée̋n d'Automobiles (ACEA) certification, the ACEA approval was stated to be more restrictive and difficult to achieve.

The second example noted was the Aureca process, which produced a fuel for diesel engines using a process developed in Spain by the Befesa Environment SA company and a department of the Industrial Engineers College in Madrid. The third process example was the Ecolube process based on the Interline technology.

Recommendations were then made for the regulation and control of the activities in each country as separate jurisdictions, and the comment was made that "for the time being, therefore, grants in the sector are essential for the activity to be attractive to the industrial system in the different countries."

The final section of this report then discussed the recycling of vegetable oils, the majority of which were apparently ending up in the sewage system. Some were being recovered to make feedstuffs for animals or to synthesize esters that are used as biofuels, or to make biodiesel and other uses. A discussion of the treatment processes for these oils then followed with conclusions and recommendations along much the same lines as for the mineral oils.

9.2.8 Taylor Nelson Sofres Report and Life Cycle Analysis Study on Re-Refining Processes

This extensive report was prepared for the European Commission, Directorate-General Environment, by Taylor Nelson Sofres Consulting and Bio Intelligence Service, following the directive 75/439/EC as amended in 87/101/EC, which stated that, among the various options for recycling of used oil, priority should be given to regeneration over its incineration (Monier and Labouze 2001).

It may be noted that both these directives have been superseded by the directive in 2008/98/EC, which incorporated used oils into a general waste management directive that established a hierarchical or stepwise structure for waste disposal that requires waste disposal at a higher level when this is technically and economically possible or desirable. The highest level on this scale is avoidance of waste generation, followed by preparation for reuse with minimal processing, then recycling to its original use or other purposes, but not including use as a fuel, and this would apply to re-refining of used oil. Recovery is the next step below recycling and includes energy recovery when used as a fuel, and finally disposal as the lowest step.

A review of activities in the members of the EU had shown that many of the countries were still using used oil as a fuel. The main objective of this study, then, was to perform a thorough techno-economic and environmental analysis of the literature available on the regeneration of used oil and its comparison with the incineration of the oil. After the publication of ISO 14040 in 1997 on life cycle analysis (LCA), four studies had been performed: one in Norway, one in France, and two in Germany, and these were also examined. The data, assumptions, and methodology of these studies were very well-documented, making this a valuable resource for comparison of the processes under the pertinent conditions at that time.

Data was collected from the GEIR (the European oil recycling organization), the member states, in interviews, and with questions regarding used oil use, collection, recycling, and disposal during the preceding year or two. The following highlights, although now dated, are interesting in establishing a baseline for that time:

- In 2000, 4.93 million tonnes (10,871 Mlbs.) of base oils were used in the EU, 65% being motor oils and 35% industrial oils
- About 50% was lost in use, leaving about 2.4 million tonnes (5292 Mlbs.) as collectible for that year, of which more than 70% was engine oil, and

of which 1.73 million tonnes (3815 Mlbs.) was actually collected, leaving some 0.67 million tonnes (1477 Mlbs.) as unaccounted for and being illegally burned or dumped

- The average collection rate in the member states was in the range of 70% to 75% of the collectible oil, varying considerably from 86% in the United Kingdom and 85% in Germany, to 47% in Spain and 37% in Greece
- In 1999 in the EU, approximately 47% of the used oil went to fuels and energy recovery, 24% to regeneration of base oils, 28% to illegal burning, and 1% to disposal
- In 1999, the regenerated oil compared with the oil collected in 2000 was 75% in Italy, 65% in Germany, 31% in France, 23% in Greece, and 0% in the United Kingdom
- The main oil burning and energy recovery methods were in cement kilns and asphalt plants
- The data was not in a consistent form, definitions and data quality varying from country to country with the situation being in a state of flux, and making a comparison across the EU difficult. The need for a more reliable form of data and record-keeping was identified as being required for the member states to give a consistent set of data, particularly for the oil regeneration area

The techno-economic analysis that followed for each of the EU member countries examined the collection costs for the used oil, the regeneration or re-refining processes and costs, tax and duty regimes, thermal cracking to make fuels, use in cement kilns and asphalt plants, before proceeding to a LCA comparison for various processes. The results may be summarized as follows:

- Plant economics were not favorable for grass-roots operations without an initial subsidy of €10 to €100 per tonne (2205 lbs.). After some of the capital costs had been paid off, the economics improved
- Regeneration plants could not compete with the fuel combustion route in many cases because the use of recycled oil as a fuel had tax advantages in many jurisdictions. This situation was expected to change due to the New Incineration Directive in the EU to take effect in 2003 for new plants, and in 2005 for old plants, which would forbid the burning of used oil. The 2008 directive, which will be discussed later, will also have had an effect in this respect
- The overall collection and delivery cost to the plant was in the range of €25 to €100 per tonne (2205 lbs.) of used oil
- The tax situation in the United Kingdom on used oil excise duty at that time was promoting the importation of used oil from other countries to be used as a fuel
- Synthetic lubricants were becoming more widely used and were decreasing the market for car engine lubricants. However, synthetic oils based on esters were less suitable for regeneration as they are less stable in the presence of alkaline caustic chemicals, as often used in the re-refining processes, and were less stable in the hydrofinishing step

- Thermal cracking to make fuel oils and lighter products were economic and required no subsidies. However, the products were unstable and required hydrotreatment or further processing to prevent discoloration and gum deposits with product degradation
- The effect of bio-oils such as sunflower or rapeseed oils in the re-refining processes was not known and would require further evaluation
- There was still some resistance to the use of re-refined lubricating oils and products, but in the United Kingdom, there was less resistance to their use in gear and hydraulic oil applications. Re-refined base oils generally sold for 10% to 25% lower prices than base oils made from virgin mineral oils
- The virgin mineral base oil supply in Europe was, at that time, in oversupply, and the longer intervals between oil changes in car engines was leading to a decrease in the market for re-refined products
- Market prices and revenues were in a very volatile situation

The economics for seven different processes and three different plant sizes were then examined in some detail, assuming a grass-roots plant in each case to be set up under a set of common assumptions, such as an expected 10% internal rate of return (IRR), and a sales price for the base oil product for each type of process. The costs and revenues for each process were then determined to see whether the re-refining process generated enough revenues to be able to pay for the used oil delivered to the gate of the re-refinery, or whether a subsidy of some sort would be required. This was measured by a *gate fee*, defined as the total costs of the process minus the revenues generated by the sale of the products. The results of this are summarized in Tables 9.1 and 9.2.

With the caveat that these plants were, in general, operating in different countries under different circumstances and with data that was not of a uniform quality and was sometimes contradictory, it was found that scale advantages did exist, depending on the size of the re-refining plant, but the larger the plant, the higher the collection costs were. It was noted that in the United Kingdom, it was expected that the advantages of scale would start to decline for plants larger than approximately 100,000 tonnes (220.5 Mlbs.) per year. The break-even point, defined as zero gate fees for the process, was generally expected to be for plants between approximately 60,000 and 80,000 tonnes (132.3 and 176.4 Mlbs.) per year of used oil, but varied from 50,000 to 80,000 tonnes (110.3 and 176.4 Mlbs.) per year for the TFE and clay finishing process to 90,000 tonnes (198.5 Mlbs.) per year for a TFE and hydrofinishing process.

The LCA, being fairly new at that time, was explained as being a *cradle-to-grave* analysis for any process from the raw material production through to the ultimate disposal of the products, and looking at the raw materials and energy inputs compared with the outputs in terms of the emissions to air, water, and solids used. According to ISO 14040:1997, subsequently replaced by ISO 14040:2006, the analysis is based on a particular model for the process and looks at several environmental effect categories. The categories considered the most reliable in this report for comparing recovery processes for used oils were:

- Consumption of fossil energy resources
- Contribution to global climate change

TABLE 9.1
TNS 2001 Report: Waste Oil Gate Fee Analysis for Various Regeneration Processes

Process	Acid–Clay	TFE + Clay					TFE + Hydro				TFE + Solvent				TDA + Clay	TDA + Hydro	PDA + Hydro
Capacity (kt/year)	100	35	50	80	160	35	50	80	85	35	35	50	80	160	100	100	57
Capital cost (million €)	34	20	25	33	50	8	43	50	47	27	31	37	44	60	45	69	42
Cost (€/t of WO)	152	242	221	198	148	237	333	275	204	289	350	308	258	148	280	304	320
Revenues (€/t of WO)	177	210	210	210	195	214	254	254	219	321	249	249	249	202	211	252	224
WO gate fee (€/t of WO)	–25	32	11	–12	–47	23	78	21	–15	–32	102	60	9	–54	68	52	96
Data source[a]	1	2	2	2	2	3	2	2	2	3	2	2	2	2	2	2	2

Source: EC-DG Environment.

Note: The WO gate fee is defined as (costs – revenues) on a Euros per tonne (2205 lbs.) of WO feed basis. A negative gate fee therefore means that revenues exceed costs and the more negative this number, the better the economics for the process.

Hydro, hydrofinishing; TFE, thin film evaporation; TDA, thermal deasphalting; WO, waste oil.

[a] Data source (as quoted in the TNS report): (1) CONCAWE (Conservation of Clean Air and Water In Europe, the oil companies' European Organization for Environment, Health and Safety). Collection and disposal of used lubricating oil. Report 5/96, 1996. Available at: http://www.concawe.be/DocShareNoFrame/docs/1/IAOBDMBBLKGNGJBAMCOHKACIVEVC7N919YBDG3BYTYP3/CEnet/docs/DLS/Rpt_96-5-2004-01722-01-E.pdf; (2) GEIR 2001: Working Group organized by TN Sofres Consulting in July 2001 with GEIR members: Viscolube, Baufeld, Dolbergen, Watco, Fuhse; (3) UK report: DETR/Oakdene Hollins Ltd. UK Waste Oil Market 2001 (DEFRA 2001).

TABLE 9.2
TNS 2001 Report: Sales Price of Base Oil Re-Refined by the Indicated Process

	Acid–Clay	TFE + Clay		TFE + Hydro	TFE + Solvent		TFE +Solvent or Hydro	TDA + Clay	TDA + Hydro	PDA + Hydro
€/tonne for re-refined base oil	250	300	300	325	320	320	296	300	325	320
Data source (see Table 9.1)	1	2	2	2	2	2	2	2	2	2

Source: EC-DG Environment.

- Contribution to regional acidifying potential
- Emission of volatile organic compounds
- Waterborne emissions
- Solid waste

The LCA analysis, according to the ISO 14040 standard, must be reviewed by an external panel of experts.

Four LCA studies were then examined in some detail, one from Norway, two from Germany, and one from France. These studies had different aims in comparing specific recovery processes with alternative routes such as incineration, for example, and the environmental impact areas chosen were specific to the studies. In addition, the assumptions made, for example, in which alternative processes were avoided by the recycling process, required some discussion with a sensitivity analysis to see how sensitive the results were to some of the assumptions made. Data for the particular recovery or incineration process used in each study were drawn from operating plants in the different countries and represented real data with its obvious advantages, as well as all the weaknesses of specific operations with the technology applied at that time. The conclusions drawn from each of the LCA studies were then summarized for each of the studies.

The overall results of the four LCA studies analyzed may again be summarized as

- There were generally fewer impacts from a regeneration plant compared with an incineration plant, although incineration in a cement kiln could have fewer environmental impacts in some categories under certain local conditions and with certain assumed scenarios
- The environmental burden of a new regeneration or incineration process by itself was generally less important than that of the process which was avoided by implementing the new process, the avoided process in the case of used oil reprocessing being the production of base oils from virgin mineral oil, or the traditional fuel or energy products
- Particularly for incineration, the bonus brought to the analysis by the avoided system depended on the particular avoided process, for instance, displaced virgin fuel oil, hydroelectric power, coal-generated or gas-generated power, or nuclear power. Apparently, regeneration would result in advantages for all environmental impacts in all scenarios if the used oil would replace non-fossil fuels such as hydroelectricity or nuclear power electricity
- The environmental impacts due to the collection and transport of used oil and primary materials are not significant compared with the impacts of the recovery processes
- Some issues were not addressed by LCA, such as noise, odor, and toxic emissions because the relative effects of toxicity had not yet been established

These results were discussed at some length, as the conditions and assumptions made were not uniform. For instance, the presence of gas cleaning or scrubbing facilities for exhaust gases from the plants was not uniform and this would have had an effect on the environmental impact of these exhaust gases for each individual case. It was therefore again emphasized that these cases represented a comparison of particular installations and processes rather than a comparison of one process system with another.

9.2.9 The UK Used Oils Market 2001

Oakdene Hollins Ltd. carried out this study for the UK Department for Environment, Food and Rural Affairs (DEFRA), and it was based on the usage of used oils in 1999 and 2000. In 1999, three refineries in the United Kingdom produced 1.09 million tonnes (2403 Mlbs.) of lubricating oils, of which 790,000 tonnes (1742 Mlbs.) were used in the country and the balance was exported. It was noted that the base oils produced were generally marketed at a price roughly twice the price of crude oil, and that there was an oversupply of base stock oils at that time in the market, particularly for the Group I oils (DEFRA 2001).

The disposal routes for the collected used oils in England and Wales in 1998 were summarized, as shown in Table 9.3, totaling some 306,734 tonnes (676.3 Mlbs.) of collected oil. It was estimated that approximately 50% of the used oil was collectible and that the collectible lubricating oil volume was approximately 391,000 tonnes (862.2 Mlbs.) per year. The short-term transfer item reflected oil in transit or storage at the time of the survey. The collected oil was recycled for use, as shown in Table 9.4, for the year 2000.

Environmental agencies estimated that approximately 475,000 tonnes (1047 Mlbs.) per year of used oil was processed, of which approximately 95,000 tonnes (209.5 Mlbs.) was imported from other European countries, and that almost all this was converted into recovered fuel oil (RFO) for use in power stations, cement kilns, and road stone users making asphalt. The use in power stations was usually in spraying it over coal being charged to the burners during start-up to improve the flame control of the process and smooth out the heating process when new coal was being charged. The

TABLE 9.3
UK DEFRA 2001 Report: England and Wales Disposal Routes in 1998 (in tonnes per year)

	Oils		Oily Wastes		
Process	**Halogenated**	**Nonhalogenated**	**Halogenated**	**Nonhalogenated**	**Total**
Incineration	165	176	516	132	989
Landfill	410	9073	159	13,496	23,138
Recycling/reuse	7621	53,818	48	10,323	71,810
Short-term transfer	3610	35,553	148	6637	45,948
Treatment	9720	110,249	477	39,497	159,943
Other	33	464	6	4403	4906
Total	21,559	209,333	1354	74,488	306,734

Source: DEFRA Oakdene Hollins.

TABLE 9.4
UK DEFRA 2001 Report: Collected Waste Oil Use in the UK in 2000

Use	Tonnes per year
Heating/furnace users	30,000
Road stone users	185,000
Cement/lime kiln users	15,000
Power station users	210,000
Quantity imported into the UK	95,000

Source: DEFRA Oakdene Hollins.

RFO had lower viscosity than the heavy fuel oil it displaced, was slightly cheaper, but caused more corrosion in the spray heads and pipes, probably due to the chlorine and sulfur content of the RFO.

Use in road stone asphalt production was being affected by the switch to natural gas fuels, and the European Waste Incineration Directive (WID), which would take full effect in 2006, was expected to phase out this disposal route. The use of recycled fuel oil in industrial furnaces, such as in steelmaking, was also expected to end with the introduction of this directive.

The study was unable to reconcile the discrepancy in the volumes of used oil collected and the reprocessed fuel oil demand, but this could have been due to the variable water contents of the used oils. The source of the collected oils is shown in Table 9.5, and of this approximately 252,000 tonnes (555.7 Mlbs.) per year was lubricating oil with another 113,000 tonnes (249.2 Mlbs.) per year probably being lubricating oil, making a collection rate of at least 80% for lubricating oils. There were also approximately 95,000 tonnes (209.5 Mlbs.) per year of oily wastes taken off ships in UK harbors in that year, leading to approximately 28,000 tonnes (61.7 Mlbs.) per year of additional reprocessed fuel oil.

TABLE 9.5
UK DEFRA 2001 Report: Collection Sources for Waste Oils in the UK

Source	Tonnes per year
Garage or service station oils	165,000
Marine oils	28,000
Waste industry	40,000
Industrial oils	147,000
Total (with 7%–8% water)	380,000

Source: DEFRA Oakdene Hollins.

It was stated that all re-refining operations in the United Kingdom had failed commercially up to that time and that the Petrus Oils plant in the West Midlands, based on the Interline process, was being dismantled for possible relocation to Germany with the addition of a hydrogenation unit. It was expected that any new plants would require a pretreatment process and a hydrogenation unit to be viable. With the increase in dispersants being added to the base oils, it was expected that the collected oils would contain more dirt, and the possible use of esters in the synthetic oils would complicate the recycling process because the esters would react with any sodium hydroxide used in the recovery process to form soaps that would have to be removed. It was, in any case, expected that the WID would force a re-examination of the three possible disposal routes of:

1. Re-refining to a base oil stock
2. Re-refining to an ultra-low diesel fuel
3. Reuse as an oil refinery feedstock

The economics of a re-refining plant were then examined, assuming a modest size of 35,000 tonnes (77.2 Mlbs.) of used oil feed per year, assuming two types of plant, the first with vacuum distillation and clay treatment, and the second with a catalytic hydrotreater replacing the clay treatment. Given the high risk perceived for such a plant, a rate of return of 15% to 20% was assumed for the investor. A product sales price of £215 to £230 per tonne (2205 lbs.) was assumed and a gate fee was calculated for the two cases as shown in Table 9.6. Because these were all negative, or uneconomic in that scenario, a case was assumed in which the plant would be built on an existing site at a capital cost saving of between £2 million and £3 million. Because the gate fee was still not very attractive, a rate of return of 10% was assumed and the gate fee for the feed oil was calculated as being in the range of −0.8p to +2.2p/L (0.264 US gallons). In this case, a positive gate fee means that the process is economic. The advantages of scale were discussed, but it was expected that these advantages would decrease for plant sizes above approximately 100,000 tonnes (220.5 Mlbs.) per year due to the larger collection area required for the used oil and the associated increase in transportation costs.

In discussions with potential buyers for the product oils, it was established that there was a negative perception problem for the re-refined oils, despite their meeting all performance criteria, and that a discount of 10% to 15% was expected for these product oils when compared with the virgin mineral oil product. There was no clear demand for the product oils in the UK—only varying degrees of resistance, but this resistance was least in the gear and hydraulic oils markets.

The four policy options that the consultant had been asked to consider were: (1) business as usual, or do nothing, (2) removal of the excise duty derogation or reduction on used oils that had been converted to fuel oil, (3) a ban on used oil combustion except in hazardous waste incinerators, and (4) a combination of these options. It should be noted that these scenarios were under an umbrella assumption of a crude oil price at that time of between $25 and $28 per barrel.

Under the first option, it was expected that the use of reprocessed fuel oil in power stations would decline as they converted to natural gas fuels. The reprocessed fuel

TABLE 9.6
UK DEFRA 2001 Report: Economics for Two Re-Refining Plants: 35,000 Tonnes per Year

Cost	Process with Clay and No Hydrogenation	Process with Hydrogenation
	New Site, 15%–25% Rate of Return	
Income	£3–5 million	£5.5–6.7 million
Capital	£8.45–10 million	£15.65–17.95 million
Operating (including 13 staff members)	£2.5–3.1 million/year	£3.2–3.5 million/year
Depreciation	£0.85–1.1 million/year	£1.6–1.8 million/year
Estimated gate fee for used oil	–4.6p to –3.4p/L	–4.7p to –6.2p/L
	Case A + Plant Built On Existing Site	
Income	£3–5 million/year	£5.5–6.7 million/year
Capital	£4.75–5.55 million	£11.15–12.45 million
Operating (including 13 staff members)	£2.5–3.1 million/year	£3.2–3.5 million/year
Depreciation	£0.46–0.56 million/year	£1.1–1.2 million/year
Estimated gate fee for used oil	–1.9p to +1.0p/L	–1.3 to –1.4p/L
	Case B + 10% Rate of Return	
Income	£3–5 million/year	£5.5–6.7 million/year
Capital	£4.75–5.55 million	£11.15–12.45 million
Operating (including 13 staff members)	£2.5–3.1 million/year	£3.2–3.5 million/year
Depreciation	£0.46–0.56 million/year	£1.1–1.2 million/year
Estimated gate fee for used oil	–1.2p to +2.2p/L	–0.8p to +2.2p/L

Source: DEFRA Oakdene Hollins.

oil would then be diverted more toward the cement kilns. Under the second option, there would be a small decrease in the volumes of reprocessed fuel oil marketed. In the third option, with the banning of used oil combustion in burners such as in cement kilns, only PCB-containing oils would be incinerated and this would promote disposal through other routes such as recycling to base oils, conversion to diesel fuel, or use as a diesel extender. Collection of the used oils would become more costly and previous reprocessed fuel oil users would switch to heavy fuel oil, coal, gas oil, or natural gas. The fourth option, a combination of the other scenarios, would be the most expected outcome and would involve power stations using reprocessed fuel oil for a limited time after 2006 when the WID was expected to take full effect, depending on whether the reprocessed fuel oil was classified as a product and not a waste material.

It was likely that the collectors would start to charge for the collection of used oil and that the majority of the oil would be routed to cement kilns after 2006 until other routes were developed. It was expected that after 2006, large-scale investment would be attracted to the sector, but that none would be focused on re-refining to make lubricating base oils. It was also forecast that of the 350 locations where used oil was being disposed of as a fuel at that time, and of which 125 were road stone quarries, this number would decrease to less than 10 and almost certainly to less than 15 with small used oil collectors being driven out of the market.

The overall conclusions of the report were summarized as

- The removal of the excise duty derogation would cause reprocessed fuel oil prices to decrease, and the WID would cause post-2006 prices to decrease further
- By 2006, it was expected that at least one major used oil collection business would be integrated with a major oil company with investment in alternative disposal routes and support for kiln feedstock, diesel fuel extender, or re-refining to base oil processes
- The technology least likely to benefit from the investment was re-refining to base oils
- Financial incentives for re-refined base oils were unlikely to be effective, given the negative perceptions involved

There were supplementary measures that could be taken, such as an agreement with a large manufacturer of lubricants to use a percentage of re-refined product in a blend, or to establish a separate set of standards for re-refined oil.

9.2.10 Process Simulation and Modeling

This article, published by Foo Chwan Yee, Rosli Mohd Yunus, and Tea Swee Sin of the Universiti Teknologi Malaysia, took laboratory results on a solvent extraction and clay treatment process and used a commercial process simulation program to model the process on a computer. The model was then used to predict the performance of a larger 5000 tonnes (11 Mlbs.) per year plant (Foo et al. 2002).

The process used solvent extraction with 2-propanol and *n*-hexane with a small amount of potassium hydroxide as a flocculating agent to form a sludge that contained much of the polar compounds, which were then separated from the oil fraction before the solvent recovery by distillation from this stream. Previous studies had shown that a mixture of 60% 2-propanol with 40% *n*-hexane was the optimal ratio for the mixed solvent. This was followed by passing the extracted oil stream through a clay adsorption step to decolorize the oil at 80°C, the clay being separated from the oil and each of the streams then being distilled to recover the solvent for recycling in the process. Small amounts of *n*-hexane and potassium hydroxide had to be added to the recycled solvent streams to bring the mixed solvent back to the required composition.

The model was built on the simulator and the computer results agreed well with the laboratory results. This model was then used to simulate a 5000 tonnes (11,200,000 lbs.) per year plant, which would treat 571 kg (1260 lbs.) per hour of used oil, requiring 39 kg (86 lbs.) per hour of the composite solvent, and 760 kg (1676 lbs.) per hour of activated clay, giving a sludge removal rate of 6.42 mass% of the used oil feed. Some data was presented on the metals removed in the adsorption process, showing high adsorption of the magnesium in particular, and a sensitivity analysis was performed on three variables: the heat duty for the third flash operation to separate the solvent from the product oil (Flash 3), and the mass flow rates for the 2-propanol and *n*-hexane. An optimal operating temperature for the Flash 3 distillation was identified as 111°C (232°F), which maximized the flow rates of the 2-propanol and *n*-hexane while minimizing the heat duty for this operation.

No information was presented on the economics of this process, and the requirement for significant quantities of clay and the energy requirements for the other distillation processes would have to be evaluated and be incorporated into an economic model. The disposal of dry sludge and dry clay would also have to be addressed.

9.2.11 Compendium of Regeneration Technologies

This report was developed by the International Centre for Science and High Technology of the United Nations Industrial Development Organization (ICS-UNIDO), which is located in Trieste, Italy, and has a mandate to transfer technology to developing and transition-economy countries to promote sustainable industrial development. As such, it represents a good overview of the used oil situation with the recycling technologies available at that time in different parts of the world (Dalla Giovanna et al. 2003).

An initial review of the lubricating oil market highlighted the following items: (1) the world lubricating oil market in the period 1996 to 1999 was for 39 million tonnes (86,000 Mlbs.) per year and (2) the market in Europe was for 5.3 million tonnes (11,700 Mlbs.) per year, of which 1.7 million tonnes (3749 Mlbs.) per year were collected for recycling. Of this, 30% was re-refined, and 70% was burned.

The processing options in increasing order of complexity were:

1. Laundering of hydraulic and cutting oils with simple filtration for solids removal
2. Reclamation of used industrial oils where they were filtered and centrifuged to produce a product to be used as mold release oil or as a base oil for chainsaw oil
3. Direct burning, for instance, in cement kilns, space heaters, or waste incinerators
4. Burning after mild re-processing of filtration and settling to be used in road stone drying and fuel blending
5. Burning after severe re-processing with flash distillation, atmospheric, or vacuum distillation
6. Re-refining back to base oil

It was noted that most re-refining plants were in East Asia, being in India, China, and Pakistan, but these were generally of small capacity of approximately 2000 tonnes (4,400,000 lbs.) per year.

The European Community had introduced various directives that affected used oil activities. These included 75/439/EEC and 87/101/EEC on used oil disposal, 96/59/EC on the disposal of polychlorinated biphenyls (PCBs) and polychlorinated terphenyls (PCTs), and 2000/76/EC on the incineration of waste. The directives required a preference for regeneration if technical, organizational, and economic constraints allowed this.

A summary of the regeneration technologies divided them into the following categories which, it should be noted again, are based on the situation at that time:

- The most advanced with proven industrial application: Mohawk and Revivoil
- Other industrial applications: Interline, Vaxon, Sotulub, acid–clay, Meinken, PROP, and others
- Prototypes, pilot plants, not yet applied in the market: Exxon, Mineraloel-Raffinerie Dollbergen (MRD), ultrafiltration
- Studies and patents: Hylube, Regelub, Recyclon, and others
- Fuel oil production: Propak, Trailblazer, the vibratory shear enhanced membrane separation (VSEP) process, Lubriclear, and others

A review of each of these was then made, with comments on the plants that had been operated up to then, and the quality of the products produced.

9.2.11.1 Implemented Processes

The Mohawk Process: A 600 bbl/day plant operating in Vancouver, Canada; the Evergreen plant in Newark, California; the Wiraswasta Gemilang, Indonesia plant of 50,000 tonnes (110.3 Mlbs.) per year, and the Southern Oil Refineries plant of 20,000 tonnes (44.1 Mlbs.) per year in Australia. The base oils produced were of good quality.

The Revivoil Process: The propane deasphalting (PDA) process developed by Viscolube S.p.A/Axens allowed 80% of the brightstock to be removed, increasing the base oil yield from 72% to 79%. Plants were operating in Pieve Fissiraga in Italy at 100,000 tonnes (220.5 Mlbs.) per year; Jellicoe, Poland, at 80,000 tonnes (176.4 Mlbs.) per year; Surabaya, Indonesia, at 40,000 tonnes (88.2 Mlbs.) per year; with two plants under construction in Spain at 20,000 tonnes (44.1 Mlbs.) per year each. BP had committed to the engineering/license for a plant in Indonesia of 20,000 tonnes (44.1 Mlbs.) per year, and UNIDO had committed to a process/license for Pakistan for 10,000 tonnes (22.1 Mlbs.) per year. With the hydrofinishing step, it was possible to produce Group II base oil products, which are accepted by API, ACEA, Mercedes Benz, and major original equipment manufacturers.

The Atomic Vacuum Co. Process: This short-path, thin film distillation/clay treatment process had a manufacturing base near Mumbai, India, and required clay regeneration or disposal. It did not use acids.

The Blonde Process of Conseco Ltd. in the Slovak Republic: This process involved the cracking and separation of cracked products using sand in a hot, fluidized bed type of reactor. One plant of 3500 tonnes (7.7 Mlbs.) per year was operating in the Slovak Republic and the product was a low-quality base oil that required additional treatment. There was expected to be high erosive wear in the equipment due to the sand.

The Cyclone Process from KTI/IFP Axens: The thin film evaporator/PDA/hydrofinishing process was being used to produce base oils in one plant in Greece.

The Dunwell Process from Dunwell Environ-tech in Hong Kong: This thin film evaporation demetallization process produced a low-quality base oil from one plant in Hong Kong that had been operating since 1993.

The Matthys–Garap Process from Matthys: This centrifugation/vacuum distillation/acid–clay process had been used in two plants in France, but the acid–clay by-product was still considered to be dangerous, the process was complicated, and it was thought that the equipment was liable to damage due to centrifugation with high-viscosity oils.

The Interline Process: Plants had been built in the United States, United Kingdom, South Korea, and Spain, but all except the plant in Spain had been closed at some time due to environmental reasons and the low quality of the oil produced. The plant in Australia was not mentioned in this report.

The KTI Relube Process: This distillation/thin film evaporation/hydrotreatment process produced good quality product oils. Plants had been built in the United States, subsequently modified to the Mohawk process; in Greece, subsequently modified to the Cyclon process; and in Tunisia, subsequently modified to the Sotulub process.

The Meinken (Meinken Engineering) Process: This was a distillation/clay treatment process in which clay was added to the dewatered oil before wiped-film evaporation and subsequent clay treatment and filtration. It produced a medium quality base oil that still contained sulfur and polynuclear aromatic hydrocarbons (PNAs). Three plants had been constructed in Germany, one in the United States, one in Taiwan, and one in Brazil.

The PROP Process: The chemical treatment/distillation/clay treatment/hydrofinishing process had seen some plants built, but these were no longer operating due to financial difficulties. The process was quite expensive, but produced good quality oil products with low metals content. Disposal of wastes was an issue.

The Acid–Clay Process from Meinken and Others: This first process was still the most widely used in the world with its low capital cost and simple processes. The product quality was very low, the plant capacities were low, and the acid sludge and spent clay were environmentally difficult to handle.

The RTI Vacuum Cyclone Distillation/Clay Treatment Process: Several acid–clay plants in the United States had converted to the RTI process, but the product quality was not very high and had no hydrotreatment step.

The Snamprogetti Process: This PDA/vacuum distillation/hydrofinishing process had one plant operating in Italy at 55,000 tonnes (121,300,000 lbs.) per year.

The Sotulub Process: One plant, originally a KTI installation, had been modified and processed 16,000 tonnes (35.3 Mlbs.) per year since 1979 in Tunisia, whereas a second plant in Kuwait processed 20,000 (44.1 Mlbs.) tonnes per year. Product quality was low.

The Vaxon Process: Originally an Enprotec fuel oil plant in Denmark with a capacity of 28,000 tonnes (61.7 Mlbs.) per year, this process was being licensed by Avista Oil of the United States. A plant of 42,000 tonnes (92.6 Mlbs.) per year was operating in Catalonia, Spain, for medium quality base oil production, and another in the Middle East with an unknown capacity.

9.2.11.2 The Other Prototype or Pilot Plant Processes

The Bechtel MP Process: This process was being used for virgin base oil production in 13 units. It had been tested in a German plant as a final bleaching/dearomatization unit.

The Entra Process: This German process uses a linear tubular reactor with careful control of the temperature and very short residence times to cause heteroatoms to be separated before the carbon–carbon bonds are broken. Acid and clay are used, as well as sodium to bind dangerous components, but PAH removal required additional treatment. One prototype plant had been operating with Südöl AG since 1988.

The Exxon Process: A pilot plant had been operated, but no industrial application is known and costs for the process have not been evaluated.

The MRD Kernsolvat Process: The report refers to the Dollbergen vacuum distillation/solvent extraction plant of 230,000 tonnes (507.2 Mlbs.) per year capacity operating near Hannover and a glass laboratory unit having been built. Removal or reduction of the polynuclear aromatic hydrocarbons content of the oil was significant.

The ROSE Process from Kellogg Brown and Root: This residue oil supercritical extraction (ROSE) process had been used for virgin base oil production. Its application to used oils was originally the Krupp–Koppers process, which was not commercialized and involved supercritical extraction by hydrocarbon gases at very high pressures.

The Ultrafiltration Process from Gerth of France: Besides ultrafiltration through a membrane, the process involved distillation and catalytic hydrotreatment. A pilot plant had been tested at Lillebonne in France in 1987, but no commercial operation was known.

9.2.11.3 The Studies and Patents Processes

The BERC Process: Goldmark Chemical Enterprises of Ghana controlled this process of distillation/solvent extraction/hydrofinishing with the use of solvents consisting of isopropanol, butanol, and 2-butanone to separate the base oil from a sludge. No industrial application was known.

The Jake Oil Process in Bolivia: The process involved distillation, followed by clay treatment and filtration. The product quality was low.

The Proterra Process from Probex Corp./Bechtel: This vacuum distillation and extraction with NMP solvent produced a Group I base oil and the first plant was being planned by Probex for construction in Wellsville near Pittsburgh with a capacity of 54 million gallons (204 ML) per year.

The Recyclon Process from Degussa/Leybold Heraeus Engineering: This chemical treatment and vacuum distillation process had no known industrial application and was a complex and high-cost process that used neither clay nor hydrogenation. It used sodium to remove polyaromatics, PCBs, and additive compounds, requiring careful handling of this material.

The Regelub Process: A distillation/thermal treatment/ultrafiltration/hydrofinishing and vacuum distillation process that was complex and used ceramic membranes, which were expensive and required frequent changing. A medium quality base oil was produced, but continuous operation seemed to be difficult for larger-sized plants.

RWE Entsorgungs AG Process: This involved treating the oil with alkaline waterglass and an aqueous solution of polyalkylene glycol to settle a sludge from the oil before subsequent hydrogenation or treatment with sodium, and was at the patent stage.

The Studies Technologies Projects Process: This was a vacuum distillation/hydrofinishing process and the comment was made that the process was based on technologies already available on the market from other licensors, and that the proposed installation had been patented by others.

The Tiqsons Technologies Inc. Process of Canada: This process was based on vacuum distillation and hydrofinishing and again seemed to be based on technologies available or patented by others. There was no evidence of an industrial application.

The UOP Hylube Process: The first plant for the DCH process was under construction in Egypt and was scheduled for startup in 2002/2003. The products were expected to be of high quality.

9.2.11.4 Fuel Production Processes

The industrially applied processes for producing fuel oil or diesel oil from used oil included the Propak, Trailblazer, Zimmark, and Soc processes. Without a stabilization process to saturate the olefins produced in the cracking process, the products would be unstable, darkening in color when exposed to oxygen and forming gums and deposits.

With the changes to legislation in Europe, North America, and elsewhere, which now require used oil to be re-refined to produce a base stock oil where this is feasible, it is expected that operators will find increasing difficulty in obtaining approval for the application of these processes.

The Propak Process from Par Excellence Developments of Canada: This process involved thermal cracking, distillation, and a Robys stabilization process to produce fuel oil or gas oil. There were some installations in Canada and one in Belgium, but the products were of poor quality, the sulfur in the oil was not reduced, and there was a residual odor in the products.

The Trailblazer Process from Texaco: This involved preflash and vacuum distillation processes to produce industrial or marine fuel and an asphalt extender. There was one plant in the United States processing 170,000 tonnes (374.9 Mlbs.) per year of used oil.

The Zimmark Process of Canada: The impurities were coagulated and precipitated at elevated temperatures but still required a settling time of days. This was followed by distillation and separation processes, which did not remove PCBs or other undesirable materials. The product was of very poor quality but the process was used in mobile units and was simple, operating on-site, and avoided the need to transport the used oil. Units were being used by the Canadian Railway System, and there are units in the United States, Mexico, and Asia.

The Springs Oil Conversion Inc. Process of Canada: After flash dewatering, the oil was thermally cracked to produce a fuel oil or gas oil of poor quality because the sulfur level was not reduced and there was a retained odor. Eighteen units in seven plants were in operation around the world. No stabilization treatment step was envisaged.

9.2.11.5 Prototypes and Other Processes for Fuels

The Environmental Oil Processing Technology Process, Nevada: The used oil was thermally cracked and produced a diesel fuel. One plant was in operation in Nampa, Idaho.

The Laser Oil Molecular Separator Process, Italy: The process used laser light energy to break up the molecules. No industrial application was known.

The Lubriclear Process from the National Center for Environmental Research: This involved carbon membrane separation of the base oils from the contaminants. Demonstration units of 75 gallons (284 L) per day had been built in the United States and Canada. Problems were expected to be the expense of the membranes and their fouling.

The Miniraff Process from Minitec Engineering GmbH of Germany: The oil is treated in a destructor unit whose function is unknown before separation and hydrotreatment of the fuel products. The first mobile unit of 500 kg (1103 lbs.) per hour was under construction.

The Polymer Membrane–Based Filtration Process from the Environmental Technologies Institute of Singapore: The oil passes through the membrane to be separated from the contaminants. Problems are the low temperature constraint due to the membrane material, high cost of the membranes, and the low quality of the product fuel oil.

The Fileas Process from the Commissariat à l'Énergie Atomique et aux Énergies Alternatives of France: The process uses supercritical carbon dioxide at high pressures to act as a solvent and reduce the viscosity of the oil, which is then filtered through a membrane, the carbon dioxide is recovered by simply lowering the pressure. The process had only operated at the pilot scale, and the cost and fouling of the membrane were concerns. Hydrotreatment would be required for the product oils and it was possible that base oils could also be produced with this process.

The VSEP Process from Dunwell Enviro-tech of Hong Kong: A vibratory membrane separation process was being operated in one unit in Hong Kong. The unit was expected to be subject to fouling, produced a low-quality product, and had a low throughput capacity.

The Electrization Process or Electrostatic Separator Process, Italy: The oil passed between electrodes maintained at a high potential difference of 10,000 to 15,000 V to separate the impurities, which are subject to induced charge from the oil which is almost electrically neutral. No industrial application was known for this process.

The final sections of this report then proceed to list web site information and used oil regenerating companies and their contact information around the world.

9.2.12 DEFRA Study on Used Oil Management in Italy and Germany

This study by Oakdene Hollins Ltd. was commissioned to provide information to the DEFRA UK authorities regarding Italian and German experience with used oil policies, and their effectiveness, to help formulate a UK policy in response to the Waste Oil Directive 75/439/EEC and its amendments from the EU (Fitzsimons and Lee 2003). The WID was implemented in the UK in 2002/2003 and was expected to be applied to existing plants from December 2005. The report was first published in 2003 and republished in 2010.

9.2.12.1 Italy

Data was provided on the six recycling plants in Italy in 2002, including one which processed only transformer oil. The changes to the Viscolube plant to produce Group

II base oils from 2003 was notable because there was little capacity for Group II base oil production in Europe at that time.

The government had established a used oil recycling program, which was run by a consortium made up of lubricant manufacturers, used oil collectors, and re-refiners. A levy on lubricant sales of €325 per tonne (2205 lbs.) at that time was used to pay a subsidy to operators of re-refining and fuel plants, which was estimated to be €94 per tonne (2205 lbs.) for base oils and €40 per tonne (2205 lbs.) for fuel oil made from used oil. In 2002, collectors had been paid €166.50 per tonne (2205 lbs.) of used oil collected and subsidies had also been paid for testing costs for the collected oil. The collection was free of charge to generators with 85 appointed collectors with legislation requiring at least 90% of the oil to be routed to re-refining and a maximum of 10% to fuel use in installations such as cement kilns. Testing was carried out on the oils being collected and no emulsions or fuel-contaminated oils were accepted by the collectors.

Prices for the re-refined base stock oils were approximately 20% below prices for virgin base stock oils, and there was competition from the Russian base oil supplier of Lukoil with similar prices. However, Viscolube was partly owned by the former state company AGIP and this had facilitated the use of the re-refined base oils in the lubricants produced by AGIP. It was estimated that approximately 50% of the lubricant manufacturers in Italy used some re-refined base stock oils. Viscolube had estimated that approximately 30% of their re-refined base stock oil production had been going to automotive use and 70% to industrial lubricant use, but with the plant upgrade to the hydrotreating facility to make more Group II base oils, the automotive use was expected to increase.

Other countries such as Australia, Spain, and Denmark had adopted systems similar to Italy's. It was felt that the system lacked competition: for instance, the used oil collector was told by the consortium where to deliver the collected oil, and subsidies were negotiated with the consortium each year. It was estimated that the annual subsidy to achieve the policy objective of giving preference to re-refining over use as a fuel in cement kilns was costing approximately €31 million per year in Italy.

9.2.12.2 Germany

The seven re-refining plants operating in Germany produced base oils that were being sold at an estimated discount of approximately 25% compared with the price of virgin base oil stocks. Before 1990, a state subsidy of approximately €80 per tonne (2205 lbs.) of used oil had been paid to companies accepting used oil as a feedstock. After the reunification of Germany in 1990, subsidies or capital grants were paid for re-refining plants located in the former East Germany, decreasing to a maximum subsidy of 20% after 2004. Also, since 2001, the government had paid a subsidy of €25 per tonne (2205 lbs.) for base oils made from used oil to operators making a financial loss, reduced by €2.5 per tonne (2205 lbs.) each year for the next 7 years until it reached zero after that.

Cement kilns and limestone plants competed with re-refiners for feedstock, cement kiln operators paying up to €70 per tonne (2205 lbs.) of used oil, but used oils had some relief from excise duty of €20 per tonne (2205 lbs.). Approximately 50,000 tonnes (110.3 Mlbs.) per year of used oil was imported from the Netherlands and the

Benelux countries. There were no public collection facilities for used oil because few Germans changed their own vehicle oil. There were also no joint ventures with base oil manufacturers or lubricant formulators, and some of the re-refiners had been marketing the re-refined lubricants under their own name. The new Puralube plant in Zeitz was expected to produce a superior product base oil when it began operations.

It was estimated that re-refining in Germany was mainly driven by facilities that had already been depreciated to almost zero, and that new capacity was being driven by capital grants and output subsidies, boosted by the enthusiasm for the Puralube operation. The increased capacity for re-refining could lead to greater imports of used oil to Germany, possibly also from the United Kingdom.

A discussion of consumer-based and producer-based market features that could affect re-refining operations identified the difference between the European market and North American market: in Europe, there was little Group II base oil production as the blending of Group I and Group III base oils was preferred to obtain better performance, whereas in North America, the production of Group II or II+ was favored. However, the first European all-hydroprocessing plant was planned for Petrola Hellas in Greece to make Group II, Group II+, and Group III base oils. The prices for Group II oils were similar to the prices for Group I oils, making a displacement of the Group I re-refiners a possibility. The gas-to-liquids process using a natural gas feed to a Fischer–Tropsch type operation was also a potential competitor with excellent quality Group III base oils, but the biggest threat to re-refining was seen to be the lower cost treatments of used oil to make fuel-type products.

The policy options and possible issues for the UK were summarized as

1. Market changes. Until March 2005, there was uncertainty regarding the effect of the WID on the market for used oil sold as RFO. Used oil was classified as a hazardous waste, but there was a possibility that an operator would develop a process to produce fuel oil that could be classified as a product and not a waste, thus bypassing the WID. In any event, the WID was expected to eliminate some of the uses for the fuels derived from used oils, and the price of used oils was expected to decrease to almost zero.
2. The removal of the derogation of duty on used oils burned as a fuel was promoted by the European Regeneration Group (GEIR), among others, but this was going to be reviewed in 2007 in Europe. The derogation could be modified so that it only supported the processing of used oils at designated re-refineries.
3. The value-added tax (VAT) could be removed from lubricants containing at least 10% of re-refined oil, this limit being the upper bound for blending oils from another source into an oil that had been already certified by the lubricant certifying bodies without requiring recertification and the additional costs thereof.
4. The possibility of capital grants for re-refining investments could be considered, similar to the grants in Germany, but there were several technologies, such as those from Avista, Viscolube, CEP, Puralube, Pesco, and Interline, among others, offering improved re-refining processes that, combined with diminishing feedstock prices, could improve re-refining economics. This could make the need for capital grants unnecessary.

5. In Italy, Denmark, Germany, France, Australia, and Catalonia in Spain, an annual output subsidy had to be negotiated by re-refinery operators with either a regulated consortium, or a regional or national government. To avoid this requirement, it could be possible to use a system of tradable permits linked to the UK vehicle end-of-life program or the carbon credits system in which the production of re-refined oil could be worth an equivalent amount of credits in the other areas.
6. Providing incentives to encourage the return of the used oils to a large, conventional refinery could perhaps be a more efficient process, but the refinery operators were reluctant to accept the used oils due to the possible problems with catalyst poisoning, increased corrosion, and contamination of wax products. Further discussions with refinery operators could be fruitful.
7. The introduction of a possible levy on lubricants to provide an annual subsidy to fund other steps was a possibility, and although a subsidy was not yet foreseen, if the markets did not respond to the WID as expected, this possible need could be anticipated.
8. The encouragement of laundering, or the simple treatment steps for oils that had not been heavily contaminated, and the intensification of use for lubricating oils was another possibility for policy development. These activities needed to be monitored and brought into the used oil re-refining discussion.

9.2.13 California LCA Study on Re-Refining and Use as a Fuel

This LCA study assessed and compared the end-of-life potential environmental effects for three management methods for used oil, namely, re-refining to a base oil, distillation to produce a marine diesel fuel oil and asphalt extender, and combustion of the untreated oil as a fuel (Boughton and Horvath 2004).

The California data for the three methods are summarized in Table 9.7 and it is seen that the majority of the used oil was burned as fuel oil, much of which was exported from California for blending with other fuel oil as a cutter stock. For the year 2002, 92 million gallons (348 ML) were collected and recycled through the three management methods studied.

For the purposes of the LCA study of the case of fuel combustion with energy recovery, the used oil was certified to meet the recycled oil standard according to the California Health and Safety Code 25250, which limits the content of certain heavy metals, total halogens, and halogenated compounds. It was generally blended with other fuel oils and it was assumed that the combustion of a blended fuel did not affect the net release of emissions with time, or in other words, the net emissions per unit of used oil consumed remained the same from an LCA perspective. Because the used oil marketed as fuel comprised less than 0.5% of the total fuel oil market in California, it was assumed that there was no significant displacement of crude refining capacity or the fuel oil market.

For the case of re-refining, the process assumed flash evaporation to remove water and light ends, a defueling step to separate gas oils, followed by a distillation to separate the lube distillate fraction from a heavy residual, and then a final hydrofinishing step for the lube oil fraction. At that time, the re-refining capacity in California was

TABLE 9.7
California Management Methods for Used Oils Generated In-State

Process	Volume Treated	Products	Wastes
Fuel oil production	48.9 million gallons (185 ML)	49.5 million gallons (187 ML) fuel oil (cutter stock)	—
Distillation	31.7 million gallons (120 ML)	15.8 million gallons (59.8 ML) asphalt flux and 13.3 million gallons (50.3 ML) MDO fuel	1.3 million gallons (4.9 ML) wastewater and 1.3 million gallons (4.9 ML) hazardous waste
Re-refining	11.4 million gallons (43.2 ML)	7.9 million gallons (29.9 ML) base stock oils and 1.9 million gallons (7.2 ML) asphalt extender	0.5 million gallons (1.9 ML) wastewater and 0.5 million gallons (1.9 ML) hazardous waste

Source: Reprinted with permission from Boughton, B., and A. Horvath. *Environ. Sci. Technol.*, 38(2), 353–358. Copyright 2004 American Chemical Society.
Note: Based on 92 million gallons (348 ML) processed in 2002 in California.

approximately 12 million gallons (45 ML) per year of used oil with a water content of 4% to 5%. For 1 L (0.264 US gallons) of dry used oil feed, the following amounts of produced and consumed components were assumed:

- Products: 0.72 L of base oil (0.2 US gallons), 0.18 L (0.05 US gallons) of asphalt, 0.06 L of gas oil (0.015 US gallons), and 0.04 L (0.10 US gallons) of light ends
- Consumed: 5.1 g of sodium hydroxide, 0.39 MJ of electricity, 1 g of hydrogen, and 4.3 MJ natural gas. The catalyst was replaced biannually, but the consumption was so small that this was neglected

It was also assumed that re-refining did not affect the overall lubricating oil market because this represented less than 5% of the California market.

For the case of distillation to produce a marine diesel oil (MDO) fraction together with an asphalt flux, the process assumed was a distillation to remove light ends and water, then a distillation to separate the heavy fuel oil as a distillate from the heavy bottoms. The capacity in California at that time was for approximately 40 million US gallons (151 ML) of used oil per year with 5% water, but the facilities generally operated at a capacity of approximately 75% due to a weak demand for the asphalt product. For each liter (0.264 US gallons) of dry used oil feed, the following produced and consumed components were assumed:

- Products: 0.44 L (0.12 US gallons) of marine diesel fuel, 0.52 L (0.14 US gallons) of asphalt flux, and 0.04 L (0.010 US gallons) of light ends
- Consumed: 2.8 g of sodium hydroxide, 0.29 MJ of electricity, and 6 MJ of natural gas

A computer software package was used to conduct the LCA and assess and compare the effects and benefits for each of the three management methods. The following components required definition in this process:

A. The Life Cycle Boundary. The study focus was on the end-of-life phase in which the used oil was being delivered for processing, and it was assumed that all preceding treatments had been equal for the three methods being compared. Dewatering of oily water was considered as pretreatment and was excluded, as were identical activities that were common to all three methods. Construction and maintenance costs for the re-refining and distillation plants were also excluded due to these being spread over very large volumes of processed oil over the 30-year life span of these facilities. Wastewater treatment on-site was included, but not water treatment of sewer discharges, which contained very little contaminants.
B. The Life Cycle Functional Unit. This was defined as the equivalent amount of products and recovered energy from 1 L (0.264 US gallons) of dry oil feed. To establish equivalent systems that enabled a comparison of the methods involved adding impacts affecting the particular process. For example, for the case of combustion of the used oil as a fuel with energy recovery, the effects from the manufacturing of base oil, asphalt tar, and gas oil from crude oil reserves were added to the effects of combustion itself.
C. Data Sources. Product effects for all products and input materials were taken from the software databases, some of this being of US origin and some of German origin. Re-refining facilities in the two countries were assumed to be equivalent for this study. In addition, a study in Vermont on heavy metals and acid gas emissions from the use of used oil as a space heating fuel provided data on emissions arising from combustion of the oil with limited or without emission controls. Used oil characteristics were gathered from California sources covering the majority of the collected used oil. Re-refining data were taken from the Evergreen facility in Newark, California, and the distillation process data came from a facility in Compton, California.
D. Temporal and Geographical Considerations. It was assumed that there were no significant differences in time, but that geographical differences might be significant. Because the majority of the fuel oil product was shipped overseas, transportation costs might affect the used oil fuel case.

The results of the study showed that the heavy metal emissions dominated the inventory, the other emissions to the air, water, and solid wastes being comparable among the cases and having little influence on the conclusions. For example, the burning of 1 L (0.264 US gallons) of fuel oil produced from used oil resulted in the emission of 800 mg of zinc and 30 mg of lead to the atmosphere. Although zinc was the main contributor to the heavy metals emissions, lead, cadmium, and copper were also significant, as well as phosphorus.

As part of the LCA analysis, the actual emission quantities of the heavy metals have to be weighted according to their effect potential in terrestrial, human

TABLE 9.8
California LCA Study: Impact Factors for Combustion Compared with Re-Refining and Distillation

Potential Environmental Impact Category	Ratio of Used Oil Combustion to Re-Refining	Ratio of Used Oil Combustion to Distillation
Terrestrial ecotoxicity (kg DCB equivalent)	150	150
Human toxicity (kg DCB equivalent)	5.7	5.7
Eutrophication(kg phosphate equivalent)	3.2	3.1
Aquatic ecotoxicity (kg DCB equivalent)	2	2
Ozone depletion (kg R11 equivalent)	1.1	1.1
Photochemical oxidant (kg ethane equivalent)	1.1	1.1
Global warming (100 years) (kg CO_2 equivalent)	0.9	0.9
Acidification (kg SO_2 equivalent)	0.5	0.5

Source: Reprinted with permission from Boughton, B., and A. Horvath. *Environ. Sci. Technol.*, 38(2), 353–358. Copyright 2004 American Chemical Society.

Note: Based on equivalent functional units of product and energy recovery with no air pollution control. DCB is dichlorobenzene, a common measure of toxicity.

toxicity, and aquatic toxicity categories. These factors vary widely in each category, cadmium being very highly weighted in terrestrial ecotoxicity potential, and chromium being highly weighted in the human toxicity category. After applying these factors to the emissions, the data shown in Table 9.8 were obtained, showing the ratios of effect characteristics for used oil combustion with little or no emission controls compared with re-refining and distillation processes and based on equivalent functional units. The values for terrestrial toxicity potential was very high at 150 times for the fuel oil case, due mainly to the zinc and cadmium emissions, and also in the human toxicity category at 5.7 times, due mainly to the lead and chromium emissions.

As a result of this study, it was recommended that the waste priorities of waste avoidance, and returning the waste to its original use, should take priority over its use as a fuel. Extending vehicle service intervals would result in lower waste production, and incentives for treatment and market support for used oil products could help reduce the effects identified in the study. Efforts to expand the re-refining and distillation capacities of plants in California could be supported by appropriate legislation and standards.

9.2.14 Preview of the Institut für Energie und Umweltforschung 2005 LCA Study

This report was issued in November 2004 by the European Re-refining Organization (GEIR) and gives a preview of and summarizes some of the results from the Institut für Energie und Umweltforschung (IFEU) report by Fehrenbach issued in February

2005 (Fehrenbach 2005). The limited success in getting member states to comply with the EEC Directives of 1975 and 1987 regarding used oils was primarily due to the lack of recycling targets in the directives (GEIR 2004). Economic factors such as return on investment and security of used oil supply were often cited as reasons for lack of implementation of the desired re-refining option, the used oil processors having to compete with the cement and limestone kiln operators for the used oil, the cement kiln fuel option being supported by a derogation of the excise duty on used oils used for this purpose in 11 of the 15 member states.

The IFEU study showed significant improvements in six environmental performance indicators: resource depletion, the greenhouse effect, acidification, nutrification, carcinogenic risk potential, and fine particle emissions. It was noted that incineration in cement kilns scores more favorably than regeneration in the area of global warming if the fuels displaced by the fuel oil were coal and petroleum coke, both of which had higher emissions than the oil and their avoidance was therefore desirable. Displacement of natural gas or conventional fuel oil use in cement kilns or other applications led to the opposite result. The long-term future of the incineration option was therefore dependent on the relative prices for the energy sources, and resource conservation and market conditions could affect the situation in the future.

It was concluded that the regeneration of the used oils resulted in a significant reduction in environmental effects when compared with the generation of lubricants from virgin mineral oils, and when compared with incineration as an alternative recovery option.

9.2.15 Nigeria Pilot Study on Used Oils

This study was sponsored by the Secretariat of the Basel Convention of Geneva and charged its Regional Centres in Egypt, Nigeria, Senegal, and South Africa to cooperate in developing a plan for environmentally sound management practices for used oil in Africa. Nigeria was chosen as a case study in developing this strategy and the study was carried out by the Basel Convention Regional Coordinating Centre in Nigeria (Bamiro and Osibanjo 2004).

The terms of reference for this study were (1) to identify the main sources and quantities of used oil in Nigeria, (2) to identify the main actors in this industry, (3) to identify the existing disposal and treatment facilities, and (4) to survey the informal sector working with used oils.

In starting with the potential volume of used oil generated in Nigeria each year, it was noted that 377 ML (99.6 million US gallons) per year of base oils were imported into the country and that there was no local base oil production. It was assumed that 80% of this went to producing lubricating oils, or approximately 300 ML (79.2 million US gallons) per year. Of this amount, approximately 50% would be lost in use, making approximately 150 ML (39.6 million US gallons), or at most 200 ML (52.8 million US gallons) per year as potentially collectible. Of this, automotive crankcase used oil was estimated to be approximately 150 ML per year (39.6 million US gallons), and industrial used oil approximately 50 ML (13.2 million US gallons) per year.

There were approximately 35 dealers in used oil in the country and they were surveyed with a questionnaire regarding the practices used. The used oil was generally stored in metal or plastic containers and the surroundings of the depots were often polluted with leakage into sewers, drainage systems, rivers, and streams. Treatment of the used oil generally involved the removal of suspended matter by settling, or removal of water by open-air heating over a fire. The uses for the used oil included direct reuse in older vehicles whose owners accepted the lower quality oil as a trade-off for the cheap price, its use as an additive in hair cream production, as a boiler fuel and weed-killer, as a mold release, in wood preservation as a termite protection measure, being mixed with grease to make gear oil, in rust protection of metals, as a wound healing application, in cooking with firewood fuel, and mixed with kerosene to make diesel fuel. With the known carcinogens in used oils, the hair cream and wound treatment uses would seem to be very dangerous options.

There was no re-refining activity in the country and the disposal practices were not organized in any systematic way, there being widespread environmental damage from used oil in the country. A study with UNIDO in 1996 had proposed the establishment of a re-refining plant based on an Italian process design, but this had not been implemented due to the escalation in the capital cost of the project and the problem of securing enough used oil supply to feed the plant. One local company had set up a research agreement and development project with the University of Lagos to develop a plant to produce fuel oil, then later diesel fuel, and in a third phase, a re-refining plant. Another company had developed and tested a pilot plant for reprocessing and re-refining with a capacity of 50 to 100 L (13.2 to 26.4 US gallons) per hour and was completing testing before marketing the technology. The Nigerian Ministry of Environment had also invited bids for an acid–clay plant and a fabrication contract had been awarded, and the hope was expressed that additional investments would be made to develop the technological backup, which was also required to make this project a success.

Regarding the legislation in place, it was noted that a 1988 decree had established the Nigerian Federal Environmental Protection Agency, now the Federal Ministry of Environment, which had issued a decree in 1991 prohibiting the discharge of oil to drains, rivers, the sea, lakes, or underground without a permit issued by the competent authorities; that collection, treatment, transportation, and final disposal of the waste was the responsibility of the industry or facility generating the waste; and if an industry or facility was likely to release gaseous, particle, liquid, or solid untreated discharges, it would have to install abatement equipment as determined by the federal agency. The comment was made that the legislation seemed to be adequate, but knowledge of the sector needed attention. From photographs in the report, apparently made during the collection of data in this regard more than 10 years after the promulgation of these requirements, it would seem that these regulations were not being applied nor enforced in many areas.

9.2.16 Nigeria Workshop on Used Oil Management in Africa

The workshop held in Lagos, Nigeria (December 6–7, 2004), was sponsored by the Basel Convention with the participation of 61 representatives from industry, government, and nongovernmental organizations (NGOs). The objectives of the workshop were:

- To present recommendations for the national plan for the environmentally sound management of used oils in Nigeria
- To present recommendations for the development of a used oil partnership in the African region
- To decide on follow-up action for the development of the partnership initiatives in the African region, and to disseminate the recommendations throughout the African region
- To decide on follow-up action for the draft national plan for the environmentally sound management of used oils

The first part of the workshop reviewed the results from the pilot study conducted in Nigeria (Bamiro and Osibanjo 2004). Articles were also presented in three technical sessions and included a description by the Director of the Basel Convention Regional Center in Pretoria, South Africa, on a workshop that had been conducted in 2002 in conjunction with the ROSE Foundation headquartered in Cape Town, which included a description of the most successful used oil recycling program in South Africa, together with the recommendations arising from that workshop. A presentation from the representative from Senegal also identified the consumption of lubricating oils in Senegal as 15,000 tonnes (33.1 Mlbs.) per year with some 1500 tonnes (3.31 Mlbs.) per year being collected for recycling. Other presentations were made by representatives from the Nigerian Standards Organization and two companies on a locally developed, skid-mounted re-refining process, which was briefly outlined, and the need for used oil management programs. The UNIDO experience was also described, including 12 techno-economic studies being conducted, 4 each in Asia, Africa (Egypt, Senegal, Kenya, and Zimbabwe), and South America.

A draft National Action Plan for Waste Oil Management in Africa was prepared by a working group of experts, focusing on job creation, poverty alleviation, and environmental protection, and identifying the components of an environmentally sound management plan for used oils in Africa, which addressed institutional coordination, awareness creation, and funding mechanisms. This also attempted to assign roles and responsibilities of stakeholders and an implementation plan, and this draft was accepted by the workshop participants, together with a set of recommendations from the workshop:

1. The creation of an enforcement arm of the Federal Ministry of Environment to deal with the regulation of used oil
2. The promotion of intersectoral collaboration between the Ministries of Environment, Health, Industry, Department of Petroleum Resources, and the Standards Organization of Nigeria
3. A review and expansion of existing legislation to identify acceptable management options for used oils, including the definition of roles and responsibilities of stakeholders
4. The strengthening of public/private partnerships through cooperation with organized occupational groups
5. The involvement of organized groups in workshops for capacity-building and training on enforcement and compliance

6. The involvement of NGOs and others in research and advocacy, awareness creation and monitoring
7. The development of a financial mechanism for used oil management
8. To establish an incentive system for used oil management initiatives
9. The provision of uniform storage facilities at strategic points close to the generators of used oil
10. To involve local know-how in reprocessing and re-refining technologies
11. To promote product responsibility, including an ecolabeling requirement for lubricating oils

The need for information dissemination, education regarding poor disposal practices, and adequate record keeping on sales and disposal of lubricating oils, together with an effective incentive system and legal framework to prevent environmental damage was also apparent (Nigeria Workshop 2004).

9.2.17 LCA Study of Re-Refining and Use as a Fuel

This intensive LCA study was commissioned by the GEIR and carried out by the IFEU GmbH of Heidelberg to compare the environmental burden of producing base lubricating oils from virgin mineral oil with the re-refining production of an equivalent quality base oil from used oil, and to compare the combustion of used oils as a fuel with the re-refining processes. Previous LCA studies had been carried out, but were considered to be no longer valid due to the following developments since the 1990s practices on which they had been based: (1) in the interim period, new regeneration and re-refining processes had been developed and implemented; (2) regulatory requirements regarding vehicle emissions had resulted in an improvement in the quality of lubricants; and (3) increasing amounts of synthetic and semisynthetic lubricants incorporated in the lubricating oils had resulted in an improvement in the quality of the used oil collected (Fehrenbach 2005).

The LCA comparisons were based on the ISO 14040 standard and examined four re-refining processes in operation in Europe, and one in the United States, including the supporting processes such as electricity generation or other fuels required, and comparing these with the production process of lube base oils from virgin mineral oil. The presence of synthetic oils in the collected used oils was also considered over a range of 0% to 30% synthetics in the used oil and covered the range from the 9% at that time to the expected value of 30% in 2030.

The environmental categories examined were:

- Resource depletion with respect to fossil energy resources in kilograms of raw oil equivalents, and each contribution was considered in terms of a characterization factor for each component in each of these categories
- Global warming. For instance, 1 kg of methane was considered as equivalent to 21 kg of CO_2 in its effect on global warming.
- Terrestrial nutrification in terms of kilograms of PO_4^{3+} equivalents
- Acidification in terms of kilograms of SO_2 equivalents

- Toxicity with respect to carcinogenic pollutants in terms of kilograms of arsenic equivalents, and fine particulates in terms of the primary particulates plus the precursors as reflected in the kilograms of PM10 value (PM10 indicating the fine particulates with a diameter of >10 μm).

The categories of aquatic and odor effects were not included as the aquatic category had been previously identified as not important with the modern technologies, as had the odor category.

Germany had been prosecuted in 1999 in the European Court of Justice and had been found guilty of not implementing the EEC Directives regarding used oils, and cases were proceeding against the UK, Northern Ireland, and Sweden, so a part of the aim of this new LCA was to determine what had changed since the LCA study in Germany in 2000, and to compare the results with that study.

Questionnaires were distributed to the companies reviewed to gather input data, these companies reflected the Cyclon, Evergreen/CEP, Hylube, MRD, and Viscolube (Revivoil) processes. The MRD installation had replaced clay treatment with a solvent extraction process. The Viscolube process was represented by operations in two plants in Italy, one in Ceccano and the other in Pieve Fissiraga, producing 130,000 tonnes (286.7 Mlbs.) per year of base oil with the process developed together with the French Axens company. The Pieve Fissiraga plant had replaced the clay process with a hydrogenation process and produced a Group II base oil product. The operating results and parameters were listed for each plant but the identification of each plant was obscured.

In considering the production of base oils from mineral oils, all products were considered and the production resources were allocated to each to come up with the share for the lubricating oil base stock. In the production of the synthetic oil components, poly α-olefins or PAOs were taken as the model for these and the resources required to produce these from ethylene from steam-cracked naphtha or natural gas condensate, polymerization to linear α-olefins, and then fractionation to obtain 1-decene used to synthesize the PAOs were determined. The functional unit assumed was 1 tonne (2205 lbs.) of collectible and re-refinable used oil, and the parameters for the five processes were averaged to form a model re-refining process.

The results of the LCA in comparing re-refining with base oil production from virgin mineral oil were summarized as

- The avoidance of environmental burden from the displaced procedures was higher than the burden from regeneration processes
- For the global warming category, CO_2 equivalent emissions are two times higher for the displaced processes than for the re-refining option
- For the nutrification category, PO_4^{3+} equivalent emissions are three to four times higher for the displaced processes
- For the acidification category and fine particulates (PM10), the level is five times higher for the displaced processes
- For the carcinogenic risk potential, the factor is 10 to 20 times higher for the displaced processes
- For fossil resource depletion, the factor was 30 times higher for the displaced processes

- The differences among the five re-refining processes were quite small
- The increasing quantity of PAOs in the used oil feedstock was mostly reflected in the global warming and nutrification categories, increasing these burdens by 20% and 35%, respectively

To obtain a meaningful comparison across the effect categories, the results were also normalized according to ISO 14042 in terms of person equivalent values (PEVs), this being the average per capita load of one inhabitant for each of the effect categories and the total used oil consumption value of 600,000 tonnes (1323 Mlbs.) per year in the EU. This identified two of the processes as being beneficial in many of the categories, due to the high yield of regenerated base oils and additional benefits from the by-products but the differences among the processes were relatively small.

A comparison was performed between the re-refining processes and direct combustion of the used oils considered for burning in cement kilns, coal power plants, steel mills, and lime kilns. The displaced fuels were difficult to define because they included hard coal, lignite, heavy fuel oil, and natural gas, and the choice of fuel tended to vary as the relative prices varied. An average fuel composition was thus assumed, consisting of a blend of the fuels generally used, and this was a point later questioned by the independent review panel.

The results for the comparison of regeneration with burning of the oil in a cement kiln showed that regeneration was clearly favored in terms of resource conservation, acidification, carcinogenic risk potential, and fine particulate emissions. In terms of global warming, used oil combustion was more beneficial when used oil substituted mainly coal and petroleum coke. The nutrification result was ambivalent, depending on the content of synthetic components in the used oil, regeneration being better for the used oil with 30% synthetics.

Comparing regeneration with combustion in coal power plants and steel works where heavy fuel oil was the displaced fuel, regeneration was always better, but with regard to global warming and nutrification, this was only the case when the synthetics were at the 30% level.

The results after normalization in terms of PEVs and the total used oil quantities showed that the global warming effect was lowered for the combustion of used oil as the displaced fuel was weighted toward coal. The combustion advantage in that case was 636 kg (1402 lbs.) of CO_2 equivalents per tonne of used oil for the zero synthetics case, and 39 kg (86 lbs.) of CO_2 equivalents for the 30% synthetics case. For the fuel oil substituting case, in terms of PEVs, regeneration was better than combustion for all categories, the global warming advantage being less for the 30% synthetics case than for the zero synthetics case.

A sensitivity analysis was also performed to look at the influence of these four factors on the overall results:

1. In preparing the feed for a lube plant in a conventional refinery, the aromatics are separated from the paraffins that feed the lube plant. Because this situation is changing due to the decrease in demand for the aromatics, which led to the production of other by-products and the aromatics increasingly

being sent to a fluid catalytic cracking treatment, the effect of shifting the cost of separating the aromatics from the paraffins as an extra burden on the paraffin stream was examined. This produced only a slight shift in favor of the regeneration option.

2. In a previous study with which the results were supposed to be compared, it had been assumed that 1 kg of product PAO needed 1 kg of raw feed material; however, considerably more than 1 kg of feed material was needed to produce 1 kg (2.21 lbs.) of PAO. When this was assumed, the burden on PAO production was increased, resulting in a mineral oil resources saving equal to the consumption of approximately 10 million European inhabitants if 600,000 tonnes (1323 Mlbs.) of used oil containing 10% synthetics was regenerated rather than being burned.
3. Other fuels such as meat and bone meal, tires, and used solvents were being increasingly used in the secondary fuel market. A level of 17% was therefore examined in the cement kiln fuel mix. This showed an increase in the regeneration advantage, the global warming effect for the 30% synthetics cases being approximately equal in this comparison.
4. The effect of distribution distances was examined because it was likely that re-refining plants would be more widely scattered than combustion plants, making the collection distances higher for the re-refining plants. Increasing the distance from the 100 km assumed previously to be 200 km, showed that this was not a very significant parameter, increasing the nutrification benefit of regeneration by 11% with the reduction in benefit in the other categories being much less.

Aside from the conclusions already listed, it was concluded that all five re-refining processes had lower environmental releases than the substituted processes. Re-refining of used oils for recovery of base oils leads to significant resource preservation and relief from environmental burdens. The priority given to re-refining by the EU was supported by the results of this LCA study.

The final step in the presentation of this report was to have it reviewed by an external panel of four independent experts. The questions raised by this panel included:

- The definition of the functional unit of 1 tonne (2205 lbs.) of used oil was questioned because the composition of used oil can vary considerably. Used oil that is unsuitable for regeneration could be burned in a cement kiln, so the functional unit for a cement kiln option could be different
- Effects on aquatic and waste treatment categories had not been included, although these had been identified previously as low impact factors
- Odor releases and solid waste disposal had not been addressed, although it was accepted that the modern re-refining processes had little effect in these areas
- The weighting factors used for the fossil fuels were questioned. For instance, coal had a weighting factor of 0.1836 kg of crude oil equivalent per kilogram of coal, and this was based on data related to the rate of depletion

and rates of consumption in Germany: these values would probably vary in other countries
- The use of an average fuel mix for the cement kilns was not correct and should have been based on the most expensive alternate of fuel oil. The IFEU response to this was that the cement kiln fuel mix varied considerably and this had, in any case, been examined in the sensitivity analysis

Several other relatively minor questions, comments, and recommendations were also listed by this panel, but with no further items that would seem to significantly affect the results of the study. The review panel thus concluded, with some minor reservations, that the study had been conducted in a way that was consistent with ISO 14040.

9.2.18 Improving Markets for Used Oils

This report was first published in 2005, based on data from 2002, and was republished in 2010 (Fitzsimons 2010). It was prepared for the Organization for Economic Cooperation and Development (OECD) by Oakdene Hollins Ltd., and was intended to provide information relevant to determining the optimal point of policy intervention in the used oil recycling industry. It included a discussion of some of the results from LCAs that had been performed up to 2002, including the Taylor Nelson Sofres (TNS) report (Monier and Labouze 2001), and indicated that the results had been inconclusive regarding the preference for used oil re-refining or incineration as a fuel, depending on which alternative fuel had been displaced. Despite this, many countries had introduced legislation to favor re-refining, often based on resource conservation and atmospheric pollution grounds, the later report from IFEU (Fehrenbach 2005) not yet being available, and the most significant environmental effect had been identified as illegal dumping of used oil. If the aim was to minimize environmental damage, the priority should thus be to maximize collection of the used oil.

The review of market failures and barriers in the recycling of used oil included the following:

- Information failures relating to waste quality. Some collectors conducted tests on collected oil before accepting it, but a rapid, mobile testing procedure was required for general use
- There was risk aversion among buyers of lubricants to using products from used oil recycling, resulting in prices often being 20% to 25% lower than the prices for virgin oil products. In addition, the costs of obtaining certification, such as that from API, could take $500,000 after the re-refining plant had been built and started production, taking perhaps 6 months to complete, during which production had to be stockpiled or sold at a cheaper price. This led to a cost of certification that could approach approximately $2 million
- Certain technical externalities associated with lubricating oil specifications. The restriction of chlorinated additives and the presence of vegetable oils in lubricating used oils had an effect on the re-refining industry

- The market power of virgin product incumbents. It was very difficult for a new re-refining operator to penetrate the existing markets and persuade buyers to switch to a product made from used oil

The policy responses in different countries had fallen into two categories: those aimed at increasing collection efficiencies and those addressing the market barriers to re-refined products. In the United Kingdom, for instance, there was no preference in the legislation for re-refining or use as a fuel, a re-refining plant having opened and closed again due to economic uncertainties, and used oil being imported for fuel use.

The situation in Italy was different, with a clear favoring of re-refining with legislation requiring at least 90% of recycled oil to be directed to re-refining and the remainder to fuel use. The 2003 rework of the subsidy system had provided a levy on lubricant sales that was used to pay a subsidy to re-refiners that was much greater than that paid to fuel producers. Testing of the used oil before it was accepted diverted unsuitable oil to cement kiln use or other fuel uses, this oil being collected at cost whereas used oil collection was otherwise free. The collaboration of the state oil company AGIP with Viscolube had provided brand recognition support for the re-refined oils to overcome some of the marketing barriers.

The Product Stewardship Arrangements for used oils in Australia provided for a levy on sales of lubricating oils, which was used to provide a subsidy for the recycled products based on a sliding scale with re-refining receiving the highest subsidy. A subsidy had also supported the construction of the new Southern Oil Refineries re-refining plant at Wagga Wagga, midway between Sydney and Melbourne. The collection efficiency issue was also being supported by subsidizing the expansion of collection points for used oil.

The system in Alberta, Canada, was based on an environmental handling charge on sales of lubricating oils with a return incentive paid to collectors for used oils, oil containers, and oil filters. The used oil was primarily used in Alberta for road asphalt and fuel burning with the focus being on maximizing collection of the used oil. There was a weakness in this system in that a lack of testing of the oils, before acceptance, would promote mixing of other components such as parts washer solvents, brake fluid and radiator antifreeze which could inflate volumes of collected fluids.

The EU Waste Directives 75/439/EEC as amended by 87/101/EEC and 91/692/EEC had given priority to re-refining over fuel use for the used oils, but the lack of implementation by the European countries had resulted in legal proceedings against 11 countries by that time. This avoidance by these countries had been attributed to several reasons, including that the re-refining process was uneconomic in their circumstances.

To promote collection, this should be made free of charge to the generators, or by payment of a subsidy, which might encourage the inflation of collected volumes by mixing other undesirable oils with the suitable used oils. A rapid test kit was needed, but rejected oils should also be collected to avoid inadvertent promotion of increased dumping of the rejected oils. The costs of switching market choice of products, such as the costs for the API tests and certification required for the re-refined products, could be subsidized.

The conclusions of this study were prefaced by the general comment that, in general, it was best to protect the environment with suitable legislation and then to let market forces decide on the best recycling route. The conclusions were then:

- It was important to target environmental policy objectives, to maximize collection of used oils and avoid uncontrolled dumping. If an environmental case could be made for preferring re-refining over fuel use, this could be done with appropriate regulations without other interventions such as subsidies. The air pollution regulations would, however, require enforcement to be effective
- The provision of information to market participants involved the highlighting of the dangers of illegal dumping of used oils, the existence of possible fines for transgressions, and the location of used oil collection points was necessary
- To discourage illegal dumping and burning of used oils, a combination of a "carrot" in the form of subsidies and refunds could be used, combined with a "stick" in the form of possible fines for flouting of the laws prohibiting this disposal route
- To maximize collection of the used oils without negatively affecting the re-refining industry, caused by mixing undesirable oils or other components with the used oil to inflate volumes, to support the development of low-cost, rapid testing equipment for the used oils
- To remove market and policy distortions by allowing incineration without emission control equipment in some areas, or imperfect enforcement of regulations. The derogation of excise duties on waste-derived fuels in the United Kingdom was scheduled to be removed in 2006
- It was important to ensure policy consistency because changes of direction in this industry were expensive and it was important to provide clear and stable incentives to investors. Efforts to increase the collection rates for the used oils should be coordinated with efforts to increase the demand or market side for the products

9.2.19 Critique of the Canadian UOMA Program Review of 2005

The Canadian National Used Oil Management Advisory Council published in 2005 the results of a Program Review of the UOMA-style programs in the Western Canadian Provinces of British Columbia, Alberta, Saskatchewan, and Manitoba (Valiante 2005). This review was carried out with a questionnaire of 24 questions sent to 400 potential stakeholders, and followed up with interviews with 30 of the 119 respondents. On the basis of this, the following three key conclusions were reached:

1. The UOMA model is a leader in program design, collection, and compensation schemes compared with 14 other global used oil management programs, and is also a world-leader in maximizing used oil collection rates of more than 75%.
2. This model was a world-leader in maximizing re-refining rates of 30%.
3. Re-refining of used oil and burning of used oil as a waste-derived fuel were environmentally equivalent and that there was no "right answer" for the appropriate end use for used oil.

This critique from the British Columbia Used Oil Management Association (BCUOMA) set out to analyze the methods used and the conclusions reached in the Program Review.

The first comment was that a survey was not an appropriate tool to obtain factually based results on a scientific basis but would return opinions on how the respondents felt about their perceptions of certain issues on which they might well be lacking hard data and insight into the ramifications of the issues. It was suggested that better alternative tools were available to determine the effect of regulatory schemes. In particular, one question was examined in some detail: "Overall, have UOMA's programs improved the collection of used oil materials in Western Canada?" After reviewing pre-UOMA and post-UOMA data, it was concluded that there was no reliable data available, even for experts, to support an opinion either way in this case, so a meaningful response to this question was impossible.

Data was then presented on used oil collection and re-refining rates in some of the European countries, which in many cases, exceeded the Canadian values that had been claimed to be exemplary. For example, France collected 81% of the available used oil in 2000 and re-refined 31% of it: Greece collected 91% of the available oil that year and re-refined 58% of it. At the time, BCUOMA was paying collectors $0.08/L (0.264 US gallons) to collect used oil, this money coming from the fees paid by the consumer in buying the new oil. The re-refiners were competing with other collectors who paid between $0.18/L and $0.24/L (0.264 US gallons) for the used oil, presumably to be supplied as a burner fuel, and the re-refined volume had dropped from approximately 35 ML (9,240,000 US gallons) per year pre-BCUOMA to approximately 24.1 ML (6,362,400 US gallons) per year post-BCUOMA. In addition, the re-refiners had previously been paid between $0.07/L and $0.09/L (0.264 US gallons) by the used oil generators to collect the oil, and their costs had increased because they had to start paying the generators $0.24/L (0.264 US gallons) to maintain the reduced processing volumes.

It was also noted that the United Kingdom did not have a formal, incentive-based, government-approved used oil collection program, and that the United Kingdom and Northern Ireland had been convicted in July 2004 by the EU court on its failure to fulfill its obligations under the 75/439/EEC Directive to give priority to the processing of used oils by regeneration.

The third contention regarding the equivalency of re-refining and burning as a fuel had apparently been based on a study that was no longer available, and the second reference supporting that view seemed to be based on data from the 1990s and did not consider many of the effect categories that a rigorous LCA would consider. The TNS study (Monier and Labouze 2001) and the IFEU study (Fehrenbach 2005) were then reviewed and the results regarding the general superiority of re-refining over use as a fuel in almost all conditions were presented and discussed.

In general, then, it was concluded that care should be taken in conducting a review of this type, that surveys were only useful in determining how people felt about their perception of an issue, and if the primary goal was the objective evaluation of the environmental effect and economic efficiency of a policy, the UOMA

Program Review was flawed and the three key conclusions could not be accepted as valid.

9.2.20 Catalan Waste Agency Position on an LCA Study

This report by Fullana (2005), a leading LCA expert, President of the Spanish LCA Society, and editor of *The International Journal of Life Cycle Assessment,* represented essentially a critique of the TNS LCA report (Monier and Labouze 2001). The work by Fullana was funded by the Waste Agency of Catalonia and the Catalan oil processor, CATOR.

A significant comment on the TNS report was that it was not correct to take outcomes from a national level study with specific boundary conditions for the LCA such as the date of the study, the technology, the market, and the fuel mix for the avoided processes and generalize these for pan-European application. Local environmental effects had to be assessed, together with economic and social effects. The best available technologies should be assessed both for the regeneration and the avoided processes, and the effect categories should be prioritized.

It was further suggested that a screening process on the influence of different parameters for the different techniques should be performed before starting more detailed assessment. This would be aimed at avoiding a large comparison matrix covering all processes in all locations under all possible substitution possibilities. This would show a range of effects for the most promising techniques, identified by environmental screening, cost–benefit analyses, or social points of view, and reduce the amount of irrelevant or low effect data to be included in the study. It was recommended that the EU work with other stakeholders in an LCA project to define the aims and execution of the project.

Specific comments included the observation that the used oil management situation in Europe was changing quickly. For instance, in Italy in 2003, 29% of the input oil to the market was regenerated compared with 18% in the TNS report of 2001, and in Catalonia, more than 65% of the used oil generated was collected and almost 100% of this was being regenerated.

Concerns were expressed regarding the non-inclusion of the quality of the used oil in the TNS report, and the data quality and methodology used in the report in which only two of the previous studies had data that was reliable under a strict LCA regime. Most of the data was from the period 1994 to 1997, and data for the avoided processes excluding base oil refining were from the years 1992 to 1993, whereas data for the base oil comparison were from the year 1996.

Future developmental estimates had not been included as they should be for decision-making and policy development studies. In relation to the sensitivity analysis in the report, it was noted that it was not clear that coal would be the only substituted fuel, and substituted fuel data from the Austrian cement industry were quoted.

In terms of the local and global environmental impacts, local or specific conditions could change the outcomes. There had been developments on including local inputs in LCA studies and these should now be included. For EU-wide policy development, it was felt that system specifications had to be representative at the EU level or at each of the regions where the policy was to be implemented.

On the omission of toxicity from the TNS study, it was expected that, even in the absence of hard data and a commonly agreed method of handling toxicity data, it should have been at least qualitatively handled. Again, recent advances had been made on including toxicity effects in LCA studies and these should be taken into account and included in future studies. Emissions to water and soils should also be included, besides the emissions to the air.

A separate quantitative analysis using LCA software had also been performed by a separate entity to examine the sensitivity of the results to some of the parameters in the TNS report and these results were presented in graphs and tables. The main conclusions from this were:

- There can be differences in the displaced fuels input in different locations
- The influence of transport distances for the used oil only became significant for distances greater than approximately 50 km. A graph was presented showing the kilograms of CO_2 equivalent as a function of transportation distance between 20 and 500 km
- The cases in the TNS report described individual technologies in individual scenarios in individual locations in Europe. Generalization to different situations and boundaries could hardly be drawn from these cases. The results might be acceptable for their specific circumstances, but should not be used in sum for reliable decision support

9.2.21 Re-Refining Processes Viewed from India

This short review of some of the re-refining processes was made by the Everest Blower Systems company of New Delhi, India (Everest 2005). It was highlighted that two re-refining processes had been used in India, the acid–clay process and vacuum distillation with clay treatment. The acid–clay process had been banned by the Central Pollution Control Board of the Government of India whereas the propane deasphalting (PDA) process was not widely used in India, the distillation process being more popular.

This company offered a skid-mounted, batch distillation version of this process, which processed a charge of 6600 L (1742.4 US gallons) in an evaporator connected to a two-stage vacuum system. The temperature was ramped up in the first stage in which a pressure of 15 to 20 torr (2–2.7 kPa) was used to separate the lighter cuts of oil. The second stage was then started with a high-vacuum pump to decrease the pressure to less than 1 torr (0.133 kPa) and the temperature was increased to between 300°C and 320°C (572°F–608°F) to complete the distillation. Each batch required approximately 8 h to complete the distillation. The products were then treated with 4% (w/v) clay to improve the color of the oil and produce a light golden-colored product for sale. The oil-saturated clay containing approximately 20% oil was generally sold to brick kiln manufacturers, but it was pointed out that a solvent cleaning process could be applied to the oily clay to regenerate the clay for reuse, enabling the use of higher quality clays and an improved quality of product oil. The disposal of the bottoms from the distillation process, apparently approximately 12% of the charge, was not addressed.

9.2.22 Used Oil Management in Scotland

This report was commissioned by the Scottish Environmental Protection Agency (Scottish EPA 2005) as a two-phase study whose primary objectives were to quantify used mineral oils in 2003 in Scotland, to identify waste management practices and infrastructure, and to assess alternative markets and technologies for the treatment and disposal of mineral oil wastes. These wastes were identified as being made up of two components: the used mineral oils together with oily wastes which consisted of other waste materials that had been contaminated with mineral oils. The first phase would be a data-gathering exercise whereas the second phase would use the data and information to assess the best practical options for the management of used mineral oils, taking into account the environmental and economic factors.

Much of the impetus for this project was provided by the EEC Waste Oil Directives 75/439/EEC as amended by 87/101/EEC and 91/692/EEC, and the WID 2000/76/EC. These directives were expected to have a major effect on the used oil recycling industry as the production of RFO and its use in coal power stations for flame control during start-up and new coal addition, the use of reprocessed fuel oil in road stone drying kilns, the disposal to landfills, and burning in small used oil burners were all expected to either be banned or severely restricted. There were no re-refining operations in either Scotland or England and this was the route to be preferred according to the directives.

The data obtained showed that in 2003, in Scotland, approximately 225,859 tonnes (498 Mlbs.) of mineral oil waste were consigned to waste management facilities, including transfer stations which collected the waste for further consignment. One of the difficulties in obtaining reliable numbers was that the waste handled by the transfer stations could be double-booked because the tracking system that would have allowed this was not in place at that time. The majority of the mineral oil wastes was in the oily waste category, being made up of offshore drilling rig waste drilling muds and other oily waste, and amounted to some 207,291 tonnes (457.1 Mlbs.). The used mineral oil component was estimated to be 15,930 tonnes (35.1 Mlbs.) and included 15,023 tonnes (33.1 Mlbs.) of used lubricating oils, which would have been suitable for re-refining: the estimated lubricating oil deliveries in 2003 were estimated at 29,000 tonnes (63.9 Mlbs.), making the collection rate approximately 46%. There was also a small component of the mineral oil wastes of 2638 tonnes (5.82 Mlbs.) that could not be assigned to either of the two main categories due to a lack of data. These numbers were also complicated by the fact that some collectors based in England collected used oil in Scotland, making the numbers probably quite a bit lower than the true values. It was estimated that 5850 tonnes (12.9 Mlbs.) of used oil originated from vehicles whose oil changes were performed at home and, of this, approximately 811 tonnes (1,790,000 lbs.) were accounted for, leaving approximately 4969 tonnes (10.96 Mlbs.) of used oil each year that was probably being disposed of through the municipal garbage system to landfills.

The main management method for the used oil was the production of reprocessed fuel oil as a cheap fuel, and this use was so popular that reprocessed fuel oil was being imported from Norway and the Netherlands to meet the demand in the United

Kingdom. The market for reprocessed fuel oil after the end of 2005 was uncertain, but the cement industry and steelmaking plants were expected to increase their consumption of this product as they had the necessary high-temperature combustion equipment that would still permit the use of reprocessed fuel oil.

Regarding the emerging practices and technologies, the possibility of using environmentally considerate lubricants made from vegetable oils, rapeseed, synthetic esters, or other biodegradable oils was recommended for further investigation, despite the higher costs of these oils. The use of these oils was already being promoted by the Scottish EPA by using them in its chain-saw and hydraulic equipment. The incentive system introduced in Australia in 2000, through which a levy on the oils sold raised the funds to pay recyclers a sliding subsidy according to a hierarchy of recycling options, was reported and recommended for further evaluation.

The final conclusions and recommendations from this first stage of the project were summarized as

- To set up a technical working group of key stakeholders to review the findings of the first stage of the project and scope the requirements for the second stage
- This group should oversee the development of best practice guidance for the management of mineral oil wastes in Scotland and liaise with others in this field
- To consider obtaining more data on mineral oil waste management, particularly cross-border movements of wastes and waste handled at multiactivity sites
- To identify resources to keep information on the locations of oil recycling banks up to date
- To assess the benefits and barriers to using environmentally considerate lubricants and promoting its use in Scotland, drawing on the experience of organizations in England and elsewhere

Although this report deals almost exclusively with the situation in Scotland for used oils, it identifies the difficulty of tracking the source and handling of the used oils, particularly when they become mixed with other waste materials, which can include drilling muds, hydraulic brake fluid, glycol from radiator coolant changes, parts washing installations, and other nonlubricating oil components. Unless the collection system for the used oil can keep these components separated, and this is a major problem where the generators tend to have a common dump tank for any waste that cannot be otherwise disposed of, the re-refining operator may have to deal with many different compositions in the used oil as fed to any re-refining process.

9.2.23 API Perspective on Re-Refining

This presentation by a representative of the API was made at the California Used Oil/HHW Conference in San Diego in April 2005 (Bachelder 2005). It contained a brief description of the treatment of used oils to produce a re-refined product, indicating that the hydrotreated recycled base oils could meet the latest

API specifications for lubricating oils such as the API SM or ILSAC GP-4 designation for gasoline engines, the API CI-4 Plus designation for diesel engines, as well as certain military specifications. This was done under the API Engine Oil Licensing and Certification System and the oils meeting these specifications were thus confirmed as meeting the manufacturer requirements for use in their engines.

There were some cost pressures on the re-refining industry due to competition for feedstock from marine fuel users, but the re-refined oil was being marketed under various brand names. There was also a market for blended re-refined and virgin base oils and some vendors were not indicating that their products contained partially recycled oil. Provided the products met the API performance criteria, the source of the base oils was not an issue.

9.2.24 Used Engine Oils: Re-Refining and Energy Recovery

This valuable contribution to the available literature on the treatment of used lubricating oils contains information on most of the re-processing options available up to late 2005, together with much of the economic data from that time. In the volatile economic world of mineral oil prices, the concomitant fluctuations in lubricating oil base stock prices, and changing process technologies and vehicle manufacturer requirements, it nevertheless provides a useful basis for comparing many of the recovery processes (Audibert 2006).

9.2.25 Design Data for a Re-Refining Plant

This book was intended to provide experimental results for process engineers and designers to enable the construction of a pilot plant and re-refining plant for Iraqi feedstock oils. The statement was made that the Iraqi lubricating oils were different from the lubricating oils in the western world and the results could not be compared with each other: the reasons for this were partly due to the high sulfur content of Iraqi crude (Awaja and Pavel 2006).

The process considered involved a dehydration step for the Iraqi used oil, followed by a solvent extraction step with an alcohol–ketone mixture before vacuum distillation. The four processes of dehydration, solvent extraction, solvent stripping, and distillation were studied in the laboratory by varying operating conditions to obtain experimental results to which polynomial models were fitted. The solvents used were butanone, 2-propanol, and 1-butanol. These data would be of interest to re-refiners using this recovery process, but the influence of various feedstock oils might require further investigation.

9.2.26 GEIR Statistics for Lubricants and Re-Refining in the EU in 2006

This survey, dated February 2008, covers the lubricants and used oil data for the EU countries for the year 2006 and is summarized in Table 9.9. There is some transboundary movement of the used oil for re-refining purposes, but much of it continued to be used for fuel purposes (GEIR 2006).

TABLE 9.9
GEIR 2006 Statistics from Questionnaire on Used Oil Collection and Utilization

EU Country	Lubricants Consumed (tonnes)	Collectible Waste Oil (%)	Collected Waste Oil (tonnes)	Amount of Each Utilization of Waste Oil (tonnes)		Which Virgin Fuel is Displaced by Burning? (e.g., heavy fuel oil)
Austria	79,000		39,596	Burning	24,700	Coal
				Burning (power plants/limeworks/steelworks)	2500	Heavy fuel oil
				Treatment to fuels		
				Re-refinery (base oil)		
				Transferred to other countries	12,396	Re-refining
Belgium	142,000		60,000	Burning (cement industry)	500	Coal
				Burning (power plants/limeworks/steelworks)		
				Treatment to fuels	25,000	Heavy fuel oil
				Transferred to other countries	15,000	Re-refining
				Others	19,500	
Bulgaria	55,000		17,000	Burning (cement industry)		
				Burning (power plants/limeworks/steelworks)		
				Recovery	1200	Heavy fuel oil
				Re-refinery (base oil)		
				Others/disposal	15,800	
Cyprus			4300	Burning (cement industry)		
				Burning (power plants/limeworks/steelworks)		
				Treatment to fuels		
				Re-refinery (base oil)		
				Transferred to others		

(continued)

TABLE 9.9 (Continued)
GEIR 2006 Statistics from Questionnaire on Used Oil Collection and Utilization

EU Country	Lubricants Consumed (tonnes)	Collectible Waste Oil (%)	Collected Waste Oil (tonnes)	Amount of Each Utilization of Waste Oil (tonnes)		Which Virgin Fuel is Displaced by Burning? (e.g., heavy fuel oil)
Czech Republic	110,689	30%	32,867	Burning (cement industry)	4800	Coal
				Burning (steelworks)	2400	Heavy fuel oil
				Burning (agriculture)	4800	Heavy fuel oil
				Burning (small furnaces)	19,881	Heavy fuel oil
				Transferred to other countries	986	Re-refining
Denmark	67,500		20,000	Burning (cement industry)	2000	Coal
				Burning (power plants/limeworks/steelworks)	2500	Heavy fuel oil
				Treatment to fuels		
				Re-refinery (base oil)	11,700	Re-refining
				Transferred to other countries	3800	Re-refining
Estonia	19,000	30%	5400	Burning (cement industry)	5400	Coal + heavy fuel oil
				Burning (power plants/limeworks/steelworks)		
				Treatment to fuels		
				Re-refinery (base oil)		
				Transferred to other countries		
Finland	79,000	30%	22,500	Burning (cement industry)	22,500	Coal + heavy fuel oil
				Burning (lime works)		
				Treatment to fuels		
				Re-refinery (base oil)		
				Transferred to other countries		

France	765,000	44%	224,759	Burning (cement industry)	78,260	50% petroleum coke, 50% heavy fuel oil
				Burning (lime works)	15,095	Heavy fuel oil
				Treatment to fuels	7214	Heavy fuel oil
				Re-refinery (base oil)	94,403	Re-refining
				Transferred to other countries	5000	Re-refining
				Others	24,787	
Germany	1,174,000		525,000 (including water)	Burning (cement industry)	70,000	Coal, other wastes
				Burning (power plants)		
				Burning (lime works/steelworks)	85,000	Heavy fuel oil
				Treatment to fuels	210,000	Heavy fuel oil
				Re-refinery (base oil)	135,000	+105,000 tonnes of imported waste oil re-refined to base oils
Greece	100,000	42,000 (t)	36,000	Burning (cement industry)		
				Burning (power plants/limeworks/steelworks)		
				Treatment to fuels		
				Re-refinery (base oil)	36,000	Re-refining
				Transferred to others		
Hungary	109,000	50%	27,823	Burning	6000	Coal
				Burning (power plants/limeworks/steelworks)		
				Burning (small furnaces)	7823	Heavy fuel oil
				Reuse (asphalt industry)	14,000	Reuse
				Transferred to others		
Ireland	38,000		20,000	Burning		
				Burning (power plants/limeworks/steelworks)		
				Treatment to fuels		
				Re-refinery (base oil)		
				Transferred to other countries		

(*continued*)

TABLE 9.9 (Continued)
GEIR 2006 Statistics from Questionnaire on Used Oil Collection and Utilization

EU Country	Lubricants Consumed (tonnes)	Collectible Waste Oil (%)	Collected Waste Oil (tonnes)	Amount of Each Utilization of Waste Oil (tonnes)		Which Virgin Fuel is Displaced by Burning? (e.g., heavy fuel oil)
Italy	542,000	49%	216,300 (including water)	Burning (cement industry)	32,800	Heavy fuel oil
				Burning (power plants)	400	Heavy fuel oil
				Treatment to fuels	1400	Heavy fuel oil
				Re-refinery (base oil)	172,600	Re-refining
				Others	800	
Latvia	37,400	30%	11,000	Burning (cement industry)	11,000	Coal + heavy fuel oil
				Burning (power plants/limeworks/steelworks)		
				Treatment to fuels		
				Re-refinery (base oil)		
				Transferred to others		
Lithuania	49,000	30%	14,000	Burning (cement industry)	14,000	Coal + heavy fuel oil
				Burning (power plants/limeworks/steelworks)		
				Treatment to fuels		
				Re-refinery (base oil)		
				Transferred to other countries		
Luxembourg	10,000		5364	Burning (cement industry)		
				Burning (power plants/limeworks/steelworks)		
				Treatment to fuels		
				Re-refinery (base oil)		
				Transferred to other countries	5364	Re-refining
Malta	4000		1200	Burning (cement industry)		
				Burning (power plants/limeworks/steelworks)		
				Treatment to fuels		

				Re-refinery (base oil)		
				Transferred to other countries		
The Netherlands	252,000		50,000	Burning (cement industry)		
				Burning (power plants/limeworks/steelworks)		
				Treatment to fuels		
				Transferred to other countries (burning)	32,000	Heavy fuel oil
				Transferred to other countries (re-refining)	18,000	Re-refining
Poland	351,000	49%	76,500	Burning (cement industry)	3000	Coal
				Burning (power plants)	10,000	Heavy fuel oil
				Treatment to fuels		
				Re-refinery (base oil)	48,500	Re-refining
				Transferred to other countries	15,000	Re-refining
Romania	130,000		27,663	Burning (cement industry)	19,129	Coal + heavy fuel oil
				Burning (power plants/limeworks/steelworks)		
				Recovery	5518	Heavy fuel oil
				Re-refinery (base oil)		
				Others/disposal	2016	
Portugal	89,000	45%	28,700	Burning (cement industry)	7100	50% pet. coke, 50% heavy fuel oil
				Burning (power plants)	10,000	Heavy fuel oil
				Re-refinery (base oil)	3400	Re-refining
				Transferred to other countries	3400	Re-refining
				Others	4800	
Slovenia	20,000		3967	Burning (cement industry)	3499	Heavy fuel oil
				Burning (power plants/limeworks/steelworks)		
				Treatment to fuels		
				Re-refinery (base oil)		
				Other	468	

(continued)

TABLE 9.9 (Continued)
GEIR 2006 Statistics from Questionnaire on Used Oil Collection and Utilization

EU Country	Lubricants Consumed (tonnes)	Collectible Waste Oil (%)	Collected Waste Oil (tonnes)	Amount of Each Utilization of Waste Oil (tonnes)		Which Virgin Fuel is Displaced by Burning? (e.g., heavy fuel oil)
Slovakia	50,000		15,000	Burning (cement industry)	6000	Coal
				Burning (power plants/limeworks/steelworks)		
				Burning (small furnaces)	9000	Heavy fuel oil
				Re-refinery (base oil)		
				Transferred to others		
Spain	545,000	40%	160,000	Burning	70,000	Heavy fuel oil
				Treatment to fuels		
				Re-refinery (base oil)	90,000	Re-refining
				Transferred to others		
Sweden	148,000	30%	45,000	Burning (cement industry)	37,000	Coal
				Burning (power plants/limeworks/steelworks)		
				Treatment to fuels		
				Re-refinery (base oil)		
				Transferred to other countries	8000	Re-refining
United Kingdom	800,000	50%	350,000	Burning (cement industry)	25,000	Gas oil
				Burning (steelworks)	185,000	Heavy fuel oil
				Treatment to fuels	30,000	CFO/heavy fuel oil
				Transferred to other countries	10,000	
				Others	20,000	Specialist oil recovery (transformer, hydraulic oil)
				SWOBS (small waste oil burners)	30,000	Heavy fuel oil

Source: Reproduced with permission from GEIR, Questionnaire on Used Oil Collection and Utilization in 2006, 6, 2006. http://www.geir-rerefining.org/documents/WO-questionnaire-EU27-final-022208.pdf (accessed on January 15, 2013).

9.2.27 A Re-Refining Process from Iran

This article indicated that the two most common reprocessing methods were still the sulfuric acid/bleaching earth process and the propane extraction/sulfuric acid/bleaching earth process, but that the environmental problems generated by the disposal of wastes had led this research group in Tehran to study a process based on dehydration and light ends distillation/vacuum distillation to produce a gas oil, base oil and residue/guard bed treatment/hydrotreatment as an alternative (Bridjanian and Sattarin 2006). The catalyst used in the hydrotreatment was a spent hydrocracking catalyst that was rejuvenated before application in the hydrotreatment and should thus reduce the costs of the hydrotreatment process over the process using fresh hydrotreatment catalyst.

The rejuvenation process was described in detail and involved naphtha washing, washing with acetic acid to remove metals deposited on the catalyst by the hydrocracking, decoking with hot air, presulfiding with hydrogen followed by gas oil containing dimethyl disulfide at elevated temperatures, followed by a sulfiding step by injecting more gas oil with the dimethyl disulfide over 12 h at 340°C.

The hydrotreatment itself was carried out at temperatures between 250°C and 370°C (482°F–698°F), pressures between 60 and 73 bar (6 and 7.3 MPa, 870 and 1059 psi), a liquid hourly space velocity between 1 and 2.3, and hydrogen purity of a minimum 70 mol%. This produced a clear and homogeneous product re-refined oil having a viscosity index of 92, a good color, and a chlorine content of 2.9 ppm, compared with the 14 ppm chlorine in the used oil.

9.2.28 Emissions Data due to Used Lubricating Oils in the UK

This study for DEFRA by a section of the UK Atomic Energy Agency was carried out to review and update the estimates for carbon emissions from the use of used lubricating oils in the United Kingdom (Norris et al. 2006). Previous studies by Oakdene Hollins Ltd. (DEFRA 2001) and others had identified the losses of lubricating oils by combustion in engines as being around 35%. Because it was expected that modern engines did not result in a combustion loss as high as that, a review was instigated to obtain a better estimate of the carbon and hence greenhouse gas emissions from this source. As the United Kingdom had been a signatory to the United Nations Framework Convention on Climate Change since 1993, it was obligated to publish data on greenhouse gas emissions and to demonstrate reductions where possible.

This study reviewed the lubricating oil consumption in the different types of engines from cars, trucks, marine or ship applications, aviation, industry, and agriculture summarizing the oil change data with associated losses in use to come up with an overall average of 22% of the annual 2004 lubricating oil volume sold of 180,000 tonnes (396.9 Mlbs.) being lost in the engines. By assuming a 25% loss number, this was judged to be a more accurate estimate than the 35% assumed previously. The loss in marine engines was higher because these users were reported to mix the used lubricating oil with the fuel oil for the engines. The change in carbon emission for each year from 1990 to 2004 was tabulated in the report.

The recalculation with the lower combustion loss in the engines resulted in a reduction of the overall carbon emissions for 2004 of some 342,470 tonnes (755.1 Mlbs.) of carbon, with 239,710 tonnes (528.6 Mlbs.) of this being from other industrial combustion in boilers, gas turbines and stationary engines, and emissions from road engines being 59,610 tonnes (131.4 Mlbs.) of carbon less than previously reported.

9.2.29 DOE Used Oil Re-Refining Study

This report was produced as a result of the US Energy Policy Act of 2005, Section 1838, which required:

> The Secretary of Energy, in consultation with the Administrator of the Environmental Protection Agency, shall undertake a study of the energy and environmental benefits of the re-refining of used lubricating oil and report to Congress within 90 days after enactment of this Act, including recommendations of specific steps that can be taken to improve collections of used lubricating oil and increase re-refining and other beneficial reuse of such oil.

The report was based on an analysis of previous studies and consultations with stakeholders, together with a review of applicable federal and state legislation (U.S. Department of Energy [USDOE] 2006).

Data regarding the use of lubricating oil in the United States and the rest of the world identified the United States as accounting for 25% of the world lubricating oil market in 2002, or 2500 million gallons (9464 ML) per year out of a total of 10,300 million gallons (39,000 ML). Approximately 60% of the US demand was for automotive use. After reviewing the sales of lubricating oil over a 10-year period, it was noted that the volumes in the United States had not changed significantly during that period, and the data from 1995 was still representative of the later situation. This indicated that the used oil potentially available for recycling was 1371 million US gallons per year (5190 ML), of which 69% was actually recovered, or 945 US million gallons per year (3577 ML). Of this recovered oil, 83% was burned as a fuel and 17% was re-refined. The used oil that had been improperly disposed of in landfills and other inappropriate ways had decreased from 426 million US gallons (1613 ML) in 1995 to 348 million gallons (1317 ML) in 2002. This was attributed to the increase in the "do-it-for-me" or DIFM vehicle oil change businesses as opposed to the "do-it-yourself" or DIY oil changes, the DIFM businesses recycling more of the used oil. There had been several presidential decrees and state requirements for federal and state departments to use re-refined oil products where this was feasible, they were reasonably available and they met vehicle manufacturer requirements.

It was noted that re-refining plants were expensive and that new investment in adding capacity was not economic at that time. Legislation varied from state to state and the oil pickup, take-back, and taxes on sales of lubricating oil to fund collection was highly variable. Although used oil for re-refining had not been classified as a hazardous waste, some of the product streams might be so classified. Care was needed in contemplating federal legislation to standardize collection and recycling operations because this might impede this activity in some jurisdictions that had been functioning well under specific local conditions. A review of worldwide re-refining operations showed that Europe was ahead of the United States in re-refining, having three times the capacity of the United States on the basis of a percentage of lubricating oil sales, and that re-refining in Japan was very limited. The importation of used oil to the United Kingdom for fuel purposes had created a shortage for re-refiners in the past, but the amendment to the Waste Framework Directive 75/442/EEC to take effect in December of 2005 had removed the priority to re-refining as LCAs had shown no clear advantage of re-refining over the use as a fuel. It should be noted that this has been superseded by the later directive of 2008.

The greatest users of used oil as a burner fuel in the United States were the hot mix asphalt plants, followed by use in space heaters and industrial boilers, with cement kilns being lower on the list of users. The review of previous studies included the study in California (Boughton and Horvath 2004), the TNS study (Monier and Labouze 2001) and the IFEU study (Fehrenbach 2005), all of which showed clear benefits of re-refining over combustion as a fuel in terms of almost all categories of energy conservation and environmental impacts. The beneficial effects of more synthetics in the used oil were noted, and the benefits of re-refining were expected to grow in the future due to these components. Some caveats were expressed in translating results from European studies to the United States. For instance, the main outlets for the used oil in combustion in Europe were in cement kilns, coal-fired power stations, and steel mills, all of which would be equipped with air pollution controls. In the United States, the corresponding industries burn a total of approximately 25% of the used oil combusted, whereas asphalt plants account for 37% and space heaters 15%.

The Sofres report (TNS 2001) had noted that, for almost all the environmental effects considered, burning in a cement kiln in which the used oil replaced fossil fuels was more favorable than burning in asphalt kilns where the displaced fuel was gas oil. Care was therefore necessary in assuming the European results would be duplicated in the United States. Assumptions regarding re-refining technologies, the type of fuel displaced and the scale and type of incineration plant can change the results obtained in the LCAs that had been carried out. A study to update the effects of synthetics, improving technologies, and changing conditions in recycling operations was advisable before clear conclusions could be drawn.

In terms of specific steps that could be taken regarding the collection and reuse of used oil, the following three were identified:

1. An information exchange activity for state and industry personnel should be established to share experience and be a focal point for updating issues. There should be a greater focus on rural and farming communities as this was seen as being a source to increase recycling quantities.
2. States that had not yet adopted the Used Oil Management Standards in 40 CFR, Part 279 from Section 3006 of the Resource Conservation and Recovery Act should be encouraged to do so.
3. For geographic areas not attaining the National Ambient Air Quality Standards, the use of used oil–fired space heaters should be reassessed. For states subsidizing space heater purchases, a study to determine whether the money would be better spent in expanding collection and financing recycling activities would be helpful.

The comments on emerging trends related to the bio-based lube products such as soy, corn, canola, and cottonseed oils indicated that the effects of these products in the used oil collected were not yet apparent and indeed not yet determined. It was felt that industrial oils such as hydraulic oils, metal-working oils, and gear oils were good targets for replacement with the more environmentally friendly bio-products, rather than the more demanding requirements in motor oils. There were some possible limitations on the use of re-refined oils in motor oil blends due to the increasing performance standards being required by the engine manufacturers, the change from ILSAC GF3 in 2000 to GF4 in 2005 and the expected GF5 in 2009/2010. However, the increasing quality of the used oils would make re-refining more attractive in the future, although this improved quality would not add to their value as a combustion fuel.

The Appendix of the report contained additional data on the re-refining technology, some economic data, and various legislation and decree examples regarding used oils.

9.2.30 Case Study for Greece of Re-Refining Benefits

This short study of the benefits of re-refining in Greece was prepared after the first reading of the new Waste Framework Directive in the European Parliament in February 2007, which had reintroduced the requirement for re-refining to be given priority in the recycling of used lubricating oils (GEIR 2007).

Approximately 100,000 tonnes (220.5 Mlbs.) per year of lubricating oils were consumed in Greece, and in July 2004, Greek legislation had banned reuse without regeneration, energy production from used oil, and the uncontrolled disposal of used oils. Under the National Collective System for Used Oils Alternative Management Program, licensed collectors were set up to collect and sell the used oil to the System, which then resold it to six regeneration plants in Greece. The largest of these, the Cyclon Hellas S.A. plant had a capacity of 40,000 tonnes (88.2 Mlbs.) per year, and was a modern plant with advanced technology. The other five plants had a capacity of approximately 15,000 (33.1 Mlbs.) tonnes per year and were scattered around the country.

Before the 2004 legislation, no more than 8000 tonnes (17.6 Mlbs.) per year of used oil had been collected in Greece. This had grown to 35,000 tonnes (77.2 Mlbs.) in 2006 and 42,000 tonnes (92.6 Mlbs.) in 2007, the latter representing approximately 42% of the lubricating oil sold in Greece. Thus, as the amounts of used oil being regenerated increased, the amounts being incinerated and being disposed of in an uncontrolled fashion had decreased.

On the basis of this success in Greece, the GEIR called for support of the provisions in the new Waste Framework Directive in Europe for the priority required for re-refining of used lubricating oils.

9.2.31 Processing Possibilities for Used Oil: Hungarian Summary Part I

The consumption of lubricating oils in the world had been changing, with consumption in Asia and Oceania increasing, whereas consumption in North America and Europe had not been changing much, the per capita consumption actually decreasing

(Baladincz et al. 2008). This had been due in part to the increase in vehicle oil change intervals and the corresponding increase in the quality of the lubricating oils being used.

A table was presented showing the consumption and used oils statistics for some of the European countries. Overall, there were 4.82 million tonnes (10,628 Mlbs.) of lubricating oils being sold in these countries, 2.21 million tonnes (4873 Mlbs.) were deemed to be collectible, and 1.78 million tonnes (3925 Mlbs.) had actually been collected for recycling, representing a collection rate of 80% in 2005. A table for the collection percentage by field of use for Hungary in 2008 was also presented, showing 70% of the motor vehicle collectible oil was actually collected, or 40% of the lubricating oils sold.

A table listed some of the main re-refining plants in some of the European countries, including six in Germany, five in Italy, three in Spain, and one each in France, the United Kingdom, Denmark, Belgium, and Greece. The contaminants expected in the used oils were discussed and a table was presented on some of the oils that were suitable for regeneration, and some that were not. The latter category included oils containing PCBs, brake fluids, metal working oils, and for reasons not further discussed, synthetic oils.

9.2.32 Improving Used Oil Recycling in California

This report was prepared by the Lawrence Livermore National Laboratory on a contract from the California Integrated Waste Management Board (CIWMB 2008). It started as a project to examine the possibility of blending used oils into the fresh crude feed to refineries, but was modified to look at the "highest and best use of used oils," which was to return them to use as a base oil stock. The objectives of the study were then:

- To increase the amount of used oil recycled to base oils
- To review the used oil market in terms of volumes sold, collected, treated, and resold
- To review various oil recycling processes and evaluate them
- To identify possible changes to incentive programs and provide recommendations for changes in policy that might be required

The sales of lubricating and industrial oils in California were approximately 150 million US gallons (568 ML) and 130 million US gallons (492 ML), respectively, in 2006. The corresponding volumes of used oil collected were approximately 85 million US gallons (322 ML) and 30 million US gallons (114 ML). California legislation, together with one other state in the United States, categorized used oil as a hazardous substance and this required state-certified recycling facilities with additional permits, reporting requirements and inspections. It was recommended that the two types of oils be kept separate in the collection process as the treatments required differed due to the different contaminants they contained. Many of the collection trucks had separate tanks for the two oils.

There were three used oil recycling facilities in California, each producing mainly one of the possible products of recycled fuel oil or reprocessed fuel oil, marine distillate oil or MDO, and re-refined base oils, the last known as closed-loop recycling. The Evergreen Oil Inc. plant at Newark rejected approximately 5% to 7% of the incoming oil as unsuitable for re-refining due mainly to high silicon and phosphorus levels, and this rejected oil was sent to a fuel blending facility. The operator had the option of selling the oil after the distillation stage as an MDO product, or sending it to the hydrotreating step to produce a base oil product. The 150N and 250N base oils had been API-certified and contained less than 100 ppm of sulfur. As stated in Chapter 7, in designations such as 150N and 250N, the higher the number, the more viscous the oil and the "N" refers to the word *neutral*, and is a throwback to the early days of refining when residual acidity from the process had to be neutralized.

Several previous reports were reviewed (Boughton and Horvath 2004; Fehrenbach 2005; USDOE 2006) and the API report, which included the comment that re-refining used oil to base oils required 50% to 85% less energy than the processes to refine base oils from virgin crude oil. The DOE 2006 statistics for the United States showed that, of the lubricating oils sold, 42% were burned in use, 40% were recovered, and 18% were improperly disposed of. Of the collected oils, 82% were combusted, and 18% were re-refined. In 2006, 40% of the oils sold in California were burned, spilled, or lost in use, 43% were collected, and 17% were unaccounted for. Of the collected oils, 48% went to fuel oil production, 24.4% to asphalt use, 10.1% to base oil production, and the rest went to waste disposal. In addition, a significant amount of used oil, or 27.7 million US gallons (105 ML), were shipped out of state to other processors.

The recycling of used oil to be blended with fresh virgin crude feed to a refinery was still seen as a possible method of treating the used oil. The problem with this method was usually the high zinc content of the used oil, which poisoned catalysts in the refining process. The process to remove the zinc before blending with the fresh feed oil were not inexpensive, and the regulatory hurdles governing refining operations in California had discouraged some processors from following this route.

The payment of incentives, introduced in California in 1992, involved the payment of $0.16 per US gal. (3.785 L) by the Certified Collection Center (CCC) for used oil brought to the CCC, and this was funded by a fee of the same amount on oil sold in California. It was felt that this incentive, which had been unchanged since 1992, was too small to be effective.

The mechanisms to increase closed-loop recycling were seen as

- Increasing the incentives paid for recycling by increasing the fee on sales of oil.
- Increasing the collection efforts for used oil, the curbside collection process having been successful where it was used.
- Encouraging the hauling of the used oil from the CCC to the California re-refinery by paying incentives to prevent the oil from being shipped out of state for a higher sales price.
- Increasing the market demand for used oil products by stressing their "green" or environmentally advantageous nature. Mandatory recycled oil content of the oils sold was another option, but influencing market forces

to achieve the desired outcome was viewed as a method preferable to legislated actions. Despite federal and state requirements to use recycled oil products in their vehicles where practicable, this was apparently not being done in many cases, and there did not seem to be any enforcement of this requirement.

- The Italian practice of payments to recyclers had been shown to work well, and the Australian system of tiered monetary incentives was also effective. The Australian product stewardship for oil (PSO) program paid \$0.50/L (0.264 US gallons) for re-refining processes, \$0.03 to \$0.07/L (0.264 US gallons) for diesel products, and nothing for recycled fuel oil. Another option would be to charge a lower sales fee on oil that contained recycled used oil.
- Increase the re-refining capacity in California. The process to obtain permits in California needed to be simplified as the permit process for the expansion of the Evergreen plant had taken 7 years to achieve.
- Subsidize the API-certified base oils produced from used oils. A range of \$0.08/gal. to \$0.20/gal. (3.785 L) had been discussed with stakeholders.

The recommendations that were then made followed the measures outlined above. They included the proposal to increase the fee on lubricating oil sales to pay for these changes, the increased fee having two components: one to cover the cost of additional collection efforts, and the other to cover the increased recycling component and incentive payments.

The Appendix to this report contained sections on the regulatory history, used oil volumes, curbside collection data, a stakeholder discussion session, and the responses to the stakeholders' comments on a number of issues related to the used oil recycling programs.

9.2.33 Processing Possibilities for Used Oil: Hungarian Summary Part II

This is the second part of a two-part review of used oil recycling from the Hungarian MOL Group (Nagy et al. 2008). The four main re-refining processes were identified as

1. Clay plus acid or solvent refining
2. Clay plus vacuum distillation
3. Vacuum distillation plus solvent refining
4. Vacuum distillation plus hydrogenation

Several of the processes were then described, including the Vaxon process operating in Germany, Denmark, Spain, and Saudi Arabia, and a table for the maximum concentration of a number of components in the feed to this process was presented. This included the maximum PCB content of 1 mg/kg, maximum sulfur content of 1.3%, maximum chloride of 1000 mg/kg, and maximum metals content of 5000 mg/kg, the yield of base oils being approximately 70%.

The Hylube or UOP DCH process was included, with its use of a guard reactor before the hydrogenation step, the Group II product base oil, and an indicated yield of base oil of greater than 90%.

The Revivoil process from the Axens/Viscolube development was in use at the Pieve Fissiraga plant in Italy, Jedlicze in Poland, Surabaya and Merak in Indonesia, Huelva and Cartagena in Spain, Hellas in Greece, and in Pakistan and Serbia.

The Dominion Oil Co. process in New Zealand and the Probex process were briefly described. The Dunwell process built by Lubrico in 1990 and by the Estate Company in Hong Kong in 1992 both had a capacity of approximately 50 tonnes (110,250 lbs.) per day, but had no hydrogenation facility. The Fetherstonhaug process from a 2002 patent involved propane extraction, followed by atmospheric and vacuum distillation and then hydrogenation. No information on the economics of these processes was provided.

The reclamation process generally involved settling, filtration or centrifugation to remove solid particles and some impurities from the used oil, to provide a heating or fuel oil for use in cement kilns, asphalt blending plants, other metalworks, coking plants and power plants. To reduce atmospheric emissions, these plants needed exhaust gas treatment, the used oil could be blended with new fuel oil to reduce the undesirable components to a level that complied with requirements, or the used oil could be pretreated in some manner. The focus at that time was on the combined reclamation of used oil and plastics.

9.2.34 Re-Refining in Europe

This report was presented by the President of GEIR at the 4th European Re-refining Congress in Brussels on November 17, 2009 (Dalla Giovanna 2009). It contained a brief overview of the GEIR with its 17 members, covering 90% of the European re-refining capacity. A table of the 27 European countries contained data for each of the lube oils consumed, collectible, and collected in 2006. For example, Germany consumed 1,174,000 tonnes (2589 Mlbs.) of lubricating oil and collected 525,000 tonnes (1158 Mlbs.) for recycling. The corresponding numbers for the second largest consumer, the United Kingdom, were 800,000 tonnes (1764 Mlbs.) sold and 350,000 tonnes (771.8 Mlbs.) collected. The total lubricating oil volume sold in these 27 countries in that year was 5.7 million tonnes (12,569 Mlbs.), 2.7 million tonnes (5954 Mlbs.) being collectible and 2.0 million tonnes (4410 Mlbs.) actually being collected. Of the 2 million tonnes (4410 Mlbs.) collected, 0.9 million tonnes (1985 Mlbs.) was burned, replacing heavy fuel oil, 0.2 million tonnes (441 Mlbs.) was burned, replacing coal, 0.7 million tonnes (1544 Mlbs.) was re-refined to produce base oils, and 0.2 million tonnes (441 Mlbs.) went to other or unknown uses.

A map and table detailed all the re-refining plants in Europe with their capacities. Of these, the following were the only ones having a re-refining process that included hydrogenation or solvent extraction, and hence, producing the higher quality base oils:

- The Dansk Olie Genbrug (DOG) plant in Denmark
- The MRD plant in Germany
- The two Viscolube Italiana plants in Italy
- The L&T Recoil plant in Finland
- The Jedlicze plant in Poland

- The Lubrica plant in Bulgaria
- The Cyclon plant in Greece

The new Waste Framework Directive 98/2008/EC had been issued in December 2008 and was expected to be passed into law by the member states of the EU. Recycling did not include the use of the products as fuels, and a five-step hierarchy was defined that passed downward from waste avoidance, through recycling to original use, to energy recovery or burning as a fuel, to disposal. Re-refining was therefore to be preferred over conversion of the used oil to a fuel, and member states were expected to promote the separate collection of the used oils, use economic instruments to promote the recycling operations, and to restrict transboundary shipments of the used oil to incineration facilities to give priority to regeneration activities.

The final section of the presentation stressed the environmental advantages of re-refining, that less energy was used to produce the base oils than was used to produce base oils from virgin crude oil, and referred again to the IFEU report (Fehrenbach 2005).

9.2.35 Solvent Extraction Process Study in Pakistan

This study covered the extraction of used oil with various solvents to precipitate a sludge containing most of the heavy metals and impurities, followed by adsorption on a packed column with an eluting solvent to improve the color of the oil (Kamal and Khan 2009). The used oil to be tested was allowed to settle for several days before being separated from the settled matter before being mixed with the solvent in a ratio of three solvent volumes to one oil volume. The sludge that formed was allowed to settle over 24 h at a temperature of 35°C (95°F), and the oil was then separated again from the sludge. The solvent was recovered from the oil by distillation.

Nine different solvents were examined for their efficacy in forming a sludge and it was found that 1-butanol was the best, followed by methyl ethyl ketone (MEK), then 1-hexanol and then 2-butanol. The *n*-heptane, *n*-hexane, methyl isobutyl ketone (MIBK), and benzene were found to be ineffective in promoting sludge formation. The other properties examined were total base number, total acid number, carbon residue, viscosities at 40°C and 100°C (104°F and 212°F), color, and flash point. As MEK was a cheap solvent with a low boiling point, making recovery easy, and its results were not much lower than the 1-butanol case, MEK was chosen for further study.

The study of the effect of settling time, settling temperature, and solvent–oil ratio identified the best settling temperature to be around 33°C (91.4°F) with 24 h settling and a ratio of solvent to oil of 3.8:1.

Decolorization of the oil was then studied with elution through a packed column, the packing being alumina, silica gel, or a local magnesite rock that was activated by heating to 500°C (932°F). Various eluting solvents were tried, the *n*-hexane, benzene, and MEK all giving satisfactory results. As *n*-hexane was cheap, it was chosen for further work. Four particle size ranges were tested for the packing and it was found that the best results were obtained with a 100- to 230-mesh magnesite packing. The magnesite could be regenerated by heating to 500°C (932°F) again, driving off the adsorbed materials. Although this procedure resulted in the use of two different solvents for the overall process, MEK and *n*-hexane, the *n*-hexane solvent elutant gave a color of 2 whereas elution with the MEK solvent gave a color of 4.

As the yield of recovered oil was approximately 94%, it was claimed that this was a better result than a distillation/clay process, which had a recovery factor of 70% to 80%, or the Mohawk process with an 82% recovery. The cheap solvents and adsorbent should make for an economic process, but further analysis would be required to properly cost an industrial process on the basis of these results. Additional chemical analysis of the recovered oil would also be required to be able to compare this process with alternative processes.

9.2.36 Policy and Technical Considerations for a Potential Re-Refining Plant in Western Australia

After a disruption to the collection and processing of used oils in Western Australia in 2006/2007, the Waste Authority in the State of Western Australia commissioned a two-part study to ensure stability in the collection and processing of these oils: this report represents Part B of the study for this authority on policy management options aimed at achieving this end, and was carried out by Oakdene Hollins Ltd. (Fitzsimons et al. 2009).

The report includes three case studies in jurisdictions that had similarities to the Western Australia situation, in each of which a levy and subsidy system was operating, similar to the Australian Federal PSO program started in the year 2000. These case studies included the Canadian Province of Alberta, in which the relatively small population and large geographic area was similar, and the European countries of Denmark and Finland, in both of which new re-refineries had been or were being built, and where Denmark had recently changed from directing most of the used oil to fuel use to the new priority of re-refining. The situation in Spain, where changes had recently been made in the regulations for used oil recycling, was also reviewed. In Western Australia, there were two used oil collectors but no re-refining operations, the oil either being burned or exported out of state.

The case study of Alberta noted that there was no central Canadian government control of used oil policies as this was handled by each province under local regulations. The Alberta Used Oil Management Association was set up in 1993 as a "delegated administrative organization" and nonprofit entity to carry out business on behalf of the Alberta Government. Funds were raised by a levy of $0.05/L (0.264 US gallons) on sales of lubricating oils, and 92% of these funds were paid to collectors of used oil. To compensate for the large transportation distances for the collected oil over a large geographic area, six collection zones had been established and the return incentive payment to the collectors varied according to the zone between $0.06/L and $0.16/L (0.264 US gallons). This system had worked well and had been adopted by four other provinces in Canada: British Columbia, Saskatchewan, Manitoba, and Quebec. The Alberta program did not favor any end use for the collected used oil, which in 2007 totaled 91 ML (24 million US gallons) and the disposition of this oil was:

- 45% to fuel for asphalt plants
- 22% to reprocessing to other lubricant products, which included fracturing fluids for the oil industry but not lubricants
- 16% recycling to base oil stock. With no re-refining plant in Alberta, the nearest re-refiner was located in Vancouver, British Columbia, more than 1000 km from Alberta
- 12% to fuel for large industrial burners
- 5% to small space heaters

Of these amounts, 44.9 ML (11.9 million US gallons) per year were used in Alberta in 2007 and 46.2 ML (12.2 million US gallons) were sent out of the Province, including 14.6 ML (3.85 million US gallons) that went to base oil production. This ability to ship the collected used oil out of the Province was seen as a desirable option in these circumstances.

The case study of the situation in Denmark noted that there were two collection systems operating in Denmark: the Mineral Oil Branch, a private sector body that provided incentives for the collection of used lubricating oils, and a market-based scheme for the lower quality oils not suitable for re-refining. Before 2002, most of the collected used oil had been burned, but a recent change in legislation had prioritized regeneration over combustion to accord with the EU directives. A levy of 0.5 DKK (in 2009, ~US$0.10) per liter (0.264 US gallons) was charged on the sales of lubricating oils suitable for recycling, which in 2007 in Denmark were 68.2 ML (18 million US gallons). Collection of used oil was free, and the levy was used to fund a payment of 1.5 DKK (~US$0.30) per liter (0.264 US gallons) for re-refined oil sold as lubricating oil. The two collectors were, to some extent, protected from competition by newcomers as any collector had to commit to collecting, free of charge and anywhere in Denmark, any used oil volume greater than 200 L (52.8 US gallons) that is suitable for re-refining: this as seen as discouraging new, smaller collectors.

The only re-refiner in Denmark, the DOG refinery at Kalundborg, had been acquired by Avista Oil, part of the Mustad Group of Houston, in 2003 but had been shut down in 2005 and the used oil collected had been sent to the MRD plant in Germany after that, the MRD plant also having been acquired by Avista Oil and converted to their Vaxon process. The DOG plant, after being shut down in 2005, had been modernized and reopened in 2008. Prior to 2008, the DOG plant had produced only industrial burner fuels, but now produced higher quality base oils, annual capacities estimated to be 24 ML (6.34 million US gallons) of base oils, 5.5 ML (1.45 million US gallons) of fuel oils, 2.5 ML (0.66 million US gallons) of heavy fuel oils, and 5 ML (1.32 million US gallons) of asphalt extender. Lower quality oils and bottoms from the distillation process at Kalundborg had been incinerated as hazardous waste. The regulations in Denmark had been revised to favor re-refining and included stricter emission standards for power plants and higher taxes on fuel oil.

Sales of lubricating oils in Denmark had been decreasing significantly: in 1986, sales were at 96 ML (25.3 million US gallons), and in 2007, they were at 68 ML (18.0 million US gallons). In 2007, the base oil production was at 13.5 ML (3.56 million US gallons) together with 2 ML (0.53 million US gallons) of heating oil from the 18.3 ML (4.82 million US gallons) of used oil collected, including water and solid contaminants. The cost of operating the Danish system was judged to be high, approximately twice the cost per liter (0.264 US gallons) of the Alberta system. The focus of the levy and subsidy system was on re-refinable oils, leaving the more problematic lower quality oils to the market-based system.

The case study of Finland described a levy system that had been introduced in 2006 at €0.0575/L (approximately US$0.90 per 0.264 US gallons) on new lubricant sales that was used to subsidize the collection and storage of used oils, but also to clean up spills of oil in the environment. Re-refining was prioritized over energy recovery use. Collection of the oil was carried out by one company to which payments were made on a "cost plus" basis, the collection being free to the generator for volumes greater than 400 L (105.6 US gallons) and a fee being charged for smaller volumes. The public could dispose of their oil at collection centers at no charge. The company was being sheltered by a 5-year contract for used oil supply and the banning of the burning of used oil in plants of less than 5 MW capacity.

A new re-refinery at the port of Hamina was scheduled to start production in the Spring of 2009, with a capacity of 60,000 tonnes (132.3 Mlbs.) per year of used oil, using a technology said to be similar to that of the Puralube plant in Germany. It was projected that about 50% of the feedstock for this plant would have to be imported, possibly from Poland or Sweden. In 2007, 77.86 ML (20.6 million US gallons) of lubricating oils had been sold in Finland, of which 31.93 ML (8.43 million US gallons) had been motor oils, and between 50 and 55 ML (13.2 and 14.5 million US gallons) were estimated to be collectible with 30 ML (7.9 million US gallons) being suitable for re-refining. Sales of lubricating oils had also been decreasing between 2000 and 2007. A levy of €0.05/L (0.264 US gallons) was charged on most new lubricating oil sales.

The situation in Spain had recently changed with the formation of the Sistema Integrado de Gestion de Aceites Usados (SIGAUS) in June, 2006, with a target of re-refining 55% of collected used oils in 2007 and 65% in 2008. With 2007 sales of lubricating oils of 520,000 tonnes (1147 Mlbs.), subsidies were being paid in October, 2008, for collection at €0.0268/L (0.264 US gallons), for re-refining at €0.09/L, and for burner fuel or bitumen products at €0.024/L (0.264 US gallons). The volume collected, mostly without charge to the generator, was 214,000 tonnes (471.9 Mlbs.) with 120,000 tonnes (264.6 Mlbs.) being re-refined; some of this was additional used oil being obtained from Portugal.

In 2008, the four re-refineries in Spain had the following annual capacities:

1. The Cator plant: 32,000 tonnes (70.6 Mlbs.)
2. The Ecolube plant: 27,000 tonnes (59.5 Mlbs.)
3. The Aurecan plant: 18,000 tonnes (39.7 Mlbs.)
4. The Aurema plant: 18,000 tonnes (39.7 Mlbs.)

Three scenarios were then discussed for policy management in the Western Australia case, based on the case studies presented. These were:

A. The baseline "do-nothing" case. Various assumptions were made regarding changes to be expected between 2008 and 2011, including volumes and prices for different products, and the effect of the collection fee that could be promoting more improper disposal of used oils. In Australia, as a whole under the PSO program, 94% of the recycled oil was being used for fuel oil purposes and 6% was being re-refined for lubricant use, but this included hydraulic and transformer oils. In almost all jurisdictions, the potential demand for burner fuel oil far exceeded the potential supply from used oil sources. Although there was some consumer reluctance to use fuel oils from used oil sources, lower prices for the used oil product made it a risk competitor for re-refiners. As the chlorine content of the used oil burner fuel was the main concern, the chlorine being converted to toxic dioxins in low-temperature combustion processes at less than 800°C (1472°F), regulatory control of burner fuels was an option.

B. The two collector companies operating in Western Australia had shown some interest in collaborating to build a local re-refining plant with a capacity of approximately 30,000 tonnes (66.2 Mlbs.) per year, and agreeing on a collection price for used oil. The cost of such a plant was estimated to be around AUD20 million (US$19 million), but depended on the technology chosen: if a plant designed to produce higher quality base oils were built, the cost could be higher but the future risk of being displaced from the market would be lower than the case in which a lower cost plant were built to produce lower quality products whose marketability was decreasing in many parts of the world. In addition, storage facilities at a port would have to be expanded for exporting used oil or the bottoms from the distillation process.

   The costs and benefits to the stakeholders were outlined, including to generators, collectors and processors, burner fuel users, and the environment. Assuming re-refining was a priority, there was a need to protect the re-refining plant from burner fuel competition, possibly with stricter atmospheric emission regulations. A government grant, linked to progression milestones, could be considered. The heavy distillation bottoms product was a risk factor and a study of its use in roofing or road asphalt, or export, could be promoted. Operating and other costs for a Sener Ecolube type plant were also presented.

C. The establishment of an out-of-state re-refining plant was a third scenario considered, a third company considering a plant with a capacity of 100,000 tonnes (220.5 Mlbs.) per year in the State of New South Wales. This plant could be partly supplied by used oil shipments from Western Australia, despite the large transportation distance involved, as shown by the Alberta case study. Assuming a market share of 25% for this new re-refinery, the costs and benefits for the stakeholders were again presented, together with overall operating estimates.

Six policy options were then outlined, including setting up a timetable to introduce new atmospheric controls for plants using waste-derived fuels, offering grants for re-refining plant and storage tank construction, instigating research into using distillation bottoms in asphalt products, and investigating the introduction of more onerous license conditions for used oil plant operators to discourage new, small competitors and provide funds to pay for the disposal of accumulated used oil or distillation bottoms in the case of the first scenario.

A sensitivity analysis was based on two key indicators: the amount of used oil self-managed by generators, and the risk of a repeat of the 2006/2007 disruption to the collection infrastructure as measured by the amount of used oil being shipped out to non-mainland burner fuel markets. Various assumptions affected the outcome of this analysis, including the market share attained by the out-of-state re-refinery in Scenario C, the growth in used oil supply from the mining sector, mainland demand for burner fuel, and the blending ratio needed for shipment of products.

The overall conclusions were as follows:

- The ability to export used oil was critical for the stability of the system in both Alberta and Denmark during periods of change
- Reducing the risk of a repeated disruption to the used oil-handling system involved an investment in re-refining
- The question arose regarding the necessity of further intervention. If the possible joint venture re-refinery of Scenario B were to go ahead, a proper business plan was needed for this. In the meantime, several policy options were available to promote progress toward either of the scenarios that involved investment in a new lube-to-lube refinery and new storage capacity for seaborne trade in used oils and products

This study again showed that there is no single solution for the re-refining of used oil that can be universally applied. A careful analysis of local conditions is supremely important to ensure that a re-refining plant has an adequate supply of suitable feedstock for a reasonable period, and that the economics for the plant will endure and provide an adequate return on investment. With the instability in oil prices in the world market, predicting future prices is a difficult process, and a risk analysis and sensitivity analysis to changes in the assumptions made provides valuable information in assessing the viability of a re-refining operation.

9.2.37 Used Oil Recycling in Taiwan

In 2004, sales of lubricating oil in Taiwan totaled 400 ML (105.6 million US gallons), and in 2005, it was at 450 ML (118.8 million US gallons). It was also estimated that vehicle lubricants accounted for 59.7% of these totals, and industrial lubricants made up 40.3% (Hsu et al. 2009). This report quotes figures for the generation of used oil from vehicles (90.6%) and from industrial lubricants (55.6%), making the volumes of used oil generated from each of these sources 240 ML (63.4 million US gallons) and 100 ML (26.4 million US gallons), respectively, in 2005. The figure 90.6% seems to be unusually high, particularly because no account seems to have been taken of the oil lost in use, this being approximately 50% in other studies. With the 14 ML (3.7 million US gallons) apparently recovered out of the total 340 ML (89.8 million US gallons) generated, the recovery rate seems to be approximately 4%. The main use for the recovered oil was 94% going to secondary oil and fuel with the remainder causing environmental pollution, presumably by uncontrolled dumping. Thus, there was much room for improvement in the management of used oils in Taiwan, and comparisons were made with recovery factors in European countries.

The various re-refining technologies were then compared with a number of criteria, including economic performance, environmental effects, and operational conditions. In conclusion, the authors indicated the need for improvement in the regulation and auditing of the used oil situation, and the need for planning and management of the recovery and sustainable utilization of the used lubricants.

9.2.38 GEIR Used Oil Statistics in Europe in 2006

This review of used oil re-refining capacity in the EU was based on data from 2006, which had been updated to 2010 (GEIR 2010). Table 9.10 summarizes these data.

9.2.39 Screening Process for a Re-Refining Plant in Turkey

The Turkish Petroleum Industry Association (PETDER) carried out this study for the Istanbul Development Agency to review the existing legislation and re-refining technologies, together with the economic and social requirements with a view to establishing a re-refining plant in Turkey. This information was being provided to local and international investors for a potential project that would align with the EU's Waste Directive 2008/98/EC (EU-WD 2008), which set up a hierarchy for waste control as previously seen. The directive, however, also allowed for life cycle considerations to be used to justify a deviation from the hierarchy, which prioritized re-refining over burning of the used oil for energy recovery. Although these LCAs were not well-known in Turkey, a study of this type was recommended to be a consideration (PETDER 2012).

It was estimated that between 400,000 and 450,000 tonnes (882 and 992 Mlbs.) of mineral oils, taken to tbe lubricating oils, were consumed annually in Turkey in 2011, generating between 200,000 and 250,000 tonnes (441 and 551 Mlbs.) of used oil. Of this, 45,000 to 50,000 tonnes (99.2–110.2 Mlbs.) were collected with the remainder going to use as fuels, blending with other fuel oils, heating purposes, or being dumped. PETDER itself collected 20,576 tonnes (45.4 Mlbs.) of used motor oil in 2011 with 9% of this going to licensed refining and regeneration facilities, 83% going to energy recovery at cement, limestone, iron, and steel factories, and 8% going to disposal facilities as hazardous waste. The refining and regeneration facilities were low-technology and the use of the used oil as a fuel was based to a large extent on illegal sales as a fuel oil.

The used oil practices in other EU countries were reviewed in terms of the fractions going to energy recovery or to re-refining facilities, and the following tracking records for waste oil were presented:

Waste oil accounted for	EU—74%	US—69%	Turkey—17%
Waste oil unaccounted for	EU—26%	US—31%	Turkey—83%

The recovery methods included burning as a fuel, and some LCAs in other countries had been inconclusive in comparing burning or re-refining as being preferable, particularly burning in high-temperature incineration plants. The high calorific value of the waste oil of 9600 kCal/kg (17,276 BTU/lbs.), the same as No. 6 fuel oil,

TABLE 9.10
GEIR Statistics on European Re-Refining Capacities

Year	Country	Company	Location	Total Used Oil Capacity (tonnes/year)	Used Oil Capacity for Base Oil Production (tonnes/year)	Products
2007	Belgium	WOS Hautrage S.A.	Hautrage	40,000	0	Gas oils, fuels
2010	Denmark	Dansk Olie Genbrug A/S	Kalundborg	40,000	40,000	Base oil + by-products
2010	Finland	L&T Recoil	Hamina	60,000	42,000	Base oil + by-products
2007	France	Compagnie Française Ecohuile	Lillebonne	125,000	125,000	Base oil + by-products
2010	Germany	Puralube GmbH	Tröglitz/Zeitz	160,000	160,000	Base oil + by-products
2010	Germany	Mineralöl-Raffinerie Dollbergen GmbH	Dollbergen	170,000	120,000	Base oil + by-products, fuels
2010	Germany	Baufeld Mineralölraffinerie Duisburg GmbH & Co. KG	Duisburg	100,000	0	Fuels
2010	Germany	Baufeld Mineralölraffinerie GmbH	Chemnitz	50,000	0	Fuels
2010	Germany	Südöl Mineralölraffinerie GmbH	Eislingen/Fils	60,000	15,000	Base oil + by-products, fuels
2010	Germany	KS Recycling GmbH & Co. KG	Sonsbeck	80,000	30,000	Base oil + by-products, fuels
2010	Germany	Horst Fuhse Mineralölraffinerie GmbH	Hamburg	100,000	0	Fuels
2010	Germany	Starke & Sohn	Niebüll	20,000	20,000	Only transformer oil
2010	Germany	Graue GmbH	Bremerhaven	25,000	0	Fuels
2010	Germany	Petrol Plus	Mannheim	25,000	0	Fuels
2010	Germany	Trafolube GmbH	Duisburg	8000	8000	Only transformer oil
2010	Greece	Cyclon Hellas S.A.	Attika	40,000	40,000	Base oil + by-products
2010	Greece	Denver S.A.	Viotia	3500	3500	Base oil + by-products

2010	Greece	Maviol	Salonica	1800	1800	Base oil + by-products
2010	Greece	Achaia Lubricants	Achaia	2500	2500	Base oil + by-products
2010	Greece	Simitzoglou	Salonica	1800	1800	Base oil + by-products
2010	Greece	Veko	Evia	1200	1200	Base oil + by-products
2010	Greece	Skamagoulis	Volos	600	600	Base oil + by-products
2010	Italy	Viscolube	Pieve Fissiraga (Lodi)	130,000	130,000	Base oil + by-products
2010	Italy	Viscolube	Ceccano (Frosinone)	84,000	84,000	Base oil + by-products
2010	Italy	Ramoil	Casalnuovo (Naples)	35,000	35,000	Base oil + by-products
2010	Italy	Siro	Corbetta (Milan)	9000	9000	Base oil + by-products
2010	Italy	PB Oil (ex Distom)	Porto Torres (Sassari)	20,000	0	Fuels
2010	Italy	Siral	Nola (Naples)	25,000	0	Fuels
2010	Netherlands	North Refinery	Farmsum	170,000	80,000	Base oils + by-products, fuels
2009	Poland	Rafineria Nafty Jedlicze	Jedlicze	80,000	33,700	Base oil + by-products
2009	Poland	Variant SA	Trzebinia	8000	6000	Base oil + by-products
2010	Spain	Tracemar	Fuenlabrada (Madrid)	36,000	36,000	Base oil + by-products
2010	Spain	Cator	Barcelona–Taragona	42,000	42,000	Base oil + by-products
2010	Spain	Tracemar–Alfaro	Alfaro	59,000	38,000	Base oil + by-products
2010	Spain	Tracemar–Huelva	Huelva	59,000	34,000	Base oil + by-products
2010	Spain	Tracemar–Murcia	Murcia	59,000	34,000	Base oil + by-products
2009	United Kingdom	Whelan Refining	Stoke-on-Trent	50,000	50,000	Base oil + by-products
			Total	1,980,400	1,223,100	

Source: Reproduced with permission from GEIR, Used Oil Capacity of European Industries, 6, 2010. http://www.geir-rerefining.org/GEIR_statistics.php (accessed on January 24, 2013).

together with the low capital costs for a relatively simple used oil treatment plant, had made this option attractive in some instances.

The available re-refining technologies were then reviewed, together with an indication of where they had been installed in other countries. In summarizing the processes, they were divided into three groups:

1. The acid–clay process
2. The processes including hydroprocessing
3. The processes including solvent extraction and no hydroprocessing

The first group, the acid–clay process, was considered to be old technology with waste disposal problems and environmental drawbacks, and was not recommended.

Generally, processes with hydroprocessing (1) needed expensive catalysts but produced Group II or II+ base oils and were not sensitive to feedstock quality, although using low-quality feedstocks could shorten the catalyst life; (2) eliminated PCBs and chlorides, as well as PNAs if high temperatures and pressures were used; (3) had higher capital and operating costs, and (4) needed a source of hydrogen or a hydrogen generation plant.

On the other hand, the processes with solvent extraction (1) had lower capital costs and could make Group II or II+ base oils if the feed oil quality was good, but were sensitive to variations in the feed properties, (2) eliminated PAH and PNA components completely, (3) preserved all synthetic oils in the feed, including the valuable PAOs, (4) operated at lower pressures and temperatures, and had a higher product yield, (5) generated smaller quantities of waste, leading to lower disposal costs, but (6) solvent costs could be high.

The fifteen recycling plants operating in Turkey were based on distillation, clay treatment, and filtration, fourteen of them operating in a batch mode, and the fifteenth scheduled to start continuous operations in the second quarter of 2012: none produced a base oil product. There were questions regarding waste disposal from these plants, the clay consumption and waste data were not clear, testing procedures for the oil and products were problematic, and the products were sold under various names, making after-sales follow-up difficult.

In compiling some recommendations for the implementation of a re-refining operation, it was noted that environmental permits were issued for a 5-year term and this would lead to uncertainties for potential investors in a re-refining plant. The main variable in determining the size of a plant was the potential feedstock volume available, which could be collected at a cost that makes the project economically viable. Three potential locations for a plant in Turkey were considered and collection costs and volumes were evaluated for each. Considering the volumes of mineral or lubricating oil sold, the likely collectible volume, and the growth over several years before the plant would become operational, a size of 80,000 tonnes (176.4 Mlbs.) per year for a plant in the Istanbul area was assumed for further evaluation.

The choice of technology needed further study, a life cycle evaluation being recommended, but more information was available to the authors on the processes using solvent extraction. Of the eleven plants in Europe examined from the *Lubes 'N' Greases* (LNG) review in 2011, eight used solvent extraction and three used hydroprocessing as shown in Table 9.11, and these produced 65% Group I base oils, and

35% Group II oils (LNG 2011). Further analysis was made on the basis of an 80,000 tonnes (176.4 Mlbs.) per year plant using solvent extraction techniques of the Avista Oil type, but it was stressed that this should not be interpreted as ruling out a hydroprocessing type plant.

Assumptions were then made with regard to the product slate, including 56% base oil yield and 15% asphalt products, the manpower needed to operate the plant and other cost items, and a 10% to 15% price discount for the products compared with current market prices for comparable products. The license fee for the technology was not included, however. It was concluded that the plant capital cost, including the required land, would be equivalent to approximately US$53 million, and the annual operating pretax profit would be approximately US$37.8 million, making the undertaking attractive to potential investors. Some data was also available to the authors from the Green Oil Co. plant in Greece, which was based on a process using hydroprocessing and had a capacity of 30,000 tonnes (66.2 Mlbs.) per year. The plant

TABLE 9.11
Some EU Re-Refinery Capacities (in tonnes per year)

	Re-Refining Plants			Capacity			
				Base Oils			
	Company	Location	Country	Group I	Group II	Group III	Total
	Processes using Solvent Extraction			**581,000**	**0**	**0**	**581,000**
1	EcoHuile	Lillebonne	France	125,000			125,000
2	Mineralölraffinerie Dollbergen 2	Dollbergen	Germany	120,000			120,000
3	Viscolube	Ceccano	Italy	84,000			84,000
4	PKN Orlen Group	Jedlicze	Poland	80,000			80,000
5	Whelan Refining	Stoke-on-Trent	UK	50,000			50,000
6	Cator	Tarragona	Spain	42,000			42,000
7	Cyclon Hellas	Attika	Greece	40,000			40,000
8	Danske Olie Genbrug	Kalundborg	Denmark	40,000			40,000
	Processes using Hydroprocessing			**20,000**	**286,000**	**0**	**306,000**
9	Puralube	Tröglitz	Germany		176,000		176,000
10	Viscolube	Pieve Fissiraga	Italy	20,000	80,000		100,000
11	Green Oil	Alexandropolis	Greece		30,000		30,000
		Total		**601,000**	**286,000**		**887,000**

Source: Annual survey of base oil refining capacity in 2012. *Lubes 'N' Greases*. LNG Publishing Company Inc., Falls Church, VA. http://www.lngpublishing.com/lngmagazine/index.cfm; PETDER, Istanbul, Turkey, Selection of the Most Appropriate Technology for Waste Mineral Oil Refining Project: Technical Research Report 2012. Turkish Petroleum Industry Association (PETDER) Contract Report, 65, 2012. http://www.petder.org.tr/admin/my_documents/my_files/2B7_ISTKATEKNIKRAPORU2012Eng.pdf (accessed on January 25, 2013). Reproduced with permission.

TABLE 9.12
Re-Refinery Capacities in 2012 (in tonnes per year)

	Re-Refining Plants			Capacity			
				Base Oils			
	Company	**Location**	**Country**	**Group I**	**Group II**	**Group III**	**Total**
		Europe					
1	Cator	Tarragona	Spain	41,600			41,600
2	Cyclon Hellas	Attika	Greece	41,600			41,600
3	Danske Olie Genbrug	Kahlundborg	Denmark	41,600			41,600
4	Eco Huile	Lillebonne	France	125,000			125,000
5	L&T Recoil Oy	Hamina	Finland		62,400		62,400
6	Mineralölraffinerie Dollbergen	Dollbergen	Germany	120,000			120,000
7	PKN Orlen Oil	Jedlicze	Poland	83,000			83,000
8	Puralube	Tröglitz	Germany		185,000		185,000
9	Viscolube	Ceccano	Italy	87,000			87,000
10	Viscolube	Pieve Fissiraga	Italy	20,000	83,000		103,000
11	Whelan	Stoke-On-Trent	England	52,000			52,000
		Total: Europe		**611,800**	**330,400**		**942,200**
		North and South America					
12	Evergreen	Newark, CA	US		60,000		60,000
13	Heartland	Columbus, OH	US		78,000		78,000
14	Heritage–Crystal Clean	Indianapolis, IN	US		104,000		104,000
15	Lwart	Lencois Paulista	Brazil		132,500		132,500
16	Safety-Kleen	Breslau, Ontario	Canada	36,400	62,400		98,800
17	Safety-Kleen	East Chicago, IL	US	41,600	218,000		259,600
		Total: North and South America		**78,000**	**654,900**		**732,900**
		Asia/Pacific/Australia					
18	Agip Lubrindo	Gempol	Indonesia	42,000			42,000
19	Wiraswasta Gemilang	Bekasi	Indonesia	52,000			52,000
		Total: Asia/Pacific/Australia		**94,000**			**94,000**
		New Plants Expected in 2013					
20	FCC Environmental	Baltimore, MD	US		104,000		104,000
21	Nexlube	Tampa, FL	US		57,000		57,000
22	UES (Avista)	Peachtree City, GA	US		62,000		62,000
		Total: New plants expected in 2013			**223,000**		**223,000**

Source: Annual survey of base oil refining capacity in 2012. *Lubes 'N' Greases*. LNG Publishing Inc. http://www.lngpublishing.com/lngmagazine/index.cfm. With permission.

Note: Only plants with hydrotreating or solvent refining processes with a capacity of at least 800 bbl/day (41,580 tonnes/year) are included.

cost had been approximately €27 million (US$35 million) and the pretax profit was estimated to be between €10 million and €11 million (US$13 million and US$14.3 million) per year.

In conclusion, the authors summarize

- The main obstacle to the establishment of a modern re-refining plant in Turkey was the illegal collection of used oil and its consumption as an illicit fuel or unqualified mineral oil in the country
- The 80,000 tonnes (176.4 Mlbs.) per year plant in the Istanbul region was economically attractive
- The two re-refining processes involving either hydroprocessing or solvent extraction needed further analysis and study before a selection was made between them

The 2012 review of base oil refining capacities around the world conducted by *Lubes 'N' Greases* magazine (LNG 2012) is summarized for the re-refining operations in Table 9.12.

9.2.40 Compendium of Recycling and Destruction Technologies for Waste Oils

This followed the 2003 compendium from the UNEP and reviewed the used oil management methods in various countries before providing an overview of the technologies themselves, followed by an outline of the Sustainability Assessment of Technologies (SAT) program developed by the International Environmental Technology Center of the UNEP (IETC/UNEP) to systematize choices and help in choosing a recycling program that would be most appropriate for any particular location. The information was aimed at national and local governments, environmental organizations, and other stakeholders, particularly in developing countries, to facilitate the choice of a recycling program for the used oil (UNEP Compendium 2012).

The used oil management procedures in several countries were reviewed, and the overall conclusion was that burning of the used oil was the most prevalent use, but re-refining was growing in application. There was a general need for better record-keeping and tracking of the fate of the used oils. There was a move toward Group II and Group III base oils, and any new plant should be cognizant of this. The consistency of the product oils was also extremely important because lubricant blenders would not easily accept a product that required constant adjustment of the blending procedure. The Indian Hazardous Waste Rules had set limits for the contaminants in used oils for recycling or reprocessing, and for fuels produced from these oils that were allowed to be burned in furnaces. Fuel oil not meeting these criteria had to be incinerated in suitable incinerators or cement kilns.

The most popular re-refining processes were then reviewed, including most of the well-known procedures, indicating that the acid–clay process was banned in many countries due to its environmental problems.

The burning of used oil was also reported, together with the U.S. EPA limits for contaminants such as cadmium, chromium, and halogen contents, and in Ohio, the

PCB content, which defined when a used oil was considered to be "on-spec" or "off-spec," the latter requiring incineration in approved facilities such as industrial furnaces or hazardous waste incinerators, which guaranteed a high enough temperature and residence time to effect the required reactions, coupled with suitable pollution control methods for the off-gases. The use as a fuel could be after mild processing such as settling, filtration, and possible blending with fresh fuel oil, or after severe processing, which included distillation and possibly hydrogenation. The thermal cracking processes were also reviewed, most of which produced a diesel oil fuel or marine diesel fuel. The disposal of hazardous wastes from the various processes often required incineration with appropriate pollution control devices or disposal in a secured landfill, and the design of these facilities was also discussed.

The choice of the best recycling program for any particular location was determined through a combination of technical, economic, environmental, and sociocultural criteria and this was facilitated with the use of the SAT program developed by the UNEP. This program was then explained in some detail, including the application to the choice of the best used oil re-refining process.

9.3 FUTURE DEVELOPMENTS

It is more than likely that increases in the level of performance of lubricating oils in engines will continue and more synthetic components such as PAOs will be used to formulate these oils in the future. This also means that the quality of the used oils will continue to increase as these make up a greater proportion of the lubricating oils. Any future re-refining operation will need to take this into account, and the question arises of whether a plant producing Group I oils will face a declining demand for its products. Although this will be true to some extent, the Group I oils will have a continuing market in other applications such as gear oils, machine oils, hydraulic fluids, and the many other uses that are not as demanding as the high-temperature environment in an internal combustion engine.

The other unresolved issue facing re-refining plant operators is the competition for feedstock with the users of used oils for fuel purposes. Although the EU, as well as some US States, the U.S. Federal Government, and other countries have given priority to the re-refining of used oils as opposed to their use as a fuel oil, many jurisdictions continue to direct most of the used oil to fuel use. The use of life cycle assessment or life cycle inventory methods provides guidance on the relative scores of the re-refining processes or fuel-burning options in the various environmental categories, but local conditions or variations can lead to arguments to override these considerations, as allowed by the EU Waste Directive 2008/98/EC (EU-WD 2008). Additional LCA studies in more areas of the world, carried out with due attention to the ISO 14040 stipulations, might provide a clearer picture of the relative benefits of re-refining and use as a fuel. Even given this clarity, there still remains the task of choosing a recycling process best suited to the local conditions and based on more than just the environmental issues, and the SAT program from the UNEP (UNEP Compendium 2012) provides a rigorous and systematized method of assessment that should be sensitive to local characteristics and provide acceptable results.

REFERENCES

Ali, M.F., Hamdan, A.J., and Rahman, F. 1995. Techno-Economic Evaluation of Waste Oil Re-Refining in Saudi Arabia, 7 p. Available at http://isi.kfupm.edu.sa/summary.aspx?sid=1586. Accessed on December 22, 2012.

Audibert, F. 2006. *Waste Engine Oils: Rerefining and Energy Recovery*. Elsevier B.V., Amsterdam, Netherlands, 323 p.

Awaja, F., and Pavel, D. 2006. *Design Aspects of Used Lubricating Oil Re-Refining*. Elsevier B.V., Amsterdam, Netherlands, 114 p.

Bachelder, D.L. 2005. Recycling Used Engine Oil by Re-Refining. *Presentation at California 2005 Used Oil/HHW Conference, San Diego, CA*, 13 p. Available at http://www.calrecycle.ca.gov/homehazwaste/Events/AnnualConf/2005/April28/Session4/DIYers/ReRefine.pdf. Accessed on December 23, 2012.

Baladincz, J., Szabó, L., Nagy, G., and Hancsók, J. 2008. Possibilities for Processing Used Lubricating Oils—Part 1, Hungarian MOL Group, 6 p. Available at http://www.mol.hu/repository/435054.pdf. Accessed on January 15, 2013.

Bamiro, O.A., and Osibanjo, O. 2004. Pilot Study of Used Oils in Nigeria. Study sponsored by the Secretariat of the Basel Convention, 62 p. Available at http://www.basel.int/Portals/4/Basel%20Convention/docs/centers/proj_activ/stp_projects/04-03.doc. Accessed on December 22, 2012.

Basel Convention Technical Guidelines (BCTG). 1995. Basel Convention Technical Guidelines on Used Oil Re-Refining or Other Re-uses of Previously Used Oil, 1995/1997, 19 p. Available at http://www.basel.int/DNNAdmin/AllNews/tabid/2290/ctl/ArticleView/mid/7518/articleID/185/Basel-Convention-Technical-Guidelines-On-Uses-Of-Previously-Used-Oil-R9.aspx. Accessed on December 20, 2012.

Boughton, B., and Horvath, A. 2004. Environmental Assessment of Used Oil Management Methods. *Environmental Science & Technology*, 38(2): 353–358. Available athttp://pubs.acs.org/doi/pdf/10.1021/es034236p. Accessed on January 14, 2013.

Bridjanian, H., and Sattarin M. 2006. *Modern Recovery Methods in Used Oil Re-Refining*. Research Institute of Petroleum Industry, Tehran, Iran, 4 p. Petroleum Coal 48(1), 40–43. Available at http://www.vurup.sk/petroleum-coal. Accessed on December 28, 2013.

California Integrated Waste Management Board (CIWMB). 2008. Improving Used Oil Recycling in California, Lawrence Livermore National Laboratory. Contract report to California Integrated Waste Management Board (CIWMB), 88 p. Available at http://www.calrecycle.ca.gov/publications/Documents/UsedOil/61008008.pdf. Accessed on January 15, 2013.

Dalla Giovanna, F. 2009. European Landscape. *Presentation at the 4th European Re-refining Congress, Brussels, November 2009,* 30 p. Available at http://www.geir-rerefining.org/geir_news.php. Accessed on January 15, 2013.

Dalla Giovanna, F., Khlebinskaia, O., Lodolo, A., and Miertus, S. 2003. Compendium of Used Oil Regeneration Technologies. Prepared for the United Nations Industrial Development Organization (UNIDO), 177 p. Available at http://institute.unido.org/documents/M8_LearningResources/ICS/95.%20Compendium%20of%20Used%20Oil%20Regeneration%20Technologies.pdf. Accessed on December 22, 2012.

Department for Environment, Food and Rural Affairs (DEFRA). 2001. UK Waste Oils Market 2001. Consulting Report for UK-DEFRA by Oakdene Hollins Ltd., 57 p. Available at http://archive.defra.gov.uk/environment/waste/topics/hazwaste/oils/pdf/wasteoils.pdf. Accessed on December 22, 2012.

European Union Waste Directive (EU-WD). 2008. European Union Waste Directive 2008/98/EC, 28 p. Available at http://eur-lex.europa.eu/LexUriServ/LexUriServ.do?uri=OJ:L:2008:312:0003:0030:EN:PDF. Accessed on January 24, 2013.

Everest. 2005. *Waste Lubricating Oil Purification and Recovery*. Everest Blower Company, India, 8 p. Available at http://www.everestblowers.com/technical-articles/Lube_Oil_Re-Refining.pdf. Accessed on December 22, 2012.

Fehrenbach, H. 2005. Ecological and Energetic Assessment of Re-refining Used Oils to Base Oils: Substitution of Primarily Produced Base Oils Including Semi-Synthetic and Synthetic Compounds. Report for GEIR by the Institut für Energie- und UmweltforschungGmbH (IFEU), Heidelberg, Germany, 104 p. Available at http://www.geir-news.com/index2.html. Accessed on December 22, 2012.

Fitzsimons, D. 2010. Improving Markets for Waste Oils, Oakdene Hollins Ltd. Consulting report to the Organization for Economic Cooperation and Development (OECD), 32 p. Available at http://www.oakdenehollins.co.uk/pdf/Used_Oil_Report_1.pdf. Accessed on January 15, 2013.

Fitzsimons, D., and Lee, P. 2003. UK Waste Oils Policy in the light of German and Italian Experience, Oakdene Hollins Ltd. Consulting report to DEFRA, 36 p. Available at http://www.oakdenehollins.co.uk/pdf/Waste_Oils_Report_2.pdf. Accessed on January 15, 2013.

Fitzsimons, D., Eatherley, D., and Rasanen, J. 2009. Analysis of Used Oil Policy Management Options, Oakdene Hollins Ltd. Consulting report to The Waste Authority, Western Australia, 90 p. Available at http://www.zerowaste.wa.gov.au/media/files/documents/analysis_used_oil_policy_management_options.pdf. Accessed on January 15, 2013.

Foo, C.Y., Rosli, M.Y., and Tea, S.S. 2002. Modeling and Simulation of Used Lubricant Oil Re-refining Process, 6 p. Available at http://kolmetz.com/pdf/Foo/WEC2002_Lube_oil_Modelling.pdf. Accessed on December 22, 2012.

Fullana, P. 2005. Catalan Waste Agency Position on the Commission Consultation on the Directive 75/439/EEC on Waste Oils: Technical Considerations, 2005, 39 p. Available at http://ec.europa.eu/environment/waste/pdf/consult/16c.pdf. Accessed on December 22, 2012.

GEIR. 2004. An Environmental Review of Waste Oils Regeneration, 5 p. Available at http://geir-rerefining.org/documents/PositionpaperGEIR161104.pdf. Accessed on December 22, 2012.

GEIR. 2006. Questionnaire on Used Oil Collection and Utilization in 2006, 6 p. Available at http://www.geir-rerefining.org/documents/WO-questionnaire-EU27-final-022208.pdf. Accessed on January 15, 2013.

GEIR. 2007. Waste Oil Regeneration: A Case Study on the Environmental and Economic Benefits to Greece, 3 p. Available at http://www.geir-rerefining.org/documents/GEIRcasestudyforGreece_2007_12_14.pdf. Accessed on January 15, 2013.

GEIR. 2010. Used Oil Capacity of European Industries, 6 p. Available at http://www.geir-rerefining.org/GEIR_statistics.php. Accessed on January 24, 2013.

Hsu, Y-L., Lee, C-H., and Kreng, V.B. 2009. Analysis and Comparison of Regenerative Technologies of Waste Lubricant, Taiwan, 15 p. Available at http://www.wseas.us/e-library/transactions/environment/2009/31-678.pdf. Accessed on January 24, 2013.

Kajdas, C. 2000. Major Pathways for Used Oil Disposal and Recycling: Part 1, September 2000, 13 p and Part 2, December 2000, 15 p. Available at http://www.tribologia.org/ptt/kaj/kaj10.htm. Also in *Tribotest Journal*, 7-7, September 2000, 61 and 7-2, December 2000, 137.

Kamal, A., and Khan, F. 2009. Effect of Extraction and Adsorption on Re-Refining of Used Lubricating Oil, 7p. Available at http://ogst.ifpenergiesnouvelles.fr/articles/ogst/pdf/2009/02/ogst08046.pdf. Accessed on January 15, 2013.

LNG. 2011. Annual survey of base oil refining capacity in 2011. *Lubes 'N' Greases*. LNG Publishing Company Inc., Falls Church, VA. Available at http://www.lngpublishing.com/lngmagazine/index.cfm.

LNG. 2012. Annual survey of base oil refining capacity in 2012. *Lubes 'N' Greases*. LNG Publishing Company Inc., Falls Church, VA. Available at http://www.lngpublishing.com/lngmagazine/index.cfm.

Nagy, G., Szabó, L., Baladincz, J., and Hancsók, J. 2010. Possibilities for Processing of Used Lubricating Oils—Part 2, Hungarian MOL Group, 8 p. Available at http://www.mol.hu/repository/610628.pdf. Accessed on January 24, 2013.

Monier, V., and Labouze, E. 2001. Critical Review of Existing Studies and Life Cycle Analysis on the Regeneration and Incineration of Waste Oils. Consulting report by Taylor Nelson Sofres Consulting, Paris, 208 p. Available at http://ec.europa.eu/environment/waste/studies/oil/waste_oil.pdf.Accessed on December 22, 2012.

New Zealand. 2000. *Used Oil Recovery, Reuse and Disposal in New Zealand.* Published by the Ministry for the Environment, Wellington, New Zealand, 55 p. Available at http://www.mfe.govt.nz/publications/waste/used-oil-recovery-dec00.pdf. Accessed on December 22, 2012.

Nigeria Workshop. 2004. Final Regional Workshop for the Development of Regional Action Plan on Environmentally Sound Management of Used Oils in Africa, Workshop hosted by the Government of Nigeria with Financial Assistance from the Secretariat of the Basel Convention, Lagos, 72 p. Available at http://www.basel.int/Portals/4/Basel%20Convention/docs/centers/proj_activ/stp_projects/04-02.pdf. Accessed December 22, 2012.

Norris, J., Stewart, R., and Passant, N. 2006. Review of the Fate of Lubricating Oils in the UK, UK-AEA Study for DEFRA, 25 p. Available at http://uk-air.defra.gov.uk/reports/cat07/0703280957_Review_of_Fate_Of_Lubricating_Oil_2005_NIR_Issue1_v1.3.1_cd4569rs.pdf. Accessed on January 15, 2013.

PETDER. 2012. Selection of the Most Appropriate Technology for Waste Mineral Oil Refining Project: Technical Research Report 2012. Turkish Petroleum Industry Association (PETDER) Contract Report, 65 p. Available at http://www.petder.org.tr/admin/my_documents/my_files/2B7_ISTKATEKNIKRAPORU2012Eng.pdf. Accessed on January 25, 2013.

Regional Activity Centre for Cleaner Production (RAC/CP). 2000. Recycling Possibilities and Potential Uses of Used Oils. Report by the Regional Activity Centre for Cleaner Production under the Mediterranean Action Plan, Barcelona, Spain, 74 p. Available at http://www.cprac.org/docs/olis_eng.pdf. Accessed on December 22, 2012.

Scottish Environmental Protection Agency (Scottish EPA), 2005. Scottish Environmental Protection Agency National Best Practice Project: Mineral Oil Wastes Final Phase 1 Report, 94 p. Available at http://www.sepa.org.uk/waste/waste_publications/idoc.ashx?docid=b652ae6f-ce7b-4bb7-be38-fe8a4324ac42&version=-1. Accessed on December 23, 2012.

Tolia, H. 1994. Evaluation of Oil Refining and Recycling Technologies. Report to US–Asia Environmental Partnership/World Environment Center, 14 p. Available at http://pdf.usaid.gov/pdf_docs/PDABL578.pdf. Accessed on December 20, 2012.

UNEP Compendium. 2012. Compendium of Recycling and Destruction Technologies for Waste Oils, prepared by a team at the Birla Institute of Management Technology, Ghaziabad, India. Available at http://www.unep.org/ietc/Portals/136/Publications/Waste%20Management/IETC%20Waste_Oils_Compendium-Full%20Doc-for%20web_Nov.2012.pdf. Accessed on January 25, 2013.

U.S. Department of Energy (USDOE). 2006. Used Oil Re-Refining Study to Address Energy Policy Act of 2005 Section 1838, U.S. Department of Energy, 121 p. Available at http://fossil.energy.gov/epact/used_oil_report.pdf. Accessed on January 15, 2013.

Valiante, U.A. 2005. A Critical Review of the Used Oil Management Association (UOMA) Program Review. Prepared by Corporate Policy Group LLP, Ontario, Canada, 27 p. Available at http://solidwastemag.com/posteddocuments/PDFs/2005/OctNov/UOMA%20Program%20Review%20Critique.pdf. Accessed on December 22, 2012.

Zachary, J.E. 1996. Re-refined Motor Oil: Overcoming the Myths, Resource Manual prepared by Community Environmental Council Gildea Resource Center, Santa Barbara, CA, 55 p. Available at http://cdm16254.contentdm.oclc.org/cdm/singleitem/collection/p178601ccp2/id/1656/rec/15. Accessed on December 22, 2012.

Glossary of Terms and Abbreviations

AAMA: The American Automobile Manufacturers Association, a trade association of automotive manufacturers with emphasis on qualification and aftermarket testing.

Abrasion: A general wearing of a surface by constant scratching, usually due to the presence of foreign matter such as dirt, grit, or metallic particles in the lubricant. It may also cause a breakdown of the material (such as the tooth surfaces of gears). Lack of proper lubrication may result in abrasion.

Abrasive Wear (cutting wear): Comes about when hard surface asperities or hard particles that have embedded themselves into a soft surface and plough grooves into the opposing harder surface, for example, a journal.

Absolute Filtration Rating: The diameter of the largest hard spherical particle that will pass through a filter under specified test conditions. This is an indication of the largest opening in the filter elements.

Absolute Pressure: The sum of atmospheric and gauge pressure.

Absolute Viscosity: A term used interchangeably with viscosity to distinguish it from either kinematic viscosity or absolute viscosity. Absolute viscosity is the ratio of shear stress to shear rate. It is a fluid's internal resistance to flow. The common unit of absolute viscosity is the poise. Absolute viscosity divided by fluid density equals kinematic viscosity. It is occasionally referred to as dynamic viscosity. Absolute viscosity and kinematic viscosity are expressed in fundamental units. Commercial viscosity such as Saybolt viscosity is expressed in arbitrary units of time, usually seconds.

Absorbent: A material having the power, capacity, or tendency to absorb.

Absorbent Filter: A filter medium that holds contaminant by mechanical means.

Absorption: The assimilation of one material into another; in petroleum refining, the use of an absorptive liquid to selectively remove components from a process stream.

AC Fine Test Dust (ACFTD): A test contaminant used to assess both filters and the contaminant sensitivity of all types of tribological mechanisms.

Accumulator: A container in which fluid is stored under pressure as a source of fluid power.

ACEA: Association des Constructeurs Européen d'Automobiles (Association of European Automobile Constructors).

Acid: In a restricted sense, any substance containing hydrogen in combination with a nonmetal or nonmetallic radical and capable of producing hydrogen ions in solution.

Acidity: In lubricants, acidity denotes the presence of acid-type constituents whose concentration is usually defined in terms of total acid number. The constituents vary in nature and may or may not markedly influence the behavior of the lubricant.

Acid Number: The quantity of base expressed in milligrams of potassium hydroxide that is required to neutralize the acidic constituents in 1 g of sample.

Acid Sludge: The residue left after treating petroleum oil with sulfuric acid for the removal of impurities. It is a black, viscous substance containing the spent acid and impurities.

Acid Treating: A refining process in which unfinished petroleum products, such as gasoline, kerosene, and lubricating oil stocks are contacted with sulfuric acid to improve their color, odor, and other properties.

Acid Value: A measure of acidity. It is normally expressed as milligrams of potassium hydroxide per gram of sample.

Activated Alumina: A highly porous material produced from dehydroxylated aluminum hydroxide. It is used as a desiccant and as a filtering medium.

Actuator: A device used to convert fluid energy into mechanical motion.

AD: Atmospheric pressure distillation.

Additive: A chemical substance added to a petroleum product to impart or improve certain properties. Common petroleum product additives are antifoam agent, antiwear additive, corrosion inhibitor, demulsifier, detergent, dispersant, emulsifier, EP additive, oiliness agent, oxidation inhibitor, pour point depressant, rust inhibitor, tackiness agent, viscosity index (VI) improver.

Additive Level: The total percentage of all additives in an oil (expressed in percentage of mass [weight] or percentage of volume).

Additive Stability: The ability of additives in the fluid to resist changes in their performance during storage or use.

Adhesion: The force or forces causing two materials such as a lubricating grease and a metal to stick together; the property of a lubricant that causes it to cling or adhere to a solid surface.

Adhesive Lubricants: Lubricants with adhesion-improving components, which are not affected by centrifugal forces.

Adhesive Wear: Is often referred to as galling, scuffing, scoring, or seizing. It happens when sliding surfaces contact one another, causing fragments to be pulled from one surface and to adhere to the other.

Adiabatic: A change occurring without loss or gain of heat.

Adiabatic Compression: Compression of a gas without extraction of heat, resulting in increased temperature. The temperature developed in compression of a gas is an important factor in lubrication because oil deteriorates more rapidly at elevated temperatures; oxidation inhibitors help prevent rapid lubricant breakdown under these conditions.

Admix: To add by mixing.

Adsorbent Filter: A filter medium primarily intended to hold soluble and insoluble contaminants on its surface by molecular adhesion.

Adsorption: Adhesion of the molecules of gases, liquids, or dissolved substances to a solid surface, resulting in relatively high concentration of the molecules at the place of contact; for example, the plating out of an antiwear additive on metal surfaces.

Adsorptive Filtration: The attraction to, and retention of particles in, a filter medium by electrostatic forces, or by molecular attraction between the particles and the medium.

ADT: Atmospheric distillation tower; the primary distillation tower of a crude distillation unit, which operates at or above atmospheric pressure.

ADU: Atmospheric distillation unit; generally, a unit for distilling crude at or above atmospheric pressure as opposed to operating under a vacuum.

AEA: UK Atomic Energy Authority.

Aeration: The state of air being suspended in a liquid such as a lubricant or hydraulic fluid.

Agglomeration: The potential of the system for particle attraction and adhesion.

AGMA: American Gear Manufacturers Associations—an organization serving the gear industry.

AGMA Lubricant Numbers: AGMA specification covering gear lubricants. The viscosity ranges of the AGMA numbers (or grades) conform to the International Standards Organization (ISO) viscosity classification system (see ISO viscosity classification system).

Air Bleeder: A device for removal of air from a hydraulic fluid line.

Air Breather: A device permitting air movement between atmosphere and the component in/on which it is installed.

Air Entrainment: The incorporation of air in the form of bubbles as a dispersed phase in the bulk liquid. Air may be entrained in a liquid through mechanical means or by release of dissolved air due to a sudden change in environment. The presence of entrained air is usually readily apparent from the appearance of the liquid (bubbly, opaque, etc.) whereas dissolved air can only be determined by analysis.

Air Motor: A device that converts compressed gas into mechanical force and motion. It usually provides rotary mechanical motion.

Air/Oil Systems: A lubrication system in which small measured quantities of oil are introduced into an air/oil mixing device that is connected to a lube line that terminates at a bearing, or other lubrication point. The air velocity transports the oil along the interior walls of the lube line to the point of application. These systems provide positive air pressure within the bearing housing to prevent the ingress of contaminants, provide cooling air flow to the bearing, and perform the lubrication function with a continuous flow of minute amounts of oil.

Air-Gap Solenoid: A solenoid that is sealed to prevent leakage of the liquid into the plunger cavity.

Alicyclic Hydrocarbons: Hydrocarbons that contain a ring of carbon atoms other than the aromatics.

Aliphatic Hydrocarbons: Hydrocarbons that have an open chain structure, as opposed to ring structures.

Alkali: Any substance having basic (as opposed to acidic) properties. In a restricted sense, it is applied to the hydroxides of ammonium, lithium, potassium, and sodium. Alkaline materials in lubricating oils neutralize acids to prevent acidic and corrosive wear in internal combustion engines.

Alkylation: The combination of an unsaturated hydrocarbon (olefin) with a saturated hydrocarbon (paraffin or isoparaffin) to form branched chain saturated hydrocarbons; may also apply to the combination of aromatic hydrocarbons with unsaturated hydrocarbons to form branched chain aromatics.

Almen EP Lubricant Tester: A journal bearing machine used for determining the load-carrying capacity or extreme pressure properties (EP) of gear lubricants.

Almen Test: A laboratory procedure used to measure extreme pressure characteristics of fluid lubricants.

Aluminum Alloy: White particles that indicate wear of aluminum component such as a casing wall.

Ambient Temperature: Temperature of the area or atmosphere around a process (not the operating temperature of the process itself).

American Automobile Manufacturers Association (AAMA): A trade association that represented car manufacturers headquartered in the United States. The AAMA disbanded on May 1, 1999. Note: On December 16, 1992, the Motor Vehicle Manufacturers Association of the United States (MVMA) changed its name to the American Automobile Manufacturers Association.

American Chemistry Council (ACC): A trade association formerly known as the Chemical Manufacturers Association (CMA) responsible for the development and administration of the Petroleum Additives Panel Product Approval Code of Practice (ACC Code).

American Petroleum Institute (API): A trade association that promotes US petroleum interests, encourages development of petroleum technology, cooperates with the government in matters of national concern, and provides information on the petroleum industry to the government and the public.

American Society for Testing and Materials (ASTM): A professional society that is responsible for the publication of test methods and the development of test evaluation techniques.

Amine: An organic compound containing basic nitrogen; may be toxic and corrosive and the lower molecular weight amines have a smell similar to ammonia.

Amp: Ampere.

Amphoteric: Possession of the quality of reacting either as an acid or as a base.

Analytical Ferrography: The magnetic precipitation and subsequent analysis of wear debris from a fluid sample. This approach involves passing a volume of fluid over a chemically treated microscope slide, which is supported over a magnetic field; permanent magnets are arranged in such a way as to create a varying field strength over the length of the substrate which causes wear debris to precipitate in a distribution with respect to size and mass over the Ferrogram. Once rinsed and fixed to the substrate, this debris deposit serves as an excellent medium for optical analysis of the composite wear particulates.

Anhydrous: The absence of water.

Aniline Point: The minimum temperature for complete miscibility of equal volumes of aniline and the sample under test (ASTM D611). A product with a high aniline point will be low in aromatics and naphthenes and, therefore, high in paraffins. Aniline point is often specified for spray oils, cleaning solvents, and thinners, in which effectiveness depends on aromatic content. In conjunction with API gravity, the aniline point may be used to calculate the net heat of combustion for aviation fuels.

ANSI: American National Standards Institute.

Antiblocking Agent: A substance, such as a finely divided solid of mineral nature, that is added to prevent the adhesion of the surfaces of films made from plastic, to each other or to other surfaces.

Antifoam Agent: One of two types of additives used to reduce foaming in petroleum products: silicone oil to break up large surface bubbles or various kinds of polymers that decrease the amount of small bubbles entrained in the oils.

Antifriction Bearing: A rolling contact type bearing in which the rotating or moving member is supported or guided by means of ball or roller elements. Does not mean without friction; usually denotes a ball or roller bearing.

Antioxidant: Additive that prolongs the induction period of a base oil in the presence of oxidizing conditions and catalyst metals at elevated temperatures. The additive is consumed and degradation products increase not only with increasing and sustained temperature but also with increases in mechanical agitation or turbulence and contamination.

Antistatic Additive: An additive that increases the conductivity of a hydrocarbon fuel to hasten the dissipation of electrostatic charges during high-speed dispensing, thereby reducing the fire/explosion hazard.

Antiwear Additives (AA): Improve the service life of tribological elements operating in the boundary lubrication regime. Antiwear compounds (e.g., ZDDP and TCP) start decomposing at 90°C to 100°C and even at a lower temperature if water (25–50 ppm) is present.

API: American Petroleum Institute: A trade association of petroleum producers, refiners, marketers, and transporters organized for the advancement of the petroleum industry by conducting research, gathering and disseminating information, and maintaining cooperation between government and the industry on all matters of mutual interest.

API Base Oil Interchangeability Guidelines: A system that reduces testing costs by permitting the interchangeable use of certain base oils without requiring a full engine and bench test program for each of the base oils. This system is described in detail in Annex E of the Guidelines.

API Certification Mark: An API Mark that remains the same for a given application (e.g., gasoline, fuel-flexible, light-duty diesel) even if a new minimum engine oil standard or standards are developed.

API Engine Service Categories: Gasoline and diesel engine oil quality levels established jointly by API, SAE, and ASTM, and sometimes called SAE or API/SAE categories; formerly called API Engine Service Classifications.

API Gravity: A gravity scale established by the American Petroleum Institute and in general use in the petroleum industry, the unit being called "the API degree." This unit is defined in terms of specific gravity (SG) as follows:

$$\text{API Gravity} = \frac{141.5}{SG} - 131.5$$

API Guidelines for SAE Viscosity-Grade Engine Testing: Guidelines established for different oil viscosity grades that allow certain engine and bench test results to be used in lieu of additional testing. These guidelines are described in detail in Annex F of the Guidelines.

API Mark: A mark licensed by API and used by oil marketers in connection with engine oil products to certify conformance with quality standards established under the API EOLCS.

API Service Symbol: An API mark that identifies specific engine oil performance levels by means of alphanumeric service categories, SAE viscosity grades, and the Energy Conserving designation as appropriate.

Apparent Viscosity: The ratio of shear stress to rate of shear of a non-Newtonian fluid such as lubricating grease or a multigrade oil, calculated from Poiseuille's equation and measured in poises. The apparent viscosity changes with changing rates of shear and temperature and must, therefore, be reported as the value at a given shear rate and temperature (ASTM D1092).

Aqueous Decontamination: Removal of a chemical or biological hazard with a water-based solution.

Aromatic: Derived from, or characterized by, the presence of the benzene ring

ARP: Aeronautically Recommended Practice.

Ash: A measure of the amount of inorganic material in lubricating oil—determined by burning the oil and weighing the residue; results are expressed as a percentage by weight.

Ash Content: Erroneous name for the ash produced from the mineral matter content of petroleum or a petroleum derived product—petroleum and petroleum products do not contain ash; more correctly, ash yield, which is the percentage by weight of residue remaining after the combustion of a sample of petroleum.

ASLE: American Society of Lubrication Engineers. Now changed to Society of Tribologists and Lubrication Engineers (STLE).

ASME: American Society of Mechanical Engineers.

Asperities: Microscopic projections on metal surfaces resulting from normal surface-finishing processes. Interference between opposing asperities in sliding or rolling applications is a source of friction, and can lead to metal welding and scoring. Ideally, the lubricating film between two moving surfaces should be thicker than the combined height of the opposing asperities.

ASTM: American Society for Testing Materials—a society for developing standards for testing petroleum and petroleum products; now known as ASTM International.

ASTM Colorimeter: Apparatus used to determine the color of lubricating oils.

ASTM Distillation: Distillations made in accordance with any of the ASTM distillation procedures.

ASTM Melting Point: The temperature at which wax first shows a minimum rate of temperature change; also known as the English melting point.

ASTM Test Monitoring Center: An entity within ASTM that monitors the calibration of engine test stands and laboratories (see Referenced Laboratory).

ASTM Viscosity: A method of specifying levels for industrial lubricants; also called the ISO viscosity classification (see ASTM D2422).

Atm: Atmosphere.

Atmospheric Pressure: Pressure exerted by the atmosphere at any specific location; sea level pressure is approximately 14.7 pounds per square inch absolute or 101.3 kPa-a.

Atomic Absorption Spectroscopy: Measures the radiation absorbed by chemically unbound atoms by analyzing the transmitted energy relative to the incident energy at each frequency. The procedure consists of diluting the fluid sample with methyl isobutyl ketone (MIBK) and directly aspirating the solution. The actual process of atomization involves reducing the solution to a fine spray, dissolving it, and finally vaporizing it with a flame. The vaporization of the metal particles depends on their time in the flame, the flame temperature, and the composition of the flame gas. The spectrum occurs because atoms in the vapor state can absorb radiation at certain well-defined characteristic wavelengths. The wavelength bands absorbed are very narrow and differ for each element. In addition, the absorption of radiant energy by electronic transitions from ground to excited state is essentially an absolute measure of the number of atoms in the flame and is, therefore, the concentration of the element in a sample.

Atomization: The conversion of a liquid into a spray of very fine droplets.

Automatic Transmission Fluid (ATF): Fluid for automatic, hydraulic transmissions in motor vehicles.

Automotive Motor Oil (Motor Oil): Oil that is used to lubricate the moving components of an internal combustion engine.

Axial-Load Bearing: A bearing in which the load acts in the direction of the axis of rotation.

Babbitt: A soft, white, nonferrous alloy bearing material composed principally of copper, antimony, tin, and lead.

Back Pressure: The pressure encountered on the return side of a system.

Background Contamination: The total of the extraneous particles that are introduced in the process of obtaining, storing, moving, transferring, and analyzing a fluid sample.

Bacteria: Microorganisms often composed of a single cell.

Bactericide: Additive included in the formulations of water-mixed cutting fluids to inhibit the growth of bacteria promoted by the presence of water, thus preventing odors that can result from bacterial action.

Baffle: A device to prevent direct fluid flow or impingement on a surface.

Ball Bearing: An antifriction rolling type bearing containing rolling elements in the form of balls.

Barrel: A unit of liquid volume of petroleum oils equal to 42 US gallons or approximately 35 Imperial gallons or 158.987 L.

Base: A material that neutralizes acids. An oil additive containing colloidally dispersed metal carbonate, used to reduce corrosive wear.

Base Number: The amount of acid, expressed in terms of the equivalent number of milligrams of potassium hydroxide, required to neutralize all basic constituents present in 1 g of sample.

Base Oil: A base stock or blend of base stocks used in an API-licensed engine oil; a liquid product consisting of mineral oil or synthetic fluid used as the primary component for various type of marketed lubricants including engine oil, automotive transmission fluid, hydraulic fluid, gear oil, metalworking oil, medicinal white oil, and grease.

Base Oil Interchangeability Guidelines: See API Base Oil Interchangeability Guidelines.

Base Stock: The base fluid, usually a refined petroleum fraction or a selected synthetic material, into which additives are blended to produce finished lubricants; a lubricant component that is produced by a single manufacturer (independent of crude source or manufacturing location), that meets the same manufacturer's specification, and that is identified by a unique formula, product identification number, or both. Base stocks may be manufactured using a variety of different processes including but not limited to distillation, solvent refining, hydrogen processing, oligomerization, esterification, and re-refining. Re-refined stock shall be substantially free from materials introduced through manufacturing, contamination, or previous use.

Base Stock Slate: A product line of base stocks that have different viscosities, but are in the same base stock grouping and from the same manufacturer.

Batch: Any quantity of material handled or considered as a "unit" in processing, that is, any sample taken from the same "batch" will have the same properties or qualities.

Bbl: Barrel.

BCUOMA: British Columbia Used Oil Management Association (Canada).

Bearing: A support or guide by means of which a moving part such as a shaft or axle is positioned with respect to the other parts of a mechanism.

Bellows Seal: A type of mechanical seal that utilizes bellows for providing secondary sealing and spring-type loading.

Bench Test: A laboratory test that measures various performance parameters of an engine oil.

Bentonite: The mineral montmorillonite, a magnesium–aluminum silicate used as a treating agent or as a component of grease.

Bernoulli's Theorem: If no work is done on or by a flowing, frictionless liquid, its energy, due to pressure, velocity, and altitude remains constant at all points along the streamline.

Beta Rating: The method of comparing filter performance based on efficiency. This is done using the Multipass Test, which counts the number of particles of a given size before and after fluid passes through a filter.

Beta Ratio: The ratio of the number of particles greater than a given size in the influent fluid to the number of particles greater than the same size in the effluent fluid, under specified test conditions (see Multipass Test).

Bevel Gear: A straight-toothed gear with the teeth cut on sloping faces and the gear shafts at an angle (normally a right angle).

BFOE: Barrels fuel oil equivalent based on net heating value of 6,050,000 Btu per BFOE.

Binder: The material used to hold the pigment of a solid lubricant system to the substrate; not to be confused with an asphaltic binder as used in road construction.

Biodegradable: The microorganism's breakdown of materials.

Bitumen: Also (incorrectly) called asphalt, pitch, or tar; occurs in nature as asphalt lakes (such as the Trinidad Asphalt Lake) and tar sands (such as the Athabasca Tar Sands or Athabasca Oil Sands in Alberta, Canada); consists of high molecular weight hydrocarbonaceous compounds that contain sulfur and nitrogen compounds.

Black Oils: Lubricants containing asphaltic materials, which impart extra adhesiveness, and are used for open gears and steel cables.

Bleeding: The separation of some of the liquid phase from a grease.

Blending: The process of mixing lubricants or components for the purpose of obtaining the desired physical or chemical properties (see Compounding).

Bloom (Florescence): The color of an oil by reflected light when this differs from its color by transmitted light.

Blow-by: Passage of unburned fuel and combustion gases past the piston rings of internal combustion engines, resulting in fuel dilution and contamination of the crankcase oil.

Boiling Point: The temperature at which a substance boils, or is converted into vapor by bubbles forming within the liquid; it varies with pressure.

Boiling Range: For a mixture of substances, such as a petroleum fraction, the temperature interval between the initial and final boiling points.

Bomb Oxidation: A test for the oxidation stability of a product obtained by sealing it in a closed container with oxygen under pressure. The decrease in pressure of the oxygen is a measure of the amount of oxidation that has occurred.

Bottoms: The product that collects in the bottom of a vessel, either during a fractionating process or while in storage.

Boundary Lubrication: Form of lubrication between two rubbing surfaces without development of a full-fluid lubricating film. Boundary lubrication can be made more effective by including additives in the lubricating oil that provide a stronger oil film, thus preventing excessive friction and possible scoring. There are varying degrees of boundary lubrication, depending on the severity of service. For mild conditions, oiliness agents may be used; by plating out on metal surfaces in a thin but durable film, oiliness agents prevent scoring under some conditions that are too severe for a straight mineral oil. Compounded oils, which are formulated with polar fatty oils, are sometimes used for this purpose. Antiwear additives are commonly used in more severe boundary lubrication applications. The more severe cases of boundary lubrication are defined as extreme pressure conditions; they are met with lubricants containing EP additives that prevent sliding surfaces from fusing together at high local temperatures and pressures.

Boyle's Law: The absolute pressure of a fixed mass of gas varies inversely as the volume, provided the temperature remains constant.

Breakdown Maintenance: Maintenance performed after a machine has failed to return it to an operating state.

Bridging: A condition of filter element loading in which contaminant spans the space between adjacent sections of a filter element, thus blocking a portion of the useful filtration.

Bright: A term generally applied to lubricating oils, meaning clear or free from moisture.

Bright Stock: A heavy residual lubricant stock with low pour point base oil fraction; used in finished blends to provide good bearing film strength, prevent scuffing, and reduce oil consumption; usually identified by its viscosity, SUS at 210°F or cSt at 100°C.

Brinelling: Permanent deformation of the bearing surfaces where the rollers (or balls) contact the races. Brinelling results from excessive load or impact on stationary bearings. It is a form of mechanical damage in which metal is displaced or upset without attrition.

Bromine Number: A test that indicates the degree of unsaturation in the test sample.

Brookfield Viscosity: Apparent viscosity in centipoise determined by Brookfield viscometer, which measures the torque required to rotate a spindle at constant speed in oil of a given temperature. Basis for ASTM D2983; used for measuring low-temperature viscosity of lubricants.

BS&W: Bottom sediment and water; the heavy material that collects in the bottom of storage tanks; composed of oil, water, and foreign matter.

BSW: See BS&W.

Btu: British thermal unit. The amount of heat required to raise the temperature of 1 lb. of water 1 degree Fahrenheit.

Bubble Point: The differential gas pressure at which the first steady stream of gas bubbles is emitted from a wetted filter element under specified test conditions.

Built-in-dirt: Material passed into the effluent stream composed of foreign materials incorporated into the filter medium.

Bulk Modulus (of elasticity): A ratio of normal stress to a change in volume. A term used in determining the compressibility of a fluid. Data for petroleum products can be found in the International Critical Tables.

Bunker C Fuel Oil: A heavy residual fuel oil used by ships, industry, and large-scale heating installations; also called Navy heavy fuel oil or No. 6 fuel oil.

Burner Oil: A clean-burning product obtained from a high-quality kerosene fraction.

Burst Pressure Rating: The maximum specified inside-out differential pressure that can be applied to a filter element without outward structural or filter–medium failure.

Bushing: A short, externally threaded connector with a smaller size internal thread.

Bypass Filtration: A system of filtration in which only a portion of the total flow of a circulating fluid system passes through a filter at any instant or in which a filter having its own circulating pump operates in parallel to the main flow.

Bypass Valve (Relief Valve): A valve mechanism that assures system fluid flow when a preselected differential pressure across the filter element is exceeded; the valve allows all or part of the flow to bypass the filter element.

C or cent.: Centigrade.

Cams: Eccentric shafts used in most internal combustion engines to open and close valves.

Capacity: The amount of contaminants a filter will hold before an excessive pressure drop is caused. Most filters have bypass valves that open when a filter reaches its rated capacity.

Capillarity: A property of a solid–liquid system manifested by the tendency of the liquid in contact with the solid to rise above or fall below the level of the surrounding liquid; this phenomenon is seen in a small-bore (capillary) tube.

Capillary Viscometer: A viscometer in which the oil flows through a capillary tube.

Carbon: A nonmetallic element—No. 6 in the periodic table. Diamonds and graphite are pure forms of carbon. Carbon is a constituent of all organic compounds. It also occurs in combined form in many inorganic substances, such as carbon dioxide, limestone, etc.

Carbon Deposit: Solid black residue in piston grooves, which can interfere with piston ring movement leading to wear or loss of power.

Carbon Residue: Coked material remaining after an oil has been exposed to high temperatures under controlled conditions.

Carbon Type: The distinction between paraffinic, naphthenic, and aromatic molecules. In relation to lubricant base stocks, the predominant type present.

Carbonyl Iron Powder: A contaminant that consists of up to 99.5% pure iron spheres.

Carcinogen: A cancer-causing substance. Certain petroleum products are classified as potential carcinogens by OSHA criteria. Suppliers are required to identify such products as potential carcinogens on package labels and Material Safety Data Sheets.

Cartridge Seal: A completely self-contained assembly including seal, gland, sleeve, mating ring, etc., usually needing no installation measurement.

Case Drain Filter: A filter located in a line conducting fluid from a pump or motor housing to a reservoir.

Case Drain Line: A line conducting fluid from a component housing to the reservoir.

Catalyst: A substance that initiates or increases the rate of a chemical reaction, without itself being used up in the process.

Catalytic Converter: An integral part of vehicle emission control systems since 1975. Oxidizing converters remove hydrocarbons and carbon monoxide (CO) from exhaust gases while reducing converters control nitrogen oxide (NO_x) emissions. Both use noble metal (platinum, palladium, or rhodium) catalysts that can be "poisoned" by lead compounds in the fuel or lubricant.

Catalytic Dewaxing: A catalytic hydrocracking process that uses catalysts such as molecular sieves to selectively hydrocrack the waxes present in hydrocarbon fractions.

Catastrophic Failure: Sudden, unexpected failure of a machine resulting in considerable cost and downtime.

Caustic: A highly alkaline substance such as sodium hydroxide.

Cavitation: Formation of an air or vapor pocket (or bubble) due to lowering of pressure in a liquid, often as a result of a solid body, such as a propeller or piston, moving through the liquid; also, the pitting or wearing away of a solid surface as a result of the violent collapse of a vapor bubble. Cavitation can occur in a hydraulic system as a result of low fluid levels that draw air into the fluid, producing tiny bubbles that expand followed by rapid implosion, causing metal erosion, and eventual pump destruction.

Cavitation Erosion: A material-damaging process that occurs as a result of vaporous cavitation. "Cavitation" refers to the occurrence or formation of gas-filled or vapor-filled pockets in flowing liquids due to the hydrodynamic generation of low pressure (below atmospheric pressure). This damage results from the hammering action when cavitation bubbles implode in the flow stream. Ultra-high pressures caused by the collapse of the vapor bubbles produce deformation, material failure, and finally, erosion of the surfaces.

Cellulose Media: A filter material made from plant fibers. Because cellulose is a natural material, its fibers are rough in texture and vary in size and shape. Compared with synthetic media, these characteristics create a higher restriction to the flow of fluids.

Centipoise (cp or cP): A unit of absolute viscosity. 1 centipoise = 0.01 poise.

Centistoke (cS or cSt): A unit of kinematic viscosity. 1 centistoke = 0.01 stoke.

Centralized Lubrication: A system of lubrication in which a metered amount of lubricant or lubricants for the bearing surfaces of a machine or group of machines are supplied from a central location.

Centrifugal Separator: A separator that removes immiscible fluid and solid contaminants that have a different specific gravity than the fluid being purified by accelerating the fluid mechanically in a circular path and using the radial acceleration component to isolate these contaminants.

Centrifuge: Equipment that removes insoluble materials by spinning a fluid at high speed; the g-force generated in a centrifuge enhances specific gravity differences between different objects or substances, effectively separating them.

CEP: Chemical Engineering Partners.

Channeling: The phenomenon observed among gear lubricants and greases when they thicken due to cold weather or other causes, to such an extent that a groove is formed through which the part to be lubricated moves without actually coming in full contact with the lubricant. A term used in percolation filtration; may be defined as a preponderance of flow through certain portions of the clay bed.

Characterization Factor: An index of feed quality used for correlating data based on physical properties; the Watson (UOP) characterization factor is defined as the cube root of the mean average boiling point in degrees Rankine divided by the specific gravity.

Chemical Manufacturers Association (CMA): See American Chemistry Council.

Chemical Stability: The tendency of a substance or mixture to resist chemical change.

Chlorinated Wax: Certain solid hydrocarbons treated with chlorine gas to form straight-chain hydrocarbons with a relatively high chlorine component. Chlorinated waxes are used primarily as polyvinyl chloride plasticizers, extreme-pressure additives for lubricants, and formulation components for many cutting fluids.

Chromatography: An analytical technique whereby a complex substance is adsorbed on a solid or liquid substrate and progressively eluted by a flow of a substance (the eluant) in which the components of the substance under investigation are differentially soluble. The eluant can be a liquid or a gas. When the substrate is filter paper and the eluant a liquid, a chromatogram of colored bands can be developed with the use of indicators. For gas chromatography, electronic detectors are normally used to indicate the passage of various components from the system.

Circulating Header System: A lubrication system having isolated lube zones wherein the lube pump runs continuously and circulates oil through the header, a return filter and back to tank during the idle period. When lubrication is required, a normal open solenoid valve in the return loop is actuated, allowing pump pressure to build. The zone valves are then sequentially opened to provide lubricant to the individual zones. Oil dispensed to the friction points is not reused, therefore, the system is a terminating type.

Circulating Lubrication: A system of lubrication in which the lubricant, after having passed through a bearing or group of bearings, is recirculated by means of a pump.

Circulating Oil: A lubrication system wherein the oil pump runs continuously and circulates oil to the friction points on a continuous basis. The oil is drained back to the tank, filtered, cooled as required, and reused.

Circulating System: A lubricating system in which oil is recirculated from a central sump to the parts requiring lubrication and then returned to the sump.

Clay Filtration: A refining process using fuller's earth (activated clay), bauxite, or other mineral to absorb minute solids from lubricating oil, as well as to remove traces of water, acids, and polar compounds.

Clay Treating: A clay adsorption process operated at elevated temperature and pressure used to neutralize or improve the color and stability of a lube base oil.

Clean: Not more than 100 particles larger than 10 μm/mL—in regard to an oil sample bottle cleanliness.

Clean Room: A facility or enclosure in which air content and other conditions (such as temperature, humidity, and pressure) are controlled and maintained at a specific level by special facilities and operating processes and by trained personnel.

Cleanable Filter: A filter element which, when loaded, can be restored by a suitable process, to an acceptable percentage of its original dirt capacity.

Cleanliness Level: A measure of relative freedom from contaminants.

Clearance Bearing: A journal bearing in which the radius of the bearing surface is greater than the radius of the journal surface.

Cleveland Open Cup: A flash point test in which the surface of the sample is completely open to the atmosphere, and which is therefore relatively insensitive to small traces of volatile contaminants.

Cloud Point: The temperature at which waxy crystals in an oil or fuel form a cloudy appearance.

Coalescer: A separator that divides a mixture or emulsion of two immiscible liquids using the interfacial tension between the two liquids and the difference in wetting of the two liquids on a particular porous medium.

Coefficient of Friction: The number obtained by dividing the friction force resisting motion between two bodies by the normal force pressing the bodies together.

Cohesion: That property of a substance that causes it to resist being pulled apart by mechanical means.

Coking: The undesirable accumulation of carbon (coke) deposits in the internal combustion engine or in a refinery plant. The process of distilling a petroleum product to dryness.

Cold Cranking Simulator: An intermediate shear rate viscometer that predicts the ability of an oil to permit a satisfactory cranking speed to be developed in a cold engine.

Cold Flow: A characteristic of plastic materials whereby they flow out of a high load area at room temperature.

Collapse: An inward structural failure of a filter element, which can occur due to abnormally high pressure drop (differential pressure) or resistance to flow.

Collapse Pressure: The minimum differential pressure that an element is designed to withstand without permanent deformation.

Combustion: Use of waste oils as fuel with the heat produced being adequately recovered.

Complex Grease: A lubricating grease thickened by a complex soap consisting of a normal soap and a complexing agent.

Compounded Oil: Petroleum oil to which other chemical substances have been added.

Compounding: The addition of fatty oils and similar materials to lubricants to impart special properties. Lubricating oils to which such materials have been added are known as compounded oils.

Compressibility: A measure of the change in volume of a sample when subjected to an increasing pressure.

Compression Ratio: In an internal combustion engine, the ratio of the volume of combustion space at bottom dead center to that at top dead center.

Conradson Carbon Residue: The residue remaining as the result of a test method used to determine the amount of carbon residue left after the evaporation and pyrolysis of the test sample at specified conditions.

Consistency: The degree to which a semisolid material such as grease resists deformation (see ASTM D217). Sometimes used qualitatively to denote viscosity of liquids.

Contaminant: Any foreign or unwanted substance that can have a negative effect on system operation, life, or reliability.

Contaminant Capacity: The weight of a specified artificial contaminant, which must be added to the influent to produce a given differential pressure across a filter at specified conditions. Used as an indication of relative service life.

Contaminant Failure: Any loss of performance due to the presence of contamination. Three types of contamination failure are abrasion, erosion and fatigue.

Contaminant Lock: A particle or fiber-induced jam caused by solid contaminants.

Contamination Control: A broad subject that applies to all types of material systems (including both biological and engineering). It is concerned with planning, organizing, managing, and implementing all activities required to determine, achieve, and maintain a specified contamination level.

Coolant: A fluid used to remove heat. See Cutting Fluid.

Copper Strip Corrosion: The gradual eating away of copper surfaces as the result of oxidation or other chemical action. It is caused by acids or other corrosive agents.

Core: The internal duct and filter media support.

Corrosion: The decay and loss of a metal due to a chemical reaction between the metal and its environment; the reaction is a transformation process in which the metal passes from its elemental form to a combined (or compound) form.

Corrosion Inhibitor: Additive for protecting lubricated metal surfaces against chemical attack by water or other contaminants. There are several types of corrosion inhibitors. Polar compounds wet the metal surface preferentially, protecting it with a film of oil. Other compounds may absorb water by incorporating it in a water-in-oil emulsion so that only the oil touches the metal surface. Another type of corrosion inhibitor combines chemically with the metal to present a nonreactive surface.

Corrosive Wear: Progressive removal of material from rubbing surface caused by a combination of chemical attack and mechanical action.

Coupling: A straight connector for fluid lines.

cP: centipoise or centiPoise, unit of dynamic viscosity.

Cracking: The process whereby large molecules are broken down by the application of heat and pressure to form smaller molecules.

Cracking Pressure: The pressure at which a pressure-operated valve begins to pass fluid.

Crankcase Oil: Lubricant used in the crankcase of the internal combustion engine.

Crown: The top of the piston in an internal combustion engine above the fire ring, exposed to direct flame impingement.

Cryogenics: The branch of physics relating to the production and effects of very low temperatures.

cSt: centistokes, unit of kinematic viscosity.

Cut: The portion or fraction of a crude oil boiling within certain temperature limits.

Cut Point: The temperature limit of a cut or fraction, usually but not limited to a true boiling point basis.

Cutting Fluid: Any fluid applied to a cutting tool to assist in the cutting operation by cooling, lubricating or other means.

Cutting Oil: A lubricant used in machining operations for lubricating the tool in contact with the workpiece, and to remove heat. The fluid can be petroleum based, water based, or an emulsion of the two. The term "emulsifiable cutting oil" normally indicates a petroleum-based concentrate to which water is added to form an emulsion which is the actual cutting fluid.

Cylinder: A device which converts fluid power into linear mechanical force and motion. It usually consists of a moveable element such as a piston and piston rod, plunger rod, plunger, or ram operating within a cylindrical bore.

Cylinder Oil: A lubricant for independently lubricated cylinders, such as those of steam engines and air compressors; also for lubrication of valves and other elements in the cylinder area. Steam cylinder oils are available in a range of grades with high viscosities to compensate for the thinning effect of high temperatures; of these, the heavier grades are formulated for super-heated and high-pressure steam, and the less heavy grades for wet, saturated, or low-pressure steam. Some grades are compounded for service in excessive moisture (see Compounded Oil). Cylinder oils lubricate on a once-through basis.

Cylinder Stock: The residuum remaining in a still after the lower-boiling constituents of crude oil have been vaporized; originally used for lubricating the cylinders of steam engines.

Deaerator: A separator that removes air from the system fluid through the application of bubble dynamics.

Deasphalted Oil: The extract or residual oil from which asphaltene and resin constituents have been removed by an extractive precipitation process called deasphalting.

Deasphalting: A process for removing asphalt from reduced crude or vacuum residua (residual oil), which utilizes the different solubility of asphaltic and nonasphaltic constituents in low-boiling hydrocarbon liquids, for example, liquid propane.

DEFRA: Department for Environment, Food, and Rural Affairs (UK).

Degas: Removing air from a liquid, usually by ultrasonic or vacuum methods.

Degradation: The progressive failure of a machine or lubricant.

Dehydrator: A separator that removes water from the system fluid.

Delamination Wear: A complex wear process in which a machine surface is peeled away or otherwise removed by forces of another surface acting on it in a sliding motion.

Demulsibility: The ability of a fluid that is insoluble in water to separate from water with which it may be mixed in the form of an emulsion.

Demulsifier: An additive that promotes oil–water separation in lubricants that are exposed to water or steam.

Density: The mass of a unit volume of a substance. Its numerical value varies with the units used.

Deplete: The depletion of additives expressed as an approximate percentage.

Deposits: Oil-insoluble materials that result from oxidation and decomposition of lube oil and contamination from external sources and engine blow-by. These can settle out on machine or engine parts. Examples are sludge, varnish, lacquer, and carbon.

Depth Filter: A filter medium that retains contaminants primarily within tortuous passages.

Depth Filter Media: Porous materials that primarily retain contaminants within a tortuous path, performing the actual process of filtration.

Dermatitis: Inflammation of the skin. Repeated contact with petroleum products can be a cause.

Desorption: Opposite of absorption or adsorption. In filtration, it relates to the downstream release of particles previously retained by the filter.

Detergent: In lubrication, either an additive or a compounded lubricant having the property of keeping insoluble matter in suspension thus preventing its deposition where it would be harmful. A detergent may also redisperse deposits already formed.

Detergent Oil: Is a lubricating oil possessing special sludge-dispersing properties usually conferred on the oil by the incorporation of special additives. Detergent oils hold formed sludge particles in suspension and thus promote cleanliness, especially in internal combustion engines. However, detergent oils do not contain "detergents" such as those used for cleaning of laundry or dishes. Also, detergent oils do not clean already "dirty" engines, but rather keep in suspension the sludge that petroleum oil forms so that the engine remains cleaner for longer periods. The formed sludge particles are either filtered out by oil filters or are drained out when the oil is changed.

Dewaxing: Removal of wax from base oil to reduce the pour point; solvent dewaxing is the process in which a number of different solvents can be used and has the following steps: feedstock is mixed with solvent and chilled, wax precipitated from solution is separated, solvent is recovered from the wax and dewaxed oil, and wax separation is accomplished by filtration, centrifuging, or settling.

Dibasic Acid Ester (Diester): Synthetic lubricant base; an organic ester, formed by reacting a dicarboxylic acid and an alcohol; properties include a high viscosity index and low volatility. With the addition of specific additives, it may be used as a lubricant in compressors, hydraulic systems, and internal combustion engines.

Dielectric Strength: A measure of the ability of an insulating material to withstand electric stress (voltage) without failure. Fluids with high dielectric strength (usually expressed in volts or kilovolts) are good electrical insulators (ASTM D877).

Differential Pressure Indicator: An indicator that signals the difference in pressure between two points, typically between the upstream and downstream sides of a filter element.

Differential Pressure Valve: A valve whose primary function is to limit differential pressure.

Directional Control Servo Valve: A directional control valve that modulates flow or pressure as a function of its input signal.

Directional Control Valve: A valve whose primary function is to direct or prevent flow through selected passages.

Dirt Capacity: The weight of a specified artificial contaminant, which must be added to the influent to produce a given differential pressure across a filter at specified conditions. Used as an indication of relative service life.

Dispersant: In lubrication, a term usually used interchangeably with detergent. An additive, usually nonmetallic (ashless), which keeps fine particles of insoluble materials in a homogeneous solution. Hence, particles are not permitted to settle out and accumulate.

Disposable: A filter element intended to be discarded and replaced after one service cycle.

Disposal: Processing or destruction of waste oils as well as their storage and tipping above or underground.

Dissolved Air: Air that is dispersed in a fluid to form a mixture.

Dissolved Gases: Those gases that enter into solution with a fluid and are neither free nor entrained gases.

Dissolved Water: Water that is dispersed in the fluid to form a mixture.

Distillation Method (ASTM D95): A method involving distilling the fluid sample in the presence of a solvent that is miscible in the sample but immiscible in water. The water distilled from the fluid is condensed and segregated in a specially designed receiving tube or tray graduated to directly indicate the volume of water distilled.

DOE: U.S. Department of Energy.

Double Seal: Two mechanical seals designed to permit a liquid or gas barrier fluid between the seals mounted back-to-back or face-to-face.

Drag: The resistance to movement caused by oil viscosity.

Dropping Point: In general, the dropping point is the temperature at which the grease passes from a semisolid to a liquid state. This change in state is typical of greases containing conventional soap thickeners. Greases containing thickeners other than conventional soaps may, without change in state, separate oil.

Drum: A container with a capacity of 55 US gallons or 208.2 L.

Dry Lubrication: The situation when moving surfaces have no liquid lubricant between them.

Dry Sump: An engine design in which oil is not retained in a pan beneath the crankshaft thus permitting splash lubrication. There may be a remote sump from which oil is recirculated, or there may be a total loss system.

Dual-Line System: A positive displacement terminating (oil or grease) lubrication system that uses two main lines supplied from a pump connected to a four-way (reverser) valve. Pressure in one main line (whereas the other is open to tank) causes the measuring piston(s) in the dual-line valve(s) to stroke in one direction, dispensing lubricant to one group of lube points. Switching the four-way (reverser) valve directs pump flow to the second main line and opens the first main line to tank. This allows pressure to build in the second main line causing the dual-line valve(s) measuring piston(s) to stroke back to their original position dispensing lubricant to a second group of lube points. The system is a parallel type and each dual-line valve operates independently of any other in the system.

Duplex Filter: An assembly of two filters with valving for selection of either or both filters.

Dust Capacity: The weight of a specified artificial contaminant, which must be added to the influent to produce a given differential pressure across a filter at specified conditions. Used as an indication of relative service life.

Dynamic Seal: A seal that moves due to axial or radial movement of the unit.

Effluent: The fluid leaving a component.

Elastohydrodynamic Lubrication: In rolling element bearings, the elastic deformation of the bearing (flattening) as it rolls, under load, in the bearing race. This momentary flattening improves the hydrodynamic lubrication properties by converting point or line contact to surface-to-surface contact.

Elastomer: A rubber or rubber-like material, both natural and synthetic, used in making a wide variety of products, such as seals and hoses. In oil seals, an elastomer's chemical composition is a factor in determining its compatibility with a lubricant.

Electrical Insulating Oil: High-quality oxidation-resistant oil refined to give long service as a dielectric and coolant for electrical equipment, most commonly transformers. An insulating oil must resist the effects of elevated temperatures, electrical stress, and contact with air, which can lead to sludge formation and loss of insulation properties. It must be kept dry because water is detrimental to dielectric strength—the minimum voltage required to produce an electric arc through an oil sample (ASTM D877).

Electrostatic Separator: A separator that removes contaminant from dielectric fluids by applying an electrical charge to the contaminant that is then attracted to a collection device of different electrical charge.

Emission Spectrometer: Works on the basis that atoms of metallic and other particular elements emit light at characteristic wavelengths when they are excited in a flame, arc, or spark. Excited light is directed through an entrance slit in the spectrometer. This light penetrates the slit, falls on a grate, and is dispersed and reflected. The spectrometer is calibrated by a series of standard samples containing known amounts of the elements of interest. By exciting these standard samples, an analytical curve can be established that gives the relationship between the light intensity and its concentration in the fluid.

Emulsibility: The ability of a non–water-soluble fluid to form an emulsion with water.

Emulsifier: Additive that promotes the formation of a stable mixture or emulsion of oil and water. Common emulsifiers are metallic soaps, certain animal and vegetable oils, and various polar compounds.

Emulsion: Intimate mixture of oil and water, generally of a milky or cloudy appearance. Emulsions may be of two types: oil-in water (where water is the continuous phase) and water-in-oil (where water is the discontinuous phase).

End Cap: A ported or closed cover for the end of a filter element.

Engine Deposits: Hard or persistent accumulation of sludge, varnish, and carbonaceous residues due to blow-by of unburned and partially burned fuel, or the partial breakdown of the crankcase lubricant. Water from the condensation of combustion products, carbon, residues from fuel or lubricating oil additives, dust, and metal particles also contribute.

Engine Oil: A lubricating agent that can be classified according to one or a combination of the viscosity grades identified in Table 1 of the most recent edition of SAE J300. Engine oils are also called motor oils. Engine oils include diesel engine oils and passenger car motor oils (PCMOs).

Engine Oil Licensing and Certification System (EOLCS): An administrative process and legally enforceable system by which API authorizes marketers of engine oil to display an API Mark or Marks on oils that meet specified industry standards, as prescribed in a formal licensing agreement.

Engine Test (also called engine sequence test or sequence test): A test of an oil's performance using a full-scale engine operating under laboratory conditions.

Entrained Air: A mechanical mixture of air bubbles having a tendency to separate from the liquid phase.

Environmental Contaminant: All material and energy present in and around an operating system, such as dust, air moisture, chemicals, and thermal energy.

EOLCS: API Engine Oil Licensing and Classification System.

EP: End-point, usually end-point of a distillation process.

EPA: U.S. Environmental Protection Agency.

EP Additives (Extreme Pressure Additives): Lubricating oil and grease additives added to prevent metal-to-metal contact in highly loaded areas. In some cases, this is accomplished by using additives that react with the metal to form a protective coating.

EP Lubricants (Extreme Pressure Lubricants): Lubricants that impart to rubbing surfaces the ability to carry appreciably greater loads than would be possible with ordinary lubricants without excessive wear or damage.

Erosion: The progressive removal of a machine surface by cavitation or by particle impingement at high velocities.

Ester: Chemical compound formed by the reaction of an organic or inorganic acid with an alcohol or with another organic compound containing the hydroxyl (–OH) radical. The reaction involves replacement of the hydrogen of the acid with a hydrocarbon group—the name of the ester indicates its derivation. For example, the ester resulting from the reaction of ethyl alcohol and acetic acid is called ethyl acetate—esters have important uses in the formulation of some petroleum additives and synthetic lubricants.

Ester Oils: Compounds of acids and alcohols used for lubrication and the production of lubricating greases.

EU: European Union.

Externally Pressurized Seal: A seal that has pressure acting on the seal parts from an external independent source of supply.

Extraction: Use of a solvent to remove edible and commercial oils from seeds (e.g., soybeans), or oils and fats from meat scraps; also, the removal of reactive components from lube distillates.

Extreme Pressure (EP) Additive: Lubricant additive that prevents sliding metal surfaces from seizing under conditions of extreme pressure. At the high local temperatures associated with metal-to-metal contact, an EP additive combines chemically with the metal to form a surface film that prevents the welding of opposing asperities, and the consequent scoring that is destructive to sliding surfaces under high loads. Reactive compounds of sulfur, chlorine, or phosphorus are used to form these inorganic films.

Extreme Pressure (EP) Property: The ability of a lubricant to reduce scuffing, scoring, and seizure of contacting bearing surfaces when applied loads are high.

Fabrication Integrity Point: The differential gas pressure at which the first stream of gas bubbles are emitted from a wetted filter element under standard test conditions.

Face Seal: A device that prevents leakage of fluids along rotating shafts. Sealing is accomplished by a stationary primary seal ring bearing against the face of a mating ring mounted on a shaft. Axial pressure maintains the contact between the seal ring and the mating ring.

False Brinelling: False brinelling of needle roller bearings is actually a fretting corrosion of the surface because the rollers are the I.D. of the bearing. Although its appearance is similar to that of brinelling, false brinelling is characterized by attrition of the steel, and the load on the bearing is less than that required to produce the resulting impression. It is the result of a combination of mechanical and chemical action that is not completely understood, and occurs when a small relative motion or vibration is accompanied by some loading, in the presence of oxygen.

Fat: Animal or vegetable oil, which will combine with an alkali to saponify and form a soap.

Fatigue Chunks: Thick three-dimensional particles exceeding 50 µm, indicating severe wear of gear teeth.

Fatigue Life: The theoretical number of revolutions (or hours of operation) a bearing will last under a given constant load and speed before the first evidence of fatigue develops on one or more of the components.

Fatigue Platelets: Normal particles between 20 and 40 µm found in gear box and rolling element bearing oil samples observed by analytical ferrography. A sudden increase in the size and quantity of these particles indicates excessive wear.

Fatigued: A structural failure of the filter medium due to flexing caused by cyclic differential pressure.

Fatty Acid: Any monobasic (one displaceable hydrogen atom per molecule) organic acid having the general formula $CH_{2n} + {}_{1}COOH$; fatty acids derived from natural fats and oils are used in the manufacture of greases and other lubricants.

Ferrography: An analytical method of assessing machine health by quantifying and examining ferrous wear particles suspended in the lubricant or hydraulic fluid.

Fiber Grease: A grease with a distinctly fibrous structure, which is noticeable when portions of the grease are pulled apart.

Film Strength: Property of a lubricant, which acts to prevent scuffing or scoring of metal parts.

Filter: Any device or porous substance used as a strainer for cleaning fluids by removing suspended matter.

Filter Aid: Materials such as diatomaceous earth or perlite minerals, which are mined, heated to remove organic materials, then ground and classified into various size ranges; useful for removal of extremely fine materials suspended in liquids and used to clarify many liquids, including water, beer, and wine.

Filter Efficiency: Method of expressing a filter's ability to trap and retain contaminants of a given size.

Filter Element: The porous device which performs the actual process of filtration.

Filter Head: An end closure for the filter case or bowl that contains one or more ports.

Filter Housing: A ported enclosure that directs the flow of fluid through the filter element.

Filter Life Test: A type of filter capacity test in which a clogging contaminant is added to the influent of a filter, under specified test conditions, to produce a given rise in pressure drop across the filter or until a specified reduction of flow is reached. Filter life may be expressed as test time required to reach terminal conditions at a specified contaminant addition rate.

Filtration: The physical or mechanical process of separating insoluble particulate matter from a fluid, such as air or liquid, by passing the fluid through a filter medium that will not allow the particulates to pass through it.

Filtration (Beta) Ratio: The ratio of the number of particles greater than a given size in the influent fluid to the number of particles greater than the same size in the effluent fluid.

Fire Point (Cleveland Open Cup): The temperature to which a combustible liquid must be heated so that the released vapor will burn continuously when ignited under specified conditions.

Fire Resistant Fluid: A fluid that is difficult to ignite and which shows little tendency to propagate flame; three common types of fire-resistant fluids are (1) water–petroleum oil emulsions, in which the water prevents burning of the petroleum constituents; (2) water–glycol fluids; and (3) nonaqueous fluids of low volatility, such as phosphate esters, silicones, and halogenated hydrocarbon-type fluids.

Fischer–Tropsch Process: The conversion of synthesis gas, a mixture of carbon monoxide and hydrogen obtained from natural gas or by gasification of other hydrocarbons such as coal, over a catalyst to form straight chain paraffins. The products range in molecular weight from light paraffins through the diesel fuel range to heavier waxes that can form the basis for the production of very high-quality lubricating base oils.

Fixed Displacement Pump: A pump in which the displacement per cycle cannot be varied.

Flash Point (Cleveland Open Cup): The temperature to which a combustible liquid must be heated to give off sufficient vapor to momentarily form a flammable mixture with air when a small flame is applied under specified conditions (ASTM D92).

Floc Point: The temperature at which wax or solids separate in an oil.

Flow Control Valve: A valve whose primary function is to control flow rate.

Flow Fatigue Rating: The ability of a filter element to resist a structural failure of the filter medium due to flexing caused by cyclic differential pressure.

Flow Rate: The volume, mass, or weight of a fluid passing through any conductor per unit of time.

Flowmeter: A device that indicates either flow rate, total flow, or a combination of both.

Fluid Compatibility: The suitability of filtration medium and seal materials for service with the fluid involved.

Foam: An agglomeration of gas bubbles separated from each other by a thin liquid film, which is observed as a persistent phenomenon on the surface of a liquid.

Foam Inhibitor: A substance introduced in a very small proportion to a lubricant or a coolant to prevent the formation of foam due to aeration of the liquid, and to accelerate the dissipation of any foam that may form.

Foaming: A frothy mixture of air and a petroleum product (e.g., lubricant, fuel oil) that can reduce the effectiveness of the product, and cause sluggish hydraulic operation, air binding of oil pumps, and overflow of tanks or sumps. Foaming can result from excessive agitation, improper fluid levels, air leaks, cavitation, or contamination with water or other foreign materials. Foaming can be inhibited with an antifoam agent. The foaming characteristics of a lubricating oil can be determined by blowing air through a sample at a specified temperature and measuring the volume of foam (ASTM D892).

FOE: Fuel oil equivalent; the heating value of a standard barrel of fuel oil equal to 6.05×10^6 Btu (6383 MJ).

Food-Grade Lubricants: Lubricants acceptable for use in meat, poultry, and other food processing equipment, applications, and plants. The lubricant types in food-grade applications are broken into categories based on the likelihood that they will contact food. The USDA created the original food-grade designations H1, H2 and H3, which is the current terminology used. The approval and registration of a new lubricant into one of these categories depends on the ingredients used in the formulation.

Force Feed Lubrication: A system of lubrication in which the lubricant is supplied to the bearing surface under pressure.

Formulation Identifier: An alphanumeric designation that permits traceability of samples in the marketplace by formulation.

Formulation/Standard Code: As defined in the ACC Code, a unique identification number that is assigned before engine testing to each candidate oil tested and that identifies the candidate's formulation, sponsor, blend, blend modification, test type, run number, testing laboratory, and test stand.

Four Ball Tester: This name is frequently used to describe either of two similar laboratory machines, the Four-Ball Wear Tester and the Four-Ball EP Tester. These machines are used to evaluate a lubricant's antiwear qualities, frictional characteristics, or load carrying capabilities. It derives its name from the four 1/2 inch steel balls used as test specimens. Three of the balls are held together in a cup filled with lubricant while the fourth ball is rotated against them.

Free Acid Points: Titration of a phosphatizing bath sample to a methyl orange end point. The point range is from 0 (weakest acid) to 14 (strongest acid); not to be confused with the pH scale.

Fretting: Wear phenomena taking place between two surfaces having oscillatory relative motion of small amplitude.

Fretting Corrosion: Can take place when two metals are held in contact and subjected to repeated small sliding, relative motions. Other names for this type of corrosion include wear oxidation, friction oxidation, chafing, and brinelling.

Friction: The resisting force encountered at the common boundary between two bodies when, under the action of an external force, one body moves or tends to move relative to the surface of the other.

FTIR: Fourier transform infrared spectroscopy; a test where infrared light absorption is used for assessing levels of soot, sulfates, oxidation, nitro-oxidation, glycol, fuel, and water contaminants.

FTIR/BRANDES: FTIR spectroscopy using the Brandes method (see also FTIR).

Fuel Dilution: The amount of raw, unburned fuel that ends up in the crankcase of an engine. It lowers an oil's viscosity and flash point, creating friction-related wear almost immediately by reducing film strength.

Full Flow Filter: A filter that, under specified conditions, filters all influent flow.

Full-Fluid Film Lubrication: Presence of a continuous lubricating film sufficient to completely separate two surfaces, as distinct from boundary lubrication. Full-fluid film lubrication is normally hydrodynamic lubrication, whereby the oil adheres to the moving parts and is drawn into the area between the sliding surfaces, where it forms a lubricating film layer at elevated pressure.

Full-flow Filtration: A system of filtration in which the total flow of a circulating fluid system passes through a filter.

FZG Four Square Gear Oil Test: Used in developing industrial gear lubricants to meet equipment manufacturers' specifications. The FZG test equipment consists of two gear sets, arranged in a four square configuration, driven by an electric motor. The test gear set is run in the lubricant at gradually increased load stages until failure, which is the point at which a 10 mg weight loss by the gear set is recorded. Also called Niemann four square gear oil test.

Galling: A form of wear in which seizing of the gear or tearing of the gear or bearing surface occurs.

Gas Turbine: An engine that uses the energy of expanding gases passing through a multistage turbine to create rotating power.

Gasohol: A blend of 10% anhydrous ethanol (ethyl alcohol) and 90% gasoline, by volume. Used as a motor fuel.

Gear: A machine part that transmits motion and force by means of successively engaging projections, called teeth. The smaller gear of a pair is called the pinion; the larger, the gear. When the pinion is on the driving shaft, the gear set acts as a speed reducer; when the gear drives, the set acts as a speed multiplier. The basic gear type is the spur gear, or straight-tooth gear, with teeth cut parallel to the gear axis. Spur gears transmit power in applications utilizing parallel shafts. In this type of gear, the teeth mesh along their full length, creating a sudden shift in load from one tooth to the next, with consequent noise and vibration. This problem is overcome by the helical gear, which has teeth cut at an angle to the center of rotation, so that the load is transferred progressively along the length of the tooth from one edge of the gear to the other. When the shafts are not parallel, the most common gear type used is the bevel gear, with teeth cut on a sloping gear face rather than parallel to the shaft. The spiral bevel gear has teeth cut at an angle to the plane of rotation, which, like the helical gear, reduces vibration and noise. A hypoid gear resembles a spiral bevel gear, except that the pinion is offset so that its axis does not intersect the gear axis; it is widely used in automobiles between the engine driveshaft and the rear axle. Offset of the axes of hypoid gears introduces additional sliding between the teeth, which, when combined with high loads, requires a high-quality EP oil. A worm gear consists of a spirally grooved screw moving against a tooth wheel; in this type of gear, where the load is transmitted across sliding, rather than rolling surfaces, compounded oils or EP oils are usually necessary to maintain effective lubrication.

Gear Oil: High-quality oil with good oxidation stability, load-carrying capacity, rust protection, and resistance to foaming, for service in gear housings and enclosed chain drives. Specially formulated industrial EP gear oils are used where highly loaded gear sets or excessive sliding action (as in worm gears) is encountered.

Gearbox (Gear Housing): A casing for gear sets that transmits power from one rotating shaft to another. A gearbox has a number of functions: it is precisely bored to control gear and shaft alignment, it contains the gear oil, and it protects the gears and lubricant from water, dust, and other environmental contaminants. Gearboxes are used in a wide range of industrial, automotive, and home machinery. Not all gears are enclosed in gearboxes; some are open to the environment and are commonly lubricated by highly adhesive greases.

GEIR: Groupement Européen de l'Industrie de Régénération (Brussels, Belgium). The European Group of the Re-Refining Industry.

Generated Contaminant: Caused by a deterioration of critical wetted surfaces and materials or by a breakdown of the fluid itself.

Graphite: A crystalline form of carbon having a laminar structure, which is used as a lubricant. It may be of natural or synthetic origin.

Gravimetric Analysis: A method of analysis whereby the dry weight of contaminant per unit volume of fluid can be measured showing the degree of contamination in terms of milligrams of contaminant per liter of fluid.

Gravity: See Specific Gravity; API Gravity.

Gravity Separation: A method of separating two components from a mixture. Under the influence of gravity, separation of immiscible phases (gas–solid, liquid–solid, liquid–liquid, solid–solid) allows the denser phase to settle out.

Grease: A lubricant composed of oil or oil thickened with a soap, soaps or other thickener to a semisolid or solid consistency.

Grease Fitting: A small fitting that connects a grease gun and the component to be lubricated. The fitting is installed by a threaded connection, leaving a nipple to which the grease gun attaches.

Grease Gun: A tool (normally hand-powered) that is used for lubrication tasks. By squeezing the trigger of the gun, grease is applied through an aperture to a specific point.

GTL (Gas-to-Liquids Process): The Fischer–Tropsch process is a catalytic conversion of synthesis gas (a mixture of carbon monoxide and hydrogen) to straight-chain paraffins that include diesel fractions and waxes. These high-quality products contain no sulfur, nitrogen, or oxygen and practically no aromatics or naphthenes, and can be used to form lubricant oils of very high quality, with properties similar to the synthetic poly α-olefin oils.

Guidelines for SAE Viscosity-Grade Engine Testing: See API Guidelines for SAE Viscosity-Grade Engine Testing.

H1 Lubricant: Food-grade lubricants used in food processing environments where there is some possibility of incidental food contact. Lubricant formulations may only be composed of one or more approved base stocks, additives, and thickeners (if grease) listed in Guidelines of Security Code of Federal Regulations (CFR) Title 21, §178.3570.

H2 Lubricant: Lubricants used on equipment and machine parts in locations where there is no possibility that the lubricant or lubricated surface contacts food. Because there is no risk of contacting food, these lubricants do not have a defined list of acceptable ingredients. They cannot, however, contain intentionally heavy metals such as antimony, arsenic, cadmium, lead, mercury, or selenium. Also, the ingredients must not include substances that are carcinogens, mutagens, teratogens, or mineral acids.

H3 Lubricant: Also known as soluble or edible oil. These are used to clean and prevent rust on hooks, trolleys, and similar equipment.

h: hour, unit of time.

Hardness: The resistance of a substance to surface abrasion.

Hard Vacuum: A term used to denote a high vacuum.

HC: Hydrocarbon.

Head: An end closure for the filter case or bowl which contains one or more ports.

Heat Exchanger: A device which transfers heat through a conducting wall from one fluid to another.

Heat Transfer Oil: A medium used for the transfer of heat at temperatures above that of steam; high-boiling petroleum oils are probably the most widely used heat transfer fluids.

Heavy Ends: The portions of a petroleum distillate fraction that are highest boiling, and therefore distill over last if the temperature is increased progressively.

Helical Gear: A cylindrical gear wheel which has slanted teeth that follow the pitch surface in a helical manner.

Housing: A ported enclosure that directs the flow of fluid through the filter element.

HVI: High viscosity index, typically from 80 to 110 VI units.

Hybrid Bearing: A bearing that consists of metal rings and ceramic balls.

Hydraulic Fluid: Fluid serving as the power transmission medium in a hydraulic system. The most commonly used fluids are petroleum oils, synthetic lubricants, oil–water emulsions, and water–glycol mixtures. The principal requirements of a premium hydraulic fluid are proper viscosity, high viscosity index, antiwear protection (if needed), good oxidation stability, adequate pour point, good demulsibility, rust inhibition, resistance to foaming, and compatibility with seal materials. Antiwear oils are frequently used in compact, high-pressure and capacity pumps that require extra lubrication protection.

Hydraulic Motor: A device which converts hydraulic fluid power into mechanical force and motion by transfer of flow under pressure. It usually provides rotary mechanical motion.

Hydraulic Oil: Oil especially suited for use as a fluid for transmitting hydraulic pressure.

Hydraulic Pump: A device which converts mechanical force and motion into hydraulic fluid power by means of producing flow.

Hydraulics: Engineering science pertaining to liquid pressure and flow.

Hydraulic System: A system designed to transmit power through a liquid medium, permitting the multiplication of force in accordance with Pascal's law, which states that "a pressure exerted on a confined liquid is transmitted undiminished in all directions and acts with equal force on all equal areas." Hydraulic systems have six basic components: (1) a reservoir to hold the fluid supply, (2) a fluid to transmit the power, (3) a pump to move the fluid, (4) a valve to regulate pressure, (5) a directional valve to control the flow, and (6) a working component—such as a cylinder and piston or a shaft rotated by pressurized fluid—to turn hydraulic power into mechanical motion. Hydraulic systems offer several advantages over mechanical systems. They eliminate complicated mechanisms such as cams, gears, and levers; are less subject to wear; are usually more easily adjusted for control of speed and force; are easily adaptable to both rotary and liner transmission of power; and can transmit power over long distances and in any direction with small losses.

Hydrocarbon: An organic compound containing carbon and hydrogen only.

Hydrocracking: A process combining cracking or pyrolysis, with hydrogenation; feedstocks can include crude oil, distillates, heavy oil, tar sand bitumen, and residua.

Hydrodynamic Lubrication: A system of lubrication in which the shape and relative motion of the sliding surfaces causes the formation of a fluid film having sufficient pressure to separate the surfaces.

Hydrofinishing: A process for treating raw extracted base stocks with hydrogen to saturate them for improved stability.

Hydrogenation: In refining, the chemical addition of hydrogen to a hydrocarbon in the presence of a catalyst; a severe form of hydrogen treating. Hydrogenation may be either destructive or nondestructive. In the former case, hydrocarbon chains are ruptured (cracked) and hydrogen is added where the breaks have occurred. In the latter, hydrogen is added to a molecule that is unsaturated with respect to hydrogen. In either case, the resulting products are highly stable. Temperatures and pressures in the hydrogenation process are usually greater than in a milder process known as hydrofining.

Hydrogen Refining: Lube oil hydrorefining and hydrocracking or severe hydrotreating processes.

Hydrolysis: Breakdown process that occurs in anhydrous hydraulic fluids as a result of heat, water, and metal catalysts (iron, steel, copper, etc.).

Hydrolytic Stability: Ability of additives and certain synthetic lubricants to resist chemical decomposition (hydrolysis) in the presence of water.

Hydrometer: An instrument for determining either the specific gravity of a liquid or the API gravity.

Hydrophilic: Compounds with an affinity for water.

Hydrophobic: Compounds that repel water.

Hydrostatic Lubrication: A system of lubrication in which the lubricant is supplied under sufficient external pressure to separate the opposing surfaces by a fluid film.

Hydro Turbine: A rotary engine whose energy is generated from moving water.

Hypoid Gear Lubricant: A gear lubricant having extreme pressure characteristics for use with a hypoid type of gear as in the differential of an automobile.

Hypoid Gears: Gears in which the pinion axis intersects the plane of the ring gear at a point below the ring gear axle and above the outer edge of the ring gear, or above the ring gear axle and below the outer edge of the ring gear.

IBP: Initial boiling point.

ICP: Inductively coupled plasma analysis.

IFEU: Institut für Energie und Umweltforschung GmbH (Institute for Energy and Environmental Research Ltd.) (Germany).

ILMA: The Independent Lubricant Manufacturers Association is a trade association of businesses engaged in compounding, blending, formulating, packaging, marketing, and distributing lubricants.

ILSAC: The International Lubricant Standardization and Approval Committee is a joint committee of AAMA and JAMA members that assists in the development of new minimum oil performance standards.

Image Analyzer: A sophisticated microscopic system involving a microscope, a television camera, a dedicated computer, and a viewing monitor similar to a television screen.

Immiscible: Incapable of being mixed without separation of phases. Water and petroleum oil are immiscible under most conditions, although they can be made miscible with the addition of an emulsifier.

Impact Sensitivity: The tendency of some materials to react with liquid oxygen when subjected to mechanical impact or vibration. This reaction is often explosive in nature.

Incompatible Fluids: Fluids that, when mixed in a system, will have a deleterious effect on that system, its components, or its operation.

Indicator: A device that provides external evidence of sensed phenomena.

Industrial Lubricant: Any petroleum or synthetic-base fluid or grease commonly used in lubricating industrial equipment, such as gears, turbines, and compressors.

Influent: The fluid entering a component.

Infrared Analysis: A form of absorption spectroscopy that identifies organic functional groups present in a used oil sample by measuring their light absorption at specific infrared wavelengths; absorbance is proportional to concentration. The test can indicate additive depletion, the presence of water, hydrocarbon contamination of a synthetic lubricant, oxidation, nitration, and glycol contamination from coolant. FTIR permits the generation of complex curves from digitally represented data.

Infrared Spectra: A graph of infrared energy absorbed at various frequencies in the additive region of the infrared spectrum. The current sample, the reference oil, and the previous samples are usually compared.

Infrared Spectroscopy: An analytical method using infrared absorption for assessing the properties of used oil and certain contaminants suspended therein (see FTIR).

Ingested Contaminants: Environmental contaminant that ingresses due to the action of the system or machine.

Ingression Level: Particles added per unit of circulating fluid volume.

Inhibitor: Any substance that slows or prevents such chemical reactions as corrosion or oxidation.

In-line Filter: A filter assembly in which the inlet, outlet, and filter element axes are in a straight line.

Inside-mounted Seal: A mechanical seal located inside the seal chamber with the pumped product's pressure at its OD.

Insolubles: Insoluble constituents—particles of carbon or agglomerates of carbon and other material. Indicates deposition or dispersant drop-out in an engine. Not serious in a compressor or gearbox unless there has been a rapid increase in these particles.

Insulating Oil: Oil used in circuit breakers, switches, transformers, and other electrical apparatus for insulating or cooling.

Intensifier: A device that converts low pressure fluid power into higher pressure fluid power.

Intercooler: A device that cools a gas between the compressive steps of a multiple stage compressor.

Interfacial Tension (IFT): The energy per unit area present at the boundary of two immiscible liquids. It is usually expressed in dynes per centimeter (ASTM D971).

Independent Lubricant Manufacturers Association (ILMA): A trade association of businesses engaged in compounding, blending, formulating, packaging, marketing, and distributing lubricants.

Interindustry Advisory Group (IAG): Provides advice to the API/Automotive Manufacturers Administrative Guidance Panel regarding the API EOLCS. The Interindustry Advisory Group consists of representatives from organizations such as Ford, General Motors, and Chrysler as well as the ACC, API, ASTM, EMA, ILMA, JAMA, PAJ, SAE, and the U.S. Army.

International Lubricant Specification Advisory Committee (ILSAC): A joint committee of Ford, General Motors, Chrysler, and JAMA members that assists in the development of new minimum oil performance standards.

Ion Exchange: A transfer of ions between two electrolytes or between an electrolyte solution and a complex. The term normally denotes the processes of purification, separation, and decontamination of aqueous and other ion-containing solutions with an insoluble (usually resinous) solid.

IP: Inspection procedure.

IP: UK Institute of Petroleum, now merged with the UK Institute of Energy to form the UK Energy Institute.

IR: Infrared spectroscopy.

IRR: Internal rate of return.

ISO: International Standards Organization, which, among many other standards, sets viscosity reference scales.

ISO Solid Contaminant Code (ISO 4406): A code assigned on the basis of the number of particles per unit volume greater than 5 and 15 μm in size. Range numbers identify each increment in the particle population throughout the spectrum of levels.

ISO Standard 4021: The accepted procedure for extracting samples from dynamic fluid lines.

ISO Viscosity Grade: A number indicating the nominal viscosity of an industrial fluid lubricant at 40°C (104°F, ASTM D2422; ISO 3448).

Japan Automobile Manufacturers Association (JAMA): A trade association that represents automobile manufacturers headquartered in Japan.

Joule: A unit of work, energy, or heat. 1 J (joule) =1 Nm (newton meter).

Journal: That part of a shaft or axle that rotates or angularly oscillates in or against a bearing or about which a bearing rotates or angularly oscillates.

Journal Bearing: A sliding type of bearing having either rotating or oscillatory motion and in conjunction with which a journal operates. In a full or sleeve type journal bearing, the bearing surface is 360° in extent. In a partial bearing, the bearing surface is less than 360° in extent, that is, 150°, 120°, etc.

Karl Fischer Reagent Method: The standard laboratory test to measure the water content of mineral base fluids. In this method, water reacts quantitatively with the Karl Fischer reagent. This reagent is a mixture of iodine, sulfur dioxide, pyridine, and methanol. When excess iodine exists, electric current can pass between two platinum electrodes or plates. The water in the sample reacts with the iodine. When the water is no longer free to react with iodine, an excess of iodine depolarizes the electrodes, signaling the end of the test.

kg: kilogram, unit of mass.

Kinematic Viscosity: The time required for a fixed amount of an oil to flow through a capillary tube under the force of gravity. The unit of kinematic viscosity is the stoke or centistoke (1/100 of a stoke). Kinematic viscosity may be defined as the quotient of the absolute viscosity in centipoises divided by the specific gravity of a fluid, both at the same temperature.

Kmol: Kilomole, unit of quantity of a substance.

kPa: Kilopascals, unit of pressure.

kPa-a: Kilopascals-absolute, a unit of pressure (pressure above absolute vacuum).

kPa-g: Kilopascals-gage, a unit of pressure (pressure above atmospheric).

KTI: Kinetics Technology International BV.

Lacquer: A deposit resulting from the oxidation and polymerization of fuels and lubricants when exposed to high temperatures; similar to, but harder than, varnish.

Laminar Flow: A flow situation in which fluid moves in parallel lamina or layers.

Laminar Particles: Particles generated in rolling element bearings, which have been flattened out by a rolling contact.

lb: Pound, a unit of mass (lb_m) or force (lb_f). Plural lbs.

LCA: Life cycle analysis or life cycle assessment.

Lead Naphthenate: A lead soap of naphthenic acids, the latter occurring naturally in petroleum.

LHSV: Liquid hourly space velocity (volumetric flow rate per hour divided by vessel volume).

License number: An identification number that is issued to a marketer on successful completion of the API licensing process and is used for audit purposes.

Licensed Fingerprint: The physical and chemical properties of a licensed formulation as described in Section 3 of Part B of the API Application for Licensure.

Light Ends: Low-boiling, volatile materials in a petroleum fraction. They are often unwanted and undesirable, but in gasoline the proportion of light ends deliberately included are used to assist low-temperature starting.

Light Obscuration: The degree of light blockage as reflected in the transmitted light impinging on the photodiode.

Lip Seal: An elastomeric or metallic seal that prevents leakage in dynamic and static applications by a scraping or wiping action at a controlled interference between itself and the mating surface.

Liquid: Any substance that flows readily or changes in response to the smallest influence. More generally, any substance in which the force required to produce a deformation depends on the rate of deformation rather than on the magnitude of the deformation.

Lithium Grease: The most common type of grease today, based on lithium soaps.

Load-carrying Capacity: Property of a lubricant to form a film on the lubricated surface, which resists rupture under given load conditions. Expressed as the maximum load that the lubricated system can support without failure or excessive wear.

Load-wear Index (LWI): Measure of the relative ability of a lubricant to prevent wear under applied loads; it is calculated from data obtained from the Four-Ball EP Method. Formerly called mean Hertz load.

LOX: An abbreviation used to denote liquid oxygen.

Lubricant: Any substance interposed between two surfaces in relative motion for the purpose of reducing the friction or the wear between them; a liquid product totally or partially consisting of mineral or synthetic oil that works to prevent metal-to-metal contact, removes contaminants, cools machine surfaces, removes wear debris, and transfers power. Lubricating oils are composed of base oils and additives.

Lubricating Grease: A solid to semifluid product of dispersion of a thickening agent in a liquid lubricant; additives imparting special properties are usually included.

Lubrication: The control of friction and wear by the introduction of a friction-reducing film between moving surfaces in contact; the lubricant used can be a fluid, solid, or plastic substance.

Lubricator: A device which adds controlled or metered amounts of lubricant into a pneumatic system.

Lubricity: Ability of an oil or grease to lubricate; also called film strength; the property of forming a lubricating film between moving surfaces, particularly when such surfaces are subject to heavy loads and rapid movements. Lubricity depends partly on wetting ability of the film-forming material. Oiliness is sometimes used as approximately equivalent to lubricity.

LVI: Low-Viscosity Index, typically below 40 VI units.

m: Meter, unit of length.

Magnetic Filter: A filter element that, in addition to its filter medium, has a magnet or magnets incorporated into its structure to attract and hold ferromagnetic particles.

Magnetic Plug: Strategically located in the flow stream to collect a representative sample of wear debris circulating in the system: for example, engine swarf, bearing flakes, and fatigue chunks. The rate of buildup of wear debris reflects degradation of critical surfaces.

Magnetic Seal: A seal that uses magnetic material (instead of springs or a bellows) to provide the closing force that keeps the seal faces together.

Magnetic Separator: A separator that uses a magnetic field to attract and hold ferromagnetic particles.

Manifold: A filter assembly containing multiple ports and integral relating components which services more than one fluid circuit.

Manifold Filter: A filter in which the inlet and outlet port axes are at right angles, and the filter element axis is parallel to either port axis.

Material Safety Data Sheet (MSDS): A publication containing health and safety information on hazardous products (including petroleum). The OSHA Hazard Communication Standard requires that an MSDS be provided by manufacturers to distributors or purchasers prior to or at the time of product shipment. An MSDS must include the chemical and common names of all ingredients that have been determined to be health hazards if they constitute 1% or greater of the product's composition (0.1% for carcinogens). An MSDS also included precautionary guidelines and emergency procedures.

Mechanical Seal: A device which works to join together systems or mechanisms to prevent leakage, contain pressure or exclude contamination.

Media Migration: Material passed into the effluent stream composed of the materials making up the filter medium.

Medium: The porous material that performs the actual process of filtration.

Metal Oxides: Oxidized ferrous particles, which are very old or have been recently produced by conditions of inadequate lubrication. Trend is important.

Metalworking Lubricant: Any lubricant, usually petroleum-based, that facilitates the cutting or shaping of metal. Basic types of metalworking lubricants are cutting and tapping fluids, drawing compounds, etc.

mg: Milligram, unit of mass.

Micrometer: See Micron.

Micron: A unit of length. One micron = 39 millionths of an inch (0.000039″). Contaminant size is usually described in microns. Relatively speaking, a grain of salt is about 60 μm and the eye can see particles to about 40 μm. Many hydraulic filters are required to be efficient in capturing a substantial percentage of contaminant particles as small as 5 μm. A micron is also known as a micrometer, and is shown as μm.

Microscope Method: A method of particle counting that measures or sizes particles using an optical microscope.

Mineral Oil: Oil derived from a mineral source, such as petroleum, as opposed to oils derived from plants and animals.

Mineral Seal Oil: A distillation fraction between kerosene and gas oil, widely used as a solvent oil in gas adsorption processes, as a lubricant for the rolling of metal foil, and as a base oil in many specialty formulations. Mineral seal oil takes its name—not from any sealing function—but from the fact that it originally replaced oil derived from seal blubber for use as an illuminant for signal lamps and lighthouses.

Miscible: Capable of being mixed in any concentration without separation of phases; for example, water and ethyl alcohol are miscible.

Mixed Film: A type of lubrication that features a combination of full-film and thin-film elements.

Mold (Release) Lubricant: A compound, often of petroleum origin, for coating the interiors of molds for glass and ceramic products. The mold lubricant facilitates removal of the molded object from the mold, protects the surface of the mold, and reduces or eliminates the need for cleaning it.

Molecular Sieve: A synthetic zeolite mineral having pores of uniform size, capable of separating molecules based on their size and structure by adsorption or sieving.

Moly: Molybdenum disulfide (MoS_2), a solid lubricant and friction reducer, colloidally dispersed in some oils and greases.

Molybdenum Disulfide: A black, lustrous powder (MoS_2) that serves as a dry-film lubricant in certain high-temperature and high-vacuum applications. It is also used in the form of pastes to prevent scoring when assembling press-fit parts, and as an additive to impart residual lubrication properties to oils and greases. Molybdenum disulfide is often called moly or moly-sulfide.

Monitoring, Enforcement, and Conformance (MEC): Aftermarket monitoring and enforcement to ensure that representation in the marketplace of API Marks to consumers and compliance with technical specifications are being adhered to, as stated in the API license agreement.

Motor: A device that converts fluid power into mechanical force and motion. It usually provides rotary mechanical motion.

Motor Bearing: A bearing that supports the crankshaft in an internal combustion engine. It is a support or guide by means of which a moving part is positioned with respect to the other parts of a mechanism.

Motor Oil (Automotive Oil): Oil that is used to lubricate the moving components of an internal combustion engine.

Motor Vehicle Manufacturers Association (MVMA): See American Automobile Manufacturers Association.

MSDS: Material Safety Data Sheet, a document required by several government agencies that typically lists the composition of a product along with hazard information, first aid measures, toxicological information, and regulatory information.

MTBF: mean time between failures.

Multigrade Oil: An oil meeting the requirements of more than one SAE viscosity grade classification, and may therefore be suitable for use over a wider temperature range than a single-grade oil.

Multipass Test: Filter performance tests in which the contaminated fluid is allowed to recirculate through the filter for the duration of the test. Contaminant is usually added to the test fluid during the test. The test is used to determine the beta ratio (q.v.) of an element.

Naphthene: A group of cyclic hydrocarbons also termed cycloparaffins; polycyclic members are also found in the higher boiling fractions.

Naphthenic: A type of petroleum fluid derived from naphthenic crude oil, containing a high proportion of closed-ring methylene groups.

Naphthenic Crude Oil: Class designation of crude oil containing predominantly naphthenes or asphaltic compounds.

NAS: National Aerospace Standard.

NASA: National Aeronautics and Space Administration.

NEC: National Electrical Code.

Needle Bearing: A rolling type of bearing containing rolling elements that are relatively long compared with their diameter.

NEMA: National Electrical Manufacturers Association.

Neutralization Number: A measure of the total acidity or basicity of an oil; this includes organic or inorganic acids or bases or a combination thereof (ASTM D974).

Newtonian Fluid: A fluid with a constant viscosity at a given temperature regardless of the rate of shear. Single-grade oils are Newtonian fluids. Multigrade oils are non-Newtonian fluids because viscosity varies with shear rate.

NFPA: National Fluid Power Association.

Nitration: Nitration products are formed during the fuel combustion process in internal combustion engines. Most nitration products are formed when an excess of oxygen is present. These products are highly acidic, form deposits in combustion areas, and rapidly accelerate oxidation.

Nitrous Oxide: A chemical compound made up of nitrogen and oxygen, N_2O. It is a liquid that turns into a gas when injected into an engine.

NLGI: National Lubricating Grease Institute; a trade association whose main interest is grease and grease technology. NLGI is best known for its system of rating greases by penetration.

NLGI Automotive Grease Classifications: Automotive lubricating grease quality levels established jointly by SAE, ASTM, and NLGI. There are several categories in two classifications: chassis lubricants and wheel bearing lubricants. Quality or performance levels within each category are defined by ASTM tests.

NLGI Consistency Grades: Simplified system established by the NLGI for rating the consistency of grease.

NMP or NM2P: N-methyl-2-pyrrolidone, a ketone used as an alternate to furfural and phenol for the extraction of lubricating oil fractions.

NMP Refining: An extraction process used to extract aromatics from lube feedstocks to improve the viscosity index and quality of lubricating oil base stock.

Nominal Filtration Rating: An arbitrary micrometer value indicated by a filter manufacturer. Due to lack of reproducibility, this rating is deprecated.

Non-Newtonian Fluid: Fluid, such as a grease or a polymer-containing oil (e.g., multigrade oil), in which shear stress is not proportional to shear rate.

Nonwoven Medium: A filter medium composed of a mat of fibers.

NORA: US National Oil Recyclers Association.

Normal Paraffin (n-Paraffin): A hydrocarbon consisting of molecules in which any carbon atom is attached to no more than two other carbon atoms; also called straight chain paraffin and linear paraffin.

Obliteration: A synergistic phenomenon of both particle silting and polar adhesion. When water and silt particles coexist in a fluid containing long-chain molecules, the tendency for valves to undergo obliteration increases.

OECD: Organization for Economic Co-operation and Development.

Off-Specification Used Oil: Used oil that does not meet any specification limits or has been mixed with hazardous waste or has ignitable characteristics not typical of lubricating oil and fails to meet the performance standard (see On-Specification Used Oil).

Oil: A greasy, unctuous liquid of vegetable, animal, mineral, or synthetic origin.

Oil Analysis: The routine activity of analyzing lubricant properties and suspended contaminants for the purpose of monitoring and reporting timely, meaningful, and accurate information on lubricant and machine condition.

Oil Change: The act of replacing dirty oil with clean oil.

Oil Consumption: The amount of lubricating fluid that is consumed by a machine, production line, plant, or company over a given time.

Oil Consumption Ratio: Annual oil purchases divided by machine charge volume. For example, if you purchased 10,000 gallons of oil in 1 year and the total amount of oil that all of your machine holds is 4200 gallons, your consumption ratio is 2.4.

Oil Drain: A large bolt or plug that secures the drain hole in the oil pan. It is generally fitted with a gasket or O-ring to prevent leakage.

Oil Filter: A device that removes the inherent or introduced impurities from the oil that lubricates an internal combustion engine.

Oil Flushing: A fluid circulation process that is designed to remove contamination and decomposition from a lubrication-based system.

Oil Marketer: Marketing organization responsible for the integrity of a brand name and the representation of the branded product in the marketplace.

Oil Mist Lubrication: A method of lubricant delivery in which oil is piped throughout the machine to desired locations and dispensed with a spray nozzle. Oil mist systems are employed to cool and lubricate many machine parts at once.

Oil Mist System: A device that delivers lubricant to multiple machine parts at once via a setup that includes piping and a spray nozzle.

Oil Oxidation: Occurs when oxygen attacks petroleum fluids. The process is accelerated by heat, light, metal catalysts, and the presence of water, acids, or solid contaminants. It leads to increased viscosity and deposit formation.

Oil Ring: A loose ring, the inner surface of which rides a shaft or journal and dips into a reservoir of lubricant from which it carries the lubricant to the top of a bearing by its rotation with the shaft.

Oil Sampling: A procedure that involves the collection of a volume of fluid from lubricated or hydraulic machinery for the purpose of performing oil analysis. Samples are typically drawn into a clean bottle which is sealed and sent to a laboratory for analytical work.

Oiler: A device for once-through lubrication. Three common types of oilers are drop-feed, wick-feed, and bottle-feed; all depend on gravity to induce a metered flow of oil to the bearing. The drop-feed oiler delivers oil from the bottom of a reservoir to a bearing one drop at a time; flow rate is controlled by a needle valve at the top of the reservoir. In a wick-feed oiler, the oil flows through a wick and drops from the end of the wick into the bearing; feed is regulated by chaining the number of strands, by increasing or lowering the oil level, or by applying pressure to the wick. In a bottle-feed oiler, a vacuum at the top of the jar keeps the fluid from running out; as tiny bubbles of air enter, the vacuum is reduced and a small amount of oil enters the bearing or is added to a reservoir from which the bearing is lubricated.

Oiliness: That property of a lubricant that produces low friction under conditions of boundary lubrication. The lower the friction, the greater the oiliness.

Oiliness Agent: An additive, usually polar in nature, used to improve the lubricity of a mineral oil. Now usually called a boundary lubrication additive.

On-Specification Used Oil: Used oil that does not exceed any specification limits, has not been mixed with hazardous waste or has had ignitable characteristics typical of lubricating oil and meets the performance standard. On-specification oil generally may be managed or burned by anyone for any legitimate oil-burning purpose (e.g., space heaters, boilers, oil furnaces) regardless of whether it is generated on-site or not. On occasion, used oil may exhibit a characteristic of a hazardous waste because of normal use and may still qualify as on-specification used oil if hazardous waste has not been added to it and if this claim can be certified (see Off-Specification Used Oil).

Open Bubble Point (Boiling Point): The differential gas pressure at which gas bubbles are profusely emitted from the entire surface of a wetted filter element under specified test conditions.

Open Gear: A gear that is exposed to the environment, rather than being housed in a protective gear box. Open gears are generally large, heavily loaded, and slow moving. They are found in such applications as mining and construction machinery, punch presses, plastic and rubber mills, tube mills, and rotary kilns. Open gears require viscous, adhesive lubricants that bond to the metal surfaces and resist run-off. Such lubricants are often called gear shields. Top-quality lubricants for such applications are specially formulated to protect the gears against the effects of water and other contaminants.

OSHA: US Occupational Safety and Health Administration.

Outside-Mounted Seal: A mechanical seal with its seal head mounted outside the seal chamber that holds the fluid to be sealed. Outside seals have the pumped fluid's pressure at their ID.

Oxidation: The reaction of oxygen with oil components—the higher the temperature of an oil, the faster oxidation reactions will take place; generally results in an increase in the total acid number (TAN) as well as an increase in viscosity—in addition, sludge and sediment will likely be formed and may separate out in the oil reservoirs, or coat the surfaces of metal components and cause a variety of mechanical problems.

Oxidation Inhibitor: Substance added in small quantities to a petroleum product to increase its oxidation resistance, thereby lengthening its service or storage life; also called antioxidant. An oxidation inhibitor may work in one of these ways: (1) by combining with and modifying peroxides (initial oxidation products) to render them harmless, (2) by decomposing the peroxides, or (3) by rendering an oxidation catalyst inert.

Oxidation Stability: Ability of a lubricant to resist natural degradation upon contact with oxygen.

P: Poise, unit of dynamic viscosity.

Pa: Pascal, unit of pressure.

PAG Synthetic Fluid: Polyalkaline glycols have excellent oxidative and thermal stability, very high VI, excellent film strength, and an extremely low tendency to leave deposits on machine surfaces. The low deposit-forming tendency is really due to two properties—the oil's ability to dissolve deposits and the fact that the oil burns cleanly. So when they are exposed to a very hot surface or subjected to microdieseling by entrained air, PAGs are less likely to leave a residue that will form deposits. PAGs may also be the only type of base oil with significantly lower fluid friction, which may allow for energy savings. The other unique property of PAGs is the ability to absorb a great deal of water and maintain lubricity. There are actually two different types of PAGs—one demulsifies and the other absorbs water. The most common applications for PAGs are compressors and critical gearing applications. The negatives of PAGs are their very high cost and the potential to be somewhat hydrolytically unstable.

Pale Oil: A petroleum lubricating or process oil refined until its color is straw to pale yellow.

PAO Synthetic Fluid: Poly α-olefins, often called synthetic hydrocarbons, are probably the most common type of synthetic base oil used today. They are moderately priced, provide excellent performance and have few negative attributes. PAO base oil is similar to mineral oil. The advantage comes from the fact that it is built, rather than extracted and modified, making it more pure. Practically all of the oil molecules are the same shape and size and are completely saturated. The potential benefits of PAOs are improved oxidative and thermal stability, excellent demulsibility and hydrolytic stability, a high VI, and very low pour point. Most of the properties make PAOs a good selection for temperature extremes—both high operating temperatures and low start-up temperatures. Typical applications for PAOs are engine oils, gear oils, and compressor oils. The negative attributes of PAOs are the price and poor solubility. The low inherent solubility of PAOs creates problems for formulators when it comes to dissolving additives. Likewise, PAOs cannot suspend potential varnish-forming degradation by-products, although they are less prone to create such material.

Paper Chromatography: A method which involves placing a drop of fluid on a permeable piece of paper and noting the development and nature of the halos, or rings, surrounding the drop through time. The roots of this test can be traced to the 1940s, when railroads used the "blotter spot" tests.

Paraffin: Any hydrocarbon identified by saturated straight (normal) or branched (iso) carbon chains; also called an alkane. The generalized paraffinic molecule can be represented by the formula C_nH_{2n+2}. Paraffins are relatively nonreactive and have excellent oxidation stability. In contrast with naphthenic oils, paraffinic lubricating oils have relatively high wax content and pour point, and generally have a high viscosity index (VI). Paraffinic solvents are generally lower in solvency than naphthenic or aromatic solvents.

Paraffinic: A type of petroleum fluid derived from paraffinic crude oil and containing a high proportion of straight chain saturated hydrocarbons. Often susceptible to cold flow problems.

Parallel Systems: Lubrication systems where the dispensing devices are connected to the main line in parallel. Each dispensing device operates independent of any other in the system.

Particle Count: The number of particles present greater than a particular micron size per unit volume of fluid often stated as particles larger than 10 μm/mL.

Particle Counter: An instrument that detects and counts particles found in a fluid such as oil.

Particle Counting: A microscopic technique that enables the visual counting of particles in a known quantity of fluid. The count identifies the number of particles present greater than a particular micron size per unit volume of fluid often stated as particles larger than 10 μm/mL.

Particle Density: An important parameter in establishing an entrained particle's potential to impinge on control surfaces and cause erosion.

Particle Erosion: Occurs when fluid-entrained particles moving at high velocity pass through orifices or impinge on metering surfaces or sharp angle turns.

Particle Impingement Erosion: A particulate wear process in which high-velocity, fluid-entrained particles are directed at target surfaces.

Particulates (Particulate Matter): Particles made up of a wide range of natural materials (e.g., pollen, dust, resins), combined with man-made pollutants (e.g., smoke particles, metallic ash); in sufficient concentrations, particulates can be a respiratory irritant.

Pascal's Law: A pressure applied to a confined fluid at rest is transmitted with equal intensity throughout the liquid and that pressure is considered to act at right angles to each surface contacted by the fluid.

Passenger car motor oils (PCMOs): Engine oils for passenger cars, light-duty trucks, and similar vehicles (see also Engine Oil).

Patch Test: A method by which a specified volume of fluid is filtered through a membrane filter of known pore structure. All particulate matter in excess of an "average size," determined by the membrane characteristics, is retained on its surface. Thus, the membrane is discolored by an amount proportional to the particulate level of the fluid sample. Visually comparing the test filter with standard patches of known contamination levels determines acceptability for a given fluid.

PCB: Polychlorinated biphenyl, a class of synthetic chemicals consisting of a homologous series of compounds beginning with monochlorobiphenyl and ending with decachlorobiphenyl. PCBs do not occur naturally in petroleum, but have been found as contaminants in used oil. PCBs have been legally designated as a health hazard, and any oil so contaminated must be handled in strict accordance with state and federal regulations.

PCT: Polychlorinated terphenyl.

PCV System: An abbreviation for "positive crankcase ventilation system." This is a system that prevents the vapors of a crankcase from being directly discharged into the atmosphere.

PCV Valve: A positive crankcase ventilation (PCV) valve is a one-way valve that ensures the continual flow and evacuation of gases from the crankcase into the engine.

Percolation: The passing of a liquid through a bed of granules or powder, for example, the slow flow of oil through a layer of decolorizing earth.

Permeability: The relationship of flow per unit area to differential pressure across a filter medium.

Petrochemical: Any chemical substance derived from crude oil or its products, or from natural gas. Some petrochemical products may be identical to others produced from other raw materials such as coal and producer gas.

Petrolatum: Soft petroleum material obtained from petroleum residua and consisting of amorphous wax and oil.

Petroleum Additives Panel Product Approval Code of Practice (ACC Code): A system developed by ACC to register and account for engine tests to help ensure that a lubricant meets a given performance specification. This system is described in detail in Appendix J of the code.

Petroleum Association of Japan (PAJ): A trade association that represents petroleum companies headquartered in Japan and promotes Japanese petroleum interests.

pH: Measure of alkalinity or acidity in water and water-containing fluids. pH can be used to determine the corrosion-inhibiting characteristic in water-based fluids. Typically, $pH > 8.0$ is required to inhibit corrosion of iron and ferrous alloys in water-based fluids.

Phenol: A white, crystalline hydroxyl (acidic) compound (C_6H_5OH) derived from benzene, used in the manufacture of phenolic resins, weed killers, plastics, disinfectants; also used in solvent extraction, a petroleum refining process. Phenol is a toxic material; skin contact must be avoided.

Phosphate Ester: Any of a group of synthetic lubricants having superior fire resistance. A phosphate ester generally has poor hydrolytic stability, poor compatibility with mineral oil, and a relatively low viscosity index (VI). It is used as a fire-resistant hydraulic fluid in high-temperature applications.

Physical and Chemical Properties: The results from several analytical tests that measure various physical characteristics and ingredients (constituents) of an engine oil.

Pigment: The solid lubricant material (such as graphite or MoS_2) used in a solid lubricant system.

Pinion: The smaller of two mating or meshing gears; can be either the driving or the driven gear.

Pitch Line: An imaginary line that divides the upper and lower halves of gear teeth while in the contact area.

Pitting: A form of extremely localized attack characterized by holes or depressions in the metal; one of the most destructive and insidious forms of corrosion. Depending on the environment and the material, a pit may take months, or even years, to become visible.

Plain Bearing: A relatively simple and inexpensive bearing typically made of two parts. A rotary plain bearing can be just a shaft running through a hole. A simple linear bearing can be a pair of flat surfaces designed to allow motion.

Pleated Filter: A filter element whose medium consists of a series of uniform folds and has the geometric form of a cylinder, cone, disc, plate, etc. Synonymous with "convoluted" and "corrugated."

PNA (Polynuclear Aromatic): Any of numerous complex hydrocarbon compounds consisting of three or more benzene rings in a compact molecular arrangement. Some types of PNAs are formed in fossil fuel combustion and other heat processes, such as catalytic cracking.

Pneumatics: Engineering science pertaining to gaseous pressure and flow.

Poise (Absolute Viscosity): A measure of viscosity numerically equal to the force required to move a plane surface of 1 cm^2 with a velocity of 1 cm/s when the surfaces are separated by a layer of fluid 1 cm in thickness. It is the ratio of the shearing stress to the shear rate of a fluid and is expressed in dyne seconds per square centimeter (dyne-s/cm^2); 1 centipoise equals 0.01 poise.

Polar Compound: A chemical compound whose molecules exhibit electrically positive characteristics at one extremity and negative characteristics at the other. Polar compounds are used as additives in many petroleum products. Polarity gives certain molecules a strong affinity for solid surfaces; as lubricant additives (oiliness agents), such molecules plate out to form a tenacious, friction-reducing film. Some polar molecules are oil-soluble at one end and water-soluble at the other end; in lubricants, they act as emulsifiers, helping form stable oil-water emulsions. Such lubricants are said to have good metal-wetting properties. Polar compounds with a strong attraction for solid contaminants act as detergents in engine oils by keeping contaminants finely dispersed.

Polishing (Bore): Excessive smoothing of the surface finish of the cylinder bore or cylinder liner in an engine to a mirror-like appearance, resulting in depreciation of ring sealing and oil consumption performance.

Polyalkylene Glycol: Mixtures of condensation polymers of ethylene oxide and water. They are any of a family of colorless liquids with high molecular weight that are soluble in water and in many organic solvents. They are used in detergents and as emulsifiers and plasticizers. PAG-based lubricants are used in diverse applications in which petroleum oil-based products do not provide the desired performance—and because they are fire-resistant and will not harm workers or the environment.

Poly α-olefins (PAOs): Synthetic base fluids for high-performance lubricants which are impurity-free and contain only well-defined hydrocarbon molecules. They offer excellent performance over a wide range of lubricating properties; PAOs are manufactured by a two-step reaction sequence from linear α-olefins, which are derived from ethylene.

Polyglycols: Polymers of ethylene or propylene oxides used as a synthetic lubricant base. Properties include very good hydrolytic stability, high viscosity index (VI), and low volatility. Used particularly in water emulsion fluids.

Polymer: A substance formed by the linkage (polymerization) of two or more simple molecules, called monomers, to form a single larger molecule having the same elements in the same proportions as the original monomers; that is, each monomer retains its structural identity. A polymer may be liquid or solid; solid polymers may consist of millions of repeated linked units. A polymer made from two or more similar monomers is called a copolymer; a copolymer composed of three different types of monomers is a terpolymer. Natural rubber and synthetic rubbers are examples of polymers. Polymers are commonly used as viscosity index improvers in multigrade oils and tackifiers in lubricating greases.

Polymerization: The chemical combination of similar-type molecules to form larger molecules.

Polyol Ester: A synthetic lubricant base, formed by reacting fatty acids with a polyol (such as a glycol) derived from petroleum. Properties include good oxidation stability at high temperatures and low volatility. Used in formulating lubricants for turbines, compressors, jet engines, and automotive engines.

Polyolefin: A polymer derived by polymerization of relatively simple olefins. Polyethylene and polyisoprene are important polyolefins.

Pore: A small channel or opening in a filter medium which allows passage of fluid.

Pore Size Distribution: The ratio of the number of effective holes of a given size to the total number of effective holes per unit area expressed as a percentage and as a function of hole size.

Porosity: The ratio of pore volume to total volume of a filter medium expressed as a percentage.

Pour Point: Lowest temperature at which an oil or distillate fuel is observed to flow, when cooled under prescribed conditions (ASTM D97). The pour point is 3°C (5°F) above the temperature at which the oil in a test vessel shows no movement when the container is held horizontally for 5 s.

Pour Point Depressant: An additive that retards the adverse effects of wax crystallization and lowers the pour point.

Pour Stability: The ability of pour point-depressed oil to maintain its original ASTM pour point when subjected to long-term storage at low temperature approximating winter conditions.

Power Unit: A combination of pump, pump drive, reservoir, controls, and conditioning components, which may be required for its application.

ppm: Parts per million (1/ppm = 0.000001); generally by weight (ppm w/w); 100 ppm = 0.01%; 10,000 ppm = 1%;

Predictive Maintenance: A type of condition-based maintenance emphasizing early prediction of failure using nondestructive techniques such as vibration analysis, thermography, and wear debris analysis.

Pressure: Force per unit area, usually expressed in pounds per square inch (psi) or newtons per square meter (Pascals).

Pressure Control Valve: A pressure control valve whose primary function is to limit system pressure.

Pressure Drop: Resistance to flow created by the element (media) in a filter. Defined as the difference in pressure upstream (inlet side of the filter) and downstream (outlet side of the filter).

Pressure Gauge (psig): Pressure differential above or below atmospheric pressure.

Pressure Indicator: An indicator that signals pressure conditions.

Pressure Line Filter: A filter located in a line conducting working fluid to a working device or devices.

Pressure Switch: An electric switch operated by fluid pressure.

Pressure, Absolute: The sum of atmospheric and gauge pressures.

Pretreatment: Usually refers to the treatment of a substrate before the application of a solid film lubricant.

Preventive Maintenance: Maintenance performed according to a fixed schedule involving the routine repair and replacement of machine parts and components.

Primary Treatment: The first step in re-refining used lubricating using dehydration in which the oil is stored to allow water and solids to separate out from the oil, then the oil is heated to approximately 120°C (248°F) in a closed vessel to boil off any emulsified water and some of the fuel diluents.

Proactive Maintenance: A maintenance strategy for stabilizing the reliability of machines or equipment. Its central theme involves directing corrective actions aimed at failure root causes, not active failure symptoms, faults, or machine wear conditions. A typical proactive maintenance regimen involves three steps: (1) setting a quantifiable target or standard relating to a root cause of concern (e.g., a target fluid cleanliness level for a lubricant), (2) implementing a maintenance program to control the root cause property to within the target level (e.g., routine exclusion or removal of contaminants), and (3) routine monitoring of the root cause property using a measurement technique (e.g., particle counting) to verify that the current level is within the target.

Processing: Operations designed to permit the reuse of waste oils, which includes both regeneration and combustion.

Process Oil: Oil that serves as a temporary or permanent component of a manufactured products. Aromatic process oils have good solvency characteristics; their applications include proprietary chemical formulations, ink oils, and extenders in synthetic rubbers. Naphthenic process oils are characterized by low pour points and good solvency properties. Paraffinic process oils are characterized by low aromatic content and light color.

Product Traceability Code: A code that permits oil samples in the marketplace to be traced by formulation, date of packaging, and source of manufacture.

Provisional License: Authority granted by the API to a marketer to permit the temporary licensing of a specific engine oil when one of the required engine tests has been declared "out of control" by ASTM. A provisional license may also be granted for an engine oil that is qualified by means of SAE viscosity-grade engine testing "read-across" from another provisionally licensed engine oil (see Chapter 2 for details).

psi: Pounds per square inch.

psia: Pounds per square inch absolute (psig + 14.696 psi); absolute, unit of pressure (pressure above total vacuum).

psid: Pounds per square inch differential.

psig: Pounds per square inch gauge (psia – 14.696 psi); unit of pressure (pressure above atmospheric).

Pump: A device that converts mechanical force and motion into hydraulic fluid power.

Pumpability: The low-temperature, low shear stress/shear rate viscosity characteristics of an oil that permit satisfactory flow to and from the engine oil pump and subsequent lubrication of moving components.

Pusher Seal: A mechanical seal in which the secondary seal is pushed along the shaft or sleeve to compensate for misalignment and face wear.

PV: The PV value is the product of load (P) on the projected bearing area (PSI) and the surface velocity in feet per minute.

Quenching Oil: Also called heat-treating oil; a high-quality, oxidation-resistant petroleum oil used to cool metal parts during their manufacture, and is often preferred to water because the oil's slower heat transfer lessens the possibility of cracking or warping of the metal. A quenching oil must have excellent oxidation and thermal stability, and should yield clean parts, essentially free of residue. In refining terms, a quenching oil is an oil introduced into high-temperature vapors of cracked petroleum fractions (see Cracking) to cool them.

Quick Disconnect Coupling: A coupling that can quickly join or separate a fluid line without the use of tools or special devices.

R&O (Rust and Oxidation Inhibited): A term applied to highly refined industrial lubricating oils formulated for long service in circulating lubrication systems, compressors, hydraulic systems, bearing housing, gear boxes, etc. The finest R&O oils are often referred to as turbine oils.

Raffinate: In solvent-refining practice, that portion of the oil that remains undissolved and is not removed by the selective solvent; the solvent lean phase.

Ramsbottom Carbon: An alternate test method for measuring the carbon residue of petroleum fractions (see Conradson carbon).

Rate of Shear: The difference between the velocities along the parallel faces of a fluid element divided by the distance between the faces.

Rated Flow: The maximum flow that the power supply system is capable of maintaining at a specific operating pressure.

Rated Pressure: The qualified operating pressure that is recommended for a component or a system by the manufacturer.

Reclamation: Treatment of used oil to separate solids and water from a variety of used oils. The methods used may include heating, filtering, dehydrating, and centrifuging. Reclaimed oil is generally used as a fuel or fuel extender.

Reclaimed Oil: Lubricating oil that is collected after use, reprocessed, and then reused as a lubricant or fuel.

Recycling: The commonly used generic term for the reprocessing, reclaiming, and regeneration (re-refining) of used oils by use of an appropriate selection of physical and chemical methods of treatment.

Reducer: A connector having a smaller line size at one end than the other.

Referenced Laboratory: An engine-testing laboratory that is monitored by the ASTM Test Monitoring Center's blind reference oil system.

Refining: A series of processes for converting crude oil and its fractions to finished petroleum products. After distillation, a petroleum fraction may undergo one or more additional steps to purify or modify it. These refining steps include thermal cracking, catalytic cracking, polymerization, alkylation, reforming, hydrocracking, hydroforming, hydrogenation, hydrogen treating, hydrofining, solvent extraction, dewaxing, deoiling, acid treating, clay filtration, and deasphalting. Refined lubricating oils may be blended with other lube stocks, and additives may be incorporated to impart special properties.

Refraction: The change of direction or speed of light as it passes from one medium to another.

Refrigeration Compressor: A special type of compressor typically used for refrigeration, heat pumping, and air conditioning. They are made to turn low-pressure gases into high-pressure and high-temperature gases. The three main types of refrigeration compressors are screw compressors, scroll compressors, and piston compressors.

Refrigerator Oil: The lubricant added to the working fluid in an expansion-type cooling unit, which serves to lubricate the pump mechanism.

Regeneration: A process whereby base oils can be produced by refining waste oils, in particular by removing the contaminants, oxidation products, and additives contained in such oils.

Remaining Useful Life: An opinion (based on data, observations, history, records, exposure, etc.) of the number of years before a fluid, system, or component will require replacement or reconditioning.

Reprocessed Oil: Lubricating oil that is collected after use, reprocessed, and refortified with additives and then reused.

Reprocessing: Treatment of used oil to remove insoluble contaminants and oxidation products such as by heating, settling, filtering, dehydrating, and centrifuging. Depending on the quality of the resultant material, this can be followed by blending with base oils and additives to bring the oil back to its original or an equivalent specification. Reprocessed oil is generally returned to its original use.

Re-refined Oil: Lubricating oil that is collected after use, reprocessed, and re-refined and then sold for reuse.

Re-refining: A process of reclaiming used lubricant oils and restoring them to a condition similar to that of virgin stocks by filtration, clay adsorption, or more elaborate methods.

Reservoir: A container for storage of liquid in a fluid power system.

Reservoir Filter: A filter installed in a reservoir in series with a suction or return line. Also known as sump filter.

Residual Dirt Capacity: The dirt capacity remaining in a service-loaded filter element after use, but before cleaning, measured under the same conditions as the dirt capacity of a new filter element.

Residuum: The highest-boiling (heaviest) components or bottoms remaining from distilling an oil, especially crude oil.

Return Line: A location in a line conducting fluid from working device to reservoir.

Return Line Filtration: Filters located upstream of the reservoir but after fluid has passed through the system's output components (cylinders, motors, etc.).

Reynolds Number: A dimensionless numerical ratio of the dynamic forces of mass flow to the shear stress due to viscosity. Flow usually changes from laminar to turbulent between Reynolds Number 2000 and 4000.

RFO: Recovered fuel oil.

Rheology: The study of the deformation and flow of matter in terms of stress, strain, temperature, and time. The rheological properties of a grease are commonly measured by penetration and apparent viscosity.

Ring Lubrication: A system of lubrication in which the lubricant is supplied to the bearing by an oil ring.

Ring Sticking: Freezing of a piston ring in its groove in a piston engine or reciprocating compressor due to heavy deposits in the piston ring zone.

Rings: Circular metallic elements that ride in the grooves of a piston and provide compression sealing during combustion. Also used to spread oil for lubrication.

Roller Bearing: An antifriction bearing comprising rolling elements in the form of rollers.

Rolling Element Bearing: A friction-reducing bearing that consists of a ring-shaped track that contains free-revolving metal balls. A rotating shaft or other part turns against such a bearing.

Rolling Oil: Oil used in hot- or cold-rolling of ferrous and nonferrous metals to facilitate feed of the metal between the work rolls, improve the plastic deformation of the metal, conduct heat from the metal, and extend the life of the work rolls. Because of the pressures involved, a rolling oil may be compounded or contain EP additives. In hot rolling, the oil may also be emulsifiable.

Roll-off Cleanliness: The fluid system contamination level at the time of release from an assembly or overhaul line. Fluid system life can be shortened significantly by full-load operation under a high fluid contamination condition for just a few hours. Contaminant implanted and generated during the break-in period can devastate critical components unless removed under controlled operating and high-performance filtering conditions.

ROSE: Recycling Oil Saves the Environment Foundation (South Africa).

ROSE Process: Residual Oil Supercritical Extraction Process (Kellogg Brown and Root).

Rotary Seal: A mechanical seal that rotates with a shaft and is used with a stationary mating ring.

Rotating Equipment: Equipment that moves liquids, solids, or gases through a system of drivers (turbines, motors, engines), driven components (compressors, pumps), transmission devices (gears, clutches, couplings) and auxiliary equipment (lube and seal systems, cooling systems, buffer gas systems).

Rotating Pressure Vessel Oxidation Test (RPVOT): The RPVOT measures an oil's oxidation stability. The oil sample is placed in a vessel containing a polished copper coil. The vessel is then charged with oxygen and placed in a bath at a constant temperature of 150°C. Stability is expressed in terms of the time it takes to achieve a pressure drop of 25.4 pounds per square inch (psi) pressure drop from maximum pressure.

Rust Inhibitor: A type of corrosion inhibitor used in lubricants to protect surfaces against rusting.

Rust Prevention Test (Turbine Oils): A test for determining the ability of an oil to aid in preventing the rusting of ferrous parts in the presence of water.

s: Second, unit of time.

SAE: Society of Automotive Engineers, an organization serving the automotive industry.

SAE Port: A straight thread port used to attach tube and hose fittings. It employs an "O" ring compressed in a wedge-shaped cavity. A standard of the Society of Automotive Engineers J514 and ANSI/B116.1.

SAE Viscosity: The viscosity classification of a motor oil according to the system developed by the Society of Automotive Engineers and now in general use. "Winter" grades are defined by viscosity measurements at low temperatures and have "W" as a suffix, whereas "Summer" grades are defined by viscosity at 100°C and have no suffix. Multigrade oils meet both a winter and a summer definition and have designations such as SAE 10W-30.

SAE Viscosity Number: An arbitrary number in a system for classifying oils, automatic transmission fluids, and differential lubricants according to their viscosities; the viscosity number does not connote quality.

Salt Base: The hydrolysis product of an alkaline material and a silicon halide.

Sample Preparation: Fluid factors that can enhance the accuracy of the particulate analysis. Such factors include particle dispersion, particle settling, and sample dilution.

Saponification Number: The amount of potassium hydroxide (KOH) that combines with 1 g of oil under specified conditions (ASTM D94). Saponification number is an indication of the amount of fatty saponifiable material in compounded oil. Caution must be used in interpreting test results if certain substances—such as sulfur compounds or halogens—are present in the oil because these also react with KOH, thereby increasing the apparent saponification number.

Saturation Level: The amount of water that can dissolve in a fluid.

Saybolt Color: A color standard for petroleum products (ASTM D156).

Saybolt Furol Viscosity: The time in seconds for 60 mL of fluid to flow through a standard Saybolt Furol viscosimeter (ASTM Method D88).

Saybolt Universal Viscosity (SUV) or Saybolt Universal Seconds (SUS): The time in seconds required for 60 cm^3 of a fluid to flow through the orifice of the Standard Saybolt Universal Viscometer at a given temperature under specified conditions (ASTM D88).

Scoring: Distress marks on sliding metallic surfaces in the form of long, distinct scratches in the direction of motion. Scoring is an advanced stage of scuffing.

Scuffing: Abnormal engine wear due to localized welding and fracture. It can be prevented through the use of antiwear, extreme-pressure, and friction modifier additives.

Scuffing Particles: Large twisted and discolored metallic particles resulting from adhesive wear due to complete lubricant film breakdown.

Seal: A device designed to prevent the movement of fluid from one area to another, or to exclude contaminants.

Seal Assembly: A group of parts, or a unitized assembly, that includes sealing surfaces, provisions for initial loading, and a secondary sealing mechanism that accommodates the radial and axial movement necessary for installation and operation.

Seal Chamber: The area between the seal chamber bore and a shaft in which a mechanical seal is installed.

Seal Face: It is either of the two lapped surfaces in a mechanical seal assembly forming the primary seal.

Seal Face Width: The radial distance from the inside edge to the outside edge of the sealing face.

Seal Swell (Rubber Swell): The swelling of rubber (or other elastomers) gaskets, or seals when exposed to petroleum, synthetic lubricants, or hydraulic fluids. Seal materials vary widely in their resistance to the effect of such fluids. Some seals are designed so that a moderate amount of swelling improves sealing action.

Sealed Motor Bearing: These bearings have rubbing seals that seal against recesses in the inner ring shoulder. They are lubricated for life. Under extreme conditions, their life can be short.

Semisolid: Any substance having the attributes of both a solid and a liquid. Similar to semiliquid but being more closely related to a solid than a liquid. More generally, any substance in which the force required to produce a deformation depends both on the magnitude and on the rate of the deformation.

Separation: Re-refining used lubricating oils, which consists of the removing of mechanical and oxidation contaminants.

Service Category: An alphanumeric code developed by API to specify a level of performance defined by ASTM D4485 and SAE J183. As new Service Categories are developed, new alphanumeric codes may be assigned.

Service Factor: A quantity that relates the actual on-stream time of a process unit to the total time available for use of the unit; frequently, a ratio of the number of actual operating days divided by 365.

Servovalve: A valve that modulates output as a function of an input command.

Settling Tank: A tank in which liquid is stored until particles suspended in the liquid sink to the bottom.

Severe Sliding: Large ferrous particles that are produced by sliding contacts. Trend is important to determine whether abnormal wear is taking place.

Severity Adjustments: Mathematically derived correction factors designed to minimize or eliminate laboratory biases. Severity adjustments are developed by the testing laboratory and confirmed by the ACC Monitoring Agency and the ASTM Test Monitoring Center.

Shear Rate: Rate at which adjacent layers of fluid move with respect to each other, usually expressed as reciprocal seconds.

Shear Stability: The ability of a lubricating grease to resist changes in consistency (hardness) during mechanical working.

Shear Stress: Frictional force overcome in sliding one "layer" of fluid along another, as in any fluid flow. The shear stress of a petroleum oil or other Newtonian fluid at a given temperature varies directly with shear rate (velocity). The ratio between shear stress and shear rate is constant; this ratio is termed viscosity of a Newtonian fluid.

Silt: Contaminant particles 5 μm and less in size.

Silting: A failure generally associated with a valve whose movements are restricted due to small particles that have wedged in between critical clearances (e.g., the spool and bore).

Single-pass Test: Filter performance tests in which contaminant that passes through a test filter is not allowed to recirculate back to the test filter.

Sintered Medium: A metallic or nonmetallic filter medium processed to cause diffusion bonds at all contacting points.

Slack Wax: The soft, oily crude wax obtained from the solvent dewaxing of paraffin distillates or lube base stocks. Slack waxes contain varying amounts of oil and must be deoiled to produce hard or finished waxes.

Sleeve Bearing: A journal bearing, usually a full journal bearing.

Sloughing Off: The release of contaminant from the upstream side of a filter element to the upstream side of the filter enclosure.

Sludge: Insoluble material formed as a result either of deterioration reactions in an oil or of contamination of an oil, or both.

Society of Automotive Engineers (SAE): An engineering society founded to develop, collect, and disseminate knowledge of mobility technology.

Solid: Any substance having a definite shape, which it does not readily relinquish. More generally, any substance in which the force required to produce a deformation depends on the magnitude of the deformation rather than on the rate of deformation.

Solid Lubricant: A solid material that provides lubrication between two relatively moving surfaces.

Solvency: Ability of a fluid to dissolve inorganic materials and polymers, which is a function of aromaticity.

Solvent: A material with a strong capability to dissolve a given substance. The most common petroleum solvents are mineral spirits, xylene, toluene, hexane, heptane, and naphtha. Aromatic-type solvents have the highest solvency for organic chemical materials, followed by naphthenes and paraffins. In most applications, the solvent disappears, usually by evaporation, after it has served its purpose. The evaporation rate of a solvent is very important in manufacturing processes.

Solvent Extraction: A refining process used to separate components (unsaturated hydrocarbons) from lube distillates to improve the oil's oxidation stability, viscosity index, and response to additives. The oil and the solvent extraction media are mixed in an extraction tower, resulting in the formation of two phases: a heavy phase consisting of the undesirable unsaturates dissolved in the solvent, and a lighter phase consisting of a high-quality oil with some solvent dissolved in it. The phases are separated and the solvent recovered from each by distillation.

Specification: A defined range for a test that must be adhered to in the manufacture of a product.

Specific Gravity: The ratio of the weight of a given volume of material to the weight of an equal volume of water.

Specific Gravity (Liquid): The ratio of the weight of a given volume of liquid to the weight of an equal volume of water.

Spectrographic Analysis: Determines the concentration of elements represented in the entrained fluid contaminant.

Spectrographic Oil Analysis Program (SOAP): Procedures for extracting fluid samples from operating systems and analyzing them spectrographically for the presence of key elements.

Spindle Oil: Light-bodied oil used principally for lubricating textile spindles and for light, high-speed machinery.

Spin-on Filter: A throwaway-type bowl and element assembly that mates with a permanently installed head.

Splash Lubrication: A system of lubrication in which parts of a mechanism dip into and splash the lubricant onto themselves or on other parts of the mechanism.

Spur Gear: This is the simplest variation of a gear. It consists of a cylinder or disk, with the teeth projecting radially. Each tooth edge is straight and aligned parallel to the axis of rotation. Such gears can be meshed together correctly only if they are fitted to parallel axles.

SSU: Saybolt Universal Seconds (or SUS), a unit of measure used to indicate kinematic viscosity, for example, SUS at 100°F.

St: Stoke, unit of kinematic viscosity.

Static Friction: The force just sufficient to initiate relative motion between two bodies under load. The value of the static friction at the instant relative motion begins is termed breakaway friction.

Static Seal: A seal between two surfaces that have no relative motion.

Stationary Seal: A mechanical seal in which the flexible members do not rotate with the shaft.

Steam Turbine: A mechanical device that extracts thermal energy from pressurized steam. The energy is converted into a rotary motion that drives a device.

Stick–slip Motion: Erratic, noisy motion characteristic of some machine ways due to the starting friction encountered by a machine part at each end of its to-and-from (reciprocating) movement. This undesirable effect can be overcome with a way lubricant, which reduces starting friction.

STLE: Society of Tribologists and Lubrication Engineers, formerly ASLE (American Society of Lubrication Engineers).

Stoke (St): Kinematic measurement of a fluid's resistance to flow defined by the ratio of the fluid's dynamic viscosity to its density.

Straight Mineral Oil: Petroleum oil containing no additives. Straight mineral oils include such diverse products as low-cost once-through lubricants and thoroughly refined white oils. Most high-quality lubricants, however, contain additives.

Straight Oil: A mineral oil containing no additives.

Strainer: A coarse filter element (pore size more than approximately 40 μm).

Suction Filter: A pump intake-line filter in which the fluid is below atmospheric pressure.

Sulfated Ash: The ash content of fresh compounded lubricating oil (ASTM D874); indicates level of metallic additives in the oil.

Sulfonate: A hydrocarbon in which a hydrogen atom has been replaced with the highly polar (SO_2OX) group, where X is a metallic ion or alkyl radical. Petroleum sulfonates are refinery by-products of the sulfuric acid treatment of white oils. Sulfonates have important applications as emulsifiers and chemical intermediates in petrochemical manufacture, and substituted sulfonates are widely used as corrosion inhibitors. Synthetic sulfonates can be manufactured from special feedstocks rather than from white oil base stocks.

Sulfur: A common natural constituent of petroleum products. Although certain sulfur compounds are commonly used to improve the EP or load-carrying properties of an oil, high sulfur content in a petroleum product may be undesirable because it can be corrosive and create an environmental hazard when burned. For these reasons, sulfur limitations are specified in the quality control of fuels, solvents, etc.

Sulfurized Oil: Oil to which sulfur or sulfur compounds have been added.

Superclean: Not more than 10 particles larger than 10 μm/mL.

Surface Fatigue Wear: The formation of surface or subsurface cracks and fatigue crack propagation. It results from cyclic loading of a surface.

Surface Filter Media: Porous materials that primarily retain contaminants on the influent face, performing the actual process of filtration.

Surface Filtration: Filtration that primarily retains contaminant on the influent surface.

Surface Tension: The contractile surface force of a liquid by which it tends to assume a spherical form and to present the least possible surface. It is expressed in dynes per centimeter, ergs per square centimeter, or newtons per meter.

Surfactant: Surface-active agent that reduces interfacial tension of a liquid. A surfactant used in a petroleum oil may increase the oil's affinity for metals and other materials.

Surge: A momentary increase of pressure in a circuit.

SUS (SSU): Saybolt Universal Seconds. A measure of lubricating oil viscosity in the oil industry. The measuring apparatus is filled with a specific quantity of oil or other fluid and its flow time through a standardized orifice is measured in seconds. Fast-flowing fluids (low viscosity) will have a low value; slow-flowing fluids (high viscosity) will have a high value.

Swarf: The cuttings and grinding fines that result from metal-working operations.

Synthetic Grease: Grease composition in which the liquid lubricant is not a mineral oil.

Synthetic Hydrocarbon: Oil molecule with superior oxidation quality tailored primarily out of paraffinic materials.

Synthetic Lubricant: A lubricant produced by chemical synthesis rather than by extraction or refinement of petroleum to produce a compound with planned and predictable properties; generally a non–petroleum-based lubricant, usually with a very narrow boiling range.

Synthetic Oil: Oil produced by synthesis (chemical reaction) rather than by extraction or refinement. Many (but not all) synthetic oils offer immense advantages in terms of high-temperature stability and low-temperature fluidity, but are more costly than mineral oils. Major advantage of all synthetic oils is their chemical uniformity.

System Pressure: The pressure that overcomes the total resistances in a system. It includes all losses as well as useful work.

Tacky: A descriptive term applied to lubricating oils and greases that seem particularly sticky or adhesive.

TAN: Total acid number.

TBN: Total base number.

Texture: The property of lubricating grease that is observed when a small separate portion of it is pressed together and then slowly drawn apart.

TFE: Thin film evaporator (see Thin Film Evaporator).

Thermal Conductivity: Measure of the ability of a solid or liquid to transfer heat.

Thermal Stability: Ability of a fuel or lubricant to resist oxidation under high-temperature operating conditions.

Thermography: The use of infrared thermography whereby temperatures of a wide variety of targets can be measured remotely and without contact. This is accomplished by measuring the infrared energy radiating from the surface of the target and converting this measurement to an equivalent surface temperature.

Thickener: The solid particles that are relatively uniformly dispersed to form the structure of lubricating grease in which the liquid is held by surface tension and other physical forces.

Thin Film Evaporator (TFE): Equipment used to create a thin film from which it is possible to evaporate or sublimate various elements; a thin film evaporator can deposit extremely thin layers of atoms or molecules onto a substrate material; consists of a vacuum chamber, a heating element, and an apparatus that holds and moves a substrate while a thin film is deposited onto it. For distillation of used oils, a thin film evaporator spreads a thin film of oil over a heated surface and the distilled oil fraction rapidly evaporates while the residue fraction falls to the bottom of the vessel. The wiped film evaporator (WFE) is a thin film evaporator in which a wiper blade or blades continuously clean the evaporating surface.

Thin Film Lubrication: A condition of lubrication in which the film thickness of the lubricant is such that the friction between the surfaces is determined by the properties of the surfaces as well as by the viscosity of the lubricant.

Thixotropy: That property of lubricating grease which is manifested by a softening in consistency as a result of shearing followed by a hardening in consistency starting immediately after the shearing is stopped.

Three-body Abrasion: A particulate wear process by which particles are pressed between two sliding surfaces.

Thrust Bearing: An axial-load bearing.

Timken EP Test: Measure of the extreme-pressure properties of a lubricating oil. The test utilizes a Timken machine, which consists of a stationary block pushed upward, by a lever arm system, against the rotating outer race of a roller bearing, which is lubricated by the product under test. The test continues under increasing load (pressure) until a measurable wear scar is formed on the block.

Timken OK Load: The heaviest load that a test lubricant will sustain without scoring the test block in the Timken test procedures (ASTM D2509 [greases] and ASTM D2782 [oils]).

Timken Ring: A standard ring used on several bench-type wear and friction testers.

Tonne: A metric tonne of 1000 kg, equivalent to 2205 lbs.

Torr: The unit of pressure that is defined in terms of a standard atmosphere and now taken as 1,013,250 dynes/cm^2 or 101.325 kPa; the torr is 1/760 atmosphere, or 1340 dynes/cm^2; 1 torr is approximately equivalent to 1 mm Hg.

Total Acid Number (TAN): The quantity of base, expressed in milligrams of potassium hydroxide, that is required to neutralize all acidic constituents present in 1 g of sample (ASTM D974). See Acid Number.

Total Base Number (TBN): The quantity of acid, expressed in terms of the equivalent number of milligrams of potassium hydroxide that is required to neutralize all basic constituents present in 1 g of sample. (ASTM D974). See Base Number.

Tribology: The science and technology of interacting surfaces in relative motion, including the study of lubrication, friction, and wear. Tribological wear is wear that occurs as a result of relative motion at the surface.

Turbidity: The degree of opacity of a fluid.

Turbine Oil: Top-quality rust- and oxidation-inhibited (R&O) oil that meets the rigid requirements traditionally imposed on steam turbine lubrication.

Quality turbine oils are also distinguished by good demulsibility, a requisite of effective oil–water separation. Turbine oils are widely used in other exacting applications for which long service life and dependable lubrication are mandatory such as certain compressors, hydraulic systems, gear drives, and other equipment. Turbine oils can also be used as heat transfer fluids in open systems, where oxidation stability is of primary importance.

Turbulent Flow: A flow situation in which the fluid particles move in a random manner; flow in which the velocity at any point varies on an erratic basis. It occurs when flow velocity exceeds a limiting value or when tube configuration irregularities preclude laminar flow.

Turbulent Flow Sampler: A sampler that contains a flow path in which turbulence is induced in the main stream by abruptly changing the direction of the fluid.

UK: United Kingdom (Britain).

Ultraclean: Not more than one particle larger than 10 μm per milliliter.

Unbalanced Seal: A mechanical seal arrangement wherein the full hydraulic pressure of the seal chamber acts to close the seal faces.

UNEP: United Nations Environmental Program.

UNIDO: United Nations Industrial Development Organization.

Unloading: The release of contaminant that was initially captured by the filter medium.

US or USA: United States of America.

U.S. DOE: United States Department of Energy.

Used Oil: A mineral-based or synthetic lubrication or industrial oil that has become unfit for the use for which they were originally intended; also referred to as Waste Oil.

U.S. EPA: United States Environmental Protection Agency.

Vacuum Dehydration: A method that involves drying or freeing of moisture through a vacuum process.

Vacuum Distillation: A distillation method that involves reducing the pressure above a liquid mixture to be distilled to less than its vapor pressure (usually less than atmospheric pressure). This causes evaporation of the most volatile liquid(s)—those with the lowest boiling points. This method works on the principle that boiling occurs when a liquid's vapor pressure exceeds the ambient pressure. It can be used with or without heating the solution.

Vacuum Pump: A device that is used to extract gas or vapor from an enclosed space, leaving behind a partial vacuum in the container.

Vacuum Separator: A separator that utilizes subatmospheric pressure to remove certain gases and liquids from another liquid because of their difference in vapor pressure.

Valve: A device which controls fluid flow direction, pressure, or flow rate.

Valve Lifter: Sometimes called a "cam follower," a component in engine designs that use a linkage system between a cam and the valve it operates. The lifter typically translates the rotational motion of the cam to a reciprocating linear motion in the linkage system.

Vapor Pressure: Pressure of a confined vapor in equilibrium with its liquid at specified temperature, and is thus a measure of a liquid's volatility.

Vapor Pressure—Reid (RVP): Measure of the pressure of vapor accumulated above a sample of gasoline or other volatile fuel in a standard bomb at 100°F (37.8°C). Used to predict the vapor locking tendencies of the fuel in a vehicle's fuel system. Controlled by law in some areas to limit air pollution from hydrocarbon evaporation while dispensing.

Variable Displacement Pump: A pump in which the displacement per cycle can be varied.

Varnish: When applied to lubrication, a thin, insoluble, nonwipeable film deposit occurring on interior parts, resulting from the oxidation and polymerization of fuels and lubricants. Can cause sticking and malfunction of close-clearance moving parts. Similar to, but softer than, lacquer.

VD: Vacuum distillation.

VDT (Vacuum Distillation Tower): Generally applies to a crude distillation tower, which operates below atmospheric pressure.

VDU (Vacuum Distillation Unit): Generally includes a VDT and associated equipment for producing distillates from the bottoms of an atmospheric distillation tower (ADT) by operating below atmospheric pressure.

VI: Viscosity index.

Viscometer (Viscosimeter): An apparatus for determining the viscosity of a fluid.

Viscosity: Measurement of a fluid's resistance to flow. The common metric unit of absolute viscosity is the poise, which is defined as the force in dynes required to move a surface 1 cm^2 in area past a parallel surface at a speed of 1 cm/s, with the surfaces separated by a fluid film 1 cm thick. In addition to kinematic viscosity, there are other methods for determining viscosity, including Saybolt universal viscosity (SUV), Saybolt Furol viscosity, Engler viscosity, and Redwood viscosity. Because viscosity varies inversely with temperature, its value is meaningless until the temperature at which it is determined is reported.

Viscosity Grade: Any of a number of systems that characterize lubricants according to viscosity for particular applications, such as industrial oils, gear oils, automotive engine oils, automotive gear oils, and aircraft piston engine oils.

Viscosity Index (VI): A commonly used measure of a fluid's change of viscosity with temperature. The higher the viscosity index, the smaller the relative change in viscosity with temperature.

Viscosity Index Improvers: Additives that increase the viscosity stability of the fluid throughout its useful temperature range. Such additives are polymers that possess thickening power as a result of their high molecular weight and are necessary for the formulation of multigrade engine oils.

Viscosity Modifier: Lubricant additive, usually a high molecular weight polymer, that reduces the tendency of an oil's viscosity to change with temperature.

Viscosity–Temperature Relationship: The manner in which the viscosity of a given fluid varies inversely with temperature. Because of the mathematical relationship that exists between these two variables, it is possible to graphically predict the viscosity of a petroleum fluid at any temperature within a limited range if the viscosities at two other temperatures are known. The charts used for this purpose are the ASTM Standard Viscosity–Temperature

Charts for liquid petroleum products, which is available in six ranges. If two known viscosity–temperature points of a fluid are located on the chart and a straight line drawn through them, other viscosity–temperature values of the fluid will fall on this line; however, values near or below the cloud point of the oil may deviate from the straight-line relationship.

Viscous: Possessing viscosity. Frequently used to imply high viscosity.

Volatility: This property describes the degree and rate at which a liquid will vaporize under given conditions of temperature and pressure. When liquid stability changes, this property is often reduced in value.

VPS (Vacuum Pipe Still): Generally includes a vacuum tower and associated equipment for the distillation of crude into lube distillates or cracking feedstocks and vacuum residua (see VDU).

Waste Oil: Any mineral-based lubrication or industrial oil that has become unfit for the use for which they were originally intended; more commonly referred to as Used Oil.

Water–Glycol Fluid: A fluid whose major constituents are water and one or more glycols or polyglycols.

Wax: Plastic, fusible, and viscous or solid substance having a characteristic luster; wax that is present in a crude oil belongs to two major varieties: paraffin wax and petrolatum.

Wax Distillate: A distillate prepared by distillation of waxy crude on a VPS; generally requires further processing including solvent refining and dewaxing to produce a lubricant base oil.

Way: Longitudinal surface that guides the reciprocal movement of a machine part.

Way Lubricant: Lubricant for the sliding ways of machine tools such as planers, grinders, horizontal boring machines, shapers, jig borers, and milling machines. A good way lubricant is formulated with special frictional characteristics designed to overcome the stick–slip motion associated with slow-moving machine parts.

Wear: The attrition or rubbing away of the surface of a material as a result of mechanical action.

Wear Debris: Particles that are detached from machine surfaces as a result of wear and corrosion; also known as wear particles.

Wear Factor (K): A proportionality constant relating radial wear (R) to load (P), velocity (V), and time (T) in the equation R = KPTV.

Wear Inhibitor: An additive that protects the rubbing surfaces against wear, particularly from scuffing, if the hydrodynamic film is ruptured.

Weld Point: The lowest applied load (in kilograms) at which the rotating ball in the Four-Ball EP test either seizes and welds to the three stationary balls, or at which extreme scoring of the three balls results.

WFE: Wiped film evaporator; equipment designed to continuously separate volatile compounds by introducing a mechanically agitated, thin film of feed material to a heated surface; the short residence time allows for efficient and reliable processing of a wide variety of high-boiling, heat-sensitive, or viscous products.

White Oil: A colorless and odorless mineral oil used in medicinal and pharmaceutical preparations and as a lubricant in food and textile industries.

Wicking: The vertical absorption of a liquid into a porous material by capillary forces.

Work Penetration: The penetration of a sample of lubricating grease immediately after it has been brought to 77°F (25°C) and then subjected to 60 strokes in a standard grease worker (ASTM D217).

Worm Gear: A gear that is in the form of a screw. The screw thread engages the teeth on a worm wheel. When rotated, the worm pulls or pushes the wheel, causing rotation.

ZDDP: An antiwear additive found in many types of hydraulic and lubricating fluids. Zinc dialkyldithiophosphate.

Appendix A: Conversion Factors

	SI	FPS or Other	Conversion Factor
Energy: *E*	J	BTU	1 BTU = 1055.1 J
			1 kcal = 4186.8 J
Force: *F*	N	lb_f	1 lb_f (32.2 lb_mft/s^2) = 4.4482 N
Torque: *T*	(1 N = 1 kg·m/s^2)	poundal	1 poundal (1 lb_mft/s^2) = 0.138255 N
	N·m	lb_f ·ft	1 lb_f ft = 1.355818 N·m
Length: *L*	m	ft	1 ft = 0.3048 m
		ins. (12″ = 1 ft)	1″ (1 inch) = 2.54 cm
Mass: *M*	kg	lb_m	1 lb_m = 0.45359 kg
			2.2046 lb_m = 1 kg
			1 slug = 14.5939 kg
Power: *W*	W	BTU/h	1 BTU/h = 0.29308 W
	(1 W = 1 J/s)	horsepower (hp)	1 hp = 745.70 W
Pressure: *P*	Pa or kPa	psi	1 psi (lb_f/in^2) = 6.895 kPa
	(1 Pa = 1 N/m^2)	(lbs/sq inch)	1 bar = 10^5 Pa
			1 standard atm = 101.325 kPa
			1 at (1 kg_f/cm^2) = 98.067 kPa
			1 mm Hg = 0.13333 kPa
			1″ H_2O = 0.24909 kPa
			1″ Hg = 3.3866 kPa
			1 lb_f/100 ft^2 = 0.4788 Pa
Temperature: *T*	K or °C	°R or °F	9°C = 5°F – 160
			–273.15°C = –459.67°F
Dynamic viscosity	Pa·s	Poise, cP	1 cP = 0.001 Pa·s
Kinematic viscosity	m^2/s	(100 cP = 1 P)	1 cSt = 1 mm^2/s = 10^{-6} m^2/s
(KV=DV/density)		centiStokes	
Volume: *V*	m^3	ft^3	1 ft^3 = 0.028317 m^3
		Imperial gallon	1 Imperial gallon = 4.5461 · 10^{-3} m^3
		US gallon	1 US gallon = 3.78541 · 10^{-3} m^3
		barrel oil (bbl)	1 bbl = 0.158987 m^3
Density: ρ	kg/m^3	lb_m/US gallon	1 lb_m/US gallon = 119.829 kg/m^3
		lb_m/ft^3	1 lb_m/ft^3 = 16.0183 kg/m^3
Universal gas	J/(mol·K)	psi·ft^3/	*R* = 10.73 psi·ft^3/(lb-moles·°R)
constant: *R*		(lb-moles·°R)	*R* = 8.3145 J/(mol·K)
Gravitational	m/s^2	ft/s^2	*g* = 32.174 ft/s^2
acceleration: *g*			*g* = 9.80665 m/s^2
			g_c = 32.174 ft/s^2

Appendix B: Lagrangian Interpolation in a Table

When it is required to obtain a value for a parameter from a set of tables, it is often found that the tables do not contain the data pair corresponding to the required values. For instance, if a value is required for parameter y corresponding to the value $x = 2.5$ and the table contains x values of integers such as 0, 1, 2, 3, 4, etc., and the y values corresponding to these x values, an interpolation must be made for the y value between the values corresponding to $x = 2$ and $x = 3$. In this case, it is useful to use the method proposed by Lagrange, known as Lagrangian interpolation, to obtain the y value corresponding to the required x value. This situation can arise, for instance, in calculating the viscosity index of an oil when using the tables contained in ASTM D2270, where values for L and H are required corresponding to the measured kinematic viscosity at 100°C.

The Microsoft Excel spreadsheet, which may be downloaded from http://www.crcpress.com/product/isbn/9781466551497, allows the interpolated value to be obtained by entering first, for example, three pairs of data for the x values (viscosity at 100°C) and the corresponding L values, choosing perhaps two table values below the x value measured for the unknown oil and one value above this x value, and clicking on the button to run the macro. The required value of L is found in cell F6. The table values for H are then entered on the sheet, replacing the L table values, and the macro is run again to obtain the required H value, again in cell F6.

In the example in the table below, it is seen that the interpolated value of y corresponding to the x value of 2.5 is $y = 4.0$. If a linear interpolation had been made between the table values corresponding to $x = 2$ and $x = 3$, a y value of 4.2 would have resulted.

It should be noted that, for the macro to run, the macro capability in Excel has to be enabled because this feature is usually disabled as a security measure.

	A	B	C	D	E	F	G	H	I
2		Macro for Lagrangian Interpolation of *x*, *y* Values from a Table							
5		N = 3			Required *x* =	2.50			
6		*x* (i)	*y* (i)		Required *y* =	4.0000			
7	1	1	1						
8	2	2	2.6		Lagrangian Interpolation				
9	3	3	5.8						
10	4								
11									
12		Note: Enter *x*, *y* values from Table in range B7:C10.							
13		Enter *x* value for the required interpolation value in cell F5							
14		Press the button to run macro: result is placed in cell F6.							
15		The order of the Lagrange polynomial is one less than the number							
16		of data pairs entered from the table, i.e., if four values of *x* and *y*							
17		are entered, the polynomial will be third order. Linear							
18		interpolation is thus obtained by entering two data pairs.							
19		Note 2: For the macro to run, macros must be enabled in Excel.							

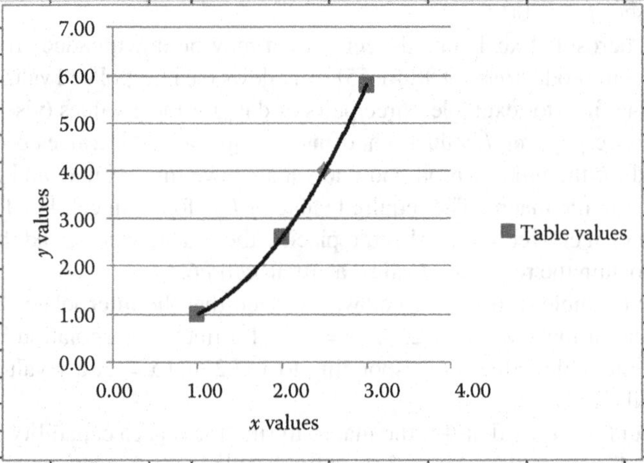
7.00
6.00
5.00
4.00
3.00
2.00
1.00
0.00
y values
0.00
1.00
2.00
3.00
4.00
x values
Table values

Appendix C: Unit Operations in a Re-Refinery

C.1 TYPICAL UNIT OPERATIONS

There are a number of chemical engineering unit operations that originate from the refining industry (Speight and Ozum 2002; Hsu and Robinson 2006; Gary et al. 2007; Speight 2014); these are applied in the re-refining of used oil and will be briefly reviewed here. The various processes developed by different operators will combine some of these unit operations in a number of different ways, ranging from the simpler processes, which are usually cheaper and result in a less valuable product, to the more complicated and expensive processes, which give a range of high-grade products. Whether a particular process is economically viable will depend in part on the size of the plant, and the capital and operating costs of these unit operations, as discussed in the chapter on the economic aspects of the processes.

The following will be a short overview of these unit operations, which form the building blocks that are put together in various ways for the different processes.

C.1.1 Filtration Processes

One of the options to remove suspended particulate matter from the recycled oil is to filter it. The impurities to be removed are usually very fine and require a filtration medium that will retain these fine particles. The boundary between microfiltration and ultrafiltration is not very distinct, and one normally considers microfiltration to be the region in which particles between 0.1 and 5 μm are retained, whereas ultrafiltration would apply to the retention of particles in the 0.001 to 1 μm range (McCabe et al. 1993). Of course, the impurities in the recycled oil are not only the particulate matter but include dissolved materials, water droplets and chemically altered hydrocarbons. Filtration of recycled oil has generally been used for the production of cleaner fuel oils or as a prestep before a distillation process. When used for pretreatment, the intention is not to remove the suspended fine solids but a coarse filter is used to remove rags and other fairly large impurities that sometimes find their way into the recycle tanks. Although membrane filter media have been used to separate oil and water mixtures, and some hydrocarbon mixtures (Melin and Rautenbach 2004; Rautenbach and Albrecht 1989), they do require periodic cleaning or replacement of the filter medium. In addition, because filtration alone does not remove any of the dissolved or chemically bound impurities, which may include significant amounts of sulfur, heavy metals, and carcinogenic compounds, the reuse of the filtrate either as a fuel or a recycled lube oil should be examined carefully.

The filter medium itself may be made of polymeric materials, sintered stainless steel, or other metals or ceramic materials, and the filtration process may be set up to involve the flow of feed material directly onto the surface of the filter material, normal to the filter surface, or in the cross-flow configuration where the feed material flows tangential to the filter medium surface. In either case, a filter cake builds up on the filter medium. In the first case, there is a gradual buildup of filter cake on the surface of the medium that gradually increases the pressure drop across the filter, which must then be periodically shut down to replace it, or a continuous cleaning device must be incorporated to scrape the filter cake off the medium. In the cross-flow case, if a sufficiently high velocity is maintained across the filter medium surface, a dynamic balance will be established in which the thickness of the filter cake is maintained by the scouring action of the feed material flowing across the surface of the filter cake. However, once the entire filter surface has reached a certain thickness, the filter must be back-flushed to remove the filter cake and restart the process.

One of the difficulties in filtering oils is the high viscosity of the oil, resulting in high pressure drops over the filter to achieve an acceptable flow rate. One way to lower the viscosity of the oil to be filtered is to raise its temperature because viscosity decreases rapidly with higher temperatures and this has led to the use of metallic filter media, and the use of ceramic membranes as in the Ceramem ultrafiltration process. Another ingenious method of lowering the viscosity of the oil to be filtered is to dissolve carbon dioxide in the oil under pressure, the dissolved carbon dioxide causing a drastic reduction in the viscosity of the oil, and then to allow the carbon dioxide to be released after the filtration process by lowering the pressure again (Rodriguez et al. 2002).

The use of a ceramic membrane is exemplified in the Ceramem process (Audibert 2006). The Ceramem filter media have usually been used for filtering water that has some contaminants in it, but tests have been carried out for its use on recycled lube oils. To meet the published standards for both recycled fuel oil and recycled lube oil base stocks, additional treatment of the filtrate would generally be necessary and the filtration processes alone have not been very popular in countries where these standards are applied.

C.1.2 Centrifugation

A variation on the filtration process, which relies on the difference in density of the particulate matter or water and the oil, is to use a centrifuge to create high centrifugal forces to discharge the impurities in an underflow from the centrifuge while the cleaned oil is collected as an overflow. This method will also not be effective for the dissolved or chemically bound components whose removal may be desired.

The centrifuge will usually consist of a cylindrical vessel that spins at a high speed, generating high inertial forces, which move the denser particles in the liquid to the periphery of the cylinder where they are continuously removed. The separation technique can, in fact, be regarded as an accelerated gravitational settling process: the closer the density of the suspended particles to the density of the liquid, the slower and more difficult the separation will be with this technique.

Centrifugation alone has not found great application in the processing of used oil but may be a step in the treatment process.

C.1.3 Solvent Extraction

The solvent extraction process for recycled oil is derived from the deasphalting process, in which a solvent such as propane or other light hydrocarbon is mixed with the feed oil, the solvent having a greater affinity for the oil fraction and forming a lighter phase, and a heavier phase containing the asphalt. As applied to recycled oil, propane has generally been found to work well and the gravity separation gives an overflow stream from which the majority of the propane can be recovered by distillation, and an underflow stream that contains the polymers, degraded additives, and heavy metals from which the residual propane can again be separated.

This type of process has also been used with a mixture of polar solvents such as hexane and alcohol, which precipitates suspended matter, and methyl pyrrolidone, which has also been used to separate aromatic hydrocarbons and hetero-organics (Compere and Griffith 2009).

C.1.4 Simple Distillation Processes

A simple distillation process will have some means of heating the incoming feed liquid to a higher temperature to evaporate the lower boiling components, which can then be collected and used in various ways. The heating is carried out at pressures near atmospheric pressure and, in its simplest form, can consist of a heating coil in a tank, the heating medium flowing through the coil could be steam or a thermal fluid that is heated in a separate boiler.

This type of simple distillation is often used as a first step on the incoming recycled oil to remove water and low boiling hydrocarbon components, these hydrocarbons, after separation from the water, can be used as a fuel elsewhere in the process. The temperatures in this case are often in the 120°C to 130°C (250°F–275°F) range, which results in the removal of almost all of the water that may be present in the feed. The use of these relatively low temperatures will generally avoid the deposition of significant amounts of solids on the heating surface, but until tests have been carried out on the particular recycled oil in any process, the possibility of fouling on the heater surfaces should always be considered.

C.1.5 Advanced Distillation Processes

The distillation processes used in a conventional petroleum refinery also find application in the re-refining process. These distillation columns are usually of two types, the most common design being the tray and bubble cap version shown in Figure C.1 or, for more specialized separations, an ordered packing type shown in Figure C.2. Heat is supplied at the bottom of the column in the reboiler section, causing the vapor phase to travel upward in the column whereas the liquid phase falls downward, the feed being supplied near the midsection of the column. A condenser system at the

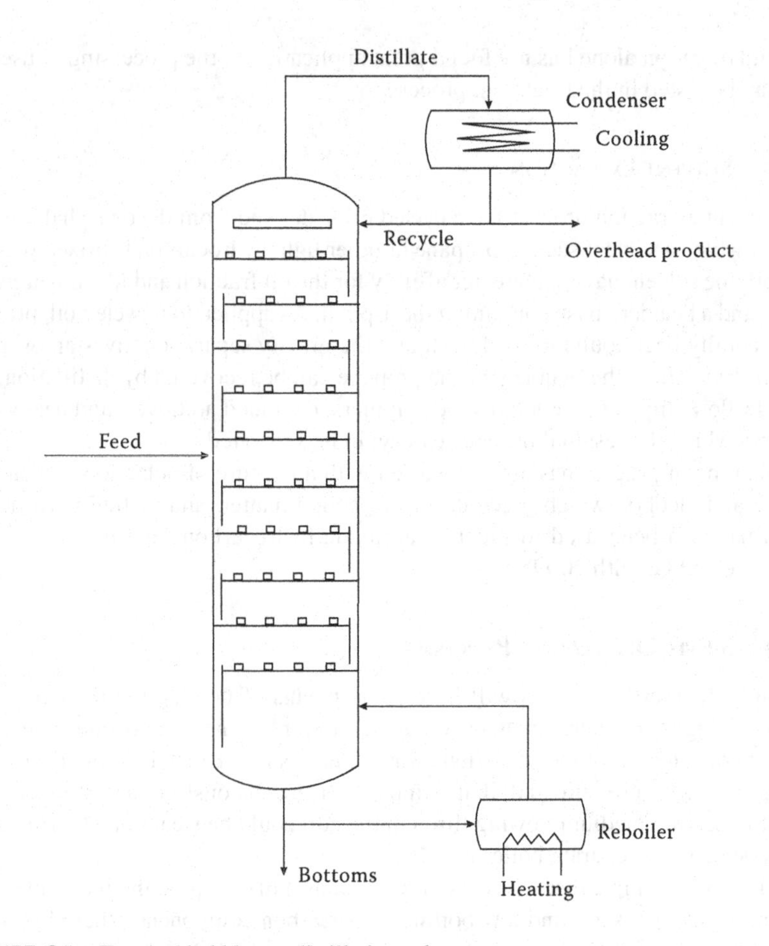

FIGURE C.1 Tray and bubble cap distillation column.

top of the column condenses the overhead stream and will usually recycle part of this stream into the top of the column to improve the separation process.

The tray and bubble cap column has a number of trays, each with many bubble caps as shown in Figure C.3, which allow the vapor phase rising through the column to bubble up through the liquid, which covers the tray: the liquid flows across the tray before falling through the downcomer to the next lower tray before flowing across the tray to the next downcomer on the opposite side of the tray. In this way, an efficient mixing of the rising vapors with the falling liquid is ensured, and heat is transferred to the liquid, which results in the evaporation of components out of the liquid. The effect of this process is to cause higher boiling components to fall to the lower part of the column while the lower boiling components are found in the upper part of the column. By arranging for off-takes at different heights on the column, different boiling range materials can be separated in the distillation process.

The packing used in the ordered packing version usually consists of a woven stainless steel mesh that is arranged in corrugated sheets in a regular pattern inside

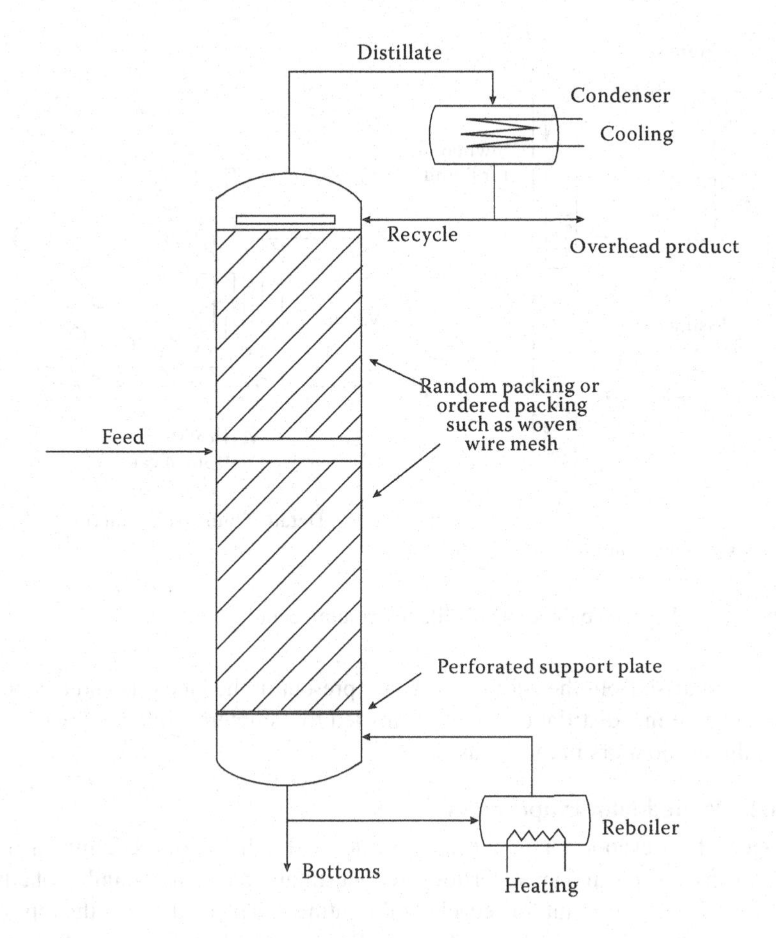

FIGURE C.2 Packed column distillation tower.

the column. This type of packing is supported at the bottom of the column on some type of grid and fills the column. Despite its success in other applications, this type of distillation column has not found great application in the refining industry.

C.1.5.1 Atmospheric and Vacuum Distillation Columns

The distillation process will often be carried out as a two-stage process. The first stage carried out with near-atmospheric pressure in the column will evaporate the lower boiling components with a boiling point up to a certain temperature. Because heating the oil to high temperatures can result in the cracking of the oil molecules to form undesirable lighter components, the second stage is operated with a low-pressure vacuum in the column, allowing components that would boil off at high temperatures under atmospheric pressure, and hence potentially cracking in the process, to boil off at a lower temperature and maintain their integrity.

One of the problems of using distillation columns with recycled lube oils is that of the deposition of solids and fouling of the internal surfaces of the column. These

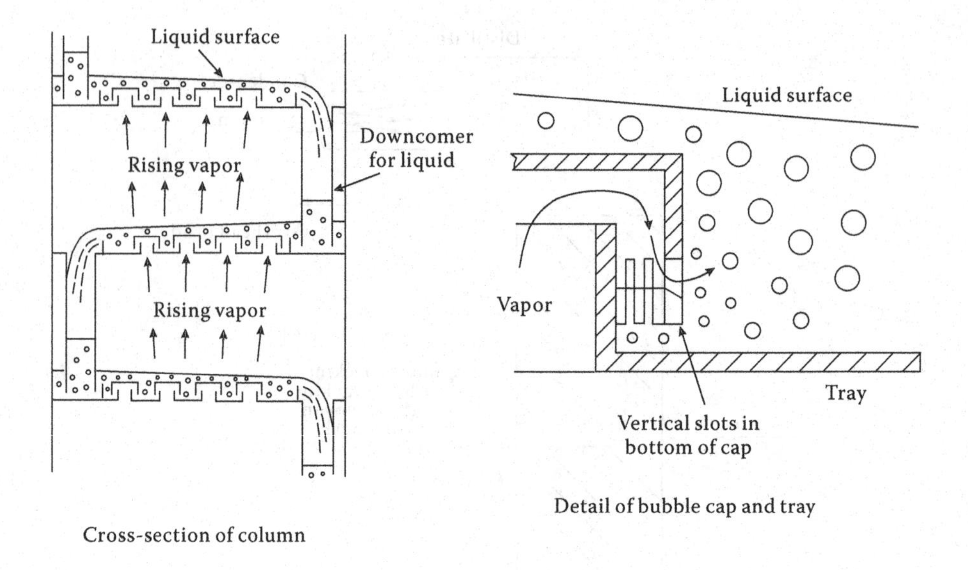

FIGURE C.3 Tray and bubble cap distillation column detail.

deposits can arise from the additives that are present in the lube oils and can be difficult to handle in a distillation column: this led to the interest in and application of wiped film evaporators in many cases.

C.1.5.2 Wiped-Film Evaporators

The wiped film evaporator is not as versatile as the distillation column in that it results in an overflow and an underflow product stream, but it has a number of advantages that can be important for recycled oil treatment. The feed enters the top of the cylindrical vessel shown in Figure C.4 and flows down the walls. A jacket around this cylindrical vessel allows the wall to be heated with a heating medium that is circulated through the jacket space, the heating medium or thermal fluid being heated in a separate heater to a carefully controlled temperature. A vertical shaft is rotated in the center of the cylindrical vessel with wiper blades being attached to this shaft. The wiper blades wipe over the inner surface of the cylindrical vessel, spreading the oil film over the heated surface, increasing the heat transfer to the oil and also continuously wiping the surface on which any deposits might tend to form. By using a thermal fluid in the jacket to heat the oil, temperatures can be accurately controlled and kept below the point where thermal cracking of the oil might become a problem with its attendant coke deposition.

Another advantage of the wiped film evaporator is realized in the short flow path design in which the evaporated vapor stream can be removed from the cylindrical vessel by having a hollow central shaft with outlet ports into the shaft allowing the evaporated vapors to escape through the shaft. This allows the evaporator to operate at a pressure that is quite uniform over the height of the vessel, with little pressure drop from the heated film to the product stream, resulting in a sharp cut separation based on the boiling points of the oil components at the pressure maintained in the vessel.

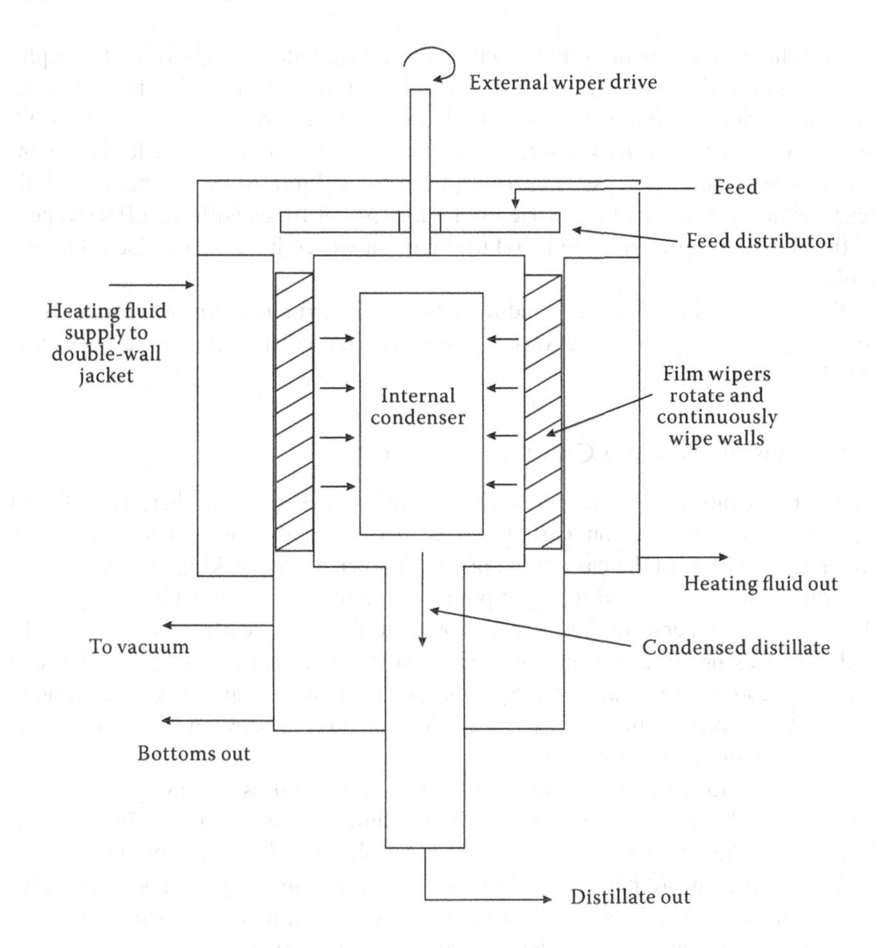

FIGURE C.4 Schematic diagram of wiped film evaporator.

The wiped film evaporators can be arranged in series with the internal pressure being adjusted in the first evaporator to give a separation at the desired temperature and pressure condition. The underflow from the first evaporator is then fed to the second evaporator, which is operated under a high vacuum to give an overflow stream of the higher boiling point components. The underflow from the second evaporator will contain the residual high-boiling components and much of the heavy metals in the recycled oil and has generally been used as an asphalt extender in applications in which it can be safely sequestered to avoid leaching out into the environment.

C.1.6 Clay Treatment

A number of mineral clays and other materials with a high adsorption affinity for asphaltic and resinous materials from an oil stream have been used in the refining industry for many years, and have mainly been used to remove the color, asphalts, and resins from a product stream.

The clay (such as bentonite) is contained in a column through which the liquid stream passes, the residence time in contact with the very fine clay particles can be of the order of minutes to hours. Periodic regeneration of the clay is required, and is done by switching the treated stream to another column while the regenerating combustion process is carried out in the column to be regenerated. This regeneration process can be carried out a number of times with the effectiveness of the clay being gradually reduced to the point where it has to be discarded and replaced.

Some of the heavy metal residues and some sulfur components can also be removed by this type of process when it is used after a distillation process for recycled lubricating oil.

C.1.7 Visbreaking and Catalytic Cracking

As the temperature of oil is increased, a point is reached where there is sufficient energy to break the carbon–carbon bonds in some of the molecules, resulting in smaller molecules. In the case of the process known as visbreaking, the intention is to reduce the viscosity and the pour points of the oil. It is a relatively mild process that occurs at temperatures between approximately 450°C and 510°C (840°F–950°F) and pressures up to approximately 5000 kPa (Speight and Ozum 2002; Hsu and Robinson 2006; Gary et al. 2007; Speight 2014) with a very short residence time in the reactor. Typically, the viscosity of waxy treated oil is between 25% and 75% of the viscosity of the feed material.

Catalytic cracking takes place at higher temperatures of 480°C to 540°C (895°F–1005°F) over a catalyst, typically in a fluidized bed with a zeolite catalyst. In this case, the aim is to break up the longer molecular chains and produce lighter products, typically producing gasoline and diesel fractions from a heavier feed material. If the process is carried out in the absence of additional hydrogen, there is a tendency to form double-bond olefins, which may be unstable.

C.1.8 Hydrogenation Processes

There are a number of processes by which hydrogen is added to the hydrocarbon compounds in the oil. The reasons for doing this include the following:

- To terminate a hydrocarbon chain that has been broken or "cracked" by adding a hydrogen atom at the end of the chain and so minimize the formation of double-bonded olefins, which can be undesirable in the product slate
- To convert unsaturated hydrocarbons such as olefins and the diolefins, which are unstable and form gums, to saturated and stable paraffins
- To remove sulfur, halides, trace metals, and nitrogen components, which combine with the hydrogen and can be separated from the oil stream

The processes vary in the severity of the pressure and temperature conditions that are applied, and catalysts can be used to promote the reactions. They include hydrotreating, hydroprocessing, hydrocracking, and hydrodesulfurization.

C.1.8.1 Hydrotreating

This is also a relatively mild process that takes place at temperatures of 315°C to 425°C (600°F–795°F) and pressures of 700 to 20,000 kPa, most often over a metal oxide catalyst of cobalt and molybdenum, and a nickel, cobalt, and molybdenum catalyst is also used. Cracking of the molecules is not promoted under these conditions. The main reactions are the saturation of olefins and the removal of mercaptans containing sulfur, and oxygen, nitrogen, and trace metals; the sulfur and nitrogen combine with the hydrogen to form hydrogen sulfide and ammonia, the metals remaining on the catalyst surface.

C.1.8.2 Hydrocracking

Hydrocracking is usually used to convert high-boiling components to products in a lower-boiling range. Where catalytic cracking treats more easily converted paraffinic gas oils, catalytic hydrocracking treats the more difficult feeds, particularly oils rich in aromatic ring compounds, which come from the distillate stream of a distillation process. The cracking process breaks the long-chain hydrocarbons into shorter-chain molecules in a hydrogen-rich atmosphere at temperatures of 290°C to 400°C (555°F–750°F) and pressures of 8 to 14 MPa or higher. The feed may be hydrotreated to saturate olefins and remove sulfur, nitrogen, and oxygen compounds before entering the hydrocracking units. The catalysts used may be silica–alumina to promote cracking and rare earths such as platinum, palladium, tungsten, or nickel, which promote hydrogenation.

C.1.8.3 Hydroprocessing

This is similar to hydrocracking in reducing the boiling range of the feed material, but usually treats the heavier components coming from the bottoms of a distillation process. The process removes metals, sulfur, and nitrogen and operates at high pressures of 14 MPa or more, and a guard reactor is used before the conversion reactors. The purpose of the guard reactor is to remove metals that would poison the catalyst, and to reduce the possibility of carbon or coke formation by cracked products. An amine absorption process can be used to remove the hydrogen sulfide from the hydrogen stream before it is recycled to the reactors.

C.1.8.4 Hydrodesulfurization

The general process for the removal of the undesirable sulfur compounds occurs to a variable extent in hydrotreating, hydrocracking, and hydroprocessing. The mercaptans, sulfides, and thiophenes (all containing sulfur) react with the hydrogen to form hydrogen sulfide, which is separated from the oil.

REFERENCES

Audibert, F. 2006. *Waste Engine Oils: Re-Refining and Energy Recovery*, Elsevier B.V., 1st Ed.

Compere, A.L., and Griffith, W.L. 2009. Third Party Evaluation of PetroTex Hydrocarbons LLC ReGen Lubricating Oil Re-Refining Process, Oak Ridge National Laboratory Report ORNL/TN-2009/093, sponsored by the U.S. Department of Energy and available at http://www.osti.gov/bridge. Accessed on December 8, 2012.

Gary, J.G., Handwerk, G.E., and Kaiser, M.J. 2007. *Petroleum Refining: Technology and Economics*, 5th Edition. CRC Press, Boca Raton, FL.

Hsu, C.S., and Robinson, P.R. (Editors) 2006. Practical Advances in Petroleum Processing Volume 1 and Volume 2. Springer Science, New York.

McCabe, W.L., Smith, J.C., and Harriott, P. 1993. *Unit Operations of Chemical Engineering* 5th Edition. McGraw-Hill Inc., New York.

Melin, T., and Rautenbach, R. 2004. *Membranverfahren*, 2nd Edition, Springer-Verlag, Heidelberg, Germany.

Rautenbach, R., and Albrecht, R. 1989. *Membrane Processes*, John Wiley & Sons Inc. Hoboken, New Jersey.

Rodriguez, C., Sarrade, S., Schrive, L., Dresch-Bazile, M., Paolucci, D., and Rios, G.M. 2002. Membrane Fouling in Cross-Flow Ultrafiltration of Mineral Oil Assisted by Pressurised CO_2. Proceedings. International Congress on Membranes and Membrane Processes (ICOM), Toulouse, France. July 7–12.

Speight, J.G. 2014. *The Chemistry and Technology of Petroleum*, 5th Edition, CRC Press, Boca Raton, FL.

Speight, J.G., and Ozum, B. 2002. *Petroleum Refining Processes*. Marcel Dekker Inc., New York.

Index

D

E

F

G

H

I

S

T

U

V

W

Z

Printed in the United States
by Baker & Taylor Publisher Services